VOLUME SEVENTY FIVE

ADVANCES IN BOTANICAL RESEARCH

Plant Microbe Interactions

ADVANCES IN BOTANICAL RESEARCH

Series Editors

VOLUME SEVENTY FIVE

ADVANCES IN BOTANICAL RESEARCH

Plant Microbe Interactions

Volume Editors

HARSH BAIS

Department of Plant and Soil Sciences,
Delaware Biotechnology Institute,
University of Delaware, Newark, USA

JANINE SHERRIER

Department of Plant and Soil Sciences,
Department of Biological Sciences,
Delaware Biotechnology Institute,
University of Delaware, Newark, USA

ELSEVIER

AMSTERDAM • BOSTON • HEIDELBERG • LONDON
NEW YORK • OXFORD • PARIS • SAN DIEGO
SAN FRANCISCO • SINGAPORE • SYDNEY • TOKYO

Academic Press is an imprint of Elsevier

Academic Press is an imprint of Elsevier
125 London Wall, London EC2Y 5AS, UK
The Boulevard, Langford Lane, Kidlington, Oxford OX5 1GB, UK
225 Wyman Street, Waltham, MA 02451, USA
525 B Street, Suite 1800, San Diego, CA 92101-4495, USA

First edition 2015

ISBN: 978-0-12-420116-3
ISSN: 0065-2296

For information on all Academic Press publications
visit our website at http://store.elsevier.com

CONTENTS

Contributors ix
Preface xi

1. Not Just Sweet Talkers: How Roots Stimulate Their Colonization by Beneficial Bacteria **1**

Pascale B. Beauregard

1. Introduction 2
2. Attraction 3
3. Nutrition 6
4. Colonization 7
5. General Effects 10
6. Compounds from the Plant That Affect the PGP Effect of Bacteria 12
7. Conclusion and Perspectives 14

References 14

2. Molecular Patterns of Rhizobacteria Involved in Plant Immunity Elicitation **21**

Martin Mariutto and Marc Ongena

1. Host Resistance Elicitation as Part of Rhizobacteria-Mediated Biocontrol 22
2. Global Aspect of Microbe-Induced Plant Immunity 24
3. MAMPs from Bacterial Pathogens and Early Physiological/Signalling Events Associated with Their Perception 25
4. Elicitors of Plant Immunity from Beneficial Bacteria 29
5. Concluding Remarks 47

Acknowledgement 48
References 48

3. Root Microbiome Assemblage is Modulated by Plant Host Factors **57**

Venkatachalam Lakshmanan

1. Introduction 58
2. Microbiome Assemblage is Influenced by the Host Genome 60
3. Microbiome Assemblage is Influenced by the Host Developmental Stage 63
4. Microbiome Assemblage is Influenced by the Host Health and Fitness 67

5. Microbiome Assemblage is Influenced by Alteration in Plant Signalling 69
6. Implications and Future Prospects 70
References 72

4. PGPR Interaction: An Ecofriendly Approach Promoting the Sustainable Agriculture System 81

Usha Bishnoi

1. Introduction 82
2. Rhizosphere, the Hot Spot for PGPR Interaction 84
3. Colonization and Competence 88
4. PGPR Rooting for the Sustainable Agriculture 89
5. Commercialization of PGPR 99
6. Conclusion 103
Acknowledgements 104
References 104

5. Human Pathogen–Plant Interactions: Concerns for Food Safety 115

Sarah M. Markland and Kalmia E. Kniel

1. Introduction: Merging Plant Science and Food Science to Address Food Safety 116
2. Human Pathogens on Plants 116
3. Human Pathogen Interactions with Phytobacteria and Rhizobacteria 120
4. Relationships Between Plant Stomata, Circadian Rhythm and Human Pathogens 123
5. Control and Prevention Strategies: Biocontrol and Plant 'Probiotics' 125
6. Attachment and Attraction Mechanisms of Human Pathogens to Plants 127
7. Research Needs and Conclusions 129
References 130

6. Early Responses in the Root–Rhizobia Interaction 137

Oswaldo Valdés-López and Georgina Hernández

1. Innate Immunity Must Be Regulated during the RN Symbiosis 139
2. Searching for Missing Regulators of the Early Responses of the RN Symbiosis 141
3. miRNAs as Regulators of the Early Events of the RN Symbiosis 143
4. Conclusions and Remarks 144
Acknowledgements 144
References 145

7. The Implications of Polyploidy for the Evolution of Signalling in Rhizobial Nodulation Symbiosis **149**

Adrian F. Powell and Jeff J. Doyle

1. Introduction 150
2. Nodulation Signalling: Determinants of Symbiotic Interactions 152
3. WGD as a Process Altering Plant Signalling, Biotic Interactions and Ecological and Geographical Ranges 158
4. Potential Implications of Polyploidy for Nodulation Signalling 162
5. Evidence of Roles for Polyploidy in the Evolution of Nodulation and Rhizobial Signalling 170
6. Current and Future Studies of Polyploidy and Nodulation: Challenges and Opportunities 175
7. Conclusion 179

References 180

8. Plant Cues and Factors Influencing the Behaviour of Beneficial Nematodes as a Belowground Indirect Defense **191**

Jared Gregory Ali and Elizabeth Davidson-Lowe

1. Nematode Diversity 193
2. A Nematode's Sensory Tool Kit 194
3. Entomopathogenic Nematodes 195
4. Indirect Defences and Herbivore-Induced Plant Cues Belowground 196
5. Sampling Belowground Herbivore-Induced Volatiles 200
6. Manipulation of Root Volatiles for Agricultural Improvement 203
7. Multitrophic Interactions and Paradox Surrounding Belowground Indirect Defences 206

Acknowledgements 209

References 209

9. Forest Canopy Precipitation Partitioning: An Important Plant Trait Influencing the Spatial Structure of the Symbiotic Soil Microbial Community **215**

Carl L. Rosier, Leslie Dean Moore, Tiehang Wu and John T. Van Stan

1. Introduction: Canopy Precipitation Partitioning and Its Potential Influence on Symbiotic Root Microbial Communities 216
2. Potential Throughfall and Stemflow Impacts on N-fixing Nodule-Forming Bacterial Signalling Pathways 219
3. Potential Throughfall and Stemflow Impacts on Endo- and Ectomycorrhizal Fungi Host Root Signalling 225

4. Future Directions for Evaluating Throughfall and Stemflow Influences on Symbiotic Microbes 230
Acknowledgement 231
References 232

10. Soluble and Volatile Metabolites of Plant Growth-Promoting Rhizobacteria (PGPRs): Role and Practical Applications in Inhibiting Pathogens and Activating Induced Systemic Resistance (ISR) 241

Gul Shad Ali, David Norman and Ashraf S. El-Sayed

1. Overview of Current Disease Management Practices 242
2. Plant Growth-Promoting Rhizobacteria 244
3. Beneficial Soluble Metabolites of PGPRs 247
4. Volatile Organic Compounds 254
5. Practical Uses in Agriculture 272
6. Outlook – Opportunities and Challenges 273

Acknowledgements 275
References 275

11. Flying Whispers of Inter-kingdom Conversation: A Complementary Perspective of Plant and Bacterial Volatile Signals 285

Gopinath Selvaraj

1. Introduction 286
2. Volatile Organic Compounds from Plants 287
3. Influence of Plant VOCs on Rhizosphere and Phyllosphere Bacteria 290
4. VOCs from Bacteria and Their Effects on Plants 292
5. Bacterial Volatiles in QS and QQ 299
6. Commonness and Distinctions among Plant and Bacterial VOCs 300
7. Future Perspectives 303

Acknowledgement 304
References 304

Subject Index *311*
Author Index *319*

CONTRIBUTORS

Gul Shad Ali
Mid-Florida Research and Education Center and Department of Plant Pathology, University of Florida/Institute of Food and Agricultural Sciences, Apopka, FL, USA

Jared Gregory Ali
Department of Entomology, Pennsylvania State University, State College, PA, USA; Department of Entomology, Michigan State University, East Lansing, MI, USA

Pascale B. Beauregard
Département de Biologie, Université de Sherbrooke, Sherbrooke, QC, Canada

Usha Bishnoi
Department of Plant and Soil Sciences, University of Delaware, Newark, DE, USA; Delaware Biotechnology Institute, Newark, DE, USA

Elizabeth Davidson-Lowe
Department of Entomology, Michigan State University, East Lansing, MI, USA

Jeff J. Doyle
Section of Plant Breeding and Genetics and Section of Plant Biology, School of Integrative Plant Science, Cornell University, Ithaca, NY, USA

Ashraf S. El-Sayed
Mid-Florida Research and Education Center and Department of Plant Pathology, University of Florida/Institute of Food and Agricultural Sciences, Apopka, FL, USA

Georgina Hernández
Centro de Ciencias Genómicas, Universidad Nacional Autónoma de México (UNAM), Cuernavaca, Morelos, Mexico

Kalmia E. Kniel
Department of Animal and Food Sciences, University of Delaware, Newark, DE, USA

Venkatachalam Lakshmanan
Department of Plant and Soil Sciences, University of Delaware, Newark, DE, USA; Delaware Biotechnology Institute, Newark, DE, USA

Martin Mariutto
Microbial Processes and Interactions Research Unit, University of Liège/Gembloux Agro-Bio Tech, Gembloux, Belgium

Sarah M. Markland
Department of Animal and Food Sciences, University of Delaware, Newark, DE, USA

Leslie Dean Moore
Department of Biology, Georgia Southern University, Statesboro, GA, USA

David Norman
Mid-Florida Research and Education Center and Department of Plant Pathology, University of Florida/Institute of Food and Agricultural Sciences, Apopka, FL, USA

Marc Ongena
Microbial Processes and Interactions Research Unit, University of Liège/Gembloux Agro-Bio Tech, Gembloux, Belgium

Adrian F. Powell
Section of Plant Biology, School of Integrative Plant Science, Cornell University, Ithaca, NY, USA

Carl L. Rosier
Boone Pickens School of Geology, Oklahoma State University, Stillwater, OK, USA; College of Agriculture and Natural Resources, University of Delaware, Newark, DE, USA

Gopinath Selvaraj
Department of Plant and Soil Sciences, University of Delaware, Newark, DE, USA

Oswaldo Valdés-López
Laboratorio de Genómica Funcional de Leguminosas, FES Iztacala, Universidad Nacional Autónoma de México (UNAM), Tlalnepantla, Estado de México, Mexico

John T. Van Stan
Department of Geology and Geography, Georgia Southern University, Statesboro, GA, USA

Tiehang Wu
Department of Biology, Georgia Southern University, Statesboro, GA, USA

PREFACE

Since the concept of beneficial symbiosis between plants and microbes was introduced almost 130 years ago, research has led to an understanding about the mechanistic complexity in these interactions. Plants associate with various benign microbes, and the last five years of work has shown significant growth in our understanding of plant-associated microbiomes. Plants grow in close association with large communities of microbes, yet comparatively little is known about the diversity of these plant-associated microbes and their interactions and effects on plant performance, crop yields and plant protection. Even though the effect of beneficial microbes on plants is discussed in the literature, there has been comparatively much less progress in elucidating the underlying mechanisms that are involved in beneficial microbes' effect(s) on plant performance and disease evasion, with the exception of the best-studied model systems. The work related to human gut microbiome revealed the functional importance of microbes in evading disease and immune response in higher animals. Since plants are more exposed to microbes than animals, it is easier to speculate that plants have evolved to facilitate more benign than pathogenic interactions. With the advances in the sequencing technology, it is now easier to find out the 'who is there' aspect of the microbiome associated with plants. The scenario related to 'what are these microbes doing' with plant associations is yet to be elucidated. Our approaches to understand plant-associated symbiotic microbiomes have changed over the last decade. Over the last 5 years or so, we have advanced our understanding of why some benign microbes associate with plants. The classical symbiotic tale of legume and *Rhizobium* species catalysed of our thinking related to how root-mediated processes play a critical role in plant–microbe interactions. Our knowledge related to root-mediated processes and mainly root exudations confirmed the functional role of root secretions as the main driving force to increase microbial associations.

Plants primarily depend on small molecular weight compounds to initiate positive or negative interactions. The knowledge pertaining to the plethora of chemicals that plants produce to initiate plant–plant and plant–microbe interactions is increased at a radicle speed. Among the most spectacular interactions mediated by a chemical exchange was the one elucidated for legume–*Rhizobium* partnerships. Over time, various interactions mediated by chemicals signifying beneficial plant–microbe interactions were elucidated. Spectacularly, there is an upsurge that plants associate with beneficial microbes under stress conditions. The overall concept of 'botanical cry's' in plants under stress is not new, but plants signalling during stress to recruit beneficial microbes adds another layer of complexity in understanding plant–microbe interactions.

The previous attempts to highlight plant–microbe communications revolved around all interactions involving both beneficial and pathogenic microbes. Here, individual chapters deal with diverse benign microbe–plants interactions. The topics include how beneficial microbes associate with plants, the interactions are usually not that symbiotic in nature. The chapters discuss how beneficial microbe hijacks plant immune system to colonize. One of the chapters talks about how beneficial microbe–plant partnerships could be utilized for sustainable and responsible agricultural practices. The shift of using beneficial microbes as biocontrol is discussed against foodborne human and plant pathogens. Chapter also discusses the different diffusible and volatile organic compounds produced by benign microbes as biocontrol agents against plant pathogens. Two chapters deal with the different factors that regulate classical legume–*Rhizobium* symbiosis. Finally, chapters discuss the involvement of other factors in rhizosphere, such as nematodes in plant communication in a tritrophic setup.

Roots and the rhizosphere have no longer remained a 'hidden half' of plants. The upsurge in understanding various rhizospheric processes has led to various scintillating discoveries in the last decade. The concept of how plants interact with beneficial microbes has taken an interesting turn. More high-throughput molecular approaches are now tested to understand the functional role of plant-associated microbiome. Plants are different than animals in the way they mediate symbiotic associations. The secondary metabolism in plants leading to production of plethora of small molecular weight compounds hold the key to drive various microbial associations. How plants use this currency towards its benefit will carve new frontiers in terms of understanding the role of plant-derived products in shaping its microbiome. In addition, the concept of 'soil health' in general has been a global point of discussion to improve sustainable crop practices. Most importantly, the overuse of fertilizers, soil salinization and microbial shifts caused by soil disturbances have further compounded the agronomic problems. The green approach of using soil microbes as biofertilizers and biocontrol agents may work as an inexpensive and sustainable to replace chemical fertilizers. The commercial use of benign microbes to improve crop yield and protection holds the future, though a note of caution should be exercised to validate and strengthen the claims of how these benign microbes improve plant growth and yield. Much of this validation would come from the lab studies that are currently lined up to understand the functional response of microbiome on plants at the molecular level. We can look forward to witnessing the next wave of novel discoveries.

Harsh Bais
Janine Sherrier

CHAPTER ONE

Not Just Sweet Talkers: How Roots Stimulate Their Colonization by Beneficial Bacteria

Pascale B. Beauregard
Département de Biologie, Université de Sherbrooke, Sherbrooke, QC, Canada
E-mail: pascale.b.beauregard@usherbrooke.ca

Contents

1. Introduction 2
2. Attraction 3
3. Nutrition 6
4. Colonization 7
5. General Effects 10
6. Compounds from the Plant That Affect the PGP Effect of Bacteria 12
7. Conclusion and Perspectives 14
References 14

Abstract

Recent studies showed that plant roots influence the composition of the bacterial population present in their rhizosphere. This observation implies that roots can affect their own colonization by various bacteria, and notably by plant growth-promoting rhizobacteria (PGPR). Since the plant-beneficial activities depend largely on the presence of PGPR on the root, it appears that the root's capacity to stimulate its own colonization is a great advantage. The exudates secreted by the root contain molecules acting as chemoattractants for many PGPR, and these molecules sometimes act specifically to attract certain cognate bacteria. In some cases, such as infection by a pathogen, the root can also secrete a higher amount of an attractant to recruit a higher quantity of PGPR. The roots exudates can also serve as a carbon source, providing energy and nutrient to the bacteria to multiply in the rhizosphere. Importantly, roots can influence biofilm formation and thus colonization of the roots. Various molecules are involved in this process such as malate, plant polysaccharides and quorum-sensing mimics. In addition, root-produced molecules can affect various cellular pathways, via modulation of transcription, translation or both. Many of these plant-stimulated pathways provide beneficial activities, such as antimicrobials and auxins production. Research on this topic is at its beginning, and most of the molecules and bacterial receptors involved are still to be found.

Advances in Botanical Research, Volume 75
ISSN 0065-2296
http://dx.doi.org/10.1016/bs.abr.2015.07.001

1. INTRODUCTION

Terrestrial plants release in the soil an important proportion of their fixed carbon, as much as 30% or even more for certain species (Haichar, Santaella, Heulin, & Achouak, 2014; Kuzyakov & Domanski, 2000). These carbon compounds are secreted as root exudates or as rhizodeposits from shedding root cells, modifying the composition of the surrounding soil. The soil under the influence of the plant root, also named the rhizosphere, is prime real estate for microbial life; its bacterial concentration can be 10–1000 fold higher than in bulk soil (Lugtenberg & Kamilova, 2009).

Many studies demonstrate that the microbial community present in a plant's rhizosphere depends at least partially on the plant species and ecotype (Berg & Smalla, 2009; Bulgarelli, Rott, & Schlaeppi, 2012; Lebeis, 2014; Lundberg, Lebeis, & Paredes, 2012; Peiffer et al., 2013; Van Overbeek & van Elsas, 2008; Wieland, Neumann, & Backhaus, 2001). The bacterial communities found in association with eight different *Arabidopsis thaliana* ecotypes differ from one another making them specific to the plant's genotype. Interestingly, chemical analysis of the exudates shows variations between ecotypes, suggesting a way by which the plant controls the assembly of the community (Micallef, Shiaris, & Colón-Carmona, 2009). The compounds present in *A. thaliana* root exudates also vary during the plant developmental stages, as do the rhizosphere microbiome. This observation suggests that *A. thaliana* emits growth stage-specific signals that modulate its root microbiome (Chaparro, Badri, & Vivanco, 2014). The plant growth stage also strongly influences the bacterial community associated with potato plants. When comparing communities associated with potato cultivars at different age, growth stage appears to have a stronger effect than the plant's genotype (Van Overbeek & van Elsas, 2008). Finally, the physiological state of the plant also influences its microbiome. The jasmonic acid signalling pathway, involved in plant response to certain stresses such as a necrophytic pathogens, can influence the composition of the associated rhizobacteria community. Its activation induces a shift in the community towards a population closely related to bacteria known to suppress plant pathogens attack, thus probably helping the plant cope with infection (Carvalhais, Dennis, Badri, et al., 2013).

As a counterpart to the beneficial environment provided by plant roots, certain free-living bacteria exert beneficial activities for the plant (reviewed in Glick, 2012; Lugtenberg & Kamilova, 2009; Vessey, 2003). Among these beneficial plant growth-promoting rhizobacteria (PGPR) are members of the bacterial genera *Azospirillum*, *Bacillus*, *Pseudomonas* and *Streptomyces*, all of which are currently used as inoculants in sustainable agriculture (Berg, 2009). The many positive effects provided by PGPR on the plant are diverse. Some of the PGPR were shown to improve plant growth and root architecture via the secretion of phytohormones and plant growth-promoting volatiles (reviewed in Bloemberg & Lugtenberg, 2001; Vacheron et al., 2013). They can also increase the bioavailability of various nutrients in the soil such as iron and phosphate. Some of them can also fix nitrogen (Mendes, Garbeva, & Raaijmakers, 2013; Richardson, Barea, McNeill, & Prigent-Combaret, 2009). Various PGPR strains can help the plant cope with abiotic stresses such as salts or heavy metals (Kloepper, Leong, Teintze, & Schroth, 1980; Mendes et al., 2013; Yang, Kloepper, & Ryu, 2009). Most importantly, PGPR help the plant resist to pathogen invasions, either directly by antagonizing the pathogens via the secretion of antimicrobial molecules, or indirectly by triggering the induced systemic resistance in the plant, a state of enhanced defensive ability (reviewed in Beneduzi, Ambrosini, & Passaglia, 2012; Berg, 2009).

In order for PGPR to exert their beneficial actions, and in particular protection against plant pathogens, they must first establish a robust colonization of the root (Bais, Fall, & Vivanco, 2004; Bianciotto, Andreotti, Balestrini, Bonfante, & Perotto, 2001; Bull, Weller, & Thomashow, 1991; Chen et al., 2013; Chin-A-Woeng, 2000; Dutta & Podile, 2010; Haggag & Timmusk, 2008). This colonization process involves several steps, which can be modulated by the plant itself. Various molecules coming from the root can act as signals or environmental cues increasing or decreasing the colonization capacities and the plant-beneficial properties of the PGPR. Here, we will take a look at the various steps taken by plant-beneficial bacteria while colonizing the root, and how the plant can modulate each of them. These various steps are summarized in Figure 1.

2. ATTRACTION

Bacterial motility is a requirement for root colonization, since nonmotile mutants of certain PGPR cannot establish themselves on plant root (reviewed in Somers, Vanderleyden, & Srinivasan, 2004). The plant can take advantage of this motility by secreting different molecules that will attract rhizobacteria via their chemotactic response (Bais, Weir, Perry, Gilroy, & Vivanco, 2006). Supporting this idea, the population of microbe

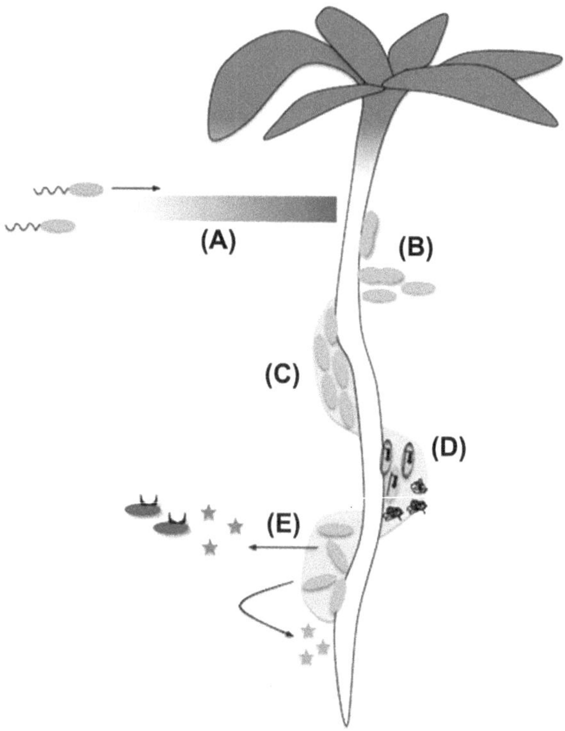

Figure 1 Schematic representation of the different steps for root colonization by plant growth-promoting rhizobacteria (PGPR) and how the plant can modulate them. (A) Root exudates secrete various molecules attracting the bacteria via chemotaxis. (B) Many of these molecules can also serve as nutrients to favour bacterial growth in the rhizosphere. (C) Once the PGPR are in contact with the root, they will establish a robust colonization via biofilm formation and matrix secretion, a process that can also be stimulated by the plant. (D) The plant also have general effects on the PGPR, both at the transcriptional and translational levels and (E) can stimulate the plant-beneficial properties of the PGPR such as antimicrobials secretion. (See colour plate)

with chemotactic potential in the rhizosphere of wheat and cowpea is less diverse than in bulk soil, suggesting that the plant exudates allow for the recruitment and thus selection of specific populations (Buchan, Crombie, & Alexandre, 2010). In experiments with model systems, a beneficial *Pseudomonas fluorescens* strain containing a mutation for the *cheA* gene and thus unable to exert chemotaxis is outcompeted by its wild-type counterpart for tomato root colonization both in soil and under gnotobiotic conditions (De Weert et al., 2002). Chemotaxis also plays a crucial role in the nitrogen-fixing PGPR *Azospirillum brasilense*, where nonflagellated and nonchemotactic mutants have a drastically reduced colonization capacity of wheat root in a gnotobiotic system (Vande Broek, Lambrecht, & Vanderleyden, 1998).

It appears likely that attraction to the root constitute the first step toward colonization for many rhizobacteria, since several studies show the chemo-attraction of various PGPR such as *P. fluorescens*, *Azospirillum* spp., *Azotobacter chroococcum* and *Bacillus* species toward whole roots exudates or seed exudates (De Weert et al., 2002; Gafny, Okon, Kapulnik, & Fischer, 1986; Heinrich & Hess, 1985; Sood, 2003; Tan et al., 2013; Yaryura et al., 2008; Zheng & Sinclair, 1996). Root exudates are a complex blend of high and low molecular weight compounds, many of which can induce chemotactic response in PGPR (Bais, Broeckling, & Vivanco, 2008). In the abundant literature on the subject, small sugars, amino acids, aromatic compounds and small organic acids are suggested to be important drivers of bacterial attraction in the rhizosphere (Bashan & Holguin, 1994; De Weert et al., 2002; Heinrich & Hess, 1985; Ling, Raza, Ma, Huang, & Shen, 2011; Neal, Ahmad, Gordon-Weeks, & Ton, 2012; Oku, Komatsu, Tajima, Nakashimada, & Kato, 2012; Parales & Harwood, 2002; Reinhold, Hurek, & Fendrik, 1985; Rudrappa, Czymmek, Paré, & Bais, 2008; Van Bastelaere et al., 1999). PGPR encodes for many chemoreceptors, which suggest a wide diversity of attracting molecules. Since most of the attractants mentioned here were discovered by a hypothesis-driven approach, it would be interesting to purify attracting compounds directly from the exudates to see if other types of molecules are involved in this process.

The exact composition of the exudates varies significantly between plants, which allow for specific recruitment of a cognate PGPR and their subsequent colonization of the root (Chaparro et al., 2013, 2014; Czarnota, Rimando, & Weston, 2003; Fan, Lane, Shenker, & Bartley, 2001; Rudrappa et al., 2008; Somers et al., 2004; Warembourg, Roumet, & Lafont, 2003). For example, an *Azospirillum lipoferum* isolated from a kallar grass rhizosphere displays strong attraction to high molecular weight compounds contained in that plant's exudates, which in turn are unable to attract an *A. brasilense* strain isolated from wheat. On the other hand, this *A. brasilense* wheat isolate strongly respond to oxalate, a significant part of wheat exudates, while the kallar grass *A. lipoferum* does not (Reinhold et al., 1985). Similarly, the *Bacillus subtilis* strain N11 isolated from the banana rhizosphere and the *Bacillus amyloliquefaciens* strain SQR9 isolated from cucumber show preferential colonization of their original host when compared to the other plant. This observation can be explained in part by the presence of fumaric acid in the banana roots exudates, which strongly attract *B. subtilis* N11 but not *B. amyloliquefaciens*. Also, citric acid present only in cucumber exudates specifically promote formation of biofilms by *B. amyloliquefaciens* SQR9, its cognate PGPR, but not by *B. subtilis* N11 (Zhang et al., 2013).

Composition of exudates and their capacity to attract various rhizobacteria can differ according to the plant's physiological state. It has been reported that root of *A. thaliana* infected with the pathogen *Pseudomonas syringae* start secreting malic acid, which in turn act as a strong attractant for the beneficial *B. subtilis* (Rudrappa et al., 2008). Similarly, cucumber roots infected by *Fusarium oxysporum* secrete more citric and fumaric acid, stimulating the chemotactic response of the protective *B. amyloliquefaciens* strain SQR9 toward the plant (Liu et al., 2014). These two examples elegantly illustrate how a plant can modulate its root-associated microbial community via attracting molecules to help it cope with environmental stresses such as pathogens.

3. NUTRITION

Most attracting molecules identified in root exudates can also be metabolized by bacteria, thus serving as a carbon source. This is particularly true for *A. brasilense* for which only metabolizable substrates can act as attractants, and their attraction potential directly correlates with their ability to support growth. With this energy taxis behaviour *A. brasilense* seeks the environment allowing its optimal growth, such as the plant rhizosphere (Alexandre, Greer, & Zhulin, 2000).

PGPR can use many of the carbon compounds secreted by the root as a carbon source. Organic acids, which are much more abundant than sugars in exudates, are the main carbon source used by *P. fluorescens* while colonizing tomato roots (Somers et al., 2004). A good root-colonizing strain of *Pseudomonas putida* can grow optimally on citric acid, succinic acid and glucose, the most dominant carbon sources present in the exudates from its host plant the grass cultivar Barmultra (Kuiper, Kravchenko, Bloemberg, & Lugtenberg, 2002). *Pseudomonas fluorescens* and *Pseudomonas chlororaphis* also show efficient growth on a 3:1 mixture of organic acid and sugars, a proportion which is typical of root exudates (Kravchenko et al., 2003). The high molecular weight fraction of root exudates, the mucilage, can also be used as a nutrient. These complex polysaccharides can support, to some extent, the growth of pea rhizosphere *P. fluorescens*, *Rhizobium leguminosarum* and *Burkholderia cepacia* strains (Knee et al., 2001). Plant polysaccharides can also be a source of galactose for *B. subtilis* while colonizing the root of *A. thaliana* (Beauregard, Chai, Vlamakis, Losick, & Kolter, 2013).

However, for certain bacterial species such as *P. fluorescens*, the capacity to utilize root secreted sugars is not the only important trait required for colonization (Lugtenberg, Kravchenko, & Simons, 1999). Using an extensive number of *P. fluorescens* strains, De La Fuente et al. did not find any link between the root colonization capacities of the strains and their growth capability on plant exudates or their predominant carbon sources (De La Fuente, Mavrodi, Thomashow, & Weller, 2007). Similarly, a study performed with a *P. putida* strain isolated from the apple tree rhizosphere could not establish a correlation between survival on the root and the efficiency to use root exudates as nutrients (Gu & Mazzola, 2001). It also appears that the amino acids present in root exudates are in concentration too low to support colonization of the root by amino acid auxotroph *P. fluorescens* (Simons, Permentier, Weger, Wijffelman, & Lugtenberg, 1997). These studies point to the fact that root exudates are definitely a nutrient source for plant-beneficial bacteria, but this mechanism is not the only way by which roots are promoting their colonization by PGPR.

4. COLONIZATION

To colonize plant roots, many plant growth-promoting rhizospheric bacteria such as *A. brasilense*, *Gluconacetobacter diazotrophicus*, *Herbaspirillum seropedicae* and *P. fluorescens* need to secrete exopolysaccharides (EPSs) (Balsanelli et al., 2010; Bianciotto et al., 2001; Meneses, Rouws, Simoes-Araujo, Vidal, & Baldani, 2011; Michiels, Vanderleyden, Gool, & Signer, 1988). Depending on the bacterial species, other extracellular polymeric substances can also be involved in colonization, as in the case of *P. putida* where synthesis of the extracellular matrix large adhesion protein F is required (Martínez-Gil, Yousef-Coronado, & Espinosa-Urgel, 2010). Secretion of these polymeric substances by bacteria is a highlight of biofilm formation, a state in which bacteria form a sessile community living embedded in a self-secreted matrix (reviewed in Flemming & Wingender, 2010). These observations thus suggest that biofilm formation is involved in root colonization, a hypothesis further supported by microscopy examination of native *Pseudomonas* species on wheat roots and of *Paenibacillus polymyxa* and *B. subtilis* on *A. thaliana* roots (Beauregard et al., 2013; Timmusk,

Grantcharova, & Wagner, 2005; Watt, Hugenholtz, White, & Vinall, 2006).

Biofilm formation is triggered by various signals that differ from one bacterial species to another. One of them is quorum sensing (QS), a cellular process by which bacteria sense their own population density via the detection of specific secreted molecules called autoinducers. QS is involved in biofilm formation by various bacteria and notably by Gram-negative ones such as *Pseudomonas*, where many of the predominant autoinducers belongs to the family of *N*-acylhomoserine lactones (Danhorn & Fuqua, 2007). Red algae, seedlings of *Medicago truncatula*, *Medicago sativa* or peas secrete compounds capable of stimulating or inhibiting QS sensing in different Gram-negative reporting bacteria. This activity seems to be caused by molecules mimicking *N*-acylhomoserine lactones (Gao, Teplitski, Robinson, & Bauer, 2003; Givskov et al., 1996; Keshavan, Chowdhary, Haines, & Gonzalez, 2005; Teplitski, Robinson, & Bauer, 2000). One could think that these plants can directly influence biofilm formation, and thus possibly root colonization, via modulation of QS signalling by beneficial or pathogenic Gram-negative bacteria found at the rhizosphere.

Biofilm formation is pinpointed in several studies as an essential trait for plant root colonization by *Bacillus* spp. (Bais et al., 2004; Beauregard et al., 2013; Chen et al., 2013; Weng, Wang, Li, Shen, & Zhang, 2013). Various signals and environmental cues, some of which are found abundantly on plant roots, trigger *B. subtilis* biofilms. Exudates from cucumber roots, infected or not with *F. oxysporum*, can stimulate biofilm formation by the beneficial *B. amyloliquefaciens* SQR9. Analysis of the root components reveals that fumarate and citrate might have a small effect on biofilm formation, but also that other unidentified molecules present in the exudates act on this bacterium (Liu et al., 2014; Zhang et al., 2013). Tomato root exudates and the various organic acids it contains (oxalate, malate, succinate, citrate and fumarate) have little to no effect on biofilm formation by the *B. amyloliquefaciens* strain T-5. However, they strongly induce swarming of the bacteria, a multicellular behaviour of coordinated and rapid movement on a solid or semisolid surface that would likely favour good colonization along the root (Tan et al., 2013).

Tomato roots exudates can trigger biofilm formation by *B. subtilis*, and further analysis of its components show that L-malate is one of the molecules involved in that process. This small organic acid is also secreted by *A. thaliana* when infected by *P. syringae*, but not when the plant is not infected. This observation suggests that tomato roots or infected *A. thaliana* roots recruit

B. subtilis via chemotaxis toward L-malate (see earlier) and then the same molecule helps promote biofilm formation, at least to some extent (Rudrappa et al., 2008). L-malate effect is specific, since D-malate or other organic acid such as oxalate, citrate or fumarate do not influence *B. subtilis* biofilm formation (Chen et al., 2012; Rudrappa et al., 2008). However, the high concentration of L-malate required to stimulate biofilm formation points to an indirect effect rather than to a direct signal. Also, this organic acid probably acts on *B. subtilis* in combination with other molecules from the exudates or from the plant surface (Beauregard et al., 2013; Chen et al., 2012; Lakshmanan & Bais, 2013). Indeed, *B. subtilis* colonization of *A. thaliana* roots is affected by various plant genes, suggesting that many plant cellular pathways are involved in inducing root colonization by *B. subtilis*, thus suggesting that cumulative effects of the plant on *B. subtilis* favour colonization (Lakshmanan & Bais, 2013; Lakshmanan, Castaneda, Rudrappa, & Bais, 2013).

Interestingly, plant polysaccharides can induce biofilm formation by the beneficial *B. subtilis* 3610, GB03 and *B. amyloliquefaciens* FZB42 strains. This discovery suggests that soon after reaching the root, possibly following attraction process via chemotaxis, *B. subtilis* starts forming a biofilm and thus colonizes the surface. Interestingly, in vitro biofilm formation by *B. subtilis* is mediated via five kinases whose activities are regulated by various intracellular and extracellular signals. However, plant polysaccharides, and especially xylan, only partially require these kinases to induce biofilm. This observation suggests the existence of an alternative cellular pathway involved in *B. subtilis* biofilm formation in the rhizosphere (Beauregard et al., 2013). A similar conclusion is drawn for *P. fluorescens*, where mutant strains defective for biofilm formation on abiotic surfaces are still able to colonize alfalfa roots to an extent similar to wild-type cells. The authors of this study conclude that the genetic program allowing biofilm formation on the rhizoplane is distinct from the one involved in abiotic surfaces, and that the two structures are not equivalent (Barahona et al., 2010). Future research on biofilm formation in the rhizoplane should take into account that these pathways might be impossible to recapitulate in vitro with our current knowledge. At the same time, it also suggests that many genes involved in the interaction with the plant have yet to be discovered, opening an exciting new area of research.

Plant molecules can induce or repress production of the biofilm matrix, but some of them can also modulate the matrix content of extracellular polymeric substances. Methyl-salicylate, a molecule synthesized in the plant's infected tissues, induces in *B. subtilis* a large increase of poly-γ-glutamic acid secretion. This polymer gives a mucoid aspect to the colonies, and thus could act as a protective capsule for *B. subtilis* against the pathogens infecting the plant and their detrimental molecules (Kobayashi, 2014). In *A. brasilense* Cd, root exudates can change the sugar composition of the EPS secreted by the bacteria, increasing the amount of arabinose and xylose units present (Fischer, Miguel, & Mori, 2003). Interestingly, an increased amount of arabinose in *A. brasilense* EPS is linked to an increase in cell aggregation, with each other and with the plant root, which could favour strong colonization (Bahat-Samet, Castro-Sowinski, & Okon, 2004; Jofré, Lagares, & Mori, 2004). Finally, some plant compounds, such as catechol, can inhibit *B. subtilis* root colonization. Catechol, a degradation product from the salicylic acid pathway, generates reactive oxygen species that strongly inhibit binding and/or biofilm formation of *B. subtilis* on the root (Rudrappa, Quinn, Stanley-Wall, & Bais, 2007). These examples illustrate that roots are not passively colonized by PGPR, but have ways to influence it, mostly positively but sometimes negatively.

5. GENERAL EFFECTS

Plant roots, and more specifically their exudates, have effects on plant growth-promoting bacteria that go beyond attraction or colonization enhancement. Indeed, they are also able to modulate an important proportion of the cell's transcriptome. Chaparro et al. recently demonstrated that *A. thaliana* exudates composition varies along plant age, and that the transcriptome of the associated microorganisms varies accordingly. Many of the mRNAs showing important variations are associated with metabolism, which correlates with the fact that the microorganisms adapt to the varying exudates metabolites composition (Chaparro et al., 2013). In *A. brasilense*, exudates of various plants, and more specifically the presence of D-galactose, L-arabinose and D-fucose in these exudates induce the expression of the *sbpA* gene. SbpA is required for chemotaxis toward these sugars as well as for optimal D-galactose uptake, thus creating a positive feedback loop allowing the optimal usage of this sugar (Van Bastelaere, De Mot, Michiels, & Vanderleyden, 1993; Van Bastelaere et al., 1999). *Azospirillum brasilense* also shows an extensive transcriptional response to auxin indole-3-acetic acid (IAA), a phytohormone produced both by that bacterium and the plant. Since many various genes such as those from secretion systems are involved in this response, it suggests that IAA acts as a signalling molecule in the plant–*Azospirillum* interaction (Van Puyvelde et al., 2011).

Flavonoids are a family of molecules secreted by roots of higher plants, which are well known for their role in endosymbiotic rhizobacteria recruitment and influence on various PGP bacteria (Badri, Weir, van der Lelie, & Vivanco, 2009). In *B. subtilis*, some flavonoids release the transcriptional inhibition mediated by the LmrA and QdoR proteins, allowing the expression of a multidrug efflux transporter (LmrB), of a flavonoids degradation enzyme (QdoI) and of a putative adenine dinucleotide (FAD)-dependent monooxygenase (YetM) (Hirooka, 2014; Hirooka et al., 2009, 2007). Similarly in *P. putida*, some flavonoids increase the expression of the TtgABC efflux pump (Terán, Krell, Ramos, & Gallegos, 2006). *Pseudomonas putida* gene expression is also influenced by DIMBOA, a secondary defence benzoxazinoids compound released by the plants during early growth stages (Frey, Schullehner, Dick, Fiesselmann, & Gierl, 2009). Exposure to this molecule induces the expression of genes involved in its degradation, in chemotaxis, as well as genes with other roles such as antibiotic production (Neal et al., 2012).

In a study using *B. amyloliquefaciens* FZB42 in pure culture, application of maize root exudates was shown to impact 8.2% of the transcriptome. Many of these genes are involved in carbon and nitrogen utilization, but there are also genes implicated in root colonization and plant-beneficial effect such as chemotaxis, motility, biofilm formation and secondary metabolites biosynthesis (Fan et al. 2012). The nutritional status of the plant can also influence the transcriptome of FZB42 via a modification of the root exudate composition. For example, exudates from nitrogen-starved plants repress genes involved in chemotaxis and motility, while phosphate-deficient maize exudates upregulate them. This correlates with the fact that *B. amyloliquefaciens* cannot help nitrogen acquisition by the plant and that there is no interest for the plant to attract it. On the other hand, FZB42 has the potential to help the plant acquire phosphate since it encodes for a phytase activity (Carvalhais, Dennis, Fan, et al., 2013). However, one should be careful in the interpretation of these transcriptomic results, since the extracellular proteome of *B. amyloliquefaciens* in response to maize exudates shows only a weak correlation between increased gene expression and increased secreted protein levels. In some cases though, genes are found to be modulated in both transcriptomic and proteomic studies, such as the biosynthesis of the plant-beneficial volatile acetoin (*alsS*) and motility (*fliD*), highlighting their importance (Kierul et al., 2014).

Bacteria other than *Bacillus* see their extracellular proteome modulated following exposition to root exudates. *Paenibacillus polymyxa* E681 have a significant number of proteins involved in metabolism, motility and bacteria–host interaction among other function that are up- or downregulated when bacteria are grown in the presence of barley plants (Seul, Park, Ryu, Lee, & Ghim, 2007). A *Bacillus cereus* strain that can promote tobacco growth shows alteration of its secreted proteome in presence of tobacco-root exudates and groundnut-root exudates. In response to the plant's exudates, *B. cereus* can modify its surface chemical composition, suggesting an adaptation that can favour strong root colonization (Dutta, Rani, & Podile, 2013). These studies demonstrate that the plant presence deeply influence the bacteria with which it associates, influencing cellular processes as diverse as secondary metabolites biosynthesis, metabolism, transport and motility.

6. COMPOUNDS FROM THE PLANT THAT AFFECT THE PGP EFFECT OF BACTERIA

Bacteria-secreted molecules mediate many of the positive effects of PGPR on the plant. Among these molecules, we find auxin IAA, a phytohormone involved in plant growth and development produced both by phytopathogenic bacteria and PGPR. Interestingly, the plant can affect PGPR's IAA production by secreting tryptophan in its root exudates. Tryptophan is a precursor of IAA biosynthesis which will activate gene transcription, upregulating IAA production in the beneficial *P. putida* and *A. brasilense* (Ona, Van Impe, Prinsen, & Vanderleyden, 2005; Patten & Glick, 2002; Rothballer, Schmid, Fekete, & Hartmann, 2005; Ryu & Patten, 2008). In *A. brasilense*, IAA synthesis is also stimulated by IAA or other natural or synthetic auxins, which can be seen as a positive feedback signal coming from the plant or the bacterium itself (Vande Broek, Lambrecht, Eggermont, & Vanderleyden, 1999).

In some cases, biosynthesis of plant-beneficial enzymes or compounds appears to be under a very complex regulation mediated by plant molecules. *Pseudomonas putida* strain UW4 encodes a 1-aminocyclopropane-1-carboxylate (ACC) deaminase, which breaks down the precursor of ethylene ACC, and thus lowers its amount in plants. A reduced amount of ethylene promotes plant growth in presence of stress. The gene encoding the ACC deaminase is activated by ACC and repressed by leucine. This double regulation system suggests a tight regulation when bacteria are in the presence of the plant since ethylene can be beneficial in some situations and detrimental in others (Cheng, Duncker, McConkey, & Glick, 2008). *Pseudomonas fluorescens* CHA0 is known to produce two antifungal compounds, 2,4-diacetylphloroglucinol (DAPG) and pyoluteorin, which production is influenced by many plant-derived molecules (De Werra, Huser, Tabacchi, Keel, & Maurhofer, 2011). A solo receptor from the LuxR family, PsoR, could be involved in this process since it was shown to bind unidentified plant molecules and then activate the transcription of DAPG biosynthetic genes, among others genes (Subramoni et al., 2011). Production of DAPG by *P. fluorescens* varies also according to plant health status. In a split root system, the DAPG biosynthesis gene *phlA* sees its expression upregulated when in contact with roots of barley but only when the other roots' half is infected with the oomycete *Pythium ultimum*. This is likely due to the increased secretion by the infected plant of vanillic, fumaric and *p*-coumaric acids that all upregulate the expression of *phlA* in *P. fluorescens* (Jousset et al., 2011). Other plants can influence DAPG production via the expression of its biosynthesis genes, since exudates from *A. thaliana* repress transcription of *phlD*, also involved in its biosynthesis, while exudates from *M. truncatula*, *Brassica napus* and *Triticum aestivum* activate it (Haichar, Fochesato, & Achouak, 2013).

Plant compounds can also modulate antibiotic secretion by Gram-positive organisms from the *Bacillus* genus. In *B. subtilis*, plant methyl-salicylate, which is synthesized in infected tissues, induces the production of the antibiotics bacilysin and fengycin via activation of the two-component system DegS-DegU, thus activating antibiosis activity (Kobayashi, 2014). Other plant factors probably also influence antibiotic secretion, since *B. amyloliquefaciens* S499 grown in contact of tomato roots produces much lower amount of iturins and fengycins relatively to surfactin than when grown in planktonic conditions. Additionally, surfactin molecules produced *in planta* are enriched in long-chain C_{14} and C_{15}. This difference in cyclic lipopeptides biosynthesis might be explained by various factors inherent to root colonization by *B. amyloliquefaciens*, such as the abundance of organic acids, important components of tomato root exudates, and oxygen availability (Nihorimbere et al., 2012).

Finally, plants can affect other beneficial aspects of PGPR. In *A. brasilense* and *Azospirillum liporefum*, binding of the lectin wheat germ agglutinin or *Datura stramonium* agglutinin on the bacterial capsule enhances transcription of genes related to nitrogen fixation, thus enhancing their nitrogenase capacity when in the presence of the root (Karpati et al., 1999). *Pseudomonas putida* strain PCL1444 degrades the pollutant naphthalene, which can be toxic to the plant. Grass cv. Barmultra can positively influence this activity since the degradation genes are most expressed in presence of salicylic acid, which could come from the plant, and of the major carbon sources present in the grass' exudates (Kuiper et al., 2002). These examples illustrate that the plant not only stimulates its own colonization by PGPR, but also favours the activation of the plant-beneficial bacterial activities.

7. CONCLUSION AND PERSPECTIVES

This chapter reviews the current knowledge on signals and/or environmental cues provided by the plant that affect PGPR behaviours. Many plant molecules, and in particular small metabolites present in root exudates, seem to influence positively PGPR. These effects are mainly about attracting the bacteria via chemotaxis, promoting root colonization via biofilm formation, providing nutrients that help bacterial growth and enhancing the bacteria plant-beneficial effects. However, research on this subject is at its very beginning, since the bacterial cellular mechanisms involved in responding to the plant signals molecules have been elucidated in only a very few cases. For example, while many molecules secreted by the plant act as attractants, few chemoreceptors binding these attractants have been identified, and their importance for colonization remains to be explored. Similarly, the molecules involved in many effects induced by the plant remain to be identified, such as the different QS mimics affecting Gram-negative or the compounds modulating antimicrobials secretion in *B. amyloliquefaciens*. A better understanding of the communication between the plant and the PGPR will open the door to engineer or select strains that are better fit to interact with plants of agricultural importance, and to broaden our knowledge of environmentally relevant interactions between bacteria and their surroundings.

REFERENCES

Alexandre, G. M., Greer, S., & Zhulin, I. (2000). Energy taxis is the dominant behavior in *Azospirillum brasilense*. *Journal of Bacteriology, 182*(21), 6042–6048.

Badri, D. V., Weir, T. L., van der Lelie, D., & Vivanco, J. M. (2009). Rhizosphere chemical dialogues: plant-microbe interactions. *Current Opinion in Biotechnology, 20*(6), 642–650.

Bahat-Samet, E., Castro-Sowinski, S., & Okon, Y. (2004). Arabinose content of extracellular polysaccharide plays a role in cell aggregation of *Azospirillum brasilense*. *FEMS Microbiology Letters, 237*, 195–203.

Bais, H. P., Broeckling, C. D., & Vivanco, J. M. (2008). Root exudates modulate plant – microbe interactions in the rhizosphere. In P. Karlovsky (Ed.), *Soil biology: Vol. 14. Secondary metabolites in soil ecology* (pp. 241–252). Berlin Heidelberg: Springer-V.

Bais, H. P., Fall, R., & Vivanco, J. (January 2004). Biocontrol of *Bacillus subtilis* against infection of *Arabidopsis* roots by *Pseudomonas syringae* is facilitated by biofilm formation and surfactin production. *Plant Physiology, 134*, 307–319.

Bais, H. P., Weir, T. L., Perry, L. G., Gilroy, S., & Vivanco, J. M. (2006). The role of root exudates in rhizosphere interactions with plants and other organisms. *Annual Review of Plant Biology, 57*, 233–266.

Balsanelli, E., Serrato, R. V., de Baura, V. A., Sassaki, G., Yates, M. G., Rigo, L. U., et al. (2010). *Herbaspirillum seropedicae* rfbB and rfbC genes are required for maize colonization. *Environmental Microbiology, 12*(8), 2233–2244.

Barahona, E., Navazo, A., Yousef-Coronado, F., Aguirre de Cárcer, D., Martínez-Granero, F., Espinosa-Urgel, M., et al. (2010). Efficient rhizosphere colonization by *Pseudomonas fluorescens* f113 mutants unable to form biofilms on abiotic surfaces. *Environmental Microbiology, 12*(12), 3185–3195.

Bashan, Y., & Holguin, G. (1994). Root-to-root travel of the beneficial bacterium *Azospirillum brasilense*. *Applied and Environmental Microbiology, 60*(6), 2120–2131.

Beauregard, P. B., Chai, Y., Vlamakis, H., Losick, R., & Kolter, R. (2013). Bacillus subtilis biofilm induction by plant polysaccharides. *PNAS, 110*(17), E1621–E1630.

Beneduzi, A., Ambrosini, A., & Passaglia, L. M. P. (2012). Plant growth-promoting rhizobacteria (PGPR): their potential as antagonists and biocontrol agents. *Genetics and Molecular Biology, 35*(4 Suppl.), 1044–1051.

Berg, G. (2009). Plant-microbe interactions promoting plant growth and health: perspectives for controlled use of microorganisms in agriculture. *Applied Microbiology and Biotechnology, 84*(1), 11–18.

Berg, G., & Smalla, K. (2009). Plant species and soil type cooperatively shape the structure and function of microbial communities in the rhizosphere. *FEMS Microbiology Ecology, 68*(1), 1–13.

Bianciotto, V., Andreotti, S., Balestrini, R., Bonfante, P., & Perotto, S. (2001). Mucoid mutants of the biocontrol strain *Pseudomonas fluorescens* CHA0 show increased ability in biofilm formation on mycorrhizal and nonmycorrhizal carrot roots. *MPMI, 14*(2), 255–260.

Bloemberg, G. V., & Lugtenberg, B. J. J. (2001). Molecular basis of plant growth promotion and biocontrol by rhizobacteria. *Current Opinion in Plant Biology, 4*(4), 343–350.

Buchan, A., Crombie, B., & Alexandre, G. M. (2010). Temporal dynamics and genetic diversity of chemotactic-competent microbial populations in the rhizosphere. *Environmental Microbiology, 12*(12), 3171–3184.

Bulgarelli, D., Rott, M., & Schlaeppi, K. (2012). Revealing structure and assembly cues for *Arabidopsis* root-inhabiting bacterial microbiota. *Nature, 488*(7409), 91–95.

Bull, C. T., Weller, D. M., & Thomashow, L. S. (1991). Relationship between root colonization and suppression of *Gaeumannomyces graminis* var. *tritici* by *Pseudomonas fluorescens* strain 2-79. *Phytopathology, 81*(9), 954–959.

Carvalhais, L. C., Dennis, P. G., Badri, D. V., Tyson, G. W., Vivanco, J. M., & Schenk, P. M. (2013). Activation of the jasmonic acid plant defence pathway alters the composition of rhizosphere bacterial communities. *PLoS One, 8*(2), e56457.

Carvalhais, L. C., Dennis, P. G., Fan, B., Fedoseyenko, D., Kierul, K., Becker, A., et al. (2013). Linking plant nutritional status to plant-microbe interactions. *PLoS One, 8*(7), e68555.

Chaparro, J. M., Badri, D. V., Bakker, M. G., Sugiyama, A., Manter, D. K., & Vivanco, J. M. (2013). Root exudation of phytochemicals in *Arabidopsis* follows specific patterns that are developmentally programmed and correlate with soil microbial functions. *PLoS One, 8*(2), e55731.

Chaparro, J. M., Badri, D. V., & Vivanco, J. M. (2014). Rhizosphere microbiome assemblage is affected by plant development. *The ISME Journal, 8*(4), 790–803.

Chen, Y., Cao, S., Chai, Y., Clardy, J., Kolter, R., Guo, J., et al. (2012). A *Bacillus subtilis* sensor kinase involved in triggering biofilm formation on the roots of tomato plants. *Molecular Microbiology, 85*(3), 418–430.

Chen, Y., Yan, F., Chai, Y., Liu, H., Kolter, R., Losick, R., et al. (2013). Biocontrol of tomato wilt disease by *Bacillus subtilis* isolates from natural environments depends on conserved genes mediating biofilm formation. *Environmental Microbiology, 15*(3), 848–864.

Cheng, Z., Duncker, B. P., McConkey, B. J., & Glick, B. R. (2008). Transcriptional regulation of ACC deaminase gene expression in *Pseudomonas putida* UW4. *Canadian Journal of Microbiology, 54*(2), 128–136.

Chin-A-Woeng, T. (2000). Root colonization by phenazine-1-carboxamide-producing bacterium *Pseudomonas chlororaphis* PCL1391 is essential for biocontrol of tomato foot and root rot. *MPMI, 13*(12), 1340–1345.

Czarnota, M., Rimando, A., & Weston, L. (2003). Evaluation of root exudates of seven sorghum accessions. *Journal of Chemical Ecology, 29*(9), 2073–2083.

Danhorn, T., & Fuqua, C. (2007). Biofilm formation by plant-associated bacteria. *Annual Review of Microbiology, 61*, 401–422.

De La Fuente, L., Mavrodi, D. V., Thomashow, L. S., & Weller, D. M. (2007). Utilization of trehalose, benzoate, valerate, and seed and root exudates by genotypes of 2,4-diacetylphloroglucinol producing *Pseudomonas fluorescens*. *Soil Biology and Biochemistry, 39*(11), 2712–2722.

De Weert, S., Vermeiren, H., Mulders, I. H. M., Kuiper, I., Hendrickx, N., Bloemberg, G. V., et al. (2002). Flagella-driven chemotaxis towards exudate components is an important trait for tomato root colonization by *Pseudomonas fluorescens*. *MPMI, 15*(11), 1173–1180.

De Werra, P., Huser, A., Tabacchi, R., Keel, C., & Maurhofer, M. (2011). Plant- and microbe-derived compounds affect the expression of genes encoding antifungal compounds in a pseudomonad with biocontrol activity. *Applied and Environmental Microbiology, 77*(8), 2807–2812.

Dutta, S., & Podile, A. R. (2010). Plant growth promoting rhizobacteria (PGPR): the bugs to debug the root zone. *Critical Reviews in Microbiology, 36*(3), 232–244.

Dutta, S., Rani, T. S., & Podile, A. R. (2013). Root exudate-induced alterations in *Bacillus cereus* cell wall contribute to root colonization and plant growth promotion. *PLoS One, 8*(10), e78369.

Fan, B., Carvalhais, L. C., Becker, A., Fedoseyenko, D., von Wirén, N., & Borriss, R. (2012). Transcriptomic profiling of *Bacillus amyloliquefaciens* FZB42 in response to maize root exudates. *BMC Microbiology, 12*, 116.

Fan, T., Lane, A., Shenker, M., & Bartley, J. (2001). Comprehensive chemical profiling of gramineous plant root exudates using high-resolution NMR and MS. *Phytochemistry, 57*, 209–221.

Fischer, S. E., Miguel, M. J., & Mori, G. B. (2003). Effect of root exudates on the exopolysaccharide composition and the lipopolysaccharide profile of *Azospirillum brasilense* Cd under saline stress. *FEMS Microbiology Letters, 219*(1), 53–62.

Flemming, H.-C., & Wingender, J. (2010). The biofilm matrix. *Nature Reviews Microbiology, 8*(9), 623–633. http://dx.doi.org/10.1038/nrmicro2415.

Frey, M., Schullehner, K., Dick, R., Fiesselmann, A., & Gierl, A. (2009). Benzoxazinoid biosynthesis, a model for evolution of secondary metabolic pathways in plants. *Phytochemistry, 70*(15–16), 1645–1651.

Gafny, R., Okon, Y., Kapulnik, Y., & Fischer, M. (1986). Adsorption of *Azospirillum brasilense* to corn roots. *Soil Biology and Biochemistry, 18*(I), 69–75.

Gao, M., Teplitski, M., Robinson, J. B., & Bauer, W. D. (2003). Production of substances by *Medicago truncatula* that affect bacterial quorum sensing. *MPMI, 16*(9), 827–834.

Givskov, M., de Nys, R., Manefield, M., Gram, L., Maximilien, R., Eberl, L., et al. (1996). Eukaryotic interference with homoserine lactone-mediated prokaryotic signalling. *Journal of Bacteriology, 178*(22), 6618–6622.

Glick, B. R. (2012). Plant growth-promoting bacteria: mechanisms and applications. *Scientifica, 2012*, 963401.

Gu, Y.-H., & Mazzola, M. (2001). Impact of carbon starvation on stress resistance, survival in soil habitats and biocontrol ability of *Pseudomonas putida* strain 2C8. *Soil Biology and Biochemistry, 33*(9), 1155–1162.

Haggag, W. M., & Timmusk, S. (2008). Colonization of peanut roots by biofilm-forming *Paenibacillus polymyxa* initiates biocontrol against crown rot disease. *Journal of Applied Microbiology, 104*(4), 961–969.

Haichar, F., Fochesato, S., & Achouak, W. (2013). Host plant specific control of 2,4-diacetylphloroglucinol production in the rhizosphere. *Agronomy, 3*(4), 621–631.

Haichar, F. E. Z., Santaella, C., Heulin, T., & Achouak, W. (2014). Root exudates mediated interactions belowground. *Soil Biology and Biochemistry, 77*, 69–80.

Heinrich, D., & Hess, D. (1985). Chemotactic attraction of *Azospirillum lipoferum* by wheat roots and characterization of some attractants. *Canadian Journal of Microbiology, 31*, 26–31.

Hirooka, K. (2014). Transcriptional response machineries of *Bacillus subtilis* conducive to plant growth promotion. *Bioscience, Biotechnology, and Biochemistry, 78*(9), 1471–1484.

Hirooka, K., Danjo, Y., Hanano, Y., Kunikane, S., Matsuoka, H., Tojo, S., et al. (2009). Regulation of the *Bacillus subtilis* divergent yetL and yetM genes by a transcriptional repressor, YetL, in response to flavonoids. *Journal of Bacteriology, 191*(11), 3685–3697.

Hirooka, K., Kunikane, S., Matsuoka, H., Yoshida, K.-I., Kumamoto, K., Tojo, S., et al. (2007). Dual regulation of the *Bacillus subtilis* regulon comprising the lmrAB and yxaGH operons and yxaF gene by two transcriptional repressors, LmrA and YxaF, in response to flavonoids. *Journal of Bacteriology, 189*(14), 5170–5182.

Jofré, E., Lagares, A., & Mori, G. (2004). Disruption of dTDP-rhamnose biosynthesis modifies lipopolysaccharide core, exopolysaccharide production, and root colonization in *Azospirillum brasilense*. *FEMS Microbiology Letters, 231*, 267–275.

Jousset, A., Rochat, L., Lanoue, A., Bonkowski, M., Keel, C., & Scheu, S. (2011). Plants respond to pathogen infection by enhancing the antifungal gene expression of root-associated bacteria. *Molecular Plant-Microbe Interactions, 24*(3), 352–358.

Karpati, E. V. A., Kiss, P., Ponyi, T., Fendrik, I., de Zamaroczym, M., & Orosz, L. (1999). Interaction of *Azospirillum lipoferum* with wheat germ agglutinin stimulates nitrogen fixation. *Journal of Bacteriology, 181*(13), 3949–3955.

Keshavan, N. D., Chowdhary, P. K., Haines, D. C., & Gonzalez, J. E. (2005). L-canavanine made by *Medicago sativa* interferes with quorum sensing in *Sinorhizobium meliloti*. *Journal of Bacteriology, 187*(24), 8427–8436.

Kierul, K., Voigt, B., Albrecht, D., Chen, X.-H., Carvalhais, L. C., & Borriss, R. (January 2014). Influence of root exudates on the extracellular proteome of the plant-growth promoting bacterium *Bacillus amyloliquefaciens* FZB42. *Microbiology, 161*, 131–147.

Kloepper, J., Leong, J., Teintze, M., & Schroth, M. (1980). Enhanced plant growth by siderophores produced by plant growth-promoting rhizobacteria. *Nature, 286*, 885–886.

Knee, E., Gong, F., Gao, M., Teplitski, M., Jones, A., Foxworthy, A., et al. (2001). Root mucilage from pea and its utilization by rhizosphere bacteria as a sole carbon source. *Molecular Plant-Microbe Interactions, 14*(6), 775–784.

Kobayashi, K. (2014). Plant methyl salicylate induces defense responses in the rhizobacterium *Bacillus subtilis*. *Environmental Microbiology*. http://dx.doi.org/10.1111/1462-2920.12613.

Kravchenko, L., Azarova, T., Leonova-Erko, E., Shaposhnikov, A. I., Makarova, N., & Tikhonovich, I. A. (2003). Root exudates of tomato plants and their effect on the growth and antifungal activity of *Pseudomonas* strains. *Microbiology, 72*(1), 37–41.

Kuiper, I., Kravchenko, L. V., Bloemberg, G. V., & Lugtenberg, B. J. J. (2002). *Pseudomonas putida* strain PCL1444, selected for efficient root colonization and naphthalene degradation, effectively utilizes root exudate components. *Molecular Plant-Microbe Interactions, 15*(7), 734–741.

Kuzyakov, Y., & Domanski, G. (2000). Carbon input by plants into the soil. *Journal of Plant Nutrition and Soil Science, 163*, 421–431.

Lakshmanan, V., & Bais, H. P. (2013). Factors other than root secreted malic acid that contributes toward. *Plant Signaling and Behavior, 8*(11), e27277.

Lakshmanan, V., Castaneda, R., Rudrappa, T., & Bais, H. P. (2013). Root transcriptome analysis of *Arabidopsis thaliana* exposed to beneficial *Bacillus subtilis* FB17 rhizobacteria revealed genes for bacterial recruitment and plant defense independent of malate efflux. *Planta, 238*(4), 657–668.

Lebeis, S. L. (June 2014). The potential for give and take in plant-microbiome relationships. *Frontiers in Plant Science, 5*, 287.

Ling, N., Raza, W., Ma, J., Huang, Q., & Shen, Q. (2011). Identification and role of organic acids in watermelon root exudates for recruiting *Paenibacillus polymyxa* SQR-21 in the rhizosphere. *European Journal of Soil Biology, 47*(6), 374–379.

Liu, Y., Zhang, N., Qiu, M., Feng, H., Vivanco, J. M., Shen, Q., et al. (2014). Enhanced rhizosphere colonization of beneficial *Bacillus amyloliquefaciens* SQR9 by pathogen infection. *FEMS Microbiology Letters, 353*(1), 49–56.

Lugtenberg, B. J. J., & Kamilova, F. (2009). Plant-growth-promoting rhizobacteria. *Annual Review of Microbiology, 63*, 541–556.

Lugtenberg, B. J. J., Kravchenko, L. V., & Simons, M. (1999). Tomato seed and root exudate sugars: composition, utilization by *Pseudomonas* biocontrol strains and role in rhizosphere colonization. *Environmental Microbiology, 1*(5), 439–446.

Lundberg, D., Lebeis, S., & Paredes, S. (2012). Defining the core *Arabidopsis thaliana* root microbiome. *Nature, 488*(7409), 86–90.

Martínez-Gil, M., Yousef-Coronado, F., & Espinosa-Urgel, M. (June 2010). LapF, the second largest *Pseudomonas putida* protein, contributes to plant root colonization and determines biofilm architecture. *Molecular Microbiology, 77*, 549–561.

Mendes, R., Garbeva, P., & Raaijmakers, J. M. (2013). The rhizosphere microbiome: significance of plant beneficial, plant pathogenic, and human pathogenic microorganisms. *FEMS Microbiology Reviews, 37*(5), 634–663.

Meneses, C. H., Rouws, L. F., Simoes-Araujo, J. L., Vidal, M. S., & Baldani, J. I. (2011). Exopolysaccharides production is required for biofilm formation and plant colonization by the nitrogen-fixing endophyte *Gluconacetobacter diazotrophicus*. *MPMI, 12*, 1448–1458.

Micallef, S. A., Shiaris, M. P., & Colón-Carmona, A. (2009). Influence of *Arabidopsis thaliana* accessions on rhizobacterial communities and natural variation in root exudates. *Journal of Experimental Botany, 60*(6), 1729–1742.

Michiels, K. W., Vanderleyden, J. O. S., Gool, A. P. V. A. N., & Signer, E. R. (1988). Isolation and characterization of *Azospirillum brasilense* loci that correct *Rhizobium meliloti* exoB and exoC mutations. *Journal of Bacteriology, 170*(11), 5401–5404.

Neal, A. L., Ahmad, S., Gordon-Weeks, R., & Ton, J. (2012). Benzoxazinoids in root exudates of maize attract *Pseudomonas putida* to the rhizosphere. *PLoS One, 7*(4), e35498.

Nihorimbere, V., Cawoy, H., Seyer, A., Brunelle, A., Thonart, P., & Ongena, M. (2012). Impact of rhizosphere factors on cyclic lipopeptide signature from the plant beneficial strain *Bacillus amyloliquefaciens* S499. *FEMS Microbiology Ecology, 79*(1), 176–191.

Oku, S., Komatsu, A., Tajima, T., Nakashimada, Y., & Kato, J. (2012). Identification of chemotaxis sensory proteins for amino acids in *Pseudomonas fluorescens* Pf0-1 and their involvement in chemotaxis to tomato root exudate and root colonization. *Microbes and Environments, 27*(4), 462–469.

Ona, O., Van Impe, J., Prinsen, E., & Vanderleyden, J. (2005). Growth and indole-3-acetic acid biosynthesis of *Azospirillum brasilense* Sp245 is environmentally controlled. *FEMS Microbiology Letters, 246*(1), 125–132.

Parales, R. E., & Harwood, C. S. (2002). Bacterial chemotaxis to pollutants and plant-derived aromatic molecules. *Current Opinion in Microbiology, 5*(3), 266–273.

Patten, C. L., & Glick, B. R. (2002). Regulation of indoleacetic acid production in *Pseudomonas putida* GR12-2 by tryptophan and the stationary-phase sigma factor RpoS. *Canadian Journal of Microbiology, 48*(7), 635–642.

Peiffer, J., Spor, A., Koren, O., Jin, Z., Tringe, S. G., Dangl, J., et al. (2013). Diversity and heritability of the maize rhizosphere microbiome under field conditions. *PNAS, 110*(16), 6548–6553.

Reinhold, B., Hurek, T., & Fendrik, I. (1985). Strain-specific chemotaxis of *Azospirillum* spp. *Journal of Bacteriology, 162*(1), 190–195.

Richardson, A. E., Barea, J.-M., McNeill, A. M., & Prigent-Combaret, C. (2009). Acquisition of phosphorus and nitrogen in the rhizosphere and plant growth promotion by microorganisms. *Plant and Soil, 321*(1–2), 305–339.

Rothballer, M., Schmid, M., Fekete, A., & Hartmann, A. (2005). Comparative in situ analysis of ipdC-gfpmut3 promoter fusions of *Azospirillum brasilense* strains Sp7 and Sp245. *Environmental Microbiology,* 7(11), 1839–1846.

Rudrappa, T., Czymmek, K. J., Paré, P. W., & Bais, H. P. (2008). Root-secreted malic acid recruits beneficial soil bacteria. *Plant Physiology, 148*(3), 1547–1556.

Rudrappa, T., Quinn, W. J., Stanley-Wall, N. R., & Bais, H. P. (2007). A degradation product of the salicylic acid pathway triggers oxidative stress resulting in down-regulation of *Bacillus subtilis* biofilm formation on *Arabidopsis thaliana* roots. *Planta, 226*(2), 283–297.

Ryu, R. J., & Patten, C. L. (2008). Aromatic amino acid-dependent expression of indole-3-pyruvate decarboxylase is regulated by TyrR in *Enterobacter cloacae* UW5. *Journal of Bacteriology, 190*(21), 7200–7208.

Seul, K., Park, S., Ryu, C., Lee, Y., & Ghim, S. (2007). Proteome analysis of *Paenibacillus polymyxa* E681 affected by barley. *Journal of Microbiology and Biotechnology, 17*(6), 934–944.

Simons, M., Permentier, H., de Weger, L. A., Wijffelman, C. A., & Lugtenberg, B. J. J. (1997). Amino acid synthesis is necessary for tomato root colonization by *Pseudomonas fluorescens* strain WCS365. *MPMI, 10*(1), 102–106.

Somers, E., Vanderleyden, J., & Srinivasan, M. (2004). Rhizosphere bacterial signalling: a love parade beneath our feet. *Critical Reviews in Microbiology, 30*(4), 205–240.

Sood, S. G. (2003). Chemotactic response of plant-growth-promoting bacteria towards roots of vesicular-arbuscular mycorrhizal tomato plants. *FEMS Microbiology Ecology, 45,* 219–227.

Subramoni, S., Gonzalez, J. F., Johnson, A., Péchy-Tarr, M., Rochat, L., Paulsen, I., et al. (2011). Bacterial subfamily of LuxR regulators that respond to plant compounds. *Applied and Environmental Microbiology,* 77(13), 4579–4588.

Tan, S., Yang, C., Mei, X., Shen, S., Raza, W., Shen, Q., et al. (2013). The effect of organic acids from tomato root exudates on rhizosphere colonization of *Bacillus amyloliquefaciens* T-5. *Applied Soil Ecology, 64,* 15–22.

Teplitski, M., Robinson, J. B., & Bauer, W. D. (2000). Plants secrete substances that mimic bacterial N-acyl homoserine lactone signal activities and affect population density-dependent behaviors in associated bacteria. *MPMI, 13*(6), 637–648.

Terán, W., Krell, T., Ramos, J. L., & Gallegos, M.-T. (2006). Effector-repressor interactions, binding of a single effector molecule to the operator-bound TtgR homodimer mediates derepression. *The Journal of Biological Chemistry, 281*(11), 7102–7109.

Timmusk, S., Grantcharova, N., & Wagner, E. G. H. (2005). *Paenibacillus polymyxa* invades plant roots and forms biofilms. *Applied and Environmental Microbiology, 71*(11), 7292–7300.

Vacheron, J., Desbrosses, G., Bouffaud, M.-L., Touraine, B., Moënne-Loccoz, Y., Muller, D., et al. (September 2013). Plant growth-promoting rhizobacteria and root system functioning. *Frontiers in Plant Science, 4*, 356.

Van Bastelaere, E., De Mot, R., Michiels, K. W., & Vanderleyden, J. (1993). Differential gene expression in *Azospirillum* spp. by plant root exudates: analysis of protein profiles by two-dimensional polyacrylamide gel electrophoresis. *FEMS Microbiology Letters, 112*, 335–341.

Van Bastelaere, E., Lambrecht, M., Vermeiren, H., Van Dommelen, A., Keijers, V., Proost, P., et al. (1999). Characterization of a sugar-binding protein from *Azospirillum brasilense* mediating chemotaxis to and uptake of sugars. *Molecular Microbiology, 32*(4), 703–714.

Van Overbeek, L., & van Elsas, J. D. (2008). Effects of plant genotype and growth stage on the structure of bacterial communities associated with potato (*Solanum tuberosum* L.). *FEMS Microbiology Ecology, 64*(2), 283–296.

Van Puyvelde, S., Cloots, L., Engelen, K., Das, F., Marchal, K., Vanderleyden, J., et al. (2011). Transcriptome analysis of the rhizosphere bacterium *Azospirillum brasilense* reveals an extensive auxin response. *Microbial Ecology, 61*(4), 723–728.

Vande Broek, A., Lambrecht, M., Eggermont, K., & Vanderleyden, J. (1999). Auxins upregulate expression of the indole-3-pyruvate decarboxylase gene in *Azospirillum brasilense. Journal of Bacteriology, 181*(4), 1338–1342.

Vande Broek, A., Lambrecht, M., & Vanderleyden, J. (1998). Bacterial chemotactic motility is important for the initiation of wheat root colonization by *Azospirillum brasilense. Microbiology, 144*(Pt 9), 2599–2606.

Vessey, J. K. (2003). Plant growth promoting rhizobacteria as biofertilizers. *Plant and Soil, 255*, 571–586.

Warembourg, F., Roumet, C., & Lafont, F. (2003). Differences in rhizosphere carbon-partitioning among plant species of different families. *Plant and Soil, 256*, 347–357.

Watt, M., Hugenholtz, P., White, R., & Vinall, K. (2006). Numbers and locations of native bacteria on field-grown wheat roots quantified by fluorescence in situ hybridization (FISH). *Environmental Microbiology, 8*(5), 871–884.

Weng, J., Wang, Y., Li, J., Shen, Q., & Zhang, R. (2013). Enhanced root colonization and biocontrol activity of *Bacillus amyloliquefaciens* SQR9 by abrB gene disruption. *Applied Microbiology and Biotechnology, 97*(19), 8823–8830.

Wieland, G., Neumann, R., & Backhaus, H. (2001). Variation of microbial communities in soil, rhizosphere, and rhizoplane in response to crop species, soil type, and crop development. *Applied and Environmental Microbiology, 67*(12), 5849–5854.

Yang, J., Kloepper, J. W., & Ryu, C. (2009). Rhizosphere bacteria help plants tolerate abiotic stress. *Trends in Plant Science, 14*(1), 1–4.

Yaryura, P. M., León, M., Correa, O. S., Kerber, N. L., Pucheu, N. L., & García, A. F. (2008). Assessment of the role of chemotaxis and biofilm formation as requirements for colonization of roots and seeds of soybean plants by *Bacillus amyloliquefaciens* BNM339. *Current Microbiology, 56*(6), 625–632.

Zhang, N., Wang, D., Liu, Y., Li, S., Shen, Q., & Zhang, R. (2013). Effects of different plant root exudates and their organic acid components on chemotaxis, biofilm formation and colonization by beneficial rhizosphere-associated bacterial strains. *Plant and Soil, 374*(1–2), 689–700.

Zheng, X., & Sinclair, J. (1996). Chemotactic response of *Bacillus megaterium* strain B153-2-2 to soybean root and seed exudates. *Physiological and Molecular Plant Pathology, 48*, 21–35.

CHAPTER TWO

Molecular Patterns of Rhizobacteria Involved in Plant Immunity Elicitation

Martin Mariutto, Marc Ongena[1]
Microbial Processes and Interactions Research Unit, University of Liège/Gembloux Agro-Bio Tech, Gembloux, Belgium
[1]Corresponding author: E-mail: Marc.Ongena@ulg.ac.be

Contents

1. Host Resistance Elicitation as Part of Rhizobacteria-Mediated Biocontrol 22
2. Global Aspect of Microbe-Induced Plant Immunity 24
3. MAMPs from Bacterial Pathogens and Early Physiological/Signalling Events Associated with Their Perception 25
   3.1 Molecular Patterns and Cognate Receptors 25
   3.2 Early Physiological Responses and Signalling 28
4. Elicitors of Plant Immunity from Beneficial Bacteria 29
   4.1 Surface-Exposed MAMPs 29
   4.2 Secreted Compounds 31
      4.2.1 Exopolysaccharides *31*
      4.2.2 Iron-Regulated Metabolites *31*
      4.2.3 Alkyl Chain-Containing Metabolites *34*
      4.2.4 Elicitors with Antibiotic Function *36*
      4.2.5 Volatiles *37*
      4.2.6 Other Elicitors *38*
   4.3 Multiple Compounds Involved in Plant Defence Elicitation by a Single PGPR Strain 38
   4.4 Host Perception of PGPR Elicitors 44
      4.4.1 Recognition at the Plant Cell Surface *44*
      4.4.2 Downstream Early Signalling and Defence-Related Events *46*
5. Concluding Remarks 47
Acknowledgement 48
References 48

Abstract

Plants possess a broad array of defences that could be actively expressed in response to the perception of pathogenic microorganisms but also following the recognition of beneficial rhizobacteria retaining the ability to stimulate systemic resistance in the host against further pathogen ingress. Induced systemic resistance (ISR) can be

Advances in Botanical Research, Volume 75
ISSN 0065-2296
http://dx.doi.org/10.1016/bs.abr.2015.07.002

globally viewed as a multistep process involving sequentially the perception by plant cells of elicitors produced by the inducing agents that initiates the phenomenon, the signal transduction that is needed to propagate the induced state systemically through the plant and the expression of defence mechanisms *sensu stricto* that limit or inhibit further pathogen penetration into the host tissues. In this review, we focus on the early molecular dialogue and provide an updated overview of the nature and diversity of the bacterial determinants acting as elicitors of ISR. Some elicitors identified from these beneficial rhizobacteria are also typically harboured by bacterial pathogens but most of them are secreted soluble or volatile compounds with very different functions for the producing cells such as siderophores, antibiotics or quorum-sensing molecules.

1. HOST RESISTANCE ELICITATION AS PART OF RHIZOBACTERIA-MEDIATED BIOCONTROL

Plant growth-promoting rhizobacteria (PGPR) are a heterogeneous group of beneficial root-associated bacteria that improve plant health and/or growth by diverse mechanisms. Some PGPR promote plant growth either by facilitating nutrient availability via fixation of atmospheric N_2, phosphate solubilization or by modulating the host's hormonal balance via the production of plant hormone-like compounds or enzymes that degrade a precursor of ethylene thus limiting the levels of this hormone in the host (Francis, Holsters, & Vereecke, 2010; Lugtenberg & Kamilova, 2009). The plant protective or so-called biocontrol activity of these bacteria relies on several traits. The first is their ability to efficiently colonize plant roots which is a prerequisite for providing protective effects notably by decreasing the availability of space (niche exclusion) or nutrients for pathogens (competition). The second is their strong antagonistic activity towards growth of various plant pathogens which is based on efficient secretion of low-size antimicrobials or hydrolytic enzymes such as *N*-acetyl glucosaminidase and glucanase (Lugtenberg & Kamilova, 2009; Manjula & Podile, 2005). Some more specific pathogen—PGPR interactions leading to direct pathogen restriction were also reported such as interference with biofilm formation, inactivation of pathogen germination factors and degradation of toxins acting as pathogenicity factors (Compant, Duffy, Nowak, Clement, & Barka, 2005; Mercado-Blanco & Bakker, 2007). A third crucial biocontrol trait of some PGPR is their ability to trigger an immune reaction in plant tissues leading to a systemically expressed resistance state that renders the host less susceptible to subsequent infection (induced systemic resistance or ISR phenomenon). ISR has been demonstrated using various phylogenetically unrelated bacterial strains including *Pseudomonas, Serratia, Burkholderia, Klebsiella, Enterobacter, Arthrobacter, Bacillus* and many others (Bakker, Djavaheri, Mercado-Blanco, & Van Loon, 2007; Barriuso, Solano, & Gutierrez Manero, 2008; De Vleesschauwer & Hofte, 2009; Kloepper, Ryu, & Zhang, 2004; Park, Kim, Lee, & Kim, 2009; Ryu et al., 2003; Compant et al., 2005). At the phenotypic level, ISR is most often associated with only partial reduction of disease incidence but the phenomenon is still of great interest from an agronomical perspective. The enhanced defensive capacity is expressed in roots as well as in leaves and is quite nonspecific regarding the nature of the infectious agent. ISR is thus effective against a broad spectrum of pathogens including virus, bacteria, fungi, oomycetes, nematodes and insects. Moreover, PGPR-mediated ISR can occur in many dicotyledonous and monocotyledonous plant species and the phenomenon is long-lasting and not conducive for development of pathogen resistance due to the multiplicity and variety of the induced defence pathways.

The ISR phenomenon may be globally viewed as a three-step process involving sequentially (1) the perception by plant cells of elicitors produced by the inducing agents that initiates the phenomenon, (2) signal transduction and establishment of the so-called primed state which allows an accelerated activation of defence responses upon pathogen attack and (3) expression of defence mechanisms *sensu stricto* that limit or inhibit further pathogen penetration into the host tissues. Phenotypically, ISR resembles the pathogen-induced systemic acquired resistance (SAR) (Fu & Dong, 2013) and is expected to share similar molecular/biochemical characteristics such as an early stimulation of locally expressed cellular/genetic events upon elicitor perception, the emission of a mobile signal transported from the site of induction to distant organs and the activation of a transduction pathway in systemic cells. However, despite major advances made over the past decades concerning the hormonal modulation and transcriptional regulation of the phenomenon (Pieterse et al., 2014; Zamioudis & Pieterse, 2012), we are still far from being able to dress a clear picture of the intricate immune-related molecular events and resistance pathways that are inducible by beneficial rhizobacteria. In this context, the objective of this review is to provide an update on the nature and diversity of PGPR molecular determinants that have been identified so far as elicitors of plant immune reactions that initiate the ISR phenomenon (De Vleesschauwer & Hofte, 2009) (Table 1).

2. GLOBAL ASPECT OF MICROBE-INDUCED PLANT IMMUNITY

Although they are sessile and seem defenceless, plants possess a strong arsenal of physical and chemical mechanisms to protect themselves against phytopathogens. The so-called basal or innate immune response is activated by the perception of conserved molecular motifs typically harboured by microbial pathogens and referred to either as pathogen-associated molecular patterns (PAMPs) or more broadly as microbe-AMPs (MAMPs). MAMPs are perceived at the plant cell surface by high-affinity pattern recognition receptors (PRRs) typically consisting in an extracellular ligand-binding domain with leucine-rich repeats (LRR), a single transmembrane domain and an intracellular serine/threonine kinase-signalling domain (Felix & Boller, 2003; Macho & Zipfel, 2014; Zipfel, 2014). They are referred to as receptor-like kinases (RLKs) or receptor-like proteins (RLPs) which are similarly structured, but lack the cytoplasmic kinase domain. A large number of genes encoding RLKs and RLPs are transcriptionally induced upon treatment with purified MAMPs, illustrating the large diversity of such perception systems and supporting their crucial role in plant defence. MAMP recognition results in convergent signalling cascades and early defence-related events leading to global transcriptional changes and ultimately to some resistant state defined as MAMP-triggered immunity (MTI) also referred as PRR-triggered immunity (PTI) (Jones & Dangl, 2006). Well-adapted microbial pathogens, however, have found ways to breach this first line of active defence by producing effectors which suppress MTI. In turn, plants have evolved perception systems to recognize the presence of these effectors and are thus able to light on a second line of defence response called effector-triggered immunity or ETI. Receptors involved in ETI are the products of R genes and most of them fall into the large family of nucleotide-binding LRR proteins. Considered as a stronger variant of PTI, ETI is generally associated with a programmed cell death called the hypersensitive response (HR) (Thomma, Nurnberger, & Joosten, 2011). The role of HR is to confine the pathogen at the infection site and to send a mobile signal in the systemic organs to induce SAR (Fu & Dong, 2013). However, the frontier between PTI and ETI seems blurred since some MAMPs were also reported to induce HR and/or systemic resistance (Thomma et al., 2011). Both MTI and ETI can thus lead to systemic resistance.

As mentioned above, ISR elicited by beneficial bacteria also relies on the perception of specific compounds secreted/harboured by the interacting microbe triggering an immune response which spreads systemically in distant tissues. Though that ISR retains some specificities compared to SAR regarding epigenetic regulation, (post)transcriptional changes and signalling, both types of systemic resistances are fine-tuned by a complex network of hormonal cross talks involving among others salicylate, jasmonate and ethylene (Pieterse, Van Der Does, Zamioudis, Leon-Reyes, & Van Wees, 2012; Zamioudis & Pieterse, 2012).

3. MAMPs FROM BACTERIAL PATHOGENS AND EARLY PHYSIOLOGICAL/SIGNALLING EVENTS ASSOCIATED WITH THEIR PERCEPTION

3.1 Molecular Patterns and Cognate Receptors

A broad array of structurally diverse MAMPs has been described originating from fungal, oomycete and bacterial pathogens. Most of these MAMPs are oligosaccharides, glycopeptides and peptides. Pathogen MAMPs and their cognate perception systems have been extensively reviewed in the past years (Beck, Heard, Mbengue, & Robatzek, 2012; Newman, Sundelin, Nielsen, & Erbs, 2013; Zipfel, 2014) and here below we summarize knowledge focussing on molecular patterns of bacterial origin. In general, the role of a specific MAMP as elicitor of plant immunity has been determined by testing the loss of function of mutants affected in its formation combined or not with the treatment of plants/cultured cells with the purified molecule. Treated plants are then tested for reduction of disease symptoms upon further infection ideally on different tissues or organs. In a complementary and more high-throughput approach, elicited tissues are also usually tested for the stimulation of typical immune-related cellular or transcriptional events. As exemplified below, this includes early events that are initiated within minutes and later responses occurring within hours like callose deposition and overexpression of defence-related genes, but collectively it results in plant resistance to pathogen attacks (Felix & Boller, 2003; Wu, Shan, & He, 2014).

Flagellin is the main subunit of the flagella essential for bacterial motility. This protein is a widespread PAMP shared by many pathogenic bacteria such as *Pseudomonas syringae* or *Xanthomonas campestris*. The perception of flagellin by the receptor FLS2 was the first described plant PAMP/PRR pair (Felix, Duran, Volko, & Boller, 1999; Gomez–Gomez & Boller, 2002). FLS2 was initially identified in the model plant *Arabidopsis thaliana* but flg22 is recognized by most higher plants and orthologues of FLS2 have already been identified in several species. In *Arabidopsis*, FLS2 recognizes flagellin by binding a conserved sequence of 22 amino acids, referred to as flg22 but plants seem able to recognize multiple epitopes within flagellin since a 28-amino acid sequence (flgII-28) induces immune responses in tomato and several *Solanaceae* species and is perceived via another RK or RLP than FLS2. FLS2 binds flg22 at the LRR ectodomain but signalling activation requires interaction with the LRR RLK BAK1 acting as coreceptor. It is assumed that such dimerization with BAK1 or related Somatic Embryogenesis Receptor Kinase proteins is necessary for full activation of most (if not all) LRR-containing PRRs involved in MAMP recognition (Antolín-Llovera et al., 2014; Böhm, Albert, Fan, Reinhard, & Nürnberger, 2014). Accordingly, beside flg22, BAK1 is required for signalling triggered by other bacterial molecular patterns (see below) like elf18, lipopolysaccharides (LPS), peptidoglycan (PGN) and bacterial cold-shock protein in *Arabidopsis* and *Nicotiana benthamiana* (Böhm et al., 2014; Chinchilla et al., 2007; Postel et al., 2010). Formation of the FLS2/BAK1 PRR heterodimer leads to phosphorylation of the receptor-like cytoplasmic kinase (RLCK) BIK1 which is released from the complex and further play a critical role in inducing reactive oxygen species (ROS) burst (activation of NADPH oxidase) and defence genes (via activation of MAPK and calcium-dependent protein kinases (CDPK) cascades) (Macho & Zipfel, 2014). Consistent with the broad role of BAK1 in PTI, BIK1 is also involved in early stages of immune signalling elicited by other MAMPs from bacterial phytopathogens such as elf18 (Zhang et al., 2010). In rice, the OsRLCK185 is phosphorylated by the BAK1-related OsCERK1 and regulates PGN-induced immune responses (Yamaguchi et al., 2013).

A second well-described PAMP/PRR system is the bacterial elongation factor EF-Tu/EFR receptor pair. The elicitor activity of EF-Tu is attributed to a highly conserved 18 amino acids epitope of the N-terminus which is recognized by the LRR-RK EFR (Kunze et al., 2004). However, elf18 perception seems restricted to the plant family *Brassicaceae* (Zipfel et al., 2006). Treatment with the 50-amino acid epitope EFa50 derived from the central region of EF-Tu induces immune responses in rice via a-yet-unknown PRR (Furukawa, Inagaki, Takai, Hirai, & Che, 2014).

PGNs are major building blocks of the cell walls of bacteria composed of polymeric alternating β(1,4)-linked N-acetylglucosamine (GlcNAc) and N-acetylmuramic acid (MurNAc) residues. PGNs provide structural integrity to the bacterial envelope and as conserved components, they also represent an excellent target for the plant immune system. A role of PGN as

MAMP triggering immune responses in various plants has been demonstrated (Bertsche, Mayer, Götz, & Gust, 2015; Erbs et al., 2008; Gust et al., 2007) and a tripartite LYM1/LYM3/CERK1 recognition system at the host plasma membrane was described in *Arabidopsis* (Willmann et al., 2011). Similar PGN perception machinery also involving LysM domain proteins (LYP4 and LYP6) has been reported for rice (Liu et al., 2012). Rather than the complex PGN matrix, it is obvious that both plants actually perceive soluble fragments released from PGN by a lysozyme-like enzyme rapidly induced in the plant following infection (Liu et al., 2014).

LPS are the major component of the outer membrane of Gram-negative bacteria and retain multiple functions like restricting membrane permeability allowing bacterial growth in unfavourable environments. The activity of LPS as MAMPs inducing innate immune responses in plants has been largely exemplified (reviewed in Erbs & Newman, 2012). They are composed by a lipid A moiety anchored in the outer leaflet of the bacterial plasma membrane, and linked to a core oligosaccharide consisting of a short series of sugars. This constitutive core region ends in the O-antigen which is composed of repeating oligorhamnans units and a strain-variable antigenic region. This O-antigen plays a key role in plant innate immunity triggered by LPS from phytopathogenic pseudomonads (Bedini et al., 2005) but works performed on purified *X. campestris* lipooligosaccharides, LPS without the O-chain, revealed that both the core oligosaccharide and the lipid A are recognized by plant cells and may both act as elicitors. A role of lipid A as MAMP of *Burkholderia cepacia* was also evidenced in *Arabidopsis* as it was found to induce transcriptional changes associated with plant defence responses. LPS eliciting activity has mostly been described in dicots, but studies in rice cells have revealed that LPS from various pathogenic bacteria, also induce defence-related gene expression in monocots, indicating that the machinery recognizing LPS is evolutionarily conserved in monocots and dicots (Desaki et al., 2006, 2012). This perception system has remained obscure until the very recent work of Ranf and colleagues who reported the key function of the bulb-type lectin S-domain-1 receptor-like kinase LORE (SD1-29) in lipid A-dependent LPS sensing by *A. thaliana* (Ranf et al., 2015).

Ax21 is a protein of yet unknown function secreted by *Xanthomonas* spp. and other pathogenic bacteria containing a 17-amino acids sulphated peptide sequence recognized as minimal epitope by the corresponding receptor XA21 in rice (Lee et al., 2009). XA21 is, together with FLS2 and EFR, among the best studied PRRs belonging to the subfamily LRR XII of non-RD class receptor kinases (Dardick & Ronald, 2006). Other MAMPs from bacterial pathogens have been identified but their corresponding PRRs are unknown. It is the case for the cold-shock protein CSP22 harbouring a conserved domain of 15 amino acids which induce immune responses in tobacco cells at subnanomolar concentrations (Felix & Boller, 2003) and for EcoRI-digested plasmid DNA fragments which induced defence responses in *Arabidopsis* (Yakushiji et al., 2009). Some bacterial PAMPs are thus only perceived by a narrow range of plants while others trigger defence responses in many host species even if there is some degree of specificity and perception efficacy for a plant family or species as in the case of flagellin (Felix & Boller, 2003; Ron & Avni, 2004).

3.2 Early Physiological Responses and Signalling

The recognition of those bacterial MAMPs by plasma membrane PRR complexes of plant cells triggers a series of early intracellular molecular changes and events. These defence-related events occur within the first seconds to minutes upon elicitor detection and represent the first transduction step leading to induction of local and further systemic resistance in the plant.

One of the earliest events occurring upon PAMP/MAMP perception is an influx of Ca^{2+} from the apoplast through the activation of membrane-associated channels causing a rapid increase in cytoplasmic Ca^{2+} concentrations. This Ca^{2+} intracellular spike may in turn serve as second messenger to promote the opening of other membrane channels (see below), or to activate CDPK leading to the stimulation of defence gene expression (Wu et al., 2014). Ca^{2+} intracellular accumulation may be quite specific regarding its amplitude, intensity, duration, frequency and location depending on the nature of the elicitation (Vadassery & Oelmüller, 2009). Early increase of Ca^{2+} intracellular concentration has been observed in response to the elicitation by various pathogen-associated MAMPs including flg22 and PGN.

Calcium import is part of a more global perturbation of ion fluxes across the plasma membrane frequently observed upon early perception of pathogen determinants. It involves K^+, NO_3^- and Cl^- effluxes and H^+ influx resulting in membrane potential depolarization and extracellular medium alkalinization. This originates, at least in part, from the activity of the plasma membrane H^+-ATPase which could be regulated by upstream Ca^{2+} signalling and protein phosphorylation. Flagellin, cold-shock protein, LPS, PGN and EF-Tu all induce medium alkalinization in various plant cell cultures within minutes albeit with slightly different amplitudes and persistence (Garcia-Brugger et al., 2006; Wu et al., 2014).

Early induced oxidative burst also plays an important role in plant–pathogen interactions and, upon perception of many MAMPs, ROS quickly accumulate at the site of infection (Nanda, Andrio, Marino, Pauly, & Dunand, 2010). This oxidative burst mainly originates from the activation of the membrane-associated NADPH oxidase (encoded by respiratory burst oxidase homologue B genes) which is activated upon phosphorylation by Ca^{2+}-regulated protein kinases and which produce superoxide anion rapidly transformed into hydrogen peroxide by the superoxide dismutase activity. ROS retain direct antimicrobial activity but also indirectly contribute to plant defence by reinforcing the cell wall and by acting as signals to induce other responses including gene transcriptional reprogramming (Torres, Jones, & Dangl, 2006). ROS accumulation also precedes the HR associated with programmed cell death during interaction between plants and avirulent pathogens.

Phosphatidic acid (PA), a key intermediate of phospholipid biosynthesis, could retain some signalling function in plant immunity and a rapid and transient accumulation of PA and its derivative diacylglycerol pyrophosphate has been observed in tomato cells upon treatment with various PAMPs, including flg22 (Testerink & Munnik, 2005; Van der Luit et al., 2000). It is assumed that PA generation is necessary for ROS production since treatment with inhibitors of the biosynthesis enzyme phospholipase C blocked the oxidative burst, which could be restored by the addition of PA. Differential PA production has also been associated with Ca^{2+} influx typical of PTI signalling (Gao et al., 2013).

4. ELICITORS OF PLANT IMMUNITY FROM BENEFICIAL BACTERIA

4.1 Surface-Exposed MAMPs

Microbial patterns such as peptidoglycans, LPS and flagellins contain immunogenic epitopes that are parts of supramolecular structures building surfaces of most bacteria either pathogenic or not for the host plant (Newman et al., 2013; Pel & Pieterse, 2013). Surprisingly, examples demonstrating a key role of these MAMPs in plant resistance elicitation by beneficial rhizobacterial strains remain limited (Table 1). Nonmotile mutants of the PGPR isolate *Pseudomonas fluorescens* WSC358 lacking flagella lost the capacity to induce ISR in *Arabidopsis* and application of purified flagella from that strain restored the resistance against *P. syringae* pv. *tomato* DC3000. However, treatment of tomato and bean plants with flagella-minus WCS358 mutants resulted in similar resistance level as induced by the wild strain, suggesting the involvement of other elicitors (Meziane, Van der Sluis, Van Loon, Höfte, & Bakker, 2005). Similarly, in an attempt to identify the elicitor of *P. fluorescens* SS101, Van de Mortel et al. (2012) bacteria-treated plants defective in flagellin receptor but the bacterium still induced significant resistance in this mutant similar to wild-type protection level, suggesting that this MAMP is not the ISR determinants in the *Arabidopsis/P. syringae* pv. *tomato* DC3000 model (Van de Mortel et al., 2012).

As stated above, PGN fragments from pathogenic bacteria were described to act as elicitors of plant innate immunity (Erbs & Newman, 2012). However, no evidence for a role of this polymer as trigger of systemic resistance by beneficial rhizobacteria has been reported so far. Interestingly, a recent study shows that only *Bacillus subtilis* PGN digests and not the native polymer PGN were able to trigger immune marker gene *FRK1* expression in *Arabidopsis* seedlings as well as medium alkalinization in cultured rice cells (Liu et al., 2014). This work also demonstrates that *in planta* generation of these immunogenic PGN fragments relies on the activity of a lysozyme-like (LYS1) enzyme which is specifically expressed upon infection. Immune defence stimulation by bacterial PGN or more exactly by its soluble fragments thus requires prior perception of the pathogen via an unknown signal. The same applies for flagellin-triggered immunity which requires the release of monomers from filamentous flagella. This could be the result of a pathogen-mediated process (Gerstel, Czapp, Bartels, & Schröder, 2009) or alternatively, mediated by inducible host factors such as proteases that could provoke shearing of flagella from bacterial surfaces (Ramos, Rumbo, & Sirard, 2004). Whatever the mechanism involved, the release of immunogenic substructures from complex matrices like flagella or PGN is therefore not likely to occur in interactions with nonpathogenic rhizobacteria.

LPS represent another MAMP common for pathogenic and noninfectious plant-associated bacteria. The first identified ISR determinant from PGPR was an LPS from the outer membrane of *P. fluorescens* WCS417r (Van Peer & Schippers, 1992). Mutants lacking the antigenic O side chain lost the capacity to induce ISR in carnation against *Fusarium oxysporum* f. sp. *raphani*. LPS-containing cell walls and pure LPS induced the same level of resistance as compared to the bacteria. In *A. thaliana*, *P. fluorescens* WCS417r and purified LPS induced a systemic resistance against *P. syringae* pv. *tomato* against *F. oxysporum* f. sp. *raphani* and *P. syringae* pv. *tomato* DC3000. LPS were then further demonstrated as ISR determinant of

P. fluorescens strains (Leeman et al., 1996), *Pseudomonas putida* strains (Meziane et al., 2005), *Chryseobacterium balustinum* AUR9 (Ramos Solano, Barriuso Maicas, Pereyra de la Iglesia, Domenech, & Gutierrez Manero, 2008), *B. cepacia* (Gerber, Zeidler, Durner, & Dubery, 2004) and *Rhizobium elti* G12 (Reitz et al., 2002). In *P. fluorescens* WCS417r, the O-antigen was identified as immunological pattern involved in resistance elicitation by LPS since bacterial mutant lacking this region lose the capacity to stimulate ISR (Van Peer & Schippers, 1992). Strain-specific variability in this saccharidic sequence seems to determine the fate of ISR-elicitation process. Indeed, LPS produced by *P. fluorescens* WCS417 induced systemic resistance in *Arabidopsis*, radish and carnation. By comparison, LPS from *P. fluorescens* WCS358 did not stimulate immunity in *Arabidopsis* and tobacco but induced the resistance in bean and tomato, and LPS synthesized by *P. fluorescens* WCS374 triggered ISR in radish but not in *Arabidopsis* (Leeman et al., 1996; Meziane et al., 2005; Steijl, Niemann, & Boon, 1999). Works performed with LPS also illustrate the importance of the dose applied to the root for induction of resistance. Treatment with 50 μg/l and 25 μg/l of purified LPS from *C. balustinum* AUR9 did not stimulate systemic defences of *Arabidopsis* against *P. syringae* pv. *tomato* DC3000. On the contrary, these LPS induced efficient resistance against the pathogenic bacteria when applied at 5 μg/l (Ramos Solano et al., 2008).

4.2 Secreted Compounds

4.2.1 Exopolysaccharides

Exopolysaccharides (EPS) are produced by many bacterial strains including pathogens and beneficials. They act as PAMPs involved in disease development, bacteria survival and plant tissue colonization for pathogen species. EPS of the symbiotic root-associated bacteria *Pantoea agglomerans* YAS34 were shown to stimulate the growth of sunflower plants (Park, Kloepper, & Ryu, 2008). EPS purified from the beneficial *Burkholderia gladioli* IN26 play a determinant role in the induction of ISR on the cucumber-*Colletotrichum orbiculare* pathosystem (Park, Kloepper et al., 2008). Recently, the immunogenic effect of *Bacillus viz.* BS4 on tomato was also linked to its production of EPS (Thenmozhi & Dinakar, 2014).

4.2.2 Iron-Regulated Metabolites

4.2.2.1 Pyoverdines/Pseudobactins

To ensure their growth in iron-limited environments, microorganisms have evolved powerful Fe^{3+}-acquisition systems based on the excretion of high-affinity iron-chelating molecules termed siderophores. PGPR siderophores are also known to antagonize pathogen populations by decreasing iron amounts in soil, resulting in nutrient competition. Pyoverdines (also called pseudobactins) are siderophores typically synthesized by fluorescent *Pseudomonas* and share the same general structure: a conserved quinoline-derived chromophore linked to a highly variable and almost strain-specific peptide sequence of 6—12 amino acids (Budzikiewicz, 2004). *Pseudomonas fluorescens* WCS358 induced ISR in various plant families and bacterial mutants deficient in pseudobactin synthesis did not induce resistance in bean, tomato, *Arabidopsis*, tobacco and *Eucalyptus*, indicating that this metabolite could act as a general ISR elicitor (Meziane et al., 2005; Van Loon, Bakker, van der Heijdt, Wendehenne, & Pugin, 2008; Ran, Li, Wu, van Loon, & Bakker, 2005). With a similar approach, the pseudobactin of *P. fluorescens* WCS374 was demonstrated to elicit ISR in rice against *Magnaporthe oryzae* (De Vleesschauwer, Djavaheri, Bakker, & Hofte, 2008). Purified pseudobactin from this strain also induced ISR in *Eucalyptus* (Ran et al., 2005). Pyoverdines also play a role in systemic resistance induced by *P. fluorescens* CHA0 in tobacco infected by tobacco necrosis virus (TNV) (Maurhofer, Hase, Meuwly, Metraux, & Defago, 1994), and in the protection afforded by *P. fluorescens* WCS374 to radish against *Fusarium* wilt (Leeman et al., 1996). However there are also experimental data talking against the involvement of pyoverdines in ISR triggering. A mutant of strain CHA0 defective in pyoverdine protected *Arabidopsis* plants as effectively as does the parent strain, suggesting that this siderophore is not involved in resistance induction in this host. Other works on resistance induced by *P. putida* BTP1 on cucumber and bean (Ongena et al., 2000, Ongena, Giger, Jacques, Dommes & Thonart, 2002) are examples showing that such a pyoverdine-mediated ISR is not involved and thus can probably not be generalized to all fluorescent *Pseudomonas*.

4.2.2.2 Salicylic Acid (SA) and SA-Derived Siderophores

The role of SA as phytohormone involved in plant immunity has been well characterized and documented (Hayat, Ali, Amara, Khalid, & Ahmed, 2010). In bacteria, SA synthesis occurs under iron-limited conditions and originates from the chorismate/isochorismate biosynthetic pathway (Kerbarh et al., 2005). SA formation is concomitant to the biosynthesis of salicylate-containing siderophores. SA production by PGPR is best exemplified with rhizosphere pseudomonads but other bacterial genera have been demonstrated to produce this compound such as *Achromobacter*, *Bacillus* and *Serratia*

(Forchetti et al., 2010; Press, Wilson, Tuzun, & Kloepper, 1997; Zhang, Moyne, Reddy, & Kloepper, 2002).

The ability of *Pseudomonas aeruginosa* 7NSK2 to elicit ISR appeared to be linked to SA production (De Meyer et al., 1999; De Meyer & Höfte, 1997). Nanogram amounts of SA produced by this strain appeared to be sufficient to trigger the phenomenon (De Meyer et al., 1999) and the KMPCH mutant of this strain was demonstrated to elicit induced resistance through the production of SA (Audenaert, Pattery, Cornelis, & Hofte, 2002). Bean and grapevine plants treated with SA-nonproducing variants of *P. aeruginosa* 7NSK2 lose the ability to activate resistance against the grey mould caused by *Botrytis cinerea* (De Meyer & Höfte, 1997; Verhagen, Trotel-Aziz, Couderchet, Hofte, & Aziz, 2010). Also quite conclusive, transformation of the non-ISR-inducing strain *P. fluorescens* P3 with SA synthesis genes conferred to the bacteria the ability to induce a systemic protection in tobacco to TNV (Maurhofer et al., 1998). Leeman et al. (1996) attributed the capacity of the pseudobactin- and LPS-negative mutants of *P. fluorescens* WCS374 and WCS417 strains to induce ISR in radish against *Fusarium* wilt to SA production under low iron availability (Leeman et al., 1996). However, SA synthesis by these *P. fluorescens* strains is not correlated with induction of resistance in other plant species and experiments with SA-deficient mutants of other natural producer strains demonstrated that SA is not their ISR determinant (Chen, Bélanger, Benhamou, & Paulitz, 1999; Djavaheri et al., 2012). For example, SA produced by *Serratia marcescens* 90−166 does not appear to be involved in the resistance induced by this strain in *Arabidopsis*, tobacco and cucumber (Press et al., 1997; Ryu et al., 2004; Zhang et al., 2002).

SA is an essential precursor or intermediate in the biosynthesis of other compounds with siderophore activity in PGPR such as pyochelin in *P. aeruginosa* (Serino et al., 1997) or pseudomonine in *P. fluorescens* (Mercado-Blanco & Bakker, 2007). ISR-eliciting activity was not clearly demonstrated for the latter compound but a role of pyochelin (together with the antibiotic pyocyanin, see below) was demonstrated in ISR triggered in tomato by *P. aeruginosa 7NSK2* (Audenaert et al., 2002).

Some reports eventually suggest that other iron-regulated bacterial products than those cited above may be involved in ISR triggering. Investigations conducted by Press and collaborators suggest that the expression of genes involved in the biosynthesis of an unidentified catechol siderophore is associated with ISR induced by *S. marcescens* strain 90−166 in cucumber (Press, Loper, & Kloepper, 2001). In summary, a clear role for PGPR-produced SA in ISR has not been clearly established and appeared to depend on the pathosystem considered. Moreover, SA production in the rhizosphere by these bacteria is not obvious as the compound is better channelled in the synthesis of other bioactive siderophores (Bakker, Ran, & Mercado-Blanco, 2014).

4.2.3 Alkyl Chain-Containing Metabolites

4.2.3.1 Lipopeptides

Another class of compounds that emerged as ISR elicitors are biosurfactants such as rhamnolipids (RLs) and lipopeptides (LPs). The potential of LPs as plant resistance inducers was demonstrated in 2007 for two different molecules synthesized by *Pseudomonas* and *Bacillus* (Ongena & Jacques, 2008; Raaijmakers, De Bruin, Nybroe, & Ongena, 2010). Tran and collaborators showed that massetolide A produced by *P. fluorescens* strain SS101 retains ISR-eliciting activity in tomato plants for the control of *Phytophthora infestans*, the causal agent of late blight (Tran, Ficke, Asiimwe, Hofte, & Raaijmakers, 2007). However, based on assays involving a massetolide A-defective mutant of SS101 or the purified molecule, it is obvious that massetolide is not involved in the induced resistance response in *Arabidopsis* against *P. syringae* pv. *tomato* (Van de Mortel et al., 2012). When applied as pure compounds, surfactin and to a lower extend, fengycin LPs formed by *Bacillus* provided a significant protective effect similar to the one induced by living cells of the producing strain. Surfactins induced significant disease reduction in bean and tomato infected with *B. cinerea* while the ISR effect of fengycins was globally lower. These results were supported by data of experiments conducted on bean and tomato showing that overexpression of surfactin biosynthetic genes in the nonproducer *B. subtilis* strain 168 was associated with a significant increase in the potential of the derivative to induce resistance in both plants. The mutant strains that overproduce fengycins did not elicit ISR in bean but activated the resistance in tomato (Ongena et al., 2007). Surfactin represents the main soluble product formed by *Bacillus amyloliquefaciens* that acts as elicitor of immunity in tobacco and tomato (Cawoy et al., 2014; Jourdan et al., 2009; Ongena et al., 2007) and was reported to induce ISR in cucumber against the cucurbit powdery mildew and in rice against *Xanthomonas oryzae* pv. *oryzicola* (Garcia-Gutierrez et al., 2013; Zhang et al., 2012). On the contrary, fengycins, but not surfactins, triggered a response in potato tuber cells (Ongena, Duby, et al., 2005). Iturins represent the third main *Bacillus* LP family that are strong antibiotics against ergosterol-containing fungi and yeasts. Treatment of tomato plants, potato tuber slices and tobacco cells with iturin did not result in plant defence induction but very recently, a PTI-inducing activity of these LPs has been reported on cotton plants and is hypothetized to act in synergy with direct inhibition to control *Verticillium* infection (Han et al., 2015). In addition, fusaricidins which are another type of cyclic LPs isolated from *Paenibacillus polymyxa*, were demonstrated to be involved in the ISR-based biocontrol of *Phytophthora* blight infection in red pepper (Lee et al., 2013).

4.2.3.2 Rhamnolipids

RLs are glycolipids produced by various bacteria species including pathogenic isolates but also some plant beneficial *Pseudomonas* and *Burkholderia* species. They are essential for bacterial surface motility and biofilm development (Chrzanowski, Ławniczak, & Czaczyk, 2012; Vatsa, Sanchez, Clement, Baillieul, & Dorey, 2010). Mono- and di-RLs have been shown to elicit plant defence responses and to induce resistance against various pathogens in grapevine (Varnier et al., 2009) and in *Arabidopsis* (Sanchez et al., 2012) associated with a central role for SA and SA-dependent plant defences.

4.2.3.3 N-Acyl-L-Homoserine Lactones

Many Gram-negative bacteria produce N-acyl-homoserine lactones (AHLs) as quorum-sensing molecules involved in cell-to-cell communication in order to monitor their population density. A role of AHLs in plant immunity triggering has been established for several bacteria such *Serratia liquefaciens* inducing resistance in tomato against *Alternaria alternata* (Schuhegger et al., 2006) or *Ensifer meliloti* in *Arabidopsis* and it is obvious that these QS compounds are readily secreted by biofilm-forming colonies on roots (Zarkani et al., 2013). Interestingly, only long-chain AHLs (oxo-C14-HSL) can induce resistance in *Arabidopsis* while shorter derivatives promote plant growth (Schenk et al., 2014). Transcriptional and biochemical studies strongly suggest that AHL-stimulated plant resistance differs in its signalling pathway from the SA-dependent SAR or the JA/ET-dependent ISR even thought that the defence mechanisms *sensu stricto* subsequently mounted by the host upon AHL treatment are similar to those already observed in SAR and ISR (Schenk et al., 2014).

4.2.3.4 N-Alkylated Benzylamine Derivative

Some bacterial metabolites with unknown function for the microbial life were identified as plant resistance elicitors. Research for identifying the determinant involved in *P. putida* BTP1-induced resistance in cucumber, bean and tomato resulted in the discovery of an n-alkylated benzylamine derivative (NABD) (Ongena, Jourdan, et al., 2005). This molecule consists in a benzylamine moiety with the quaternary ammonium substituted by methyl, ethyl and a long alkyl group. NABD production is stimulated during iron-limited condition, but its role as siderophores was not determined. The plant-immunity eliciting function of NABD was established on the basis of treatment of bean roots with the pure compound that provided a similar level of protection compared to the producing strain and by showing that a BTP1 derivative affected in NABD synthesis was also impaired in its efficacy to stimulate ISR. Treatment of bean and cucumber roots with pure NABD elicited similar protective effect than treatment with living bacteria cells (Ongena et al., 2008). In tomato, the pure elicitor induced a lower protective effect than observed with the producing strain *P. putida* BTP1, suggesting the synthesis of an additional ISR determinant. Perception of NABD by root cells may lead to systemic expression of the oxylipin pathway involved in disease reduction triggered by BTP1 via accumulation of LOX-formed antifungal hydroperoxides (Mariutto et al., 2014, 2013).

4.2.4 Elicitors with Antibiotic Function

The direct control of plant pathogens is mainly based on the production of antibiotics by beneficial rhizobacteria. Interestingly, some of these PGPR antibiotics were also shown to elicit ISR, making them promising tools for biocontrol with dual action on pathogen populations. It is notably the case for 2,4-diacetylphloroglucinol (DAPG) formed by various *P. fluorescens* strains (Weller et al., 2012) as evidenced by the activity of the pure molecule and based on the complete loss of capacity of mutants affected in DAPG synthesis to induce ISR in *Arabidopsis* infected by *Peronospora parasitica* (Iavicoli, Boutet, Buchala, & Metraux, 2003) and in tomato against the pathogen nematode *Meloidogyne javanica* (Siddiqui, Shaukat, Khan, & Ali, 2003).

Pseudomonas aeruginosa 7NSK2 produces other metabolites than SA and pyochelin potentially active in ISR triggering such as the phenazine-type antibiotic pyocyanin. An SA-producing but pyocyanin nonproducing mutant of this strain did not induce ISR in tomato against *B. cinerea*. In addition, treatment of tomato roots with SA and pyochelin-deficient but pyocyanin-producing mutants of 7NSK2 did not result in enhanced immunity. Coinoculation of both mutants restored the resistance at the same level as the parental strain, suggesting a synergic effect of the elicitors for the induction of ISR in tomato (Audenaert et al., 2002). Pyocyanin-minus

7NSK2 mutants, but not pyochelin nor pyoverdine-deficient mutants, lost their ability to induce ISR in rice against the hemibiotroph *Particularia grisea*. Very low concentration of pure pyocyanin induced disease protection at the same extent than 7NSK2. However this metabolite and the bacteria triggered susceptibility to the necrotroph *Rhizoctonia solani* and *Cochliobolus miyabeanus* in rice (De Vleesschauwer, Cornellis, & Hofte, 2006). Interestingly, pyocyanin nonproducing mutants protected the plant against rice sheath blight caused by *R. solani*, suggesting that defence mechanisms induced by pyocyanin antagonized the resistance triggered by the other 7NSK2 elicitors (De Vleesschauwer et al., 2006).

4.2.5 Volatiles

Some studies have provided insight into the role of bacterial volatile organic compounds (VOCs) as initiators of resistance and defence responses in plants. Exposure to VOCs from *B. subtilis* was efficient to activate ISR in *Arabidopsis* seedlings and chemical analysis of the bacterial volatile emissions suggested a key role played by (2R,3R)-butanediol (Ryu et al., 2003, 2004). Pharmaceutical application of synthetic 2,3-butanediol showed similar disease-protection capacity compared to a volatile extract collected from strain GB03, which was also comparable to that induced by direct inoculation of the bacterium (Ryu et al., 2004). On another hand, colonization of tobacco roots by the PGPR *Pseudomonas chlororaphis* O6 induced ISR against the pathogens *Erwinia carotovora* and *P. syringae* pv. *tabaci*. Two different elicitors seemed to be specifically involved in defence against each pathogen. Rhizobacterial mutants impaired in butanediol production lost the capacity to trigger resistance to *E. carotovora* in tobacco. Treatment of plants with pure 2R,3R-butanediol increased protection against *E. carotovora* at the same level as the producing bacterium. On the contrary, butanediol-induced resistance was not effective against *P. syringae* pv. *tabaci* (Han, Anderson, et al., 2006, Han, Lee, et al., 2006). The structurally similar volatile acetoin released by root-colonizing *B. subtilis* FB17 triggered ISR in *A. thaliana* (Rudrappa et al., 2010).

Bacterial VOCs with longer hydrocarbon chains were also identified as elicitors of ISR such as tridecane emitted by *P. polymyxa* E681 (Lee et al., 2012), and hexadecane (Park, Lee, Kloepper, & Ryu, 2013) which conferred protection to *Arabidopsis* from infection of a biotrophic pathogen, *P. syringae* pv. *maculicola* and a necrotrophic pathogen, *Pectobacterium carotovorum* subsp *carotovorum*. The protective effect of *Bacillus cereus* strain C1L was associated with the synthesis of the volatile dimethyl disulphide

(Huang et al., 2012). Application of this compound as soil drench induced ISR in tobacco and corn respectively against *B. cinerea* and *Cochliobolus heterostrophus*. Choi and collaborators also recently reported that the 3-pentanol produced by *B. amyloliquefaciens* IN937a stimulate ISR in pepper against *Xanthomonas axonopodis* and *Cucumber mosaic virus* (Choi, Song, Yi, & Ryu, 2014).

4.2.6 Other Elicitors

Investigations for ISR metabolites involved in protection induced by *P. chlororaphis* O6 against *P. syringae* led to the identification of the 4-(aminocarbonyl)-phenylacetate. However, on the contrary of other PGPR elicitors which are active at concentration lower than 1 mM, high amount (68 mM) of phenylacetate should be applied to plants to induce an efficient protection (Park, Kim, et al., 2008), suggesting the involvement of other determinants.

Klebsiella oxytoca C1036 is a rhizobacterium capable of inducing resistance in tobacco against the soft-rot pathogen *Pseudomonas carotovorum* subsp. *carotovorum* SCC1 and butyl 2-pyrrolidone-5-carboxylate was suggested as the metabolite mainly responsible for this ISR activity even if application at millimolar concentration was necessary to significantly suppress the disease symptoms (Park et al., 2009). Fractioning of a cell-free culture of *Bacillus* sp. BS107 led to the isolation and determination of its elicitor 2-amino benzoic acid was demonstrated to be involved in resistance induced of in tobacco against *P. carotovorum* (Yang et al., 2011). The plant growth-promoting rhizobacterium *Ochrobactrum lupini* KUDC1013 elicited ISR in tobacco against soft rot disease caused by *P. carotovorum* subsp. *carotovorum*. Beside LPS and flagella that were effective in ISR elicitation, other substances present in the crude cell-free supernatant such as phenylacetic acid, 1-hexadecene and linoleic acid also retained host resistance stimulating activity (Sumayo, Hahm, & Ghim, 2013) (Table 1).

4.3 Multiple Compounds Involved in Plant Defence Elicitation by a Single PGPR Strain

Some PGPR strains were shown to stimulate ISR in various plant families. However, the metabolite involved in host defence elicitation depends on the plant species (Table 2). In *P. fluorescens* WCS358, LPS-defective mutants lost the capacity to elicit ISR in bean and tomato, but were as effective as the wild-type strain in *Arabidopsis*, tobacco and *Eucalyptus*. By comparison, flagellin from WCS358 induced ISR in *Arabidopsis* but not in the four other

Table 1 Plant Growth-Promoting Rhizobacteria Products Identified as Elicitors of Plant Immunity and Systemic Resistance

Elicitor	Origin	Host Plant	References
Cell Surface Components			
Flagellin	*Pseudomonas putida, Pseudomonas fluorescens*	*Arabidopsis*	Meziane et al. (2005), Ran et al. (2005)
Lipopolysaccharides	*P. fluorescens, Burkholderia cepacia, Rhizobium elti, Chryseobacterium balustinum*	*Arabidopsis*, tomato, potato, tobacco, carnation	Van Peer, Niemann and Schippers (1991), Duijff, Gianinazzi-Pearson and Lemanceau (1997), Meziane et al. (2005), Coventry and Dubery (2001), Reitz et al. (2002), Erbs and Newman (2003), Gomez–Gomez et al. (2002), Ramos Solano et al. (2008)
Peptidoglycan fragments	*Bacillus subtilis*	*Arabidopsis*	Liu et al. (2014)
Secreted Compounds			
Iron-Regulated Metabolites			
Pyoverdines/pseudobactins	*P. fluorescens, P. putida*	*Eucalyptus*, tobacco, tomato, *Arabidopsis*, rice	Meziane et al., (2005), DeVleeschauwer et al. (2008)
Salicylic acid	*P. fluorescens, P. putida*	Tomato, tobacco, bean	Leeman et al. (1996), Audenaert et al. (2002), De Meyer et al. (1999), Bigirimana and Hofte (2002)
Pyochelin	*Pseudomonas aeruginosa*	Tomato	Audenaert et al. (2002)
Alkyl Chain-Containing Molecules			
Massetolide	*P. fluorescens*	Tomato	Tran et al. (2007)
Surfactin, fengycin	*B. subtilis, Bacillus amyloliquefaciens*	Tomato, bean	Ongena et al. (2007), Jourdan et al. (2009)
Iturin	*B. subtilis*	Cotton	Han et al. (2015)
Fusaricidin	*Paenibacillus polymyxa*	Red-pepper	Lee et al. (2013)

(Continued)

Table 1 Plant Growth-Promoting Rhizobacteria Products Identified as Elicitors of Plant Immunity and Systemic Resistance—cont'd

Elicitor	Origin	Host Plant	References
N-Acyl Benzylamine derivative	*P. putida*	Bean, cucumber	Ongena, Jourdan, et al., (2005), Ongena et al. (2008)
Rhamnolipids	*P. aeruginosa*	Grapevine, *Arabidopsis*	Varnier et al. (2009), Sanchez et al. (2012)
N-acyl homoserine lactone	*Serratia liquefaciens*, *Ensifer meliloti*	Tomato, *Arabidopsis*	Schuhegger et al. (2006), Schenk et al. (2014)
Compounds with Antibiotic Function			
2,4-Diacetylphloroglucinol	*P. fluorescens*	*Arabidopsis*, tomato	Iavicoli et al. (2003), Siddiqui and Shaukat (2003), Weller et al. (2012)
Pyocyanine	*P. aeruginosa*	Rice	De Vleesschauwer et al. (2006)
Volatiles			
Acetoin	*B. subtilis*	*Arabidopsis*	Rudrappa et al. (2010)
Butanediol	*B. subtilis*, *B. amyloliquefaciens*	*Arabidopsis*, tobacco	Heil and Ton (2008), Ryu et al. (2004)
Tridecane	*P. polymyxa*	*Arabidopsis*	Lee et al. (2012)
Hexadecane	*P. polymyxa*	*Arabidopsis*	Park et al. (2013)
3-pentanol	*B. amyloliquefaciens*	Pepper	Choi et al. (2014)
Dimethyl disulfide	*Bacillus cereus*	Tobacco	Huang et al. (2012)
1-hexadecene	*Ochrobactrum lupini*	Tobacco	Sumayo et al. (2013)
Other Elicitors			
Phenylacetic acid	*O. lupini*	Tobacco	Sumayo et al. (2013)
2-Amino benzoic acid	*Bacillus* sp.	Tobacco	Yang et al. (2011)
Butyl 2-pyrrolidone-5-carboxylate	*Klebsiella oxytoca*	Tobacco	Park et al. (2009)
4-(Aminocarbonyl) phenylacetate	*Pseudomonas chlororaphis*	Tobacco	Park, Kim, et al. (2008)
Exopolysaccharides	*Burkholderia gladioli*, *Bacillus* sp.	Cucumber, tomato	Park, Kloepper, et al. (2008), Thenmozhi and Dinakar (2014)

Table 2 Elicitors Involved in Induced Systemic Resistance Triggered by Specific Plant Growth-Promoting Rhizobacteria (PGPR) Strains in Various Pathosystems.

PGPR	Plant	Pathogen	Elicitor	References
Pseudomonas fluorescens CHA0	Tobacco	Tobacco mosaic virus (TMV)[b]	Pyoverdine + unidentified metabolite(s)	Maurhofer et al. (1994)
	Arabidopsis	*Peronospora parasitica*[b]	2,4-diacetylphloroglucinol	Iavicoli et al. (2003)
	Tomato	*Meloidogyne javanica*[b]	2,4-diacetylphloroglucinol	Siddiqui and Shaukat (2003)
	Arabidopsis	*Hyaloperonospora arabidopsidis*[b]	2,4-diacetylphloroglucinol	Iavicoli et al. (2003)
	Arabidopsis	*Pseudomonas syringae* pv. *tomato*[hb]	2,4-diacetylphloroglucinol	Weller et al. (2012)
Pseudomonas aeruginosa 7NSK2	Tobacco	TMV[b]	SA	De Meyer et al. (1999)
	Bean	*Botrytis cinerea*[n]	Pyochelin + pyocyanin + SA	De Meyer and Höfte (1997)
	Bean	*Colletotrichum lindemuthianum*[hb]	SA	Bigirimana and Hofte (2002)
	Tomato	*B. cinerea*[n]	Pyochelin + pyocyanin + SA	Audenaert et al. (2002)
	Rice	*Magnaporthe grisea*[hb]	Pyocyanin	De Vleesshauwer et al. (2006)
	Grapevine	*B. cinerea*[n]	SA + maybe pyocyanin and/ or pyoverdine	Verhagen et al. (2010)
Pseudomonas putida BTP1	Cucumber	*Colletotrichum lagenarium*[n]	NABD	Ongena et al. (2008)
	Bean	*B. cinerea*[n]	NABD	
	Tomato	*B. cinerea*[n]	NABD + unidentified metabolite(s)	
P. fluorescens WCS358	Bean	*B. cinerea*[n] and *C. lindemuthianum*[hb]	LPS, pseudobactin	Meziane et al. (2005)
	Arabidopsis	*P. syringae* pv. *tomato*[hb]	Pseudobactin, flagellin	
	Tomato	*B. cinerea*[n]	LPS, pseudobactin	
	Tobacco	*Erwinia carotovora*[n]	Pseudobactin	Van Loon et al. (2008)
	Eucalyptus	*Ralstonia solanacearum*	Pseudobactin	Ran et al. (2005)

(*Continued*)

Table 2 Elicitors Involved in Induced Systemic Resistance Triggered by Specific Plant Growth-Promoting Rhizobacteria (PGPR) Strains in Various Pathosystems.—cont'd

PGPR	Plant	Pathogen	Elicitor	References
P. fluorescens WCS374	Radish	*Fusarium oxysporum* f. sp. *raphani*[hb]	Pseudobactin + LPS	Leeman et al. (1995), Leeman et al. (1996)
	Arabidopsis	*P. syringae* pv. *tomato* DC3000[hb]	Flagella	Djavaheri et al. (2012)
	Rice	*Magnaporthe oryzae*[hb]	Pseudobactin	De Vleesschauwer et al. (2008)
	Eucalyptus	*R. solanacearum*	Pseudobactin + unidentified metabolite	Ran et al. (2005)
P. fluorescens WCS417r	Radish	*F. oxysporum* f. sp. *raphani*[hb]	LPS	Van Peer and Schippers (1992), Leeman et al. (1995)
	Arabidopsis	*P. syringae* pv. *tomato*[hb]	LPS + unidentified metabolite(s)	Leeman et al. (1996)
	Carnation	*F. oxysporum* f. sp. *raphani*[hb]	LPS	Van Peer and Schippers (1992), Leeman et al. (1995)
P. fluorescens SS101	Tomato	*Phytophthora infestans*	Massetolide A	Tran et al. (2007)
	Arabidopsis	*P. syringae* pv. *tomato* DC3000	Unidentified metabolite	
Pseudomonas chlororaphis O6	Tobacco	*E. carotovora*[n]	2R,3R-butanediol	Han, Lee et al., (2006)
	Tobacco	*P. syringae* pv. *tabaci*[b]	4-(Aminocarbonyl) phenylacetate	Park, Kim et al., (2008)
	Arabidopsis	*P. syringae* pv. *tomato*[b]	2R, 3R-butanediol	Rudrappa et al. (2010)
Bacillus amyloliquefaciens S499	Tomato	*B. cinerea*[n]	Surfactin + fengycin	Ongena et al. (2007)
	Bean	*B. cinerea*[n]	Surfactin	
	Cucumber	*C. lagenarium*	Unidentified not heat-stable envelope component(s)	Ongena, Jourdan, et al. (2005)

b, biotroph; hb, hemibiotroph; n, necrotroph; LPS; lipopolysaccharides; SA, salicylic acid; NABD, n-alkylated benzylamine derivative.

plant species. In addition, mutants deficient in synthesis of pseudobactin did not induce resistance in the five plant species, indicating that this siderophore could act as a more general ISR elicitor (Meziane et al., 2005; Van Loon et al., 2008; Ran et al., 2005). In a similar approach, various metabolites produced by *P. fluorescens* WCS374 were shown to specifically trigger ISR in different plants. LPS-deficient WCS374 mutants completely lost the capacity to induce ISR in radish against *Fusarium* wilt, confirming their crucial role as ISR determinants. In addition, unlike WCS358, the pseudobactin of WCS374 triggered a systemic resistance in radish (Leeman et al., 1996) but also in rice against *M. oryzae* (De Vleesschauwer et al., 2008). Pseudobactin defective/SA producing and pseudobactin producing/SA defective mutants did not protect *Arabidopsis* plants against the *turnip crinkle virus* (TCV), while pseudobactin/SA-producing strains induced ISR. These data suggested that separately pseudobactin and SA are unable to stimulate *Arabidopsis* immunity against the TCV, and are both required for WCS374-triggered ISR in this pathosystem. On the contrary, bacterial mutants disrupted in pseudobactin or/and SA production induced similar protection as the parental strain in the same plant but infected by *P. syringae* pv. *tomato* (Djavaheri et al., 2012). In the same pathosystem, flagella-lacking mutants of WCS374 failed to induce resistance but purified flagella triggered a systemic protection. Purified pseudobactin from WCS374 induced ISR in *Eucalyptus*. However, the pseudobactin-minus mutants induced a similar protection than the wild-type strain, suggesting that, in addition to this siderophore, another unknown determinant(s) may be involved in the elicitation of the WCS374-mediated ISR in *Eucalyptus* (Ran et al., 2005).

Other examples of ISR induction by specific elicitors in different plant species are available as in the case of *P. fluorescens* CHA0. Production of the antibiotic DAPG is important for ISR triggering by this isolate in *Arabidopsis* and in tomato against the nematode *M. javanica* (Iavicoli et al., 2003; Siddiqui, Shaukat, Khan, et al., 2003). However, other determinants seem to be involved in resistance induced by CHA0 in tobacco plants against the TNV. Indeed, *gacA* mutants (suppression of general antibiotic production) induced the resistance to the same extent as the wild-type strain while the pyoverdine-mutant partially lost the faculty to protect tobacco (Maurhofer et al., 1994). In the case of *P. aeruginosa* 7NSK2, SA is the main elicitor active to protect bean and grapevine against grey mould caused by *B. cinerea* (De Meyer & Höfte, 1997; Verhagen et al., 2010) and is involved in resistance of bean to *Colletotrichum lindemuthianum* (Bigirimana & Hofte, 2002). By contrast, 7NSK2-induced resistance in tomato against

B. cinerea mostly rely on a synergistic effect of SA, pyochelin and pyocyanin (Audenaert et al., 2002) while this last antibiotic but not pyochelin nor pyoverdine acts as stimulator of defence in rice (De Vleesschauwer et al., 2006).

4.4 Host Perception of PGPR Elicitors

4.4.1 Recognition at the Plant Cell Surface

The perception of MAMPs from bacterial pathogens which involves specific host-encoded pattern recognition receptors is being described in its intricate molecular mechanisms (Macho & Zipfel, 2014). As they share similar immunogenic epitopes, recognition of microbial patterns like peptidoglycans, LPS and flagellins harboured by beneficial rhizobacteria should obviously be mediated via the same receptor-dependent process (Pieterse et al., 2014). For example, specific features in the substructure of LPS seemed to determine the fate of ISR induced by some PGPR. Indeed, LPS produced by WCS417 induced systemic resistance in *Arabidopsis*, radish and carnation. By comparison, LPS from WCS358 were not ISR determinant in *Arabidopsis*, but induced the resistance in bean and tomato, and LPS synthesized by WCS374-triggered ISR in radish but not in *Arabidopsis*. In those cases, differential elicitation of ISR may be associated with the specificity of plant receptors to recognize specific LPS epitopes.

PGPR elicitors other than these structural MAMPs are quite specific and present a wide variability in their chemical nature, including linear alkanes, amphiphilic compounds, SA-related metabolites, proteins, etc. Even if no clues have been provided so far, these compounds could be perceived by individual receptors since structural similarities are not really obvious within these metabolites. Results obtained by comparing the activity of pure benzylamine with that of NABD in ISR assays performed on bean and cucumber suggest that the aromatic amino part of the molecule is important for its biological activity (Ongena et al., 2008). SA and DAPG also contain an aromatic phenolic group and thus such phenyl-derived moieties could constitute a general motif widely recognized by specific plant cell receptors. Additional experiments are required to appreciate the relative importance of such structural traits by testing multiple naturally coproduced or chemically synthesized derivatives.

LPs may be sensitized by plant tissues via a less specific mechanism than high-affinity PRR. On the basis of absence of refractory effect and of competition between active and inactive homologues, and persistence of activity on protease-pretreated tobacco cells, Henry and collaborators suggested that surfactin is not perceived by a proteic receptor but that the perception is better driven by a lipid-mediated process in the plasma membrane (Henry, Deleu, Jourdan, Thonart, & Ongena, 2011). Indeed, due to their amphiphilic nature and their surfactant properties, these LPs readily interact with phospholipid bilayers with the potential to create some disturbance or channelling in the plasma membrane. This may, in turn, activate a cascade of molecular events leading to defensive responses. It is noteworthy that such membrane perturbation should remain limited since the pure compounds had no toxic effect on plant health at the concentration used (Jourdan et al., 2009). Recent data obtained with *Arabidopsis* also support that, from a mechanistic viewpoint, surfactin-induced immunity appears quite specific compared to other MAMPs like flagellin. For instance, surfactin does not consistently trigger MTI-related early events such as ROS burst in *Arabidopsis* root tissues and the transcriptional response in terms of defence marker genes is weak compared to the response observed upon flagellin recognition. The LP thus acts by priming *Arabidopsis* plants to mount efficient defensive lines and to elevate its resistance towards further pathogen ingress but, in contrast with other MAMPs, by avoiding major genetic reprogramming and fitness costs (Debois et al., 2015).

The strain and host-specific effect of pyoverdines in ISR may be explained as far as the peptide chain is involved in the perception process by plant cells because of the structural differences between naturally occurring pyoverdines. Actually, there is no partial sequence shared by the three active pyoverdines from strains WCS358, WCS374 and CHA0. Testing a wider range of heterogeneous pyoverdines on the same plant is required to evaluate whether some amino acid sequences may represent epitopes perceived by specific receptors in the membrane of root cells. An alternative to direct recognition of pyoverdines and other siderophores via specific receptors by the host plant arose from the key role of the root-specific transcription factor MYB72 for the onset of ISR in *Arabidopsis* elicited by *P. fluorescens* WCS417r or *P. putida* WCS358r (Segarra, Van Der Ent, Trillas, & Pieterse, 2009). *MYB72* and other marker genes are specifically induced in roots under iron-limited conditions and therefore siderophore-induced iron stress in the root's environment as indirect mechanism underlying ISR activation by beneficial rhizobacteria has emerged (Pieterse et al., 2014). However, siderophore mutants of *P. fluorescens* WCS417r were still able to induce *MYB72* and other iron uptake genes like *FRO2* and *IRT1* and thus iron-deficiency response is activated even though plants do not physically experience iron limitation (Pieterse et al., 2014). In addition, whether MYB72-dependent ISR triggered by fluorescent pseudomonads is strain-specific or not is questionable since other isolates than WCS358 such as *P. fluorescens* SS101 were able to induce a systemic protection against *P. syringae* pv. *tomato* DC3000 in *myb72 Arabidopsis* mutants (Van de Mortel et al., 2012).

Concerning volatile elicitors, the experimental procedures were conducted by exposing whole plants to bacterial volatiles in closed Petri dishes. Nothing is known about their site of perception and they could be perceived by roots or by above parts of the plant. The first hypothesis implies the generation of an internal mobile signal emitted from roots to aerial organs. In the second case, volatiles must be produced in sufficient quantities to be perceived by leaves in open environment and the resistance should be defined as local induced resistance because elicited tissues are directly in contact with the elicitor (Choudhary & Johri, 2009). Farag and colleagues also hypothesized that volatile elicitors, such as 2,3-butanediol, may be translocated from roots to shoots *in planta* because they are structurally similar to signal alcohols released during insect and wound damage (Farag, Zhang, & Ryu, 2013).

4.4.2 Downstream Early Signalling and Defence-Related Events

The way PGPR elicitors are perceived at the plasma membrane level is not clear but still there is a clear convergence in the early steps of downstream signalling with MTI. Treatment of tobacco cells respectively with LPS-containing cell wall and flagellin of *P. fluorescens* WCS417, WCS358 and WCS374, as well as pseudobactin isolated from strain WCS374 induced a similar rapid, transient and monophasic Ca^{2+} influx (Van Loon et al., 2008). Comparatively, LPS produced by the PGPR *B. cepacia* induced a biphasic increase of cytosolic Ca^{2+} concentration (Gerber et al., 2004). Stimulation of tobacco cells with purified surfactin from *Bacillus* was also related to Ca^{2+} influx (Jourdan et al., 2009). Elicitors of *P. fluorescens* WSC358 and WCS417 inactive at inducing defence gene expression in tobacco cells still triggered a strong Ca^{2+} influx but not downstream early events (Van Loon et al., 2008). So in many instances, chemical inhibition of calcium influx by Ca^{2+} chelator and/or via ion channel inhibition prevented the occurrence of downstream ISR-linked events.

LPS from *B. cepacia* and surfactin formed by *B. amyloliquefaciens* induced a single transient oxidative burst in tobacco cells peaking at 10 min and then returning to basal level within 30–40 min after treatment (Gerber et al., 2004; Jourdan et al., 2009). LPS from WCS417, LPS and flagellin from WCS358, pseudobactin and LPS from WCS374 triggered a comparable oxidative burst with similar kinetics (Van Loon et al., 2008). By comparison, LPS and cryptogein respectively from pathogenic *X. campestris* and *Phytophthora cryptogea* induced a later and more prolonged ROS accumulation. Inactive pseudobactin of WCS358 and WCS417 strains did not induce ROS production (Meyer, Puhler, & Niehaus, 2001; Van Loon et al., 2008).

Extracellular alkalinization was also reported to be induced by various PGPR determinants such as LPS from *B. cepacia* in tobacco cells (Gerber et al., 2004). The LP surfactin and ISR elicitors of *Pseudomonas* strains such as flagella also triggered a transient pH increase within minutes after treatment (Jourdan et al., 2009; Van Loon et al., 2008). Inhibition of the NADPH oxidase activity by diphenyleneiodonium partially decreased extracellular alkalinization caused by surfactin in tobacco cells (Jourdan et al., 2009). On the contrary, this inhibitor had no effect on the pH increase induced by LPS/flagellin-containing cell wall of *P. fluorescens* WCS358 in tobacco cells (Van Loon et al., 2008). *Burkholderia phytofirmans* PsJN, which did not activate oxidative burst, triggered a transient and extracellular alkalinization in grapevine cells (Bordiec et al., 2011).

5. CONCLUDING REMARKS

The molecular basis underpinning plant perception of the rather limited number of MAMPs identified from pathogenic bacteria is being well deciphered. This PRR-based recognition is of unsuspected complexity involving at least coreceptors, regulatory proteins substrates that link PRR activation to the induction of early signalling components and negative regulators. By contrast, even if the capacity of PGPRs to induce plant immunity partly relies on similar molecular patterns, it also depends on the emission of a vast array of structurally diverse low-molecular weight metabolites. More than 20 PGPR elicitors have been characterized so far as ISR inducers but for most of them, the mechanisms involved in the recognition at the plant cell surface are not known. Nevertheless, there should be some convergence between pathogen-induced and PGPR-mediated plant immunity since common early defence events are stimulated downstream of the first interaction at the plasma membrane level. However, in most cases, a given PGPR elicitor seems not to be universal and induces protection in specific plant species. This feature suggests that plant perception of the ISR determinants should retain some specificity. Also it appears that each ISR elicitor may trigger a specific signalling pathway, leading to expression of specific defence mechanisms effective against a limited spectrum of pathogens. Corollary, it suggests the importance of using appropriate resistance-inducing PGPR for controlling specific pathogen populations. By contrast, some studies also highlighted the production of several elicitors by a same PGPR which could therefore activate different ISR pathways in parallel, resulting in resistance against a broad spectrum of pathogens with different life style.

ACKNOWLEDGEMENT

M.O. is senior research associate at the FRS-FNRS (Fonds National de la Recherche Scientifique, Belgium). Work in the MO's lab received financial support from the INTERREG program France-Wallonie-Vlaanderen and from the program Action de Recherches Concertées (FIELD project) from the University of Liège.

REFERENCES

Antolín-Llovera, M., Petutsching, E. K., Ried, M. K., Lipka, V., Nürnberger, T., Robatzek, S., et al. (2014). Knowing your friends and foes — plant receptor-like kinases as initiators of symbiosis or defence. *New Phytologist, 204*, 791—802.

Audenaert, K., Pattery, T., Cornelis, P., & Hofte, M. (2002). Induction of systemic resistance to *Botrytis cinerea* in tomato by *Pseudomonas aeruginosa* 7NSK2: role of salicylic acid, pyochelin, and pyocyanin. *Molecular Plant-Microbe Interactions, 15*, 1147—1156.

Bakker, P., Djavaheri, M., Mercado-Blanco, J., & Van Loon, L. (2007). Analysis of determinants of *Pseudomonas fluorescens* WCS374r involved in induced systemic resistance in *Arabidopsis thaliana. Phytopathology, 97*, S7.

Bakker, P. A. H. M., Ran, L. X., & Mercado-Blanco, J. (2014). Rhizobacterial salicylate production provokes headaches! *Plant and Soil, 382*, 1—16.

Barriuso, J., Solano, B. R., & Gutierrez Manero, F. J. (2008). Protection against pathogen and salt stress by four plant growth-promoting rhizobacteria isolated from *Pinus* sp. on *Arabidopsis thaliana. Phytopathology, 98*, 666—672.

Beck, M., Heard, W., Mbengue, M., & Robatzek, S. (2012). The INs and OUTs of pattern recognition receptors at the cell surface. *Current Opinion in Plant Biology, 15*, 367—374.

Bedini, E., De Castro, C., Erbs, G., Mangoni, L., Dow, J. M., Newman, M. A., et al. (2005). Structure-dependent modulation of a pathogen response in plants by synthetic O-antigen polysaccharides. *Journal of the American Chemical Society, 127*, 2414—2416.

Bertsche, U., Mayer, C., Götz, F., & Gust, A. A. (2015). Peptidoglycan perception-sensing bacteria by their common envelope structure. *International Journal of Medical Microbiology, 305*, 217—223.

Bigirimana, J., & Hofte, M. (2002). Induction of systemic resistance to *Colletotrichum lindemuthianum* in bean by a benzothiadiazole derivative and rhizobacteria. *Phytoparasitica, 30*, 159—168.

Böhm, H., Albert, I., Fan, L., Reinhard, A., & Nürnberger, T. (2014). Immune receptor complexes at the plant cell surface. *Current Opinion in Plant Biology, 20*, 47—54.

Bordiec, S., Paquis, S., Lacroix, H., Dhondt, S., Ait Barka, E., Kauffmann, S., et al. (2011). Comparative analysis of defence responses induced by the endophytic plant growth-promoting rhizobacterium *Burkholderia phytofirmans* strain PsJN and the non-host bacterium *Pseudomonas syringae* pv. pisi in grapevine cell suspensions. *Journal of Experimental Botany, 62*, 595—603.

Budzikiewicz, H. (2004). Bacterial catecholate siderophores. *Mini-Reviews in Organic Chemistry, 1*, 163–168.
Cawoy, H., Mariutto, M., Henry, G., Fisher, C., Vasilyeva, N., Thonart, P., et al. (2014). Plant defense stimulation by natural isolates of *Bacillus* depends on efficient surfactin production. *Molecular Plant-Microbe Interactions, 27*, 87–100.
Chen, C., Bélanger, R., Benhamou, N., & Paulitz, T. (1999). Role of salicylic acid in systemic resistance induced by *Pseudomonas* spp. against *Pythium aphanidermatum* in cucumber roots. *European Journal of Plant Pathology, 105*, 477–486.
Chinchilla, D., Zipfel, C., Robatzek, S., Kemmerling, B., Nürnberger, T., Jones, J. D. G., et al. (2007). A flagellin-induced complex of the receptor FLS2 and BAK1 initiates plant defence. *Nature, 448*, 497–500.
Choi, H. K., Song, G. C., Yi, H. S., & Ryu, C. M. (2014). Field evaluation of the bacterial volatile derivative 3-pentanol in priming for induced resistance in pepper. *Journal of Chemical Ecology, 40*, 882–892.
Choudhary, D. K., & Johri, B. N. (2009). Interactions of *Bacillus* spp. and plants—with special reference to induced systemic resistance (ISR). *Microbiology Research, 164*, 493–513.
Chrzanowski, Ł., Ławniczak, Ł., & Czaczyk, K. (2012). Why do microorganisms produce rhamnolipids? *World Journal of Microbiology and Biotechnology, 28*, 401–419.
Compant, S., Duffy, B., Nowak, J., Clement, C., & Barka, E. A. (2005). Use of plant growth-promoting bacteria for biocontrol of plant diseases: principles, mechanisms of action, and future prospects. *Applied and Environmental Microbiology, 71*, 4951–4959.
Coventry, H. S., & Dubery, I. A. (2001). Lipopolysaccharides from *Burkholderia cepacia* contribute to an enhanced defensive capacity and the induction of pathogenesis-related proteins in *Nicotianae tabacum*. *Physiological and Molecular Plant Pathology, 58*, 149–158.
Dardick, C., & Ronald, P. (2006). Plant and animal pathogen recognition receptors signal through non-RD kinases. *PLoS Pathogens, 2*, e2.
De Meyer, G., Capieau, K., Audenaert, K., Buchala, A., Metraux, J. P., & Hofte, M. (1999). Nanogram amounts of salicylic acid produced by the rhizobacterium *Pseudomonas aeruginosa* 7NSK2 activate the systemic acquired resistance pathway in bean. *Molecular Plant-Microbe Interactions, 12*, 450–458.
De Meyer, G., & Höfte, M. (1997). Salicylic acid produced by rhizobacterium *Pseudomonas aeruginosa* 7NSK2 induces resistance to leaf infection by *Botrytis cinerea* on bean. *Phytopathology, 87*, 588–603.
De Vleesschauwer, D., Cornellis, P., & Hofte, M. (2006). Redox-active pyocyanin secreted by *Pseudomonas aeruginosa* 7NSK2 triggers systemic resistance to *Magnaporthe grisea* but enhances *Rhizoctonia solani* susceptibility in rice. *Molecular Plant-Microbe Interactions, 19*, 1406–1419.
De Vleesschauwer, D., Djavaheri, M., Bakker, P. A., & Hofte, M. (2008). *Pseudomonas fluorescens* WCS374r-induced systemic resistance in rice against *Magnaporthe oryzae* is based on pseudobactin-mediated priming for a salicylic acid-repressible multifaceted defense response. *Plant Physiology, 148*, 1996–2012.
De Vleesschauwer, D., & Hofte, M. (2009). Rhizobacteria-induced systemic resistance. *Advances in Botanical Research, 51*, 223–281.
Debois, D., Fernandez, O., Franzil, L., Jourdan, E., de Brogniez, A., Willems, L., et al. (2015). Plant polysaccharides initiate underground cross talk with bacilli by inducing synthesis of the immunogenic lipopeptide surfactin. *Environmental Microbiology Reports, 7*, 570–582.
Desaki, Y., Miya, A., Venkatesh, B., Tsuyumu, S., Yamane, H., Kaku, H., et al. (2006). Bacterial lipopolysaccharides induce defense responses associated with programmed cell death in rice cells. *Plant and Cell Physiology, 47*, 1530–1540.
Desaki, Y., Otomo, I., Kobayashi, D., Jikumaru, Y., Kamiya, Y., Venkatesh, B., et al. (2012). Positive crosstalk of MAMP signaling pathways in rice cells. *PLoS One, 7*, e51953.

Djavaheri, M., Mercado-Blanco, J., Versluis, C., Meyer, J. M., Loon, L. C., & Bakker, P. A. (2012). Iron-regulated metabolites produced by *Pseudomonas fluorescens* WCS374r are not required for eliciting induced systemic resistance against *Pseudomonas syringae* pv. tomato in *Arabidopsis*. *Microbiology Open, 1*, 311–325.

Duijff, B. J., Gianinazzi-Pearson, V., & Lemanceau, P. (1997). Involvement of the outer membrane lipopolysaccharides in the endophytic colonization of tomato roots by biocontrol *Pseudomonas fluorescens* strain WCS417r. *New Phytologist, 135*, 325–334.

Erbs, G., & Newman, M. A. (2003). The role of lipopolysaccharides in induction of plant defence responses. *Molecular Plant Pathology, 4*, 421–425.

Erbs, G., & Newman, M. A. (2012). The role of lipopolysaccharide and peptidoglycan, two glycosylated bacterial microbe-associated molecular patterns (MAMPs), in plant innate immunity. *Molecular Plant Pathology, 13*, 95–104.

Erbs, G., Silipo, A., Aslam, S., De Castro, C., Liparoti, V., Flagiello, A., et al. (2008). Peptidoglycan and muropeptides from pathogens *Agrobacterium* and *Xanthomonas* elicit plant innate immunity: structure and activity. *Chemistry & Biology, 15*, 438–448.

Farag, M. A., Zhang, H., & Ryu, C. M. (2013). Dynamic chemical communication between plants and bacteria through airborne signals: induced resistance by bacterial volatiles. *Journal of Chemical Ecology, 39*, 1007–1018.

Felix, G., & Boller, T. (2003). Molecular sensing of bacteria in plants – the highly conserved RNA-binding motif RNP-1 of bacterial cold shock proteins is recognized as an elicitor signal in tobacco. *Journal of Biological Chemistry, 278*, 6201–6208.

Felix, G., Duran, J. D., Volko, S., & Boller, T. (1999). Plants have a sensitive perception system for the most conserved domain of bacterial flagellin. *Plant Journal, 18*, 265–276.

Forchetti, G., Masciarelli, O., Izaguirre, M. J., Alemano, S., Alvarez, D., & Abdala, G. (2010). Endophytic bacteria improve seedling growth of sunflower under water stress, produce salicylic acid, and inhibit growth of pathogenic fungi. *Current Microbiology, 61*, 485–493.

Francis, I., Holsters, M., & Vereecke, D. (2010). The Gram-positive side of plant-microbe interactions. *Environmental Microbiology, 12*, 1–12.

Fu, Z. Q., & Dong, X. (2013). Systemic acquired resistance: turning local infection into global defense. *Annual Review of Plant Biology, 64*, 839–863.

Furukawa, T., Inagaki, H., Takai, R., Hirai, H., & Che, F. S. (2014). Two distinct EF-Tu epitopes induce immune responses in rice and *Arabidopsis*. *Molecular Plant-Microbe Interactions, 27*, 113–124.

Gao, X., Chen, X., Lin, W., Chen, S., Lu, D., Niu, Y., et al. (2013). Bifurcation of *Arabidopsis* NLR immune signaling via Ca^{2+}-dependent protein kinases. *PLoS Pathogens, 9*, e1003127.

Garcia-Brugger, A., Lamotte, O., Vandelle, E., Bourque, S., Lecourieux, D., Poinssot, B., et al. (2006). Early signaling events induced by elicitors of plant defenses. *Molecular Plant-Microbe Interactions, 19*, 711–724.

Garcia-Gutierrez, L., Zeriouh, H., Romero, D., Cubero, J., de Vicente, A., & Perez-Garcia, A. (2013). The antagonistic strain *Bacillus subtilis* UMAF6639 also confers protection to melon plants against cucurbit powdery mildew by activation of jasmonate- and salicylic acid-dependent defence responses. *Microbial Biotechnology, 6*, 264–274.

Gerber, I. B., Zeidler, D., Durner, J., & Dubery, I. A. (2004). Early perception responses of *Nicotiana tabacum* cells in response to lipopolysaccharides from *Burkholderia cepacia*. *Planta, 218*, 647–657.

Gerstel, U., Czapp, M., Bartels, J., & Schröder, J. M. (2009). Rhamnolipid-induced shedding of flagellin from *Pseudomonas aeruginosa* provokes hBD-2 and IL-8 response in human keratinocytes. *Cellular Microbiology, 11*, 842–853.

Gomez-Gomez, L., & Boller, T. (2002). Flagellin perception: a paradigm for innate immunity. *Trends in Plant Sciences, 7*, 251–256.

Gust, A. A., Biswas, R., Lenz, H. D., Rauhut, T., Ranf, S., Kemmerling, B., et al. (2007). Bacteria-derived peptidoglycans constitute pathogen-associated molecular patterns triggering innate immunity in *Arabidopsis*. *Journal of Biological Chemistry, 282*, 32338–32348.

Han, Q., Wu, F., Wang, X., Qi, H., Shi, L., Ren, A., et al. (2015). The bacterial lipopeptide iturins induce *Verticillium dahliae* cell death by affecting fungal signalling pathways and mediate plant defence responses involved in pathogen-associated molecular pattern-triggered immunity. *Environmental Microbiology, 17*, 1166–1188.

Han, S. H., Anderson, A. J., Yang, K. Y., Cho, B. H., Kim, K. Y., Lee, M. C., et al. (2006). Multiple determinants influence root colonization and induction of induced systemic resistance by *Pseudomonas chlororaphis* O6. *Molecular Plant Pathology, 7*, 463–472.

Han, S. H., Lee, S. J., Moon, J. H., Park, K. H., Yang, K. Y., Cho, B. H., et al. (2006). GacS-dependent production of 2R, 3R-butanediol by *Pseudomonas chlororaphis* O6 is a major determinant for eliciting systemic resistance against *Erwinia carotovora* but not against *Pseudomonas syringae* pv. tabaci in tobacco. *Molecular Plant-Microbe Interactions, 19*, 924–930.

Hayat, R., Ali, S., Amara, U., Khalid, R., & Ahmed, I. (2010). Soil beneficial bacteria and their role in plant growth promotion: a review. *Annals of Microbiology, 60*, 579–598.

Heil, M., & Ton, J. (2008). Long-distance signalling in plant defence. *Trends in plant science, 13*, 264–272.

Henry, G., Deleu, M., Jourdan, E., Thonart, P., & Ongena, M. (2011). The bacterial lipopeptide surfactin targets the lipid fraction of the plant plasma membrane to trigger immune-related defence responses. *Cellular Microbiology, 13*, 1824–1837.

Huang, C. J., Tsay, J. F., Chang, S. Y., Yang, H. P., Wu, W. S., & Chen, C. Y. (2012). Dimethyl disulfide is an induced systemic resistance elicitor produced by *Bacillus cereus* C1L. *Pest Management Science, 68*, 1306–1310.

Iavicoli, A., Boutet, E., Buchala, A., & Metraux, J. P. (2003). Induced systemic resistance in *Arabidopsis thaliana* in response to root inoculation with *Pseudomonas fluorescens* CHA0. *Molecular Plant-Microbe Interactions, 16*, 851–858.

Jones, J. D. G., & Dangl, J. L. (2006). The plant immune system. *Nature, 444*, 323–329.

Jourdan, E., Henry, G., Duby, F., Dommes, J., Barthelemy, J. P., Thonart, P., et al. (2009). Insights into the defense-related events occurring in plant cells following perception of surfactin-type lipopeptide from *Bacillus subtilis*. *Molecular Plant-Microbe Interactions, 22*, 456–468.

Kerbarh, O., Bulloch, E. M. M., Payne, R. J., Sahr, T., Rébeillé, F., & Abell, C. (2005). Mechanistic and inhibition studies of chorismate-utilizing enzymes. *Biochemical Society Transactions, 33*, 763–766.

Kloepper, J., Ryu, C., & Zhang, S. (2004). Induced systemic resistance and promotion of plant growth by *Bacillus* spp. *Phytopathology, 94*, 1259–1266.

Kunze, G., Zipfel, C., Robatzek, S., Niehaus, K., Boller, T., & Felix, G. (2004). The N terminus of bacterial elongation factor Tu elicits innate immunity in *Arabidopsis* plants. *Plant Cell, 16*, 3496–3507.

Lee, B., Farag, M. A., Park, H. B., Kloepper, J. W., Lee, S. H., & Ryu, C. M. (2012). Induced resistance by a long-chain bacterial volatile: elicitation of plant systemic defense by a C13 volatile produced by *Paenibacillus polymyxa*. *PLoS One, 7*, e48744.

Lee, S. H., Cho, Y. E., Park, S. H., Balaraju, K., Park, J. W., Lee, S. W., et al. (2013). An antibiotic fusaricidin: a cyclic depsipeptide from *Paenibacillus polymyxa* E681 induces systemic resistance against *Phytophthora* blight of red-pepper. *Phytoparasitica, 41*, 49–58.

Lee, S. W., Han, S. W., Sririyanum, M., Park, C. J., Seo, Y. S., & Ronald, P. C. (2009). A type I-secreted, sulfated peptide triggers XA21-mediated innate immunity. *Science, 326*, 850–853.

Leeman, M., den Ouden, E. M., van Pelt, J. A., Dirkx, F. P. M., Steijl, H., Bakker, P., et al. (1996). Iron availability affects induction of systemic resistance to Fusarium wilt of radish by *Pseudomonas fluorescens*. *Phytopathology, 86*, 149–155.

Leeman, M., van Pelt, J. A., den Ouden, F. M., Heinsbrock, M., Bakker, P. A. H. M., & Schippers, B. (1995). Induction of systemic resistance against *Fusarium* wilt of radish by lipopolysaccharides of *Pseudomonas fluorescens*. *Phytopathology, 85*, 1021–1027.

Liu, B., Li, J. F., Ao, Y., Qu, J., Li, Z., Su, J., et al. (2012). Lysin motif-containing proteins LYP4 and LYP6 play dual roles in peptidoglycan and chitin perception in rice innate immunity. *Plant Cell, 24*, 3406–3419.

Liu, X., Grabherr, H. M., Willmann, R., Kolb, D., Brunner, F., Bertsche, U., et al. (2014). Host-induced bacterial cell wall decomposition mediates pattern-triggered immunity in *Arabidopsis*. *eLife, 3*, e01990.

Lugtenberg, B., & Kamilova, F. (2009). Plant-growth-promoting rhizobacteria. *Annual Review of Microbiology, 63*, 541–556.

Macho, A. P., & Zipfel, C. (2014). Plant PRRs and the activation of innate immune signaling. *Molecular Cell, 54*, 263–272.

Manjula, K., & Podile, A. R. (2005). Production of fungal cell wall degrading enzymes by a biocontrol strain of *Bacillus subtilis* AF 1. *Indian Journal of Experimental Biology, 43*, 892–896.

Mariutto, M., Fauconnier, M. L., Ongena, M., Laloux, M., Wathelet, J. P., du Jardin, P., et al. (2014). Reprogramming of fatty acid and oxylipin synthesis in rhizobacteria-induced systemic resistance in tomato. *Plant Molecular Biology, 84*, 455–467.

Mariutto, M., Fernandez, O., Thonart, P., Clément, C., Dommes, J., Dorey, S., et al. (2013). Comparison of plant defense elicitor properties of amphiphilic compounds from plant-associated bacteria. *IOBC-WPRS Bulletin, 89*, 223–226.

Maurhofer, M., Hase, C., Meuwly, P., Metraux, J. P., & Defago, G. (1994). Induction of systemic resistance of tobacco to Tobacco Necrosis Virus by the root-colonizing *Pseudomonas fluorescens* Strain CHA0-Influence of the *GacA* gene and of pyoverdine production. *Phytopathology, 84*, 139–146.

Maurhofer, M., Reimmann, C., Schmidli-Sacherer, P., Heeb, S., Haas, D., & Défago, G. (1998). Salicylic acid biosynthesis genes expressed in *Pseudomonas fluorescens* strain P3 improve the induction of systemic resistance in tobacco against tobacco necrosis virus. *Phytopathology, 88*, 678–694.

Mercado-Blanco, J., & Bakker, P. (2007). Interactions between plants and beneficial *Pseudomonas* spp.: exploiting bacterial traits for crop protection. *Antonie Van Leeuwenhoek International Journal of General and Molecular Microbiology, 92*, 367–389.

Meyer, A., Puhler, A., & Niehaus, K. (2001). The lipopolysaccharides of the phytopathogen *Xanthomonas campestris* pv. campestris induce an oxidative burst reaction in cell cultures of *Nicotiana tabacum*. *Planta, 213*, 214–222.

Meziane, H., Van Der Sluis, I., Van Loon, L. C., Höfte, M., & Bakker, P. A. H. M. (2005). Determinants of *Pseudomonas putida* WCS358 involved in inducing systemic resistance in plants. *Molecular Plant Pathology, 6*, 177–185.

Nanda, A. K., Andrio, E., Marino, D., Pauly, N., & Dunand, C. (2010). Reactive oxygen species during plant-microorganism early interactions. *Journal of Integrative Plant Biology, 52*, 195–204.

Newman, M. A., Sundelin, T., Nielsen, J. T., & Erbs, G. (2013). MAMP (microbe-associated molecular pattern) triggered immunity in plants. *Frontiers in Plant Science, 4*, 139.

Ongena, M., Daayf, F., Jacques, P., Benhamou, N., Paulitz, T., & Bélanger, R. (2000). Systemic induction of phytoalexins in cucumber in response to treatments with fluorescent pseudomonads. *Plant Pathology, 49*, 523–530.

Ongena, M., Duby, F., Jourdan, E., Beaudry, T., Jadin, V., Dommes, J., et al. (2005). *Bacillus subtilis* M4 decreases plant susceptibility towards fungal pathogens by increasing host resistance associated with differential gene expression. *Applied Microbiology and Biotechnology, 67*, 692–698.

Ongena, M., Giger, A., Jacques, P., Dommes, J., & Thonart, P. (2002). Study of bacterial determinants involved in the induction of systemic resistance in bean by *Pseudomonas putida* BTP1. *European Journal of Plant Pathology, 108*, 187–196.

Ongena, M., & Jacques, P. (2008). *Bacillus* lipopeptides: versatile weapons for plant disease biocontrol. *Trends in Microbiology, 16*, 115–125.

Ongena, M., Jourdan, E., Adam, A., Paquot, M., Brans, A., Joris, B., et al. (2007). Surfactin and fengycin lipopeptides of *Bacillus subtilis* as elicitors of induced systemic resistance in plants. *Environmental Microbiology, 9*, 1084–1090.

Ongena, M., Jourdan, E., Adam, A., Schaefer, M., Budzikiewicz, H., & Thonart, P. (2008). Amino acids, iron, and growth rate as key factors influencing production of the *Pseudomonas putida* BTP1 benzylamine derivative involved in systemic resistance induction in different plants. *Microbial Ecology, 55*, 280–292.

Ongena, M., Jourdan, E., Schafer, M., Kech, C., Budzikiewicz, H., Luxen, A., et al. (2005). Isolation of an n-alkylated benzylamine derivative from *Pseudomonas putida* BTP1 as elicitor of induced systemic resistance in bean. *Molecular Plant-Microbe Interactions, 18*, 562–569.

Park, H. B., Lee, B., Kloepper, J. W., & Ryu, C. M. (2013). One shot-two pathogens blocked: exposure of *Arabidopsis* to hexadecane, a long chain volatile organic compound, confers induced resistance against both *Pectobacterium carotovorum* and *Pseudomonas syringae*. *Plant Signaling and Behavior, 8*, e24619.1–e24619.3.

Park, K., Kloepper, J. W., & Ryu, C. M. (2008). Rhizobacterial exopolysaccharides elicit induced resistance on cucumber. *Journal of Microbiology and Biotechnology, 18*, 1095–1100.

Park, M. R., Kim, Y. C., Lee, S., & Kim, I. S. (2009). Identification of an ISR-related metabolite produced by rhizobacterium *Klebsiella oxytoca* C1036 active against soft-rot disease pathogen in tobacco. *Pest Management Science, 65*, 1114–1117.

Park, M. R., Kim, Y. C., Park, J. Y., Han, S. H., Kim, K. Y., Lee, S. W., et al. (2008). Identification of an ISR-related metabolic produced by *Pseudomonas chlororaphis* O6 against the wildfire pathogen *Pseudomonas syringae* pv. *tabaci* in Tobacco. *Journal of Microbiology and Biotechnology, 18*, 1659–1662.

Pel, M. J. C., & Pieterse, C. M. J. (2013). Microbial recognition and evasion of host immunity. *Journal of Experimental Botany, 64*, 1237–1248.

Pieterse, C. M. J., Van Der Does, D., Zamioudis, C., Leon-Reyes, A., & Van Wees, S. C. M. (2012). Hormonal modulation of plant immunity. *Annual Review of Cell and Developmental Biology, 28*, 489–521.

Pieterse, C. M. J., Zamioudis, C., Berendsen, R. L., Weller, D. M., Van Wees, S. C. M., & Bakker, P. A. H. M. (2014). Induced systemic resistance by beneficial microbes. *Annual Review of Phytopathology, 52*, 347–375.

Postel, S., Küfner, I., Beuter, C., Mazzotta, S., Schwedt, A., Borlotti, A., et al. (2010). The multifunctional leucine-rich repeat receptor kinase BAK1 is implicated in *Arabidopsis* development and immunity. *European Journal of Cell Biology, 89*, 169–174.

Press, C. M., Loper, J. E., & Kloepper, J. W. (2001). Role of iron in rhizobacteria-mediated induced systemic resistance of cucumber. *Phytopathology, 91*, 593–598.

Press, C. M., Wilson, M., Tuzun, S., & Kloepper, J. W. (1997). Salicylic acid produced by *Serratia marcescens* 90-166 is not the primary determinant of induced systemic resistance in cucumber or tobacco. *Molecular Plant-Microbe Interactions, 10*, 761–768.

Raaijmakers, J., De Bruin, I., Nybroe, O., & Ongena, M. (2010). Natural functions of cyclic lipopeptides from *Bacillus* and *Pseudomonas*: more than surfactants and antibiotics. *FEMS Microbiology Reviews, 34*, 1037–1062.

Ramos, H. C., Rumbo, M., & Sirard, J. C. (2004). Bacterial flagellins: mediators of pathogenicity and host immune responses in mucosa. *Trends in Microbiology, 12*, 509–517.

Ramos Solano, B., Barriuso Maicas, J., Pereyra de la Iglesia, M. T., Domenech, J., & Gutierrez Manero, F. J. (2008). Systemic disease protection elicited by plant growth promoting

rhizobacteria strains: relationship between metabolic responses, systemic disease protection, and biotic elicitors. *Phytopathology, 98*, 451–457.
Ran, L. X., Li, Z. N., Wu, G. J., van Loon, L. C., & Bakker, P. (2005). Induction of systemic resistance against bacterial wilt in *Eucalyptus urophylla* by fluorescent *Pseudomonas* spp. *European Journal of Plant Pathology, 113*, 59–70.
Ranf, S., Gisch, N., Schäffer, M., Illig, T., Westphal, L., Knirel, Y. A., et al. (2015). A lectin S-domain receptor kinase mediates lipopolysaccharide sensing in *Arabidopsis thaliana. Nature Immunology, 16*, 426–433.
Reitz, M., Oger, P., Meyer, A., Niehaus, K., Farrand, S. K., Hallmann, J., et al. (2002). Importance of the O-antigen, core-region and lipid A of rhizobial lipopolysaccharides for the induction of systemic resistance in potato to *Globodera pallida. Nematology, 4*, 73–79.
Ron, M., & Avni, A. (2004). The receptor for the fungal elicitor ethylene-inducing xylanase is a member of a resistance-like gene family in tomato. *Plant Cell, 16*, 1604–1615.
Rudrappa, T., Biedrzycki, M., Kunjeti, S., Donofrio, N., Czymmek, K. J., Paré, P., et al. (2010). The rhizobacterial elicitor acetoin induces systemic resistance in *Arabidopsis thaliana. Communicative and Integrative Biology, 3*, 8.
Ryu, C., Farag, M., Hu, C., Reddy, M., Paré, P., & Kloepper, J. (2003). Volatiles produced by PGPR elicit plant growth promotion and induced resistance in *Arabidopsis*. In *Proceedings of the 6th International PGPR Workshop.*
Ryu, C. M., Farag, M. A., Hu, C. H., Reddy, M. S., Kloepper, J. W., & Pare, P. W. (2004). Bacterial volatiles induce systemic resistance in *Arabidopsis. Plant Physiology, 134*, 1017–1026.
Sanchez, L., Courteaux, B., Hubert, J., Kauffmann, S., Renault, J. H., Clément, C., et al. (2012). Rhamnolipids elicit defense responses and induce disease resistance against biotrophic, hemibiotrophic, and necrotrophic pathogens that require different signaling pathways in *Arabidopsis* and highlight a central role for salicylic acid. *Plant Physiology, 160*, 1630–1641.
Schenk, S. T., Hernández-Reyes, C., Samans, B., Stein, E., Neumann, C., Schikora, M., et al. (2014). N-acyl-homoserine lactone primes plants for cell wall reinforcement and induces resistance to bacterial pathogens via the salicylic acid/oxylipin pathway. *Plant Cell, 26*, 2708–2723.
Schuhegger, R., Ihring, A., Gantner, S., Bahnweg, G., Knappe, C., Vogg, G., et al. (2006). Induction of systemic resistance in tomato by N-acyl-L-homoserine lactone-producing rhizosphere bacteria. *Plant Cell and Environment, 29*, 909–918.
Segarra, G., Van Der Ent, S., Trillas, I., & Pieterse, C. M. J. (2009). MYB72, a node of convergence in induced systemic resistance triggered by a fungal and a bacterial beneficial microbe. *Plant Biology, 11*, 90–96.
Serino, L., Reimann, C., Visca, P., Beyeler, M., DellaChiesa, V., & Haas, D. (1997). Biosynthesis of pyochelin and dihydroaeruginoic acid requires the iron-regulated pchDCBA operon in *Pseudomonas aeruginosa. Journal of Bacteriology, 179*, 248–257.
Siddiqui, M. A., & Shaukat, S. S. (2003). Suppression of root-knot disease by Pseudomonas fluorescens CHA0 in tomato: importance of bacterial secondary metabolite, 2,4-diacetylpholoroglucinol. *Soil Biology & Biochemistry, 35*, 1615–1623.
Siddiqui, I. A., Shaukat, S. S., Khan, G. H., & Ali, N. I. (2003). Suppression of *Meloidogyne javanica* by *Pseudomonas aeruginosa* IE-6S(+) in tomato: the influence of NaCl, oxygen and iron levels. *Soil Biology & Biochemistry, 35*, 1625–1634.
Steijl, H., Niemann, G., & Boon, J. (1999). Changes in chemical composition related to fungal infection and induced resistance in carnation and radish investigated by pyrolysis mass spectrometry. *Physiological Molecular Plant Pathology, 55*, 297–311.
Sumayo, M., Hahm, M. S., & Ghim, S. Y. (2013). Determinants of plant growth-promoting *Ochrobactrum lupini* KUDC1013 involved in induction of systemic resistance against

Pectobacterium carotovorum subsp. *carotovorum* in tobacco leaves. *Plant Pathology Journal, 29*, 174–181.
Testerink, C., & Munnik, T. (2005). Phosphatidic acid: a multifunctional stress signaling lipid in plants. *Trends in Plant Science, 10*, 368–375.
Thenmozhi, P., & Dinakar, S. (2014). Exopolysaccharides (EPS) mediated induction of systemic resistance (ISR) in *Bacillus – Fusarium oxysporum* f.sp. *lycopersici* pathosystem in tomato (var. PKM-1). *International Journal of Current Microbiology and Applied Sciences, 3*, 8.
Thomma, B. P., Nurnberger, T., & Joosten, M. H. (2011). Of PAMPs and effectors: the blurred PTI-ETI dichotomy. *Plant Cell, 23*, 4–15.
Torres, M. A., Jones, J. D. G., & Dangl, J. L. (2006). Reactive oxygen species signaling in response to pathogens. *Plant Physiology, 141*, 373–378.
Tran, H., Ficke, A., Asiimwe, T., Hofte, M., & Raaijmakers, J. M. (2007). Role of the cyclic lipopeptide massetolide A in biological control of *Phytophthora infestans* and in colonization of tomato plants by *Pseudomonas fluorescens*. *New Phytologist, 175*, 731–742.
Vadassery, J., & Oelmüller, R. (2009). Calcium signaling in pathogenic and beneficial plant microbe interactions: what can we learn from the interaction between *Piriformospora indica* and *Arabidopsis thaliana*. *Plant Signaling & Behavior, 4*, 1024–1027.
Van de Mortel, J. E., de Vos, R. C. H., Dekkers, E., Pineda, A., Guillod, L., Bouwmeester, K., et al. (2012). Metabolic and transcriptomic changes induced in *Arabidopsis* by the rhizobacterium *Pseudomonas fluorescens* SS101. *Plant Physiology, 160*, 2173–2188.
Van der Luit, A. H., Piatti, T., Van Doorn, A., Musgrave, A., Felix, G., Boller, T., et al. (2000). Elicitation of suspension-cultured tomato cells triggers the formation of phosphatidic acid and diacylglycerol pyrophosphate. *Plant Physiology, 123*, 1507–1515.
Van Loon, L. C., Bakker, P., van der Heijdt, W. H. W., Wendehenne, D., & Pugin, A. (2008). Early responses of tobacco suspension cells to rhizobacterial elicitors of induced systemic resistance. *Molecular Plant-Microbe Interactions, 21*, 1609–1621.
Van Peer, R., Niemann, G. J., & Schippers, B. (1991). Induced resistance and phytoalexin accumulation in biological-control of Fusarium-wilt of carnation by *Pseudomonas* sp strain WCS417R. *Phytopathology, 81*, 728–734.
Van Peer, R., & Schippers, B. (1992). Lipopolysaccharides of plant-growth-promoting *Pseudomonas* sp strain WCS417R induce resistance in carnation to Fusarium-wilt. *Netherlands Journal of Plant Pathology, 98*, 129–139.
Varnier, A. L., Sanchez, L., Vatsa, P., Boudesocque, L., Garcia-Brugger, A., Rabenoelina, F., et al. (2009). Bacterial rhamnolipids are novel MAMPs conferring resistance to *Botrytis cinerea* in grapevine. *Plant Cell and Environment, 32*, 178–193.
Vatsa, P., Sanchez, L., Clement, C., Baillieul, F., & Dorey, S. (2010). Rhamnolipid biosurfactants as new players in animal and plant defense against microbes. *International Journal of Molecular Sciences, 11*, 5095–5108.
Verhagen, B. W., Trotel-Aziz, P., Couderchet, M., Hofte, M., & Aziz, A. (2010). *Pseudomonas* spp.-induced systemic resistance to *Botrytis cinerea* is associated with induction and priming of defence responses in grapevine. *Journal of Experimental Botany, 61*, 249–260.
Weller, D. M., Mavrodi, D. V., Van Pelt, J. A., Pieterse, C. M. J., Van Loon, L. C., & Bakker, P. A. H. M. (2012). Induced systemic resistance in *Arabidopsis thaliana* against *Pseudomonas syringae* pv. tomato by 2,4-diacetylphloroglucinol-producing *Pseudomonas fluorescens*. *Phytopathology, 102*, 403–412.
Willmann, R., Lajunen, H. M., Erbs, G., Newman, M. A., Kolb, D., Tsuda, K., et al. (2011). *Arabidopsis* lysin-motif proteins LYM1 LYM3 CERK1 mediate bacterial peptidoglycan sensing and immunity to bacterial infection. *Proceedings of the National Academy of Sciences of the United States of America, 108*, 19824–19829.

Wu, S., Shan, L., & He, P. (2014). Microbial signature-triggered plant defense responses and early signaling mechanisms. *Plant Science, 228*, 118–126.
Yakushiji, S., Ishiga, Y., Inagaki, Y., Toyoda, K., Shiraishi, T., & Ichinose, Y. (2009). Bacterial DNA activates immunity in *Arabidopsis thaliana*. *Journal of General Plant Pathology, 75*, 227–234.
Yamaguchi, K., Yamada, K., Ishikawa, K., Yoshimura, S., Hayashi, N., Uchihashi, K., et al. (2013). A receptor-like cytoplasmic kinase targeted by a plant pathogen effector is directly phosphorylated by the chitin receptor and mediates rice immunity. *Cell Host and Microbe, 13*, 347–357.
Yang, S. Y., Park, M. R., Kim, I. S., Kim, Y. C., Yang, J. W., & Ryu, C. M. (2011). 2-Aminobenzoic acid of *Bacillus* sp. BS107 as an ISR determinant against *Pectobacterium carotovorum* subsp. *carotovorum* SCC1 in tobacco. *European Journal of Plant Pathology, 129*, 371–378.
Zamioudis, C., & Pieterse, C. M. J. (2012). Modulation of host immunity by beneficial microbes. *Molecular Plant-Microbe Interactions, 25*, 139–150.
Zarkani, A. A., Stein, E., Röhrich, C. R., Schikora, M., Evguenieva-Hackenberg, E., Degenkolb, T., et al. (2013). Homoserine lactones influence the reaction of plants to rhizobia. *International Journal of Molecular Sciences, 14*, 17122–17146.
Zhang, J., Li, W., Xiang, T., Liu, Z., Laluk, K., Ding, X., et al. (2010). Receptor-like cytoplasmic kinases integrate signaling from multiple plant immune receptors and are targeted by a *Pseudomonas syringae* effector. *Cell Host and Microbe, 7*, 290–301.
Zhang, R., Liu, Y., Luo, C., Wang, X., Liu, Y., Qiao, J., et al. (2012). *Bacillus amyloliquefaciens* Lx-11, a potential biocontrol agent against rice bacterial leaf streak. *Journal of Plant Pathology, 94*, 11.
Zhang, S., Moyne, A. L., Reddy, M. S., & Kloepper, J. W. (2002). The role of salicylic acid in induced systemic resistance elicited by plant growth-promoting rhizobacteria against blue mold of tobacco. *Biological Control, 25*, 288–296.
Zipfel, C. (2014). Plant pattern-recognition receptors. *Trends in Immunology, 35*, 345–351.
Zipfel, C., Kunze, G., Chinchilla, D., Caniard, A., Jones, J. D. G., Boller, T., et al. (2006). Perception of the bacterial PAMP EF-Tu by the receptor EFR restricts *Agrobacterium*-mediated transformation. *Cell, 125*, 749–760.

CHAPTER THREE

Root Microbiome Assemblage is Modulated by Plant Host Factors

Venkatachalam Lakshmanan[*,§]
*Department of Plant and Soil Sciences, University of Delaware, Newark, DE, USA
[§]Delaware Biotechnology Institute, Newark, DE, USA
E-mail: venki@udel.edu

Contents

1. Introduction 58
2. Microbiome Assemblage is Influenced by the Host Genome 60
3. Microbiome Assemblage is Influenced by the Host Developmental Stage 63
4. Microbiome Assemblage is Influenced by the Host Health and Fitness 67
5. Microbiome Assemblage is Influenced by Alteration in Plant Signalling 69
6. Implications and Future Prospects 70
References 72

Abstract

In recent years, the term plant microbiome inhabiting the root interior of healthy plants, as well as the rhizosphere, has received substantial attention, since it influences nutrient availability, plant growth promotion and disease suppression. The community structure and diversity of plant-associated microbiome is well characterized in modern systems such as *Arabidopsis*, maize and rice using deep rRNA gene sequencing techniques. Increasing evidence indicates that composition of root exudates secreted by host plants can shape the soil microbiome. The molecular communication fluctuates according to the plant genotype, plant development stage, biotic/abiotic stress, proximity to neighbouring species, management techniques and many other factors which may shape the rhizospheric microbiome. Plants secretes blends of organic compounds that are differentially produced at distinct stages of development to help orchestrate rhizosphere microbiome assemblage. A comprehensive understanding of the mechanisms that govern selection and activity of microbial communities by plant roots will provide new opportunities to increase crop production. Hence, this chapter focuses mainly on existing knowledge based on the root microbiome, its functional importance and potential influence of host plant to the establishment of its rhizospheric associates.

Advances in Botanical Research, Volume 75
ISSN 0065-2296
http://dx.doi.org/10.1016/bs.abr.2015.09.004

1. INTRODUCTION

The rhizospheric microbiome is the dynamic community of microorganisms associated with plant roots, a possible determinant of plant fitness and productivity and has received substantial attention in the recent years. Now scientists are taking a census of plant—microbes—which includes distinct microbial communities live inside roots, on leaves and within flowers, and all in all have an estimated three to six orders of magnitude greater genetic diversity than their plant hosts. This second genome, much like the human microbiome, scientists and farmers alike think represents the next big thing in agriculture. The study of plant microbiome by new techniques has significantly improved our understanding of the structure and specificity of the plant microbiome. The microbes living in, on, or surrounding an organism are called microbiota and comprise the microbiome and encompass bacteria, fungi, and archaea (Dridi, Raoult, & Drancourt, 2011; Mendes et al., 2011). The composition of community structure and abundance at rhizosphere which is the narrow zone of soil, can contain up to 10^{11} microbial cells per gram root (Egamberdieva, 2007) and more than 30,000 prokaryotic species (Mendes et al., 2011) and bulk soil of same environment is having only 10^{6}—10^{9} cells per gram of the soil (Bulgarelli, Schlaeppi, Spaepen, van Themaat, & Schulze-Lefert, 2013). The current knowledge of rhizosphere microbiome structure, their assemblage, their function and their promising biotechnological potential was summarized by Berendsen, Pieterse, and Bakker (2012), Bakker, Berendsen, Doornbos, Wintermans, and Pieterse (2013), Mendes, Garbeva, and Raaijmakers (2013). Turner, James, and Poole (2013), Lakshmanan, Selvaraj, and Bais (2014), Lebeis et al. (2015) and Edwards et al. (2015) reviewed more specifically the plant microbiome and plant health relationship. For instance, the soil microbiome suppresses plant diseases by imposing physiological restrictions on pathogens establishing and infecting plant tissues (Kumar et al., 2012; Mendes et al., 2011; Rudrappa et al., 2010; Rudrappa, Czymmek, Paré, & Bais, 2008; Weller, Raaijmakers, Gardener, & Thomashow, 2002). Equally, the soil microbiome also imparts a certain degree of resistance to the system against 'invaders', thereby linking the diversity of the microbiome to its intrinsic ability to restrict or inhibit the survival of exogenous organisms (van Elsas et al., 2012). Several other functions have been attributed by plant microbiome, for instance, their ability to provide nutrients, such as (1) phosphorus solubilization and nitrogen fixation; (2) their support of nutrient uptake from the soil, as described for the mycorrhiza and (3) their capacity to promote plant protection by hindering agents of plant stresses, such as infection by pathogens and pests (Hallmann, Quadt-Hallmann, Mahaffee, & Kloepper, 1997; Mendes et al., 2013, 2011; Quecine et al., 2014). Thus, manipulation of plant microbiome increases agricultural production, reduces chemical agrochemical inputs resulting in more sustainable agricultural practices (Adesemoye, Torbert, & Kloepper, 2009 and reviewed by Bakker et al., 2013; Berendsen et al., 2012; Lakshmanan et al., 2014; Perez-Jaramillo, Mendes, & Raaijmakers, 2015).

Traditionally, the composition of rhizosphere microbial community was characterized by isolating and culturing microbes on different media and growth conditions. However, only a very small fraction of the microbial community was identified as culturable microorganisms, which only represent 1–10% of microorganisms inhabiting soil (Philippot, Raaijmakers, Lemanceau, & van der Putten, 2013). Deeper and more comprehensive research should be conducted to further explore the composition of rhizosphere. Recently, amplicon sequencing of 16S ribosomal RNA (rRNA) genes and internal transcribed spacer sequences has been used for in-depth analysis of soil bacterial and fungal communities (Schoch et al., 2012). This method could provide a large number of parallel reads at a reasonable cost and in a relatively short time period and offer an opportunity to achieve a high-throughput and deeper insight into the composition of rhizosphere microbial communities (Binladen et al., 2007; Yu, Grishkan, & Steinberger, 2013). Over the last 5 years, several root microbiomes including *Arabidopsis* (Bulgarelli et al., 2012; Lundberg et al., 2012; Lundberg, Yourstone, Mieczkowski, Jones, & Dangl, 2013; Schlaeppi, Dombrowski, Oter, Ver Loren van Themaat, & Schulze-Lefert, 2014), rice (Edwards et al., 2015; Ikeda et al., 2011), maize (Peiffer et al., 2013), sugarcane (Yeoh et al., 2015), sweet potato (Shakya et al., 2013) and domesticated barley (Bulgarelli et al., 2015) were established using high-throughput sequencing.

The microbiome is an active component of the host and also responsive to changes in environmental (biotic and abiotic) conditions. More generally, bacterial communities of above- and belowground differ in both their diversity and structure. The composition of microbial communities of the rhizosphere seems to be determined primarily by the combined influence of various biotic and abiotic factors. Biotic and abiotic factors such as soil type, seasons, plant developmental stage, proximity to root, root architecture, plant species and cultivars can affect the structure of microbial communities in the rhizosphere (Castellanos, Dohrmann, Imfeld, Baumgarte, & Tebbe, 2009; reviewed by Berg & Smalla, 2009; Buee, De Boer, Martin, van Overbeek, & Jurkevitch, 2009; Lakshmanan et al., 2014). Plant species can select specific microbial populations from the bulk soil by root exudates excretion (reviewed by Bais, Weir, Perry, Gilroy, & Vivanco, 2006; Haichar et al., 2008; Huang et al., 2014). Differences in the composition of root exudates but also of the root architecture are assumed to affect the relative abundance of soil microorganisms in the vicinity of the root (Somers, Vanderleyden, & Srinivisam, 2004). In several studies, the influence not only of the plant species but also of the plant genotypes on the rhizosphere microbial communities was demonstrated (reviewed by Andreote et al., 2009; Berg & Smalla, 2009; van Overbeek & van Elsas, 2008; Smalla et al., 2001; Weinert et al., 2011). Taken together, these environmental factors should lead to predictable spatial patterns in microbial community structure within and between individual plants. Correspondingly, there is some evidence that this appears to be the case.

This chapter focuses on the description of plant microbiomes by not only describing the compartments where the microbes live (rhizosphere, endosphere and phyllosphere) but by also discussing the importance of host-derived factors that determine the interactions between plants and microorganisms. Due to the availability of data, most recent studies focus on bacterial fractions of the microbiome.

2. MICROBIOME ASSEMBLAGE IS INFLUENCED BY THE HOST GENOME

Recent rhizospheric microbiome studies reveal that there are highly complex and dynamic interactions between a plant and its microbiome assemblage. Moreover, plant host specificity is well understood in the case of phytopathogenic interaction with fungi or bacteria (Raaijmakers, Paulitz, Steinberg, Alabouvette, & Moënne-Loccoz, 2009) and symbiotic interaction with bacteria and legumes (Long, 1989). Studies using high-throughput 16S rRNA gene sequencing showed that host genome may modulate their rhizosphere microbiome in a host-dependent way. Each plant species promotes a particular set of rhizosphere microbes (Haichar et al., 2008; Ofek, Voronov-Goldman, Hadar, & Minz, 2014; Turner, Ramakrishnan, et al., 2013). The composition of their rhizosphere microbial assemblage appears to increase microbiome assemblage with an increase in the phylogenetic distance between plant species (Bouffaud, Poirier, Mulle, & Moënne-Loccoz, 2014; Pongsilp, Nimnoi, & Lumyong, 2012; Wieland, Neumann, & Backhaus, 2001). Recently, the microbiota of three *Arabidopsis thaliana* ecotypes along with its sister species *Arabidopsis halleri* and *Arabidopsis lyrata*, as well as *Cardamine hirsuta*, which diverged from the former ~35 Mya was surveyed their under controlled environmental conditions and of *A. thaliana* and *C. hirsuta* in two natural habitats. The major finding is that the diversification of the root microbiota of the tested host species is largely quantitative, host species-specific ecological adaptation and phylogenetic distance of the hosts could contribute to microbiota diversification (Schlaeppi et al., 2014). For example, two species *A. thaliana* and *A. lyrata* occur in similar habitats whereas *A. halleri* has evolved a distinctive lifestyle resulted in the selection of a distinctive microbiota with habitat specific manner (Schlaeppi et al., 2014). Not only different plant species, but also different genotypes of the same species may differ in their rhizosphere microbiome composition. For example, a recent experiment with the establishment of rhizosphere communities in three cultivars of potato (*Solanum tuberosum*) grown in two distinct field sites revealed that only 4% of operational taxonomic units were dependent on the host genotype by 40% soil-specific abundance (Weinert et al., 2011). Interestingly, three different potato cultivars showed differences in microbes belonging to the families of bacteria that have been studied extensively for their ability to control plant pathogens and a portion of the detected OTUs of Streptomycetaceae responded in a cultivar-dependent manner (Weinert et al., 2011). This genotype-dependent microbiome assemblage on sweet potato (*Ipomoea batatas*) is further confirmed by another study, which showed that plant age and genotype influenced the root microbiome (Marques et al., 2014). They also found that *Sphingobium*, *Pseudomonas*, *Acinetobacter*, *Stenotrophomonas* and *Chryseobacterium* were enriched on the low-starch genotype as compared to two high-starch genotypes (Marques et al., 2014). In a similar way, another study involving three distinct genotypes wheat, pea and oat (*Avena sativa*), plants were grown for 4 weeks in similar bulk soil and the microbiomes were evaluated. Where pea and oat are often used as break crops in rotation schemes to promote yield of crops such as wheat (Seymour, Kirkegaard, Peoples, White, & French, 2012). Interestingly, the microbiomes were found to be different from each other, with a profound change in the balance of prokaryotes and eukaryotes between pea and oat plant. Oat and pea exerted strong selection on eukaryotes, whereas selection by wheat was much weaker (Turner, Ramakrishnan, et al., 2013).

This combined information suggests that the host genotypes at species level may have some control over the microbiome assemblage and functional diversity at the rhizosphere.

On the other hand, Bulgarelli et al. (2012) used high-resolving 16S rRNA gene pyrosequencing of microbiome and identified only one OTU of the bacterial root endophyte community that showed significantly different quantitative enrichment between the two *Arabidopsis* ecotypes tested. Similar way, Lundberg et al. (2012) also used 16S rRNA gene sequencing of microbiome collected from eight ecotypes of *Arabidopsis* and identified 12 OTUs among 778 detectable OTUs, which are host genotype-dependent. In these two studies, the endophytic microbiome of *Arabidopsis* exhibits some genotype-dependent variation within the species and an increased variation when other related species are examined (Bulgarelli et al., 2012; Lundberg et al., 2012, 2013; Schlaeppi et al., 2014). An important finding of these studies is that soil type and the respective soil bacterial biomes had a greater influence than the host genotype on the composition of root endophyte communities (mainly Actinobacteria, Bacteroidetes and Proteobacteria). A significant but weak host genotype-dependent effect acts in the selection of *Arabidopsis* root-inhabiting bacterial communities. However, it is unclear whether microbiota divergence is greater in host species belonging to other plant families and whether the process of domestication, which gave rise to modern cultivated plants (Abbo et al., 2014) and which cannot be studied in *A. thaliana*.

A recent study in root microbiota of soil-grown wild, traditional and modern accessions of barley using a 16S rRNA gene pyrosequencing revealed that host innate immune system as well as the root metabolism is relevant host factors for bacterial recruitment (Bulgarelli et al., 2015). The same investigation found that the host genotype accounts for approximately 5.7% of the variance in the rhizosphere microbiome composition. Another study in 27 inbred lines (genetic variation in a single species) of cultivated maize examined microbiome variation across different sites and found a large variation arising from geographical location between three different states in the United States and a relatively smaller dependence on the genotype (Peiffer et al., 2013). In another study, Shakya et al. (2013) showed that a high percentage (more than 90%) of OTUs specific to sampled mature *Populus deltoides* trees from two watersheds of North Carolina and Tennessee in two seasons (spring and fall) had the dominant phyla Proteobacteria (56.1%), Actinobacteria (17.5%) and Acidobacteria (10%). However, dominance of Proteobacteria was replaced by Actinobacteria from spring to fall in the

Tennessee samples. This study further attempted to investigate the impact of host genotype and phenotype on microbiome community structure, where more than 40% variation of factors was statistically unexplained and only approximately 20% was significantly contributed in two habitats of the tree belowground: rhizosphere and endosphere. Thus, the composition of root microbiota depends more on interactions with the ecological habitat than with host species. In addition, microbe–microbe interactions are also a driving force in the taxonomic differentiation of the root-associated bacterial assemblages.

In contrast, microbial communities of sugarcane were profiled at two field sites of under low and standard fertilizer application revealed that previously reported sugarcane diazotrophs belonging to the genera *Beijerinckia*, *Acetobacter* and *Klebsiella* were below in detection in all samples their relative abundance was also not correlated with nitrogen treatment (Yeoh et al., 2015). This investigation suggested that passive manipulation of native diazotrophic populations by reducing nitrogen fertilizer rates is ineffective and does not generate conditions conducive of biological nitrogen fixation. These studies suggested that the composition and structure of plant microbiomes and the factors that affect them. However, in recent years microbiome profiling was carried out using deep 16S rRNA gene sequencing of root and corresponding soil samples of numerous plant species including cultivated food crops (Bulgarelli et al., 2012; Edwards et al., 2015; Lundberg et al., 2012; Schlaeppi et al., 2014), different sampling protocols and low-resolution profiling methods make it difficult to reexamine and compare these for the existence of a conserved core microbiota between plant species. However, these recent developments on plant root microbiome studies posed open-ended questions like what are the host factors/specific genes/alleles that are involved in root endophytic as well as rhizospheric microbiome assemblage. Thus, the dissection of the molecular mechanisms underlying plant–microbe community associations at the root–soil interface will be a crucial step toward the rational exploitation of the microbiome for agricultural purposes.

3. MICROBIOME ASSEMBLAGE IS INFLUENCED BY THE HOST DEVELOPMENTAL STAGE

Next to genotype-specific effects, also the plant developmental stage is a strong driver shaping the rhizobial microbiome community structure. Previously, plant age effects on the composition of bacterial communities in the rhizosphere were demonstrated in several studies (Buee et al., 2009; Herschkovitz et al., 2005; Heuer, Kroppenstedt, Lottmann, Berg, & Smalla, 2002; Lerner et al., 2006; Smalla et al., 2001), and the plant growth developmental stage was considered a main factor affecting bacterial communities in the potato (Marques et al., 2014; van Overbeek & van Elsas, 2008), soybean (Sugiyama, Ueda, Zushi, Takase, & Yazaki, 2014), *Arabidopsis* rhizosphere (Chaparro, Badri, & Vivanco, 2014). In case of phytopathogenic interaction studies, age-related resistance is well documented and correlated with plant evelopmental stage. For example, younger plant of maize is more susceptible to common rust (*Puccinia sorghi*) than mature plants (Abedon & Tracy, 1996). Similarly, later stages of *Arabidopsis* showed age-related defense strategies against bacterial and fungal pathogens (An & Mou, 2011; Bednarek, 2012; Clay, Adio, Denoux, Jander, & Ausubel, 2009; Millet et al., 2010; Rogers, Glazebrook, & Ausubel, 1996). In another study, *Arabidopsis* transitions from the vegetative to the floral phase correlate with resistance to *Pseudomonas syringae* (Kus, Zaton, Sarkar, & Cameron, 2002; Rusterucci et al., 2005) and biotrophic pathogen *Hyaloperonospora arabidopsidis* (Carviel et al., 2009, 2014; Rusterucci et al., 2005). For instance, the phytotoxin coronatine (*P. syringae*) and several *Hpa* effectors (*H. arabidopsidis*) secretion contributes to the suppression of salicylic acid (SA), in young plants thus promoting disease susceptibility. At later stages of development, plants gain competence and high levels of SA, which may act as an antimicrobial agent in the intercellular space and are resistant to *Pst and HPa* infection (Carella, Wilson, & Cameron, 2015). In plant—bacterial symbiotic interaction per se, soybean roots secrete isoflavones such as daidzein and genistein into the rhizosphere, a small region around the roots, where these compounds signal rhizobia to form nodules on the roots (Weston & Mathesius, 2013). In addition to signalling rhizobia to induce the NodD gene, isoflavones play various roles in biological communications with soil microbes (Sugiyama & Yazaki, 2014; Weston & Mathesius, 2013). In soybean, *Bradyrhizobium*, *Bacillus* and *Stenotrophomonas* were more abundant at the flowering stage as compared to vegetative and mature stages (Sugiyama, Ueda, Zushi, et al., 2014). This symbiosis research leads to host genome and it is developmentally control.

Root growth, root exudate and its composition changes are highly associated with plant developmental stage and are also important factor for microbial community interaction. Through automated ribosomal intergenic spacer analysis or denaturing gradient gel electrophoresis demonstrated that plant microbial communities change in response to plant development gradient (Baudoin, Benizri, & Guckert, 2002; Houlden, Timms-Wilson, Day, & Bailey, 2008; Micallef, Channer, Shiaris, & Colón-Carmona, 2009; Mougel et al., 2006). For example, in *Arabidopsis* early plant development was more distinct to the bulk soil and that this difference decreased with plant age (Micallef, Channer, et al., 2009), in potato microbiomes disappeared as the plants aged (Inceoglu, Al-Soud, Salles, Semenov, & van Elsas, 2011) and in soybean early-reproductive growth stages produced more complex microbial communities than late-stage plants (Xu et al., 2009). These studies produce a fingerprint of the community structure but not of its members' identity. Recent studies using rRNA gene pyrosequencing attempted that the molecular cross talk between plants and the soil microbes is largely orchestrated by alteration in root exudates, not only at the one compound–one microbe level, but at the community level (Chaparro, Badri, & Vivanco, 2014; Marques et al., 2014; Sugiyama, Ueda, Zushi, et al., 2014). For instance, the structure and function of the rhizospheric bacterial community associated with *Arabidopsis* at four different plant development stages (seedling, vegetative, bolting and flowering) were analyzed and showed that there were no significant differences in bacterial community structure (Chaparro, Badri, & Vivanco, 2014). Interestingly, the microbial community at the seedling stage was found to be distinct from the other developmental time points (Chaparro, Badri, & Vivanco, 2014). In the last decade, evidence demonstrating root exudates mediate rhizospheric interactions and microbial composition have been extensively studied (Badri, Chaparro, Zhang, Shen, & Vivanco, 2013; Broeckling, Broz, Bergelson, Manter, & Vivanco, 2008; Chaparro, Badri, Bakker, et al., 2014; Doornbos, Loon, & Bakker, 2012; Micallef, Channer, et al., 2009; Micallef, Shiaris, et al., 2009). It was further demonstrated that the change in specific compound of root exudates shown dramatic shifts in the composition of the microbial community (Eilers, Lauber, Knight, & Fierer, 2010; Shi et al., 2011). Recently found that compounds such as phenolics or amino acids more readily influence soil rhizosphere microbial communities (Chaparro, Badri, Bakker, et al., 2014; Chaparro, Badri, & Vivanco, 2014). For example, fractions of root exudates containing phenolic and phenolic related compounds when applied to the soil (in the absence of the plant) significantly modulated the soil microbiome (Badri et al., 2013) and also *Arabidopsis* mutant that secretes more phenolic compounds than sugars showed that it cultured more beneficial microbes such as Plant Growth Promoting Rhizobacteria (PGPR) and those involved in N-fixation such as bradyrhizobia and cyanobacteria.

In another study, the rhizosphere bacterial communities of bulk and rhizospheric soils at different growth (initial, vegetative, flowering, and mature) of soybeans were analyzed using both culture-dependent and 16S rRNA gene metagenomics. It was reported that several species of potential beneficial rhizobacteria like *Bradyrhizobium* (nitrogen fixation; Barbour, Hattermann, & Stacey, 1991), *Stenotrophomonas maltophilia* (reduces nematode densities in soils; Insunza, Alstrom, & Eriksson, 2002), *Enterobacter cloacae* (produces indole-3-acetic acid and solubilize phosphate; Richardson, Barea, McNeill, & Prigent-Combaret, 2009; de Souza et al., 2013), *Sphingomonas* sp. (found to produce indole-3-acetic acid; de Vasconcellos & Cardoso, 2009) were elevated in the rhizosphere of soybean (Sugiyama, Ueda, Zushi, et al., 2014). Moreover ICHG, rhizosphere bacterial communities contained potential PGPR genera were highly abundant and growth stage-specific (Sugiyama, Ueda, Zushi, et al., 2014). Also reported that, ICHG expression was increased vegetative stages of soybean and suggest the secretion of the isoflavone (malonyl) glucosides that may have accumulated in vacuoles. The influences of alteration in composition of exudates during soybean growth have been reported to correlate during the growth of *Arabidopsis* (Chaparro, Badri, Bakker, et al., 2014; Chaparro, Badri, & Vivanco, 2014; Sugiyama et al., 2015). Another microbial fungal communities, however, no significant effects of the soybean growth stage were detected (Sugiyama, Ueda, Takase, & Yazaki, 2014). These effects, however, are not general as some studies highlighted a stronger selective rhizosphere effect at young plant growth stages (Chaparro, Badri, & Vivanco, 2014; Gomes et al., 2001; Jin et al., 2009; Micallef, Channer, et al., 2009; Xu et al., 2009) while other studies documented stronger selective effects on the microbiome at flowering (Inceoglu, Salles, Van Overbeek, & Van Elsas, 2010; Smalla et al., 2001). The developmentally and nutritionally regulated secretion of isoflavone and its effect on microbial communities need further research to propose as a general per se.

From the above mentioned studies, it will be concluded that plant growth stage will influence root physiology and change the quality and quantity of root exudates; consequently, these changes exert a selection on root-associated microorganisms at different growth stages (Aulakh, Wassmann, Bueno, Kreuzwieser, & Rennenberg, 2001; Chaparro, Badri, Bakker, et al., 2014; Chaparro, Badri, & Vivanco, 2014; Dunfield & Germida, 2003; Houlden et al., 2008; Sugiyama, Ueda, Takase, et al., 2014; Sugiyama, Ueda, Zushi, et al., 2014). Even within the rhizosphere, there are smaller microenvironments that are created by root exudation.

There are certain areas of the root that exude more compounds than others and these are the areas where bacteria will accumulate (Fukui, Schroth, Hendson, Hancock, & Firestone, 1994). The composition of the secretions also differs along the root length and this results in unique bacterial communities along the root length (Ofek, Hadar, & Minz, 2011). As more research is completed on the identity and role of root exudates corresponds with plant developmental stage, we gain a better understanding of how plants communicate in the rhizosphere. Understanding the composition of rhizosphere microbial communities during growth in the field could provide a basis for optimizing agricultural utilization of rhizosphere microbes.

4. MICROBIOME ASSEMBLAGE IS INFLUENCED BY THE HOST HEALTH AND FITNESS

Plants are sessile and interact with their surroundings by use of root exudates as chemical cues (Badri, Weir, van der Lelie, & Vivanco, 2009; De-la-Pena et al., 2010; De-la-Pena, Lei, Watson, Sumner, & Vivanco, 2008). The low molecular weight compounds are more abundant in exudates and include amino acids, organic acids, phenolic compounds, simple sugars and other small secondary metabolites (Bais et al., 2006). Root exudate composition and concentrations change depending upon the environment in which a plant is growing, including soil edaphic and biological factors like disease, stress and even the microbial community (Flores, Vivanco, & Loyola-Vargas, 1999; Lankau, 2011; Micallef, Shiaris, et al., 2009). In recent years, changes in specific compounds such as L-malic acid, citric acid and fumaric acid and colonization of specific microbes are well documented (Lakshmanan et al., 2012; Lakshmanan & Bais, 2013; Liu et al., 2014; Rudrappa et al., 2008; Tan et al., 2013). Plants use these compounds as a chemical cues to attract beneficial soil microorganisms and establish interactions (Neal, Ahmad, Gordon-Weeks, & Ton, 2012). For example, benzoxazinoids, found in the root exudates of maize, attract plant-beneficial rhizobacteria (Neal et al., 2012). Similarly, *p*-coumaric acid increased pathogenic fungal taxa while vanillic acid secretion promoted beneficial rhizobacteria *Bacillus* sp. colonization in cucumber (*Cucumis sativus*) plant (Zhou & Wu, 2012, 2013).

Alteration in plant health like insect infestation, foliar pathogenic infection of plants release compounds such as hormones, exogenous volatile organic compounds and modify the composition of root secretion (for review, see Erb, Lenk, Degenhardt, & Turlings, 2009). For example, under herbivory, wounding of plant tissues by insect feeding triggers the release of volatile signals that attract natural enemies of insects (Kessler & Baldwin, 2001). Recently, the first evidence of the recruitment of beneficial root microbes after aboveground herbivory and pathogenic bacterial attack was shown: aerial aphid feeding and pathogenic microbial attack increased the population of the nonpathogenic rhizobacterium *Bacillus subtilis* in the rhizosphere of sweet pepper (*Capsicum annuum*) (Lee, Lee, & Ryu, 2012). Rudrappa et al. (2008) demonstrated that foliar infection by *P. syringae* in *A. thaliana* influences the expression of root malate transporter that potentially influences exudation of malic acid in the rhizosphere that, in turn, attracts the beneficial rhizobacterium *B. subtilis* (Lakshmanan et al., 2012; Rudrappa et al., 2008). Malic acid attracted a beneficial strain of *B. subtilis*, which then colonized the rhizoplane of the same plant, forming a biofilm that protected roots against further aggression from the phytopathogen (Rudrappa et al., 2008). These findings were further validated by Chen et al. (2013), where exudates of tomato roots strongly stimulated *B. subtilis* biofilm formation ex planta, and that an abundant small molecule in the exudates, L-malic acid, was able to stimulate biofilm formation. In an another study, root infection of *Fusarium oxysporum* f. sp. *cucumerinum* in cucumber plants resulted in increased root secretion of fumaric acid and citric acid, and enhanced chemotaxis, leading to abundant colonization and biofilm formation by *Bacillus amyloliquefaciens* SQR9 (Liu et al., 2014). Tricarboxylic acids, such as malic acid and citrate, are suitable carbon sources for many microorganisms (López-Bucio, de La Vega, Guevara-García, & Herrera-Estrella, 2000; for review, see Pineda, Zheng, van Loon, Pieterse, & Dicke, 2010). Carbon enrichment of the rhizosphere, especially carboxylate excretion and acidification at the root surface, might have a strong impact on structuring rhizospheric microbial communities (Marschner, Neumann, Kania, Weiskopf, & Lieberei, 2002). These studies showed that even a minor modification in plant roots could have important repercussions for soil microbial communities. Recent development in plant-beneficial microbe interaction studies revealed that root exudates play a role in establishing specific interactions with particular microbes in the rhizosphere (legume's symbiotic interaction with rhizobia, interaction of plants with mycorrhizae, and PGPR) (Bais et al., 2006; Nagahashi & Douds, 2000; Prithiviraj, Paschke, & Vivanco, 2007; Rudrappa et al., 2008). These studies reveal a new type of interaction and raise the question of how multiple herbivory/pathogen attacks would affect the colonization of root-associated microbes. Moreover, the above-said studies deal with one-to-one interactions (plant–microbe), but in nature, plants are exposed to numerous beneficial as well as pathogenic microbes in the soil. So, it is important to understand these rhizospheric interactions at the microbial community level.

Recent reports showed that the specific compound in *root exudates* plays an important *role* in *shaping* the *rhizosphere microbiome* (Badri et al., 2013). For example, an *Arabidopsis* ABC transporter mutant that secreted more phenolics than sugars compared with the wild type caused significant changes to the natural microbial community (Badri et al., 2009). Likewise, the secretion of sugars and sugar alcohols, which may help to orchestrate the assemblage of the rhizospheric microbiome on roots (Chaparro, Badri, Bakker, et al., 2014). Similarly, Badri et al. (2013) observed that the addition of distinct blends of natural chemicals derived from *Arabidopsis* root exudates added to the soil produced distinct rhizosphere microbial communities. In addition, the specific role of root exudates in the shaping of the rhizosphere is further confirmed by showing different groups of natural compounds derived from plant root exudates synergistically modifying the root microbiome (Badri et al., 2013).

5. MICROBIOME ASSEMBLAGE IS INFLUENCED BY ALTERATION IN PLANT SIGNALLING

Two commonly studied signalling pathways in plants are induced systemic resistance (ISR) jasmonic acid (JA)/ethylene (ET)-dependent ISR and SA-dependent systemic acquired resistance (SAR) (Glazebrook, 2005). In natural environment, plants are under continuous attack by various organisms such as foliar feeding insects and pathogens and mount multiple responses that have systemic effects on belowground microbes. Other plant hormones, such as abscisic acid (ABA), cytokinin, gibberellin and auxin also function as modulators of the hormone-signalling backbone (Giron, Frago, Glevarec, Pieterse, & Dicke, 2013; Meldau, Erb, & Baldwin, 2012; Pieterse, Van der Does, Zamioudis, Leon-Reyes, & Van Wees, 2012; Robert-Seilaniantz, Grant, & Jones, 2011). Feeding habits of herbivores determines the induction of JA/SA signalling (Erb et al., 2009; Soler, Erb, & Kaplan, 2013; Stam et al., 2014; Thaler, Humphrey, & Whiteman, 2012). Upon foliar feeding by herbivory or pathogenic infection, plants respond in several ways and activate defense in distal parts like root and may affect belowground rhizospheric microbial community (Doornbos, Geraats, Kuramae, Van Loon, & Bakker, 2011). The feeding of sap-sucking pests like whiteflies or aphids on pepper plants and induced SA-dependent SAR and JA-dependent ISR signalling throughout the plants that is aerial foliar parts as well as belowground roots (Lee, Farag, et al., 2012; Yang et al., 2011) and further showed that aphid feeding elicits plant to attract beneficial bacterial *B. subtilis* GB03 populations in the rhizosphere of sweet pepper (*C. annuum*) (Lee, Lee, et al., 2012). Moreover, upregulation of JA signalling in whole plant by repeated wounding and foliar application of JA in *Medicago truncatula* leads to enhanced colonization of beneficial mycorrhizal colonization (Landgraf, Schaarschmidt, & Hause, 2012). For instance, activation of JA defense pathway by application of methyl jasmonate (MeJA) significantly altered the composition of the rhizosphere community but did not directly influence bulk soil microbial communities (Carvalhais et al., 2013). Further reported that JA signalling mutants *myc2* and *med2* released lower amounts of asparagine, ornithine and tryptophan, and *med25* released less glutamic acid and changes in the abundance of rhizosphere microbial populations. This study provides evidence to suggest that root exudates released in response to changes in JA signalling may influence the microbiome abundance (Carvalhais et al., 2015). At the same time, *Arabidopsis* plants with altered SA signalling have root microbiomes that differ in the relative abundance of specific bacterial families as compared with those of wild type (Lebeis et al., 2015). Alternatively, application of plant hormones such as SA, MeJA, ABA and ET to *A. thaliana* grown in a model soil (potting mix) showed no changes to the belowground bacterial richness and evenness. Relative to the control, all hormone treatment led to more than double the abundance of a *Limnobacter* population; ABA led to a fourfold increase in a *Cellvibrio* population, ET led to a c. 70% decrease in an *Algoriphagus* population, and MeJA and ABA applications halved the abundance of a *Massilia* population (Carvalhais et al., 2013; Carvalhais, Dennis, & Schenk, 2014).

6. IMPLICATIONS AND FUTURE PROSPECTS

The composition and structure of plant microbiome and its functional influence on host plant health need better exploration. However, recent studies have focused on evaluating belowground microbial community diversity using deep rRNA gene sequencing technique, its influence on host structure has advanced our understanding of this exciting interaction (Bulgarelli et al., 2012; Chaparro, Badri, & Vivanco, 2014; Lundberg et al., 2012; Peiffer et al., 2013; for review, see Chaparro, Sheflin, Manter,

& Vivanco, 2012; Huang et al., 2014; Lakshmanan et al., 2014; Perez-Jaramillo et al., 2015; Spence & Bais, 2013; Turner, James, et al., 2013). Moreover, these recent microbiome analyses tried merely to identify its structure and complexity rather than to determine molecular understanding of these microbial assemblages and its effect on altering the plant phenome, which is essential to explore toward its utilization (Figure 1). Recently, Edwards et al. (2015) attempted to dissect assembling process of the rice root microbiome. They found that microbiome acquisition from soil is rapid, namely that rice plants begin to assemble an endospheric microbiome within a day after transplantation from sterile media to soil, and the relative level approaches steady state within 2 weeks and showed that the 13-day endosphere and rhizoplane microbiomes were most similar to the older, 42-day, microbiomes. Further, it has been proposed that plants might assemble their microbiomes in two steps: with the first step involving a general recruitment to the vicinity of the root and a second step for entry inside the root that involves species-specific genetic factors (Bulgarelli et al., 2013; Edwards et al., 2015). Thus, microbiome assemblage on endosphere, rhizosphere and rhizoplane responding to multiple signals from the plant coordinates the assembly of the root microbiome. Furthermore, scientific attempts are required to expand our molecular understanding of the

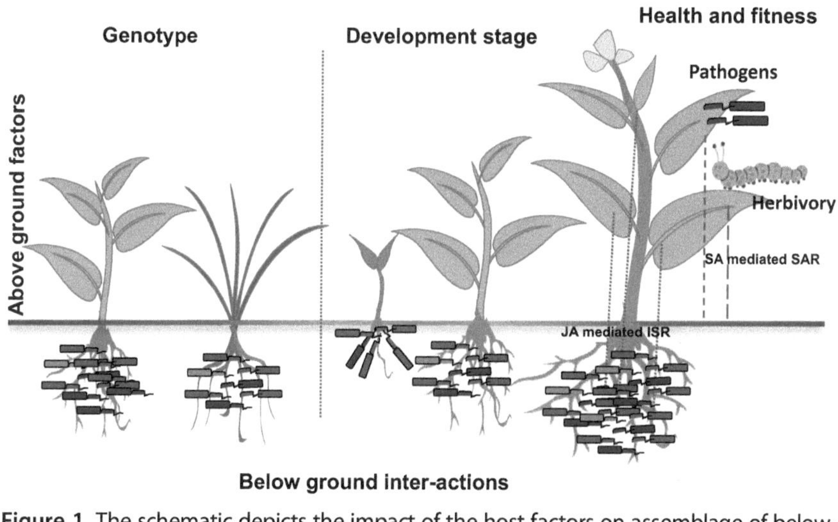

Figure 1 The schematic depicts the impact of the host factors on assemblage of below-ground rhizospheric microbiome in plants. It is shown that host genotype, developmental stage and plant health may alter the plant signalling and composition of root secretions which in turn facilitates rhizospheric composition of microbes. (See colour plate)

plant–microbiome interaction and its impact on plant health and agricultural productivity. In this way, the development of microbiome and microbe-driven biologicals as rational agriculture input might result in the next-generation agriculture, resulting in a more sustainable system for crop production and food security.

REFERENCES

Abbo, S., Pinhasi van-Oss, R., Gopher, A., Saranga, Y., Ofner, I., & Peleg, Z. (2014). Plant domestication versus crop evolution: a conceptual framework for cereals and grain legumes. *Trends in Plant Science, 19*, 351–360.

Abedon, B. G., & Tracy, W. F. (1996). Corn grass 1 of Maize (*Zea mays* L.) delays development of adult plant resistance to common rust (*Puccinia sorghi* Schw.) and european corn borer (*Ostrinia nubilalis* Hubner). *Journal of Heredity, 87*, 219–223.

Adesemoye, A. O., Torbert, H. A., & Kloepper, J. W. (2009). Plant growth-promoting rhizobacteria allow reduced application rates of chemical fertilizers. *Microbial Ecology, 58*, 921–929.

An, C., & Mou, Z. (2011). Salicylic acid and its function in plant immunity. *Journal of Integrative Plant Biology, 53*, 412–428.

Andreote, F. D., Araújo, W. L., Azevedo, J. L., vanElsas, J. D., Rocha, U. N., & van Overbeek, L. (2009). Endophytic colonization of potato (*Solanum tuberosum* L.) by a novel competent bacterial endophyte, *Pseudomonas putida* strain P9, and the effect on associated bacterial communities. *Applied and Environmental Microbiology, 75*, 3396–3406.

Aulakh, M. S., Wassmann, R., Bueno, C., Kreuzwieser, J., & Rennenberg, H. (2001). Characterization of root exudates at different growth stages of ten rice (*Oryza sativa* L.) cultivars. *Plant Biology, 3*, 139–148.

Badri, D. V., Chaparro, J. M., Zhang, R., Shen, Q., & Vivanco, J. M. (2013). Application of natural blends of phytochemicals derived from the root exudates of *Arabidopsis* to the soil reveal that phenolic related compounds predominantly modulate the soil microbiome. *Journal of Biological Chemistry, 288*, 4502–4512.

Badri, D. V., Weir, T. L., van der Lelie, D., & Vivanco, J. M. (2009). Rhizosphere chemical dialogues: plant–microbe interactions. *Current Opinion in Biotechnology, 20*, 642–650.

Bais, H. P., Weir, T. L., Perry, L. G., Gilroy, S., & Vivanco, J. M. (2006). The role of root exudates in rhizosphere interactions with plants and other organisms. *Annual Review of Plant Biology, 57*, 233–266.

Bakker, P. A., Berendsen, R. L., Doornbos, R. F., Wintermans, P. C., & Pieterse, C. M. (2013). The rhizosphere revisited: root microbiomics. *Frontiers in Plant Science, 4*, 165.

Barbour, W. M., Hattermann, D. R., & Stacey, G. (1991). Chemotaxis of *Bradyrhizobium japonicum* to soybean exudates. *Applied and Environmental Microbiology, 57*, 2635–2639.

Baudoin, E., Benizri, E., & Guckert, A. (2002). Impact of growth stage on the bacterial community structure along maize roots, as determined by metabolic and genetic fingerprinting. *Applied Soil Ecology, 19*, 135–145.

Bednarek, P. (2012). Chemical warfare or modulators of defence responses – the function of secondary metabolites in plant immunity. *Current Opinion in Plant Biology, 15*, 407–414.

Berendsen, R. L., Pieterse, C. M. J., & Bakker, P. A. H. M. (2012). The rhizosphere microbiome and plant health. *Trends in Plant Science, 17*, 478–486.

Berg, G., & Smalla, K. (2009). Plant species and soil type cooperatively shape the structure and function of microbial communities in the rhizosphere. *FEMS Microbiology Ecology, 68*, 1–13.

Binladen, J., Gilbert, M. T., Bollback, J. P., Panitz, F., Bendixen, C., Nielsen, R., et al. (2007). The use of coded PCR primers enables high-throughput sequencing of multiple homolog amplification products by 454 parallel sequencing. *PLoS One, 2*, e197.

Bouffaud, M.-L., Poirier, M.-A., Mulle, D., & Moënne-Loccoz, Y. (2014). Root microbiome relates to plant host evolution in maize and other Poaceae. *Environmental Microbiology, 16*, 2804–2814.

Broeckling, C. D., Broz, A. K., Bergelson, J., Manter, D. K., & Vivanco, J. M. (2008). Root exudates regulate soil fungal community composition and diversity. *Applied and Environmental Microbiology, 74*, 738–744.

Buee, M., De Boer, W., Martin, F., van Overbeek, L., & Jurkevitch, E. (2009). The rhizosphere zoo: an overview of plant-associated communities of microorganisms, including phages, bacteria, archaea, and fungi, and of some of their structuring factors. *Plant and Soil, 321*, 189–212.

Bulgarelli, D., Garrido-Oter, R., Münch, P. C., Weiman, A., Dröge, J., Pan, Y., et al. (2015). Structure and function of the bacterial root microbiota in wild and domesticated barley. *Cell Host and Microbe, 17*, 392–403.

Bulgarelli, D., Rott, M., Schlaeppi, K., van Themaat, E. V. L., Ahmadinejad, N., Assenza, F., et al. (2012). Revealing structure and assembly cues for *Arabidopsis* root-inhabiting bacterial microbiota. *Nature, 488*, 91–95.

Bulgarelli, D., Schlaeppi, K., Spaepen, S., van Themaat, E. V. L., & Schulze-Lefert, P. (2013). Structure and functions of the bacterial microbiota of plants. *Annual Review of Plant Biology, 64*, 807–838.

Carella, P., Wilson, D. C., & Cameron, R. K. (2015). Some things get better with age: differences in salicylic acid accumulation and defense signaling in young and mature *Arabidopsis*. *Frontiers in Plant Science, 5*, 775.

Carvalhais, L. C., Dennis, P. G., Badri, D. V., Kidd, B. N., Vivanco, J. M., & Schenk, P. M. (2015). Linking jasmonic acid signaling, root exudates, and rhizosphere microbiomes. *Molecular Plant-Microbe Interactions, 28*, 1049–1058.

Carvalhais, L. C., Dennis, P. G., Badri, D. V., Tyson, G. W., Vivanco, J. M., Schenk, P. M., et al. (2013). Activation of the jasmonic acid plant defence pathway alters the composition of column break rhizosphere bacterial communities. *PLoS One, 8*, e56457.

Carvalhais, L. C., Dennis, P. G., & Schenk, P. M. (2014). Plant defense inducers rapidly influence the diversity of bacterial communities in a potting mix. *Applied Soil Ecology, 84*, 1–5.

Carviel, J. L., Al-Daoud, F., Neumann, M., Mohammad, A., Provart, N. J., Moeder, W., et al. (2009). Forward and reverse genetics to identify genes involved in the age-related resistance response in *Arabidopsis thaliana*. *Molecular Plant Pathology, 10*, 621–634.

Carviel, J. L., Wilson, D. C., Isaacs, M., Carella, P., Catana, V., Golding, B., et al. (2014). Investigation of intercellular salicylic acid accumulation during compatible and incompatible *Arabidopsis-Pseudomonas syringae* interactions using a fast neutron-generated mutant allele of EDS5 identified by genetic mapping and whole-genome sequencing. *PLoS One, 9*, e88608.

Castellanos, T., Dohrmann, A. B., Imfeld, G., Baumgarte, S., & Tebbe, C. C. (2009). Search of environmental descriptors to explain the variability of the bacterial diversity from maize rhizospheres across a regional scale. *European Journal of Soil Biology, 45*, 383–393.

Chaparro, J. M., Badri, D. V., Bakker, M. G., Sugiyama, A., Manter, D. K., & Vivanco, J. M. (2014). Root exudation of phytochemicals in *Arabidopsis* follows specific patterns that are developmentally programmed and correlate with soil microbial functions. *Plos One, 8*, e55731.

Chaparro, J. M., Badri, D. V., & Vivanco, J. M. (2014). Rhizosphere microbiome assemblage is affected by plant development. *The ISME Journal, 8*, 790–803.

Chaparro, J. M., Sheflin, A. M., Manter, D. K., & Vivanco, J. M. (2012). Manipulating the soil microbiome to increase soil health and plant fertility. *Biology and Fertility of Soils, 48*, 489–499.
Chen, Y., Yan, F., Chai, Y., Liu, H., Kolter, R., Losick, R., et al. (2013). Biocontrol of tomato wilt disease by *Bacillus subtilis* isolates from natural environments depends on conserved genes mediating biofilm formation. *Environmental Microbiology, 15*, 848–864.
Clay, N. K., Adio, A. M., Denoux, C., Jander, G., & Ausubel, F. M. (2009). Glucosinolate metabolites required for an *Arabidopsis* innate immune response. *Science, 323*, 95–101.
De-la-Pena, C., Badri, D. V., Lei, Z., Watson, B. S., Brandao, M. M., Silva-Filho, M. C., et al. (2010). Root secretion of defense-related proteins is development-dependent and correlated with flowering time. *Journal of Biological Chemistry, 285*, 30654–30665.
De-la-Pena, C., Lei, Z., Watson, B. S., Sumner, L. W., & Vivanco, J. M. (2008). Root-microbe communication through protein secretion. *Journal of Biological Chemistry, 283*, 25247–25255.
Doornbos, R. F., Geraats, B. P. J., Kuramae, E. E., Van Loon, L. C., & Bakker, P. A. H. M. (2011). Effects of jasmonic acid, ethylene, and salicylic acid signaling on the rhizosphere bacterial community of Arabidopsis thaliana. *Molelcular Plant Microbe Interaction, 24*, 395–407.
Doornbos, R., Loon, L., & Bakker, P. H. M. (2012). Impact of root exudates and plant defense signaling on bacterial communities in the rhizosphere. A review. *Agronomy for Sustainable Development, 32*, 227–243.
Dridi, B., Raoult, D., & Drancourt, M. (2011). Archaea as emerging organisms in complex human microbiomes. *Anaerobe, 17*, 56–63.
Dunfield, K. E., & Germida, J. J. (2003). Seasonal changes in the rhizosphere microbial communities associated with field-grown genetically modified canola (*Brassica napus*). *Applied and Environmental Microbiology, 69*, 7310–7318.
Edwards, J., Johnson, C., Santos-Medellín, C., Lurie, E., Podishetty, N. K., Bhatnagar, S., et al. (2015). Structure, variation, and assembly of the root-associated microbiomes of rice. *Proceedings of the National Academy of Sciences, 112*, E911–E920.
Egamberdiyeva, D. (2007). The effect of plant growth promoting bacteria on growth and nutrient uptake of maize in two different soils. *Applied Soil Ecology, 36*, 184–189.
Eilers, K. G., Lauber, C. L., Knight, R., & Fierer, N. (2010). Shifts in bacterial community structure associated with inputs of low molecular weight carbon compounds to soil. *Soil Biology and Biochemistry, 42*, 896–903.
van Elsas, J. D., Chiurazzi, M., Mallon, C. A., Elhottovā, D., Krištůfek, V., & Salles, J. F. (2012). Microbial diversity determines the invasion of soil by a bacterial pathogen. *Proceedings of the National Academy of Sciences of the United States of America, 109*, 1159–1164.
Erb, M., Lenk, C., Degenhardt, J., & Turlings, T. C. J. (2009). The underestimated role of roots in defense against leaf attackers. *Trends in Plant Science, 14*, 653–659.
Flores, H. E., Vivanco, J. M., & Loyola-Vargas, V. M. (1999). 'Radicle' biochemistry: the biology of root-specific metabolism. *Trends in Plant Science, 4*, 220–226.
Fukui, R., Schroth, M. N., Hendson, M., Hancock, J. G., & Firestone, M. K. (1994). Growth patterns and metabolic activity of *Pseudomonas* in sugar beet spermospheres: relationship to pericarp colonization by *Pythium ultimum*. *Phytopathology, 84*, 1331–1338.
Giron, D., Frago, E., Glevarec, G., Pieterse, C. M. J., & Dicke, M. (2013). Cytokinins as key regulators in plant–microbe–insect interactions: connecting plant growth and defence. *Functional Ecology, 27*, 599–609.
Glazebrook, J. (2005). Contrasting mechanisms of defense against biotrophic and necrotrophic pathogens. *Annual Review of Phytopathology, 43*, 205–227.
Gomes, N. C. M., Heuer, H., Schonfeld, J., Costa, R., Mendonca-Hagler, L., & Smalla, K. (2001). Bacterial diversity of the rhizosphere of maize (*Zea mays*) grown in tropical soil studied by temperature gradient gel electrophoresis. *Plant and Soil, 232*, 167–180.

Haichar, F. Z., Marol, C., Berge, O., Rangel-Castro, J. I., Prosser, J. I., Balesdent, J., et al. (2008). Plant host habitat and root exudates shape soil bacterial community structure. *The ISME Journal, 2*, 1221–1230.

Hallmann, J., Quadt-Hallmann, A., Mahaffee, W. F., & Kloepper, J. W. (1997). Bacterial endophytes in agricultural crops. *Canadian Journal of Microbiology, 43*, 895–914.

Herschkovitz, Y., Lerner, A., Davidov, Y., Rothballer, M., Hartmann, A., Okon, Y., et al. (2005). Inoculation with the plant growth promoting rhizobacterium *Azospirillum brasilense* causes little disturbance in the rhizosphere and rhizoplane of maize (*Zea mays*). *Microbial Ecology, 50*, 277–288.

Heuer, H., Kroppenstedt, R. M., Lottmann, J., Berg, G., & Smalla, K. (2002). Effects of T4 lysozyme release from transgenic potato roots on bacterial rhizosphere communities are negligible relative to natural factors. *Applied and Environmental Microbiology, 68*, 1325–1335.

Houlden, A., Timms-Wilson, T. M., Day, M., & Bailey, M. J. (2008). Influence of plant developmental stage on microbial community structure and activity in the rhizosphere of three field crops. *FEMS Microbiology Ecology, 65*, 193–201.

Huang, X. F., Chaparro, J. M., Reardon, K. F., Zhang, R., Shen, Q., & Vivanco, J. M. (2014). Rhizosphere interactions: root exudates, microbes, and microbial communities. *Canadian Journal of Botany, 92*, 267–275.

Ikeda, S., Okubo, T., Takeda, N., Banba, M., Sasaki, K., Fujihara, S., et al. (2011). The genotype of the calcium/calmodulin-dependent protein kinase gene (CCaMK) determines bacterial community diversity in rice roots under paddy and upland field conditions. *Applied and Environmental Microbiology, 77*, 4399–4405.

Inceoglu, O., Al-Soud, W. A., Salles, J. F., Semenov, A. V., & van Elsas, J. D. (2011). Comparative analysis of bacterial communities in a potato field as determined by pyrosequencing. *PLoS One, 6*, e23321.

Inceoglu, Ö., Salles, J. F., Van Overbeek, L., & Van Elsas, J. D. (2010). Effect of plant genotype and growth stage on the ß-proteobacterial community associated with different potato cultivars in two fields. *Applied and Environment Microbiology, 76*, 3675–3684.

Insunza, V., Alstrom, S., & Eriksson, K. B. (2002). Root bacteria from nematicidal plants and their biocontrol potential against trichodorid nematodes in potato. *Plant and Soil, 241*, 271–278.

Jin, J., Wang, G. H., Liu, X. B., Liu, J. D., Chen, X. L., Herbert, S. J., et al. (2009). Temporal and spatial dynamics of bacterial community in the rhizosphere of soybean genotypes grown in a black soil. *Pedosphere, 19*, 808–816.

Kessler, A., & Baldwin, I. T. (2001). Defensive function of herbivore-induced plant volatile emissions in nature. *Science, 291*, 2141–2144.

Kumar, A. S., Lakshmanan, V., Caplan, J. L., Powell, D., Czymmek, K. J., Levia, D. F., et al. (2012). Rhizobacteria *Bacillus subtilis* restricts foliar pathogen entry through stomata. *The Plant Journal, 72*, 694–706.

Kus, J. V., Zaton, K., Sarkar, R., & Cameron, R. K. (2002). Age related resistance in *Arabidopsis* is a developmentally regulated defense response to *Pseudomonas syringae*. *The Plant Cell, 14*, 479–490.

Lakshmanan, V., & Bais, H. P. (2013). Factors other than root secreted malic acid that contributes toward *Bacillus subtilis* FB17 colonization on *Arabidopsis* roots. *Plant Signaling and Behavior, 8*, e27277.

Lakshmanan, V., Kitto, S. L., Caplan, J. L., Hsueh, Y. H., Kearns, D. B., Wu, Y. S., et al. (2012). Microbe-associated molecular patterns-triggered root responses mediate beneficial rhizobacterial recruitment in *Arabidopsis*. *Plant Physiology, 160*, 1642–1661.

Lakshmanan, V., Selvaraj, G., & Bais, H. P. (2014). Functional soil microbiome: belowground solutions to an aboveground problem. *Plant Physiology, 166*, 689–700.

Landgraf, R., Schaarschmidt, S., & Hause, B. (2012). Repeated leaf wounding alters the colonization of *Medicago truncatula* roots by beneficial and pathogenic microorganisms. *Plant Cell and Environment, 35*, 1344–1357.

Lankau, R. A. (2011). Intraspecific variation in allelochemistry determines an invasive species' impact on soil microbial communities. *Oecologia, 165*, 453–463.

Lebeis, S. L., Paredes, S. H., Lundberg, D. S., Breakfield, N., Gehring, J., McDonald, M., et al. (2015). Salicylic acid modulates colonization of the root microbiome by specific bacterial taxa. *Science, 349*, 860–864.

Lee, B., Farag, M. A., Park, H. B., Kloepper, J. W., Lee, S. H., & Ryu, C. M. (2012). Induced resistance by a long-chain bacterial volatile: elicitation of plant systemic defense by a C13 volatile produced by *Paenibacillus polymyxa*. *PLoS One, 7*, e48744.

Lee, B. L., Lee, S., & Ryu, C. M. (2012). Foliar aphid feeding recruits rhizosphere bacteria and primes plant immunity against pathogenic and non-pathogenic bacteria in pepper. *Annals of Botany, 110*, 281–290.

Lerner, A., Herschkovitz, Y., Baudoin, E., Nazaret, S., Moenne-Loccoz, Y., Yaacov, O., et al. (2006). Effect of *Azospirillum brasilense* on rhizobacterial communities analyzed by denaturing gradient gel electrophoresis and automated intergenic spacer analysis. *Soil Biology and Biochemistry, 38*, 1212–1218.

Liu, Y., Zhang, N., Qiu, M., Feng, H., Vivanco, J. M., Shen, Q., et al. (2014). Enhanced rhizosphere colonization of beneficial *Bacillus amyloliquefaciens* SQR9 by pathogen infection. *FEMS Microbiology Letters, 353*, 49–56.

Long, S. R. (1989). Rhizobium-legume nodulation: life together in the underground. *Cell, 56*, 203–214.

López-Bucio, J., de La Vega, O. M., Guevara-García, A., & Herrera-Estrella, L. (2000). Enhanced phosphorus uptake in transgenic tobacco plants that over produce citrate. *Nature Biotechnology, 18*, 450–453.

Lundberg, D. S., Lebeis, S. L., Paredes, S. H., Yourstone, S., Gehring, J., Malfatti, S., et al. (2012). Defining the core *Arabidopsis thaliana* root microbiome. *Nature, 488*, 86–90.

Lundberg, D. S., Yourstone, S., Mieczkowski, P., Jones, C. D., & Dangl, J. L. (2013). Practical innovations for high-throughput amplicon sequencing. *Nature Methods, 10*, 999–1002.

Marques, J. M., da Silva, T. F., Vollu, R. E., Blank, A. F., Ding, G. C., Lucy, S., et al. (2014). Plant age and genotype affect the bacterial community composition in the tuber rhizosphere of field-grown sweet potato plants. *FEMS Microbiology Ecology, 88*, 424–435.

Marschner, P., Neumann, G., Kania, A., Weiskopf, L., & Lieberei, R. (2002). Spatial and temporal dynamics of the microbial community structure in the rhizosphere of cluster roots of white lupin (*Lupinus albus* L.). *Plant and Soil, 246*, 167–174.

Meldau, S., Erb, M., & Baldwin, I. T. (2012). Defence on demand: mechanisms behind optimal defense patterns. *Annals of Botany, 110*, 1503–1514.

Mendes, R., Garbeva, P., & Raaijmakers, J. M. (2013). The rhizosphere microbiome: significance of plant beneficial, plant pathogenic, and human pathogenic microorganisms. *FEMS Microbiology Reviews, 37*, 634–663.

Mendes, R., Kruijt, M., de Bruijn, I., Dekkers, E., van der Voort, M., Schneider, J. H., et al. (2011). Deciphering the rhizosphere microbiome for disease-suppressive bacteria. *Science, 332*, 1097–1100.

Micallef, S. A., Channer, S., Shiaris, M. P., & Colón-Carmona, A. (2009). Plant age and genotype impact the progression of bacterial community succession in the *Arabidopsis* rhizosphere. *Plant Signaling and Behavior, 4*, 777–780.

Micallef, S. A., Shiaris, M. P., & Colon-Carmona, A. (2009). Influence of *Arabidopsis thaliana* accessions on rhizobacterial communities and natural variation in root exudates. *Journal of Experimental Botany, 60*, 1729–1742.

Millet, Y. A., Danna, C. H., Clay, N. K., Songnuan, W., Simon, M. D., Werck-Reichhart, D., et al. (2010). Innate immune responses activated in *Arabidopsis* roots by microbe associated molecular patterns. *The Plant Cell, 22*, 973–990.

Mougel, C., Offre, P., Ranjard, L., Corberand, T., Gamalero, E., Robin, C., et al. (2006). Dynamic of the genetic structure of bacterial and fungal communities at different developmental stages of *Medicago truncatula* Gaertn.cv. Jemalong line J5. *New Phytologist, 170*, 165–175.

Nagahashi, G., & Douds, D. D. (2000). Partial separation of root exudate components and their effects upon the growth of germinated spores of AM fungi. *Mycological Research, 104*, 1453–1464.

Neal, A. L., Ahmad, S., Gordon-Weeks, R., & Ton, J. (2012). Benzoxazinoids in root exudates of maize attract *Pseudomonas putida* to the rhizosphere. *PLoS One, 7*, e35498.

Ofek, M., Hadar, Y., & Minz, D. (2011). Colonization of cucumber seeds by bacteria during germination. *Environmental Microbiology, 13*, 2794–2807.

Ofek, M., Voronov-Goldman, M., Hadar, Y., & Minz, D. (2014). Host signature effect on plant root-associated microbiomes revealed through analyses of resident vs. active communities. *Environmental Microbiology, 16*, 2157–2167.

van Overbeek, L., & van Elsas, J. D. (2008). Effects of plant genotype and growth stage on the structure of bacterial communities associated with potato (*Solanum tuberosum* L.). *FEMS Microbiology Ecology, 64*, 283–296.

Peiffer, J. A., Spor, A., Koren, O., Jin, Z., Tringe, S. G., Dangl, J. L., et al. (2013). Diversity and heritability of the maize rhizosphere microbiome under field conditions. *Proceedings of the National Academy of Sciences of the United States of America, 110*, 6548–6553.

Perez-Jaramillo, J. E., Mendes, R., & Raaijmakers, J. M. (2015). Impact of plant domestication on rhizosphere microbiome assembly and functions. *Plant Molecular Biology*, 1–10. http://dx.doi.org/10.1007/s11103-015-0337-7.

Philippot, L., Raaijmakers, J. M., Lemanceau, P., & van der Putten, W. H. (2013). Going back to the roots: the microbial ecology of the rhizosphere. *Nature Reviews Microbiology, 11*, 789–799.

Pieterse, C. M. J., Van der Does, D., Zamioudis, C., Leon-Reyes, A., & Van Wees, S. C. (2012). Hormonal modulation of plant immunity. *Annual Review of Cell and Developmental Biology, 28*, 489–521.

Pineda, A., Zheng, S. J., van Loon, J. J. A., Pieterse, C. M. J., & Dicke, M. (2010). Helping plants to deal with insects: the role of beneficial soil-borne microbes. *Trends in Plant Science, 15*, 507–514.

Pongsilp, N., Nimnoi, P., & Lumyong, S. (2012). Genotypic diversity among rhizospheric bacteria of three legumes assessed by cultivation-dependent and cultivation-independent techniques. *World Journal of Microbiology and Biotechnology, 28*, 615–626.

Prithiviraj, B., Paschke, M. W., & Vivanco, J. M. (2007). Root communication: the role of root exudates. *Encyclopedia of Plant and Crop Science, 1*, 1–4.

Quecine, M. C., Araújo, W. L., Tsui, S., Parra, J. R. P., Azevedo, J. L., & Pizzirani-Kleiner, A. A. (2014). Control of *Diatraea saccharalis* by the endophytic *Pantoea agglomerans* 33.1 expressing cry1Ac7. *Archives of Microbiology, 196*, 227–234.

Raaijmakers, J. M., Paulitz, T. C., Steinberg, C., Alabouvette, C., & Moënne-Loccoz, Y. (2009). The rhizosphere: a playground and battlefield for soilborne pathogens and beneficial microorganisms. *Plant and Soil, 321*, 341–361.

Richardson, A. E., Barea, J. M., McNeill, A. M., & Prigent-Combaret, C. (2009). Acquisition of phosphorus and nitrogen in the rhizosphere and plant growth promotion by microorganisms. *Plant and Soil, 321*, 305–339.

Robert-Seilaniantz, A., Grant, M., & Jones, J. D. (2011). Hormone crosstalk in plant disease and defense: more than just jasmonate-salicylate antagonism. *Annual Review of Phytopathology, 49*, 317–343.

Rogers, E. E., Glazebrook, J., & Ausubel, F. N. (1996). Mode of action of the *Arabidopsis thaliana* phytoalexin camalexin and its role in *Arabidopsis*-pathogen interactions. *Molecular Plant-Microbe Interactions, 9*, 748–757.

Rudrappa, T., Biedrzycki, M. L., Kunjeti, S. G., Donofrio, N. M., Czymmek, K. J., Paré, P. W., et al. (2010). The rhizobacterial elicitor acetoin induces systemic resistance in *Arabidopsis thaliana. Communicative and Integrative Biology, 3*, 130–138.

Rudrappa, T., Czymmek, K. J., Paré, P. W., & Bais, H. P. (2008). Root-secreted malic acid recruits beneficial soil bacteria. *Plant Physiology, 148*, 1547–1556.

Rusterucci, C., Zhao, Z., Haines, K., Mellersh, D., Neumann, M., Cameron, R. K., et al. (2005). Age-related resistance to *Pseudomonas syringae* pv. tomato is associated with the transition to flowering in *Arabidopsis* and is effective against *Peronospora parasitica. Physiological and Molecular Plant Pathology, 66*, 222–231.

Schlaeppi, K., Dombrowski, N., Oter, R. G., Ver Loren van Themaat, E., & Schulze-Lefert, P. (2014). Quantitative divergence of the bacterial root microbiota in *Arabidopsis thaliana* relatives. *Proceedings of the National Academy of Sciences of the United States of America, 111*, 585–592.

Schoch, C. L., Seifert, K. A., Huhndorf, S., Robert, V., Spouge, J. L., André Levesque, C., et al. (2012). Nuclear ribosomal internal transcribed spacer (ITS) region as a universal DNA barcode marker for Fungi. *Proceedings of the National Academy of Sciences of the United States of America, 109*, 6241–6246.

Seymour, M., Kirkegaard, J. A., Peoples, M. B., White, P. F., & French, R. J. (2012). Break-crop benefits to wheat in Western Australia—insights from over three decades of research. *Crop and Pasture Science, 63*, 1–16.

Shakya, M., Gottel, N., Castro, H., Yang, Z. K., Gunter, L., Labbé, J., et al. (2013). A multi-factor analysis of fungal and bacterial community structure in the root microbiome of mature *Populus deltoides* trees. *PLoS One, 8*, e76382.

Shi, S., Richardson, A. E., O'Callaghan, M., DeAngelis, K. M., Jones, E. E., Stewart, A., et al. (2011). Effects of selected root exudate components on soil bacterial communities. *FEMS Microbiology Ecology*, 77, 600–610.

Smalla, K., Wieland, G., Buchner, A., Zock, A., Parzy, J., Kaiser, S., et al. (2001). Bulk and rhizosphere soil bacterial communities studied by denaturing gradient gel electrophoresis: plant dependent enrichment and seasonal shifts. *Applied and Environmental Microbiology, 67*, 4742–4751.

Soler, R., Erb, M., & Kaplan, I. (2013). Long distance root–shoot signaling in plant–insect community interactions. *Trends in Plant Science, 18*, 149–156.

Somers, E., Vanderleyden, J., & Srinivisam, M. (2004). Rhizosphere bacterial signaling: a love parade beneath our feet. *Critical Reviews in Microbiology, 30*, 205–240.

de Souza, R., Beneduzi, A., Ambrosini, A., da Costa, P. B., Meyer, J., Vargas, L. K., et al. (2013). The effect of plant growth-promoting rhizobacteria on the growth of rice (*Oryza sativa* L.) cropped in southern Brazilian fields. *Plant and Soil, 366*, 585–603.

Spence, C., & Bais, H. (2013). Probiotics for plants: rhizospheric microbiome and plant fitness. In F. J. de Bruijn (Ed.), *Molecular microbial ecology of the rhizosphere* (Vols. 1 and 2, pp. 713–721). Hoboken, NJ: John Wiley & Sons. http://dx.doi.org/10.1002/9781118297674.ch67.

Stam, J. M., Kroes, A., Li, Y., Gols, R., van Loon, J. J., Poelman, E. H., et al. (2014). Plant interactions with multiple insect herbivores: from community to genes. *Plant Biology, 65*, 689–713.

Sugiyama, A., Ueda, Y., Takase, H., & Yazaki, K. (2014). Pyrosequencing assessment of rhizosphere fungal communities from a soybean field. *Canadian Journal of Microbiology, 60*, 687–690.

Sugiyama, A., Ueda, Y., Zushi, T., Takase, H., & Yazaki, K. (2014). Changes in the bacterial community of soybean rhizospheres during growth in the field. *Plos One, 9*, e100709.

Sugiyama, A., Yamazaki, Y., Yamashita, K., Takahashi, S., Nakayama, T., & Yazaki, K. (2015). Developmental and nutritional regulation of isoflavone secretion from soybean roots. *Bioscience, Biotechnology, and Biochemistry*. http://dx.doi.org/10.1080/09168451.2015.1062714.

Sugiyama, A., & Yazaki, K. (2014). Flavonoids in plant rhizospheres: secretion, fate and their effects on biological communication. *Plant Biotechnology, 31*, 431–443.

Tan, S., Yang, C., Mei, X., Shen, S., Raza, W., Shen, O., et al. (2013). The effect of organic acids from tomato root exudates on rhizosphere colonization of *Bacillus amyloliquefaciens* T5. *Applied and Soil Ecology, 64*, 15–22.

Thaler, J. S., Humphrey, P. T., & Whiteman, N. K. (2012). Evolution of jasmonate and salicylate signal crosstalk. *Trends in Plant Science, 17*, 260–270.

Turner, T. R., James, E. K., & Poole, P. S. (2013). The plant microbiome. *Genome Biology, 14*, 209–219.

Turner, T. R., Ramakrishnan, K., Walshaw, J., Heavens, D., Alston, M., Swarbreck, D., et al. (2013). Comparative metatranscriptomics reveals kingdom level changes in the rhizosphere microbiome of plants. *The ISME Journal, 7*, 2248–2258.

de Vasconcellos, R. L. F., & Cardoso, E. J. B. N. (2009). Rhizospheric streptomycetes as potential biocontrol agents of *Fusarium* and *Armillaria* pine rot and as PGPR for *Pinus taeda*. *BioControl, 54*, 807–816.

Weinert, N., Piceno, Y., Ding, G. C., Meincke, R., Heuer, H., Berg, G., et al. (2011). PhyloChip hybridization uncovered an enormous bacterial diversity in the rhizosphere of different potato cultivars: many common and few cultivar-dependent taxa. *FEMS Microbiology Ecology, 75*, 497–506.

Weller, D. M., Raaijmakers, J. M., Gardener, B. B. M., & Thomashow, L. S. (2002). Microbial populations responsible for specific soil suppressiveness to plant pathogens. *Annual Review of Phytopathology, 40*, 309–348.

Weston, L. A., & Mathesius, U. (2013). Flavonoids: their structure, biosynthesis and role in the rhizosphere, including allelopathy. *Journal of Chemical Ecology, 39*, 283–297.

Wieland, G., Neumann, R., & Backhaus, H. (2001). Variation of microbial communities in soil, rhizosphere, and rhizoplane in response to crop species, soil type, and crop development. *Applied and Environmental Microbiology, 67*, 5849–5854.

Xu, Y., Wang, G., Jin, J., Liu, J., Zhang, Q., & Liu, X. (2009). Bacterial communities in soybean rhizosphere in response to soil type, soybean genotype, and their growth stage. *Soil Biology and Biochemistry, 41*, 919.

Yang, J. W., Yi, H. S., Lee, B., Lee, S., Ghim, S. Y., & Ryu, C. M. (2011). Whitefly infestation of pepper plants elicits defence responses against bacterial pathogens in leaves and roots and changes the below-ground microflora. *Journal of Ecology, 99*, 46–56.

Yeoh, Y. K., Paungfoo-Lonhienne, C., Dennis, P. G., Robinson, N., Ragan, M. A., Schmidt, S., et al. (2015). The core root microbiome of sugarcanes cultivated under varying nitrogen fertilizer application. *Environmental Microbiology*. http://dx.doi.org/10.1111/1462-2920.12925.

Yu, J., Grishkan, I., & Steinberger, Y. (2013). Micro fungal community diversity in *Zygophyllum dumosum* and *Hammada scoparia* root zones in the northern Negev Desert. *Journal of Basic Microbiology, 53*, 390–400.

Zhou, X., & Wu, F. (2012). p-Coumaric acid influenced cucumber rhizosphere soil microbial communities and the growth of *Fusarium oxysporum* f.sp. cucumerinum Owen. *PLoS One, 7*, e48288.

Zhou, X., & Wu, F. (2013). Artificially applied vanillic acid changed soil microbial communities in the rhizosphere of cucumber (*Cucumis sativus* L.). *Canadian Journal of Soil Science, 93*, 13–21.

CHAPTER FOUR

PGPR Interaction: An Ecofriendly Approach Promoting the Sustainable Agriculture System

Usha Bishnoi[*,§]
[*]Department of Plant and Soil Sciences, University of Delaware, Newark, DE, USA
[§]Delaware Biotechnology Institute, Newark, DE, USA
E-mail: bishnoi@udel.edu

Contents

1. Introduction 82
2. Rhizosphere, the Hot Spot for PGPR Interaction 84
3. Colonization and Competence 88
4. PGPR Rooting for the Sustainable Agriculture 89
   4.1 PGPR as Biofertilizers 89
      4.1.1 Nitrogen Fixation *89*
      4.1.2 Phosphorus Solubilization *90*
      4.1.3 Absorption of Iron *91*
      4.1.4 Phytostimulators *92*
   4.2 PGPR as Biopesticides 96
   4.3 PGPR as Bioremediators 97
5. Commercialization of PGPR 99
6. Conclusion 103
Acknowledgements 104
References 104

Abstract

Soil, the dynamic and valuable natural resource harbouring a vast collection of microorganisms, is vital for the production of food and fibre, in addition to fulfilling critical roles in the maintenance of global nutrient balance and ecosystem function. Plant growth promoting rhizobacteria (PGPR), a diverse group of soil bacteria, are key components of soil–plant systems, where they are engaged in an intense network of interactions in the rhizosphere, thus affecting the plant growth and yield. Agricultural productivity rests on the foundation of microbial diversity in the soil, and in recent years, PGPR have emerged as an important and promising tool for sustainable agriculture. PGPR promote plant growth and development directly or indirectly, either by releasing plant growth regulators/phytohormones or other biologically active substances, altering endogenous levels of phytohormones, enhancing the availability

Advances in Botanical Research, Volume 75
ISSN 0065-2296
http://dx.doi.org/10.1016/bs.abr.2015.09.006

and uptake of nutrients through fixation and mobilization, or reducing the harmful effects of pathogenic microorganisms on plants and/or by employing multiple mechanisms of action. Besides offering an economically attractive and ecologically sound route for augmenting the nutrient supply and protecting against soilborne pathogens, PGPR are the key players in efforts to preserve soil fertility. In this chapter, the different mechanisms of plant growth promotion rendered by the PGPR are discussed briefly, and the need for the development and commercialization of these PGPR for sustainable agriculture is emphasized.

1. INTRODUCTION

Agriculture, the science or the practice of cultivating plants, animals and other life forms, is undoubtedly one of the factors that propelled human civilization and development. It is difficult to determine precisely where, when and how agriculture started; and the views of origin are diverse, ranging from mythological to ecological. Development of agriculture did not just happen overnight and instantly spread to all corners of the world; it is an evolutionary process that eventually transformed plants from being independent, wild progenitors into fully dependent, domesticated cultivars with the concomitant evolution of agricultural economics (Zeder, 2009). This relationship between humans, the Earth and food sources further affirms soil as the foundation of agriculture, and the vital role played by the microbes in sustaining our farmed and natural ecosystems only adds urgency to their study.

The utilization of bacteria to stimulate plant growth in agriculture has been practiced for millennia, and the idea of using bacteria in land stewardship to sustain land productive for future generations is not new. The Middle Eastern farmers practiced crop rotation in 6000 BC without understanding the chemistry or the important roles played by the bacteria, alternately planting legumes and cereal. More recently in human history, Hellriegel and Wilfarth (1888) studied the rhizosphere root colonization and suggested the ability of soil bacteria to convert atmospheric N_2 into plant-usable forms, and the establishment of legumes on cultivated lands resulted in improved soil fertility (Chew, 2002). Besides crop rotation and intercropping, the mixing of different soils as a means of remedying defects and adding 'heart' to the soil is a well-known process (Tisdale & Nelson, 1975). It was Kloepper and Schroth (1978), who coined the term 'plant growth promoting rhizobacteria (PGPR)' for these beneficial microbes. Since then, the research on PGPR has made great strides and PGPR now represent a wide variety of soil bacteria which, when grown in association with a host plant, result in stimulation of growth of the host. PGPR thus constitute an indispensable part of the rhizosphere biota, and they stimulate the plant growth by a plethora of mechanisms. Genera of PGPR include *Pseudomonas, Bacillus, Azospirillum, Agrobacterium, Azotobacter, Arthrobacter, Alcaligenes, Serratia, Rhizobium, Enterobacter, Burkholderia, Beijerinckia, Klebsiella, Clostridium, Vario-vovax, Xanthomonas* and *Phyllobacterium* (Lucy, Reed, & Glick, 2004; Lugtenberg, Chin A-Woeng, & Bloemberg, 2002; Weller & Thomashow, 1994). Among these, *Pseudomonas* and *Bacillus* are the most widely reported PGPR (Podile & Kishore, 2006).

The PGPR are defined by three distinct characteristics: (1) they must be able to colonize the root; (2) they must survive and multiply in microhabitats associated with the root surface, in competition with other microbiota, at least for the time needed to express their plant promotion/protection activities and (3) they must promote plant growth. Plant treatments with PGPR may also increase the percentage of seed germination, seedling vigour, emergence, plant stand, root and shoot growth, total biomass of the plants, seed weight, grains, fodder and fruit yields or decrease the time to flowering (van Loon, Bakker, & Pieterse, 1998; Zahran, 2001). Though the exact mechanisms involved in growth promotion for many microbes are still unclear, various mechanisms have been suggested (Figure 1; Vacheron et al., 2013). Bacterially mediated phytohormone production is the most likely explanation for PGPR activity in the absence of pathogens (Brown, 1974; Tien, Gaskins, & Hubbell, 1979), while siderophore production by PGPR is considered more important for plant growth stimulation when other potentially deleterious microorganisms are present in the rhizosphere (Kloepper, Leong, Teintze, & Schroth, 1980).

PGPR play a pivotal role in plant function by influencing plant physiology and development, either directly or indirectly. Direct stimulation includes biological nitrogen fixation (Zahran, 2001), producing or changing the concentrations of phytohormones such as auxins, cytokinins, gibberellins (GA) (Vacheron et al., 2013; Tien et al., 1979) or ethylene (Glick, Karaturovic, & Newell, 1995), solubilizing minerals like phosphorus and iron (Delvasto, Valverde, Ballester, Igual, & Munoz, 2006; Rodríguez, Fragai, González, & Bashan, 2006), production of siderophores and enzymes, and induction of systemic resistance (van Loon, 2007). The indirect stimulation of plant growth is fundamentally related to biocontrol, including antibiotic production, chelation of Fe in the rhizosphere, synthesis of extracellular enzymes to hydrolyze fungal cell wall and competition

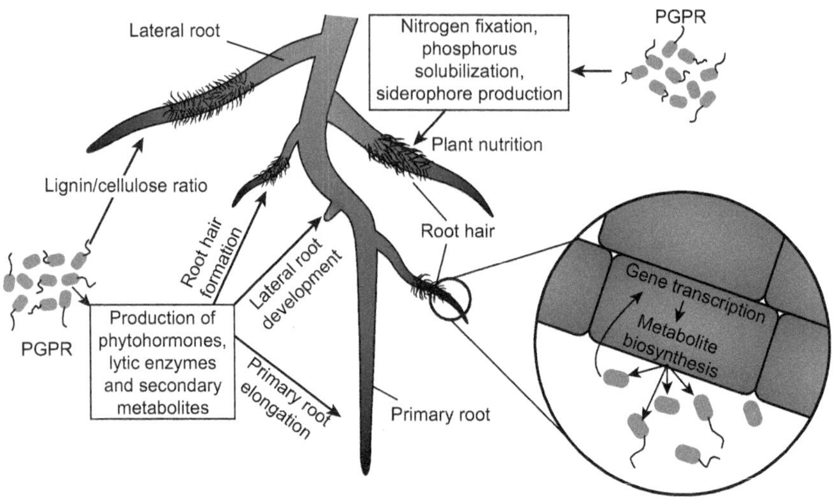

Figure 1 The possible mode of actions used by plant growth promoting rhizobacteria (PGPR) towards growth promotion in plants. *Adapted from Vacheron et al. (2013).* (See colour plate)

for niches within the rhizosphere (van Loon, 2007). PGPR strains, particularly, *Pseudomonas fluorescens* and *Bacillus subtilis* are often studied as the most promising candidates of indirect stimulation (Damayanti, Pardede, & Mubarik, 2007). PGPR may utilize more than one mechanisms to enhance plant growth, as experimental evidence suggests that the plant growth stimulation is the net result of multiple mechanisms that may be activated simultaneously (Martinez-Viveros, Jorquera, Crowley, Gajardo, & Mora, 2010).

Thus, based on their mechanism of action, PGPR can be categorized into three general forms such as biofertilizer, phytostimulator and biopesticide (Table 1).

2. RHIZOSPHERE, THE HOT SPOT FOR PGPR INTERACTION

Rhizosphere, the narrow zone of soil specifically influenced by the root system (Dobbelaere, Venderleyden, & Okon, 2003), is a dynamic and versatile environment of intensive plant–microbe interactions for extracting essential macro- and micronutrients from a confined nutrient pool. This zone is rich in nutrients for microbes when compared to the bulk soil, and this is well reflected by the number of bacteria that are present around the roots of

Table 1 Different Modes of Action Considered for Growth Promotion by PGPR (Plant Growth Promoting Rhizobacteria)

Term	Definition	Mechanism	References
Biofertilizer	A substance which contains live microorganisms which when applied to the seed, plant or the soil, colonizes the rhizosphere or the interior of the plant and promotes the growth through increased supply/or availability of primary nutrients to the host plant	Biological nitrogen fixation	Vessey (2003) Somers, Vanderleyden, and Srinivasan (2004)
		Utilization of insoluble forms of phosphorus	Fuentes-Ramirez and Caballero Mellato (2005)
Phytostimulator	Microorganism with the ability to produce or change the concentration of growth regulators such as IAA, GA, cytokinins and ethylene	Production of phytohormones (auxins, cytokinins and GA)	Lugtenberg et al. (2002) Somers et al. (2004)
Biopesticides	Microorganisms that promote plant growth through the control of phytopathogenic agents, mainly for the production of antibiotics and antifungal metabolites	Production of antibiotics and production of hydrolytic enzymes	Vessey (2003) Somers et al. (2004)
		Competitive exclusion, acquired and induced systemic resistance	Chandler, Davidson, Grant, Greaves, and Tatchell (2008)

GA, gibberellins.

the plants, generally 10 to 100 times higher than in the bulk soil (Weller & Thomashow, 1994). The plant roots exude a huge diversity of organic nutrients (organic acids, phytosiderophores, sugars, vitamins, amino acids, nucleosides, mucilage) and signals that attract the microbial populations, especially those able to metabolize plant-exuded compounds and proliferate in this habitat (Badri, Weir, van der Lelie, & Vivanco, 2009; Bais, Weir, Perry, Gilroy, & Vivanco, 2006; Drogue, Combes-Meynet, Mënne-Loccoz, Wisniewski-Dyé, & Prigent-Combaret, 2013; Lakshmanan, Selvaraj, & Bais, 2014). The corresponding microbial community associated to plant roots is referred as the rhizo-microbiome (Chaparro et al., 2013), and its composition is distinctive from that of the microbial community of the surrounding bulk soil, a direct consequence of bacterial competition for nutrients liberated in the vicinity of plant roots (Bulgarelli, Schlaeppi, Spaepen, Ver Loren van Themaat, & Schulze-Lefert, 2013; Chaparro et al., 2013). Thus, in contrast to other microenvironments of plant, the rhizosphere is characterized by a high microbial abundance and activities. *Via* root exudation, roots may regulate the soil microbial community in their immediate vicinity, deter or respond to herbivores, encourage beneficial symbioses, change the chemical and physical properties of the soil or inhibit the growth of competing plant species (Bais, Park, Weir, Callaway, & Vivanco, 2004). Since composition of root exudate changes along the length of the root system, according to stages of plant development and to plant genotype, the rhizo-microbiome composition also differs accordingly (Berg & Smalla, 2009; Bulgarelli et al., 2013; Chaparro et al., 2013). The specific composition of root exudates may create a niche that influences which microorganisms are to colonize the rhizosphere, thereby altering the composition and diversity of microorganisms colonizing the rhizosphere in a plant-specific manner (Grayston, Wang, Campbell, & Edwards, 1998). Thus, the plant species, plant developmental stage and soil type are the key factors governing the composition of rhizosphere microbial communities prevailing in a particular ecological niche (Berendsen, Pieterse, & Bakker, 2012; Berg & Smalla, 2009; Broeckling, Broz, Bergelson, Manter, & Vivanco, 2008). The rhizosphere is of central importance not only for plant nutrition, health and quality, but it also plays a critical role in the microorganism-driven carbon sequestration. Carbon sequestration, of course, has an important role in ecosystem functioning and nutrient cycling in terrestrial ecosystems. Thus, the rhizosphere has emerged as a versatile and dynamic ecological environment of intensive plant—microbe interactions affecting the plant growth, and serving as a protection shield against the soilborne pathogens.

In the rhizosphere, plants and microorganisms are permanently interacting in a continuum ranging from deleterious to beneficial and based on their effects on the plant growth can be classified into beneficial, deleterious and neutral groups (Dobbelaere et al., 2003). In negative interactions, the phytopathogenic rhizobacteria produce phytotoxic substances such as hydrogen cyanide or ethylene, thus, exhibiting negative influence on the growth and physiology of the plants. Apart from these deleterious bacteria, there are PGPR that exert a positive impact on plant growth by direct mechanisms, such as solubilization of nutrients, nitrogen fixation, production of growth regulators, etc., or by indirect mechanisms such as stimulation of mycorrhizae development, competitive exclusion of pathogens or removal of phytotoxic substances (Bashan & de-Bashan, 2010). Neutral interactions, on the other hand, do not benefit or harm the plant directly, but the microbes' presence is vital for the soil fertility (Brimecombe, De Leij, & Lynch, 2007). Based on their degree of association with the plant root cells, PGPR can be further classified into extracellular plant growth promoting rhizobacteria (ePGPR) and intracellular plant growth promoting rhizobacteria (iPGPR) (Martinez-Viveros et al., 2010). The ePGPR may exist in the rhizosphere, on the rhizoplane or in the spaces between the cells of root cortex; whereas, iPGPR locate generally inside the specialized nodular structures of root cells. The bacterial genera such as *Agrobacterium, Arthrobacter, Azotobacter, Azospirillum, Bacillus, Burkholderia, Caulobacter, Chromobacterium, Erwinia, Flavobacterium, Micrococcus, Pseudomonas and Serratia* belong to ePGPR (Gray & Smith, 2005). The iPGPR includes the endophytes (*Allorhizobium, Azorhizobium, Bradyrhizobium, Mesorhizobium* and *Rhizobium* of the family *Rhizobiaceae*) and *Frankia* species both of which can symbiotically fix atmospheric N_2 with the higher plants (Verma, Yadav, & Tiwari, 2010).

ePGPR represent a wide variety of soil bacteria that may induce plant growth by direct or indirect modes of action (Glick, 1995). The direct modes include (1) the production of phytohormones (Vacheron et al., 2013) and stimulatory bacterial volatiles (Ryu et al., 2003); (2) the lowering of plant ethylene levels (Glick, Penrose, & Li, 1998); (3) the improvement of plant nutrient status by either making available macro- and micronutrients from insoluble sources (Delvasto et al., 2006; Rodríguez et al., 2006) or by non-symbiotic nitrogen fixation (Boddey, Urquiaga, Alves, & Reis, 2003) and (4) the stimulation of disease-resistance mechanisms such as induced systemic resistance (ISR) (van Loon, 2007). Indirect effects arise when ePGPR act as biocontrol agents leading to reduced diseases (Compant et al., 2005), when they stimulate other beneficial symbioses (Burdman,

Jurkevitch, & Okon, 2000) or when they protect the plant by degrading xenobiotics in inhibitory contaminated soils (Jacobsen, 1997).

3. COLONIZATION AND COMPETENCE

Colonization and competence is a prerequisite of PGPR towards plant growth promotion. In all successful plant–microbe interactions, the competence to colonize plant habitats is important (Kamilova, Validov, Azarova, Mulders, & Lugtenberg, 2005; Lugtenberg et al., 2002). Single bacterial cells can attach to surfaces and, after cell division and proliferation, form dense aggregates commonly referred to as macrocolonies or biofilms (Bloemberg & Lugtenberg, 2004; Rudrappa, Czymmek, Par, & Bais, 2008). Steps of colonization include attraction, recognition, adherence, invasion (only endophytes and pathogens), colonization and growth. Plant roots initiate crosstalk with soil microbes by producing signals that the microbes recognize, which in turn, produce signals that initiate plant colonization (Berg, 2009). PGPR reach root surfaces by active motility facilitated by flagella and are guided by chemotactic responses (Pinton, Varanini, & Nannipieri, 2007). In most cases, the population of many PGPR inoculants actually declines progressively in time after inoculation from 10e7–10e9 cells per gram dry soil to 10e5–10e6 cells per gram dry soil after 2–3 weeks (DeFlaun & Gerba, 1993). Rhizosphere competence of biocontrol agents thus involves effective root colonization combined with the ability to survive and proliferate along growing plant roots over a large time period, in the presence of the indigenous microflora (Lugtenberg & Dekkers, 1999). Given the relative abundance of substrates in the rhizosphere, the efficiency of nutrient uptake and catabolism by bacteria is a key factor in competitiveness (Chin A-Woeng, Bloemberg, & Lugtenberg, 2003). Besides, the characteristic for rapid growth mediated by effective colonization, the rhizobacteria deploy many other metabolic strategies like the extracellular conversion of glucose to gluconic acid and 2-ketogluconic acid which enables some bacteria, including several species of *Pseudomonas* to sequester glucose effectively thus giving a competitive advantage over other microorganisms that lack the ability to utilize these compounds (Gottschalk, 1986).

Similarly, competition for trace elements, such as iron, copper, zinc, manganese, also occurs in the soil. Iron, for example, is an essential growth element for all living organisms and the scarcity of its bioavailable form in soil habitats results in an intense competition (Loper et al., 1997). Iron siderophores, low molecular weight compounds with high iron affinity are produced by some microorganisms to solubilize and competitively acquire ferric ion under iron-limiting conditions, thereby making iron unavailable to other organisms which cannot grow for lack of it (Crowley, 2006; Loper et al., 1997). Suppression of soilborne plant pathogens by siderophore-producing Pseudomonads has been reported (Buysens, Heungens, Poppe, & Hofte, 1996; Weger et al., 1988), and the competition for nutrients and niches between pathogens and beneficial microbes is an important factor for limiting disease incidence and severity (Kamilova et al., 2005).

4. PGPR ROOTING FOR THE SUSTAINABLE AGRICULTURE

The PGPR interactions are vital not only for the growth and productivity of the plants but are also imperative to our planet's health and functioning. The role of PGPR as biofertilizers, phytostimulators, biopesticides and bioremediators is discussed briefly:

4.1 PGPR as Biofertilizers

Next to water and temperature, nutrients are the environmental factor that most strongly constrains terrestrial plant growth. PGPR promote the plant growth by increasing the accessibility or uptake of nutrients from a confined nutrient pool in the soil/rhizosphere.

4.1.1 Nitrogen Fixation

Nitrogen is an essential element for all forms of life; a basic requisite for synthesizing nucleic acids, proteins and other organic nitrogenous compounds. Although, there is about 78% N_2 in the atmosphere, it is highly inert and unavailable to growing plants. Atmospheric N_2 is converted into plant-usable forms by a process called biological N_2 fixation (BNF) in which nitrogen fixing microorganisms transform elemental nitrogen into ammonia using a complex enzyme system known as nitrogenase (Kim & Rees, 1994). Nitrogen fixing organisms are generally categorized as (1) symbiotic N_2-fixing bacteria including members of the family rhizobiaceae (*Rhizobium, Sinorhizobium, Bradyrhizobium, Mesorhizobium* and *Azorhizobium*, collectively termed rhizobia) which forms symbiosis with leguminous plants (Zahran, 2001) and nonleguminous trees (e.g. *Frankia*) and (2) non-symbiotic (free living, associative and endophytes) nitrogen fixing forms such as cyanobacteria (*Anabaena, Nostoc*), *Azospirillum, Azotobacter, Gluconacetobacter diazotrophicus* and *Azocarus,* etc. (Bhattacharyya & Jha, 2012). The signalling processes (Long, 2001), the evolutionary history (Henson, Watson, & Barnum, 2004) and the molecular aspects determining host specificity in the rhizobial–legume symbiosis (Young, Mutch, Ashford, Zézé, & Mutch, 2003), have been reviewed. BNF imparts 180×106 metric tons per year globally, out of which symbiotic nitrogen fixation contributes 80% and the remaining comes from free living nitrogen fixation. BNF thus represents an economically beneficial and environmentally sound alternative to replace the high regimens of chemical fertilizers used in the present day agriculture (Adesemoye, Torbert, & Kloepper, 2009).

4.1.2 Phosphorus Solubilization

Phosphorus (P) is one of the major essential macronutrients for plant growth and development, and low availability of P to crop plants is a worldwide issue and on 30–40% of world arable lands crop yield is reduced due to P availability (Vance, Uhde-Stone, & Allan, 2003). P in soil is present in two main insoluble forms: mineral forms, such as apatite, hydroxyapatite and oxyapatite, and organic forms including inositol phosphate (soil phytate), phosphomonoesters, phosphodiesters and phosphotriesters. Solubilization and mineralization of P by phosphate-solubilizing bacteria (PSB) are among the most important bacterial physiological traits in soil biogeochemical cycles (Jeffries, Gianinazzi, Perotto, Turnau, & Barea, 2003), as well as in plant growth promotion by PGPR (Richardson, 2001; Rodriguez et al., 2006). Bacterial genera like *Azotobacter, Bacillus, Beijerinckia, Burkholderia, Enterobacter, Erwinia, Flavobacterium, Microbacterium, Pseudomonas, Rhizobium* and *Serratia* are reported as the most significant PSB (Bhattacharyya & Jha, 2012). Besides, examples of some widely reported P-solubilizing microbial species intimately associated with a large number of agricultural crops like potato, tomato, wheat, radish, pulses, etc. are *Azotobacter chroococcum, Bacillus circulans, Cladosporium herbarum, Bradyrhizobium japonicum, Enterobacter agglomerans, Pseudomonas chlororaphis, Pseudomonas putida* and *Rhizobium leguminosarum* (Antoun, Beauchamp, Goussard, Chabot, & Lalande, 1998; Cattelan, Hartel, & Fuhrmann, 1999; Chabot, Beauchamp, Kloepper, & Antoun, 1998). Typically, the solubilization of inorganic phosphorus occurs as a consequence of the action of low molecular weight organic acids which are synthesized by various soil bacteria (Zaidi, Khan, Ahemad, & Oves, 2009). Conversely, the mineralization of organic phosphorus occurs through the synthesis of a variety of different phosphatases, catalyzing the hydrolysis of phosphoric esters (Glick, 2012). Phosphate solubilization and mineralization can coexist in the same bacterial strain (Tao, Tian, Cai, & Xie, 2008), besides providing P to the plants, the PS bacteria also augment the growth of plants by stimulating the efficiency of BNF, enhancing the availability of other trace elements (such as iron, zinc). The possibility of enhancing P uptake of crops by inoculation with P-solubilizing strains of PGPR presents a promising approach towards recovering the reservoirs of insoluble phosphorus from the soil and thus minimizing the external application of phosphate fertilizers to the soil.

4.1.3 Absorption of Iron

Iron is an essential micronutrient for plants as it serves as a cofactor of many enzymes with redox activity and it is required in a number of major physiological processes like N_2 fixation, photosynthesis, respiration, etc. To meet their iron requirement, microorganisms and plants have evolved specific mechanisms to chelate insoluble iron through the release of siderophores and uptake of iron–siderophore complexes through specific outer membrane receptor proteins (Sharma & Johri, 2003). These siderophores can be of different types: hydroxamates, phenol-catecholates and carboxylates (Podile & Kishore, 2006). The synthesis of siderophores in bacteria is induced by the low level of Fe^{3+} and in acid soil, where solubility and availability grow, their protective effect comes down. Microbial siderophores in the rhizosphere are frequently associated with biocontrol activities and not with plant nutrition (Vessey, 2003). Iron is essential for cellular growth and metabolism, such that Fe acquisition through siderophore production plays an essential role in determining the competitive fitness of bacteria to colonize plant roots and to compete for iron with other microorganisms in the rhizosphere (Crowley, 2006). Siderophore-producing PGPR can prevent the proliferation of pathogenic microorganisms by sequestering Fe^{3+} in the area around the root since these bacterial iron chelators are thought to sequester the limited supply of iron available in the rhizosphere making it unavailable to pathogenic fungi, thereby restricting their growth (Loper & Henkels, 1999). Recent studies have clearly shown that the iron nutrition of the plant influences the rhizosphere microbial community structure (Yang & Crowley, 2000). Iron competition in Pseudomonads has been intensively studied and the role of the pyoverdine siderophore produced by many *Pseudomonas* species has been clearly demonstrated in the control of *Pythium* and *Fusarium* species, either by comparing the effects of purified pyoverdine with synthetic iron chelators or through the use of pyoverdine minus mutants (Duijff, De Kogel, Bakker, & Schippers, 1994). The dynamics of iron competition in the rhizosphere are often complex. For example, some siderophores can only be used by the bacteria that produce them (Ongena et al., 1999), whereas others can be used by many different bacteria (Loper & Henkels, 1999). Different environmental factors can also influence the quantity of siderophores produced (Duffy & Défago, 1999). Various studies have isolated siderophore-producing bacteria belonging to the genera *Bradyrhizobium, Pseudomonas, Rhizobium, Serratia* and *Streptomyces* (Boopathi & Rao, 1999; Kuffner, Puschenreiter, Wieshammer, Gorfer, & Sessitsch, 2008) from the rhizosphere. Positive effects on plant growth have been reported after the inoculation of siderophore-producing *Pseudomonas, Rhizobium* and *Azospirillum* in iron-limited conditions (Carrillo-Castañeda et al., 2002).

4.1.4 Phytostimulators

Phytohormones are organic compounds that in extremely low concentrations influence biochemical, physiological and morphological processes in plants, and their synthesis is finely regulated (Fuentes-Ramirez & Caballero-Mellado, 2005). Studies have demonstrated that the PGPR can stimulate plant growth through the production of auxins (indole acetic acid) (Spaepen, Vanderleyden, & Remans, 2007), GA (Bottini, Cassán, & Piccoli, 2004) and cytokinins (Timmusk, Nicander, Granhall, & Tillberg, 1999), or by regulating the high levels of endogenous ethylene in the plant (Glick et al., 1998).

4.1.4.1 Indole-3-Acetic Acid

Microbial synthesis of the phytohormone auxin (indole-3-acetic acid/IAA) has been known for a long time. It is reported that 80% of microorganisms isolated from the rhizosphere of various crops possess the ability to synthesize and release auxins as secondary metabolites (Patten & Glick, 2002). Generally, IAA secreted by rhizobacteria interferes with many plant developmental processes because the endogenous pool of plant IAA may be altered by the acquisition of IAA that has been secreted by soil bacteria (Glick, 2012; Spaepen et al., 2007). Evidently, IAA also acts as a reciprocal signalling molecule affecting gene expression in several microorganisms. Consequently, IAA plays a very important role in rhizobacteria–plant interactions (Spaepen & Vanderleyden, 2011). Moreover, downregulation of IAA as signalling is associated with the plant defence mechanisms against a number of phytopathogenic bacteria as evidenced in enhanced susceptibility of plants to the bacterial pathogen by exogenous application of IAA or IAA produced by the pathogen (Spaepen & Vanderleyden, 2011). This diversity of function is reflected by the extraordinary complexity of IAA biosynthetic, transport and signalling pathways (Santner, Calderon-Villalobos, & Estelle, 2009). Generally, IAA affects plant cell division and differentiation; stimulates seed and tuber germination; increases the rate of xylem and root development; controls processes of vegetative growth; initiates lateral and adventitious root formation; affects photosynthesis, biosynthesis of various metabolites and resistance to stressful conditions. IAA produced by rhizobacteria interfere the above physiological processes of plants by changing the plant auxin pool. Exogenous IAA controls a wide variety of processes in plant development and plant growth, low concentrations of IAA can stimulate primary root elongation, whereas high IAA levels stimulate the formation of lateral roots, decrease primary root length and increase root hair formation (Patten & Glick, 2002; Spaepen et al., 2007; Vacheron et al., 2013). Moreover, bacterial IAA increases root surface area and length, and thereby provides the plant greater access to soil nutrients. Also, rhizobacterial IAA loosens plant cell walls and as a result facilitates an increasing amount of root exudation that provides additional nutrients to support the growth of rhizosphere bacteria (Glick, 2012). Thus, rhizobacterial IAA is identified as an effector molecule in plant–microbe interactions, both in pathogenesis and phytostimulation (Spaepen & Vanderleyden, 2011). IAA is usually synthesized by rhizobacteria from tryptophan, which is found at different concentrations in root exudates according to plant genotype (Kamilova et al., 2006). The ability to synthesize IAA has been detected in many rhizobacteria as well as in pathogenic, symbiotic and free-living bacterial species (Tsavkelova, Klimova, Cherdyntseva, & Netrusov, 2006). At present, auxin-synthesizing rhizobacteria are the most well-studied phytohormone producers (Spaepen et al., 2007; Tsavkelova et al., 2006). These rhizobacteria synthesize IAA from tryptophan by different pathways, phytopathogenic bacteria mainly use the indole acetamide pathway to synthesize IAA, whereas the acid indole pyruvic pathway is used by the PGPR (Patten & Glick, 2002). Among PGPR species, *Azospirillum* is one of the best studied IAA producers

(Dobbelaere, Croonenborghs, Thys, Vande Broek, & Vanderleyden, 1999). Other IAA-producing bacteria belonging to *Aeromonas, Azotobacter, Bacillus, Burkholderia, Enterobacter, Pseudomonas* and *Rhizobium* genera have been isolated from different rhizosphere soils (Spaepen et al., 2007).

4.1.4.2 Gibberellin

GA are another group of phytohormones that are involved in all phases of plant growth and development from germination to senescence. However, the most prominent physiological effect of GA is in shoot elongation. Some other plant growth-related functions of GA include overcoming dormancy and dwarfism in plants, inducing flowering in some photoperiodically sensitive and other low-temperature-dependent plants and contributing to fruit ripening. Several soil microbes are known to produce GA and GA-like substances (Spaepen & Vanderleyden, 2011). Like auxins and cytokinins, GA mainly act in combination with other hormones. Production of GA has been documented in several PGPR belonging to *Achromobacter xylosoxidans, Acinetobacter calcoaceticus, Azospirillum* spp., *Azotobacter* spp., *Bacillus* spp., *Herbaspirillum seropedicae, Gluconobacter diazotrophicus* and rhizobia (Bottini et al., 2004; Dodd, Zinovkina, Safronova, & Belimov, 2010) The mechanism of plant growth stimulation by GA is still obscure. Fulchieri, Lucangeli and Bottini (1993) speculate that GA increase root hair density in root zones involved in nutrient and water uptake.

4.1.4.3 Cytokinin

Cytokinins stimulate plant cell division, control root meristem differentiation, induce proliferation of root hairs, but inhibit lateral root formation and primary root elongation (Riefler, Novak, Strnad, & Schmülling, 2006). Inoculation of plants with bacteria producing cytokinin has been shown to stimulate shoot growth and reduce the root to shoot ratio (Arkhipova et al., 2007). They play an important role in nodule development and formation. Symbiotic N_2-fixing bacteria, *Rhizobium*, free-living N_2-fixing bacteria, *Azospirillum* and *Azotobacter,* and mycorrhizal fungus, *Rhizopogon roseulus*, are known to produce cytokinins in the rhizosphere along with other growth-promoting substances. Cytokinin production (especially zeatin) has been documented in various PGPR-like *Arthrobacter giacomelloi, Azospirillum brasilense, B. japonicum, Bacillus licheniformis, P. fluorescens* and *Paenibacillus polymyxa* (Per-rig et al., 2007; Timmusk et al., 1999).

4.1.4.4 Regulating Plant Ethylene Levels

Ethylene, at high concentrations, induces defoliation and cellular processes that lead to inhibition of stem and root growth as well as premature senescence, all of which lead to reduced crop performance (Li, Saleh-Lakha, & Glick, 2005). Under different types of environmental stress, such as cold, drought, flooding, infections with pathogens, presence of heavy metals, the plants respond by synthesizing 1-aminocyclopropane-1-carboxylate (ACC), which is the precursor for ethylene (Glick, 2012). Some of the ACC is secreted into the rhizosphere and is readsorbed by the roots, where it is converted into ethylene. This accumulation of ethylene leads to a downward spiral effect, as poor root growth leads to a diminished ability to acquire water and nutrients, which, in turn, leads to further stress. Thus, PGPR with the ability to degrade ACC in the rhizosphere can help to break this downward cycle and re-establish a healthy root system that is needed to cope with environmental stress. The primary mechanism rhizobacteria utilize to degrade ethylene involves enzyme ACC deaminase. This enzyme can diminish or prevent some of the harmful effects of the high ethylene levels (Glick et al., 1998). The ACC deaminase acts on ACC, an immediate ethylene precursor in higher plants, degrading this chemical to alpha-ketobutyrate and ammonium (Glick et al., 1998; Grichko & Glick, 2001; Mayak, Tirosh, & Glick, 2004). Rhizosphere bacteria with ACC deaminase activity belonging to the *Achromobacter* (Govindasamy, Senthilkumar, Gaikwad, & Annapurna, 2008), *Azospirillum* (Li et al., 2005), *Bacillus* (Ghosh, Penterman, Little, Chavez, & Glick, 2003), *Enterobacter* (Li et al., 2005), *Pseudomonas* (Govindasamy et al., 2008) and *Rhizobium* (Duan, Miiller, Charles, Vesely, & Glick, 2009) genera have been isolated from different soils. Various studies have demonstrated that plants treated with PGPR bacteria that produce ACC deaminase have increased their resistance to environmental stress, e.g. Grichko and Glick (2001) inoculated tomato seeds with the ACC deaminase expressing bacteria *Enterobacter cloacae* and *P. putida* and recorded an increase in plant resistance on 55 days of age to 9 consecutive days of flooding. Ghosh et al. (2003) found ACC deaminase activity in three *Bacillus* species (*B. circulans* DUC1, *Bacillus firmus* DUC2 and *Bacillus globisporus* DUC3), which stimulated root elongation of *Brassica campestri* plants. Mayak et al. (2004) evaluated tomato plants inoculated with the bacterium *Achromobacter piechaudii* under water and saline stress conditions and reported a significant increase in fresh and dry weight of inoculated plants.

4.2 PGPR as Biopesticides

Diseases are very common in plants and are responsible for the loss of approximately one-third of the crop yield worldwide (Lugtenberg et al., 2002). PGPR with biocontrol traits can be considered as an alternative to the high doses of pesticides applied on crops to deter the pathogens and reduce the disease severity. A large number of mechanisms used by PGPR are involved in biocontrol such as direct antagonism via production of antibiotics, siderophores, HCN, hydrolytic enzymes (chitinases, proteases, lipases, etc.), or indirect mechanisms in which the biocontrol organisms act as a probiotic by competing with the pathogen for a niche (Lugtenberg & Kamilova, 2009). Siderophores, bacteriocins and antibiotics production are the most effective and well-known strategies employed by the PGPR to minimize or prevent the phytopathogenic proliferation, and several PGPR have been identified as potential ISR elicitors, for their ability to induce resistance in plants towards pathogenic fungi, bacteria and viruses (van Loon, 2007; van Loon et al., 1998). Many bacterial determinants induce ISR, i.e. flagella, siderophores (pycholin and pyocyanin), LPS, salicylic acid (van Loon, 2007), cyclic lipopeptides (Ongena et al., 2007), N-acyl homoserine lactone molecules (Shuhegger et al., 2006), the bacterial volatile 2,3-butanediol produced by *Bacillus* spp. (Ryu et al., 2003) and antibiotics such as Phl (Iavicoli, Boutet Buchala, & Metraux, 2003).

Moreover, ISR involves jasmonate and ethylene signalling within the plant and these hormones stimulate the host plant's defence responses against a variety of plant pathogens (Glick, 2012). Many rhizobacteria have been reported to produce antifungal metabolites like HCN, phenazines, pyrrolnitrin, 2, 4-diacetylphloroglucinol, pyoluteorin, viscosinamide and tensin (Bhattacharyya & Jha, 2012). *Azospirillum brasilense* cells contain a low molecular weight compound that inhibits germination and growth of the radical of Egyptian broom rape seeds (*Orobanche aegyptiaca*), a specific weed parasite of sunflower (Dadon, Nun, & Mayer, 2004). Similarly, *Azospirillum* spp. inhibited germination of the parasitic striga weed (witchweed) seeds (*Striga hermonthica*) that infest fields of tropical sorghum, thereby promoting growth of sorghum (Bouillant et al., 1997). Among the many defence mechanisms, the induction of resistance in plants (ISR) through the application of PGPR strains seems to have transgressed boundaries or limitations to any particular groups of pathogens, and it is effective against a broad range of pathogens like bacterial, viral, nematodes, arthropods, etc. (Kloepper, Ryu, & Zhang, 2004). PGPR role as biocontrol agents against various crop diseases is shown in Table 2. With the advent of 'omics'-technologies, the developments in the next generation bio-products seems even more promising (Berg, Zachow, Müller, Philipps, & Tilcher, 2013).

4.3 PGPR as Bioremediators

Bioremediation is recognized as an important tool to restore contaminated sites, reforest eroded areas and restore degraded ecosystems. Although PGPR are mainly considered for promoting the plant growth and for the biocontrol of plant diseases, much attention has recently been paid on bioremediation using PGPR (Huang, El Alawi, Penrose, Glick, & Greenberg, 2004; Narasimhan, Basheer, Bajic, & Swarup, 2003). Due to the sensitivity and the sequestration ability of the microbial communities to heavy metals, microbes have been successfully used for bioremediation (Burd, Dixon, & Glick, 2000; Ma, Rajkumar, & Freitas, 2009a). In contrast with inorganic compounds, microorganisms can degrade and even mineralize organic compounds in association with plants (Saleh, Huang, Greenberg, & Glick, 2004). PGPR capable of degrading certain kind of organic pollutant, such as polychlorinated biphenyls, have been isolated from a range of sites and the pathways, and encoding genes have been studied (Brazil et al., 1995). Several endophytic bacteria to degrade some pollutants (i.e. explosives, herbicides or hydrocarbons) have been reported (Germaine, Liu, Cabellos, Hogan, & Ryan, 2006). Also, endophytic bacteria resistant to high concentrations of heavy metals, benzene, toluene, ethylbenzene and xylenes, trichloroethylene or polyaromatic hydrocarbons have been identified (Doty, 2008; Moore et al., 2006). Several effective methods have been developed to improve the degradation efficiency and the tolerance of bacteria to contaminants in soils, and the combined use of PGPR and specific contaminant-degrading bacteria can successfully remove complex contaminants (Huang, El Alawi, Gurska, Glick, & Greenberg, 2005). The application of certain rhizobacteria can increase the uptake of Ni from soils by changing its phase (Abou-Shanab, Angle, & Chaney, 2006). PGPR capable of producing siderophores that chelate iron as well as other heavy metals like cadmium, lead, nickel, arsenic, aluminium, magnesium, zinc, copper, cobalt and strontium can help in adsorbing/absorbing heavy metals from

Table 2 Plant Growth Promoting Rhizobacteria (PGPR) as Biopesticides/Biocontrol Agents Against Various Plant Diseases

PGPR	Crop	Disease	References
Azospirillum strains SPS2, WBPS1 and Z2-7	Rice	Rice blast	Naureen Zakira, Hafeez Adam, Roberts Fauzia, and Michael (2009)
Bacillus amyloliquefaciens 937a, *Bacillus subtilis* 937b, *Bacillus pumilus* SE34	Tomato	Tomato mottle virus	Murphy et al. (2000)
B. amyloliquefaciens strain IN937a, *B. pumilus* strain SE34, *B. subtilis* strain IN937b	Cucumber	Cucumber mosaic virus	Zehnder, Yao, Murphy, Sikora, and Kloepper (2000)
Bacillus cereus strains B101R, B212R, A068R	Tomato	Foliar diseases	Silva et al. (2004)
Bacillus spp. strains BB11, FH17	Bell pepper	Blight of bell pepper	Jiang, Guo, Li, Qi, and Guo (2006)
B. pumilus strain INR7	Cucumber	Bacterial wilt	Zehnder, Murphy, Sikora, and Kloepper (2001)
B. pumilus strain SE34	Tobacco	Blue mold	Zhang, Moyne, Redddy, and Kloepper (2002)
B. subtilis strain GBO3, *B. pumilus* strain INR7, *B. pumilus* strain T4	Pearl millet	Downy mildew	Niranjan et al. (2003)
B. subtilis strain ME488	Cucumber, Pepper	Soilborne pathogens	Chung et al. (2008)
Burkholderia strains MBf21 and MBf15	Maize	Maize rot	Hernandez-Rodriguez, Heydrich-Perez, Acebo-Guerrero, Velazquez-del Valle, and Hernandez-Lauzardo (2008)
Enterobacter sp.	Chickpea	*Fusarium avenaceum*	Hynes, Leung, Hirkala, and Nelson (2008)
Paenibacillus polymyxa strain E681	Sesame	Fungal disease	Ryu, Kim, Choi, Kim, and Park (2006)
Pseudomonas sp.	White clover Medicago	*Acyrthosiphon kondoi*	Kempster, Scott, and Davies (2002)
Pseudomonas fluorescens strain CHAO + chitin bioformulations	Banana	Banana bunchy top virus	Kavino, Harish, Kumar, Saravanakumar, and Samiyappan (2010)
Streptomyces marcescens strain 90–116	Tobacco	Blue mold	Zhang et al. (2002)

the soil, besides plant growth promotion and phytopathogen suppression. Some notable PGPR strains used in bioremediation are listed in Table 3.

5. COMMERCIALIZATION OF PGPR

Since the discovery of rhizobia in 1886 (Hellriegel & Wilfarth, 1888), Rhizobium inoculants have been commercially produced worldwide, mainly in the developed countries (Catroux, Hartmann, & Revellin, 2001; Deaker, Roughley, & Kennedy, 2004). In 1897, a bacteriological fertilizer based *on B. subtilis* was marketed for inoculation of cereals under the proprietary name Alinit by Bayer AG (Killian et al., 2000). During the early 1950s, research findings from China, Russia and several other western countries further prompted the potential use of microbes to be explored for plant disease management (Backman, Wilson, & Murphy, 1997) and opened new vistas to use PGPR as an alternative to chemical pesticides for the management of soilborne pathogens (Dunleavy, 1955; Kloepper, 1993). Owing to the potential of PGPR, the first commercial product of *B. subtilis* was introduced during 1985 by Gustafson, Inc. (Plano, Texas) in the United States and the strains of *B. subtilis* A-13, GB03, GB07 were sold for the management of soilborne pathogens under the trade names of Quantum, Kodiak and Epic, respectively (Broadbents et al., 1977). In the United States, 60–75% of cotton, peanut, soya bean, corn, vegetables and small grain crops are now treated with commercial products of *B. subtilis* for the management of soilborne pathogens (Nakkeeran, Kavitha, Chandrasekar, Renukadevi, & Fernando, 2006). PGPR have been largely applied in 48 different crops over 3.35 million ha in China (Wenhua & Hetong, 1997) and productivity gains as high as 23% in sweet potatoes and 22.5% in potatoes were reported in addition to 85.5% and 80.3%, reduction levels of disease caused by *Xanthomonas oryzae* and *Glomerella cingulata,* respectively. PGPR-based commercial products used for combating various diseases in crops are listed in Table 4.

Among several PGPR strains, *Bacillus*-based products gained momentum for commercialization since *Bacillus* spp. produce endospores which can tolerate extremes of conditions such as temperature, pH and exposure to pesticides and fertilizers (Backman et al., 1997). In the mid-1990s in the United States, *B. subtilis* started to be used as seed dressing, with registration in more than seven crops and application to more than 2 million ha (Backman et al., 1997). Eighteen different commercial products of *Bacillus*

Table 3 Plant Growth Promoting Rhizobacteria (PGPR) Interaction with Plants Towards Bioremediation

PGPR	Crop	Heavy Metal	References
Achromobacter xylosoxidaus Ax10	*Brassica juncea*	Cu	Ma et al. (2009a)
Azospirillum brasilense cd, *Enterobacter cloacae* CAL 2, *Pseudomonas putida* UW3	*Festuca arundinacea*	Polycyclic aromatic hydrocarbons	Huang et al. (2004)
Azotobacter chroococcum HKN-5, *Bacillus megaterium* HKP-1, *Bacillus mucilaginosus* HKK-1	*B. juncea*	Pb and Zn	Wu, Wood, Mulchandani, and Chen (2006)
Azospirillum lipoferum strains, *A. brasilense* strains	*Triticum aestivum*	Crude oil	Muratova et al. (2005)
Bacillus sp. PSB10	*Cicer arietinum*	Cr	Wani and Khan (2010)
Bacillus subtilis SJ-101	*B. juncea*	Ni	Zaidi, Usmani, Singh, and Musarrat (2006)
Bradyrhizobium sp. RM8	*Brassica napus*	Ni and Zn	Wani, Khan, and Zaidi (2007)
Brevibacillus B-1	*Trifolium repens*	Zn	Vivas, Biro, Ruiz-Lozano, Barea, and Azcon (2006)
E. cloacae CAL 2, *E. cloacae* UW4	*Festuca arundinacea*	Total petroleum hydrcarbons	Huang et al. (2005)
Kluyvera ascorbata SUD 165, SUD 165/26	*B. juncea, Lycopersicon esculentum*	Ni, Pb and Zn	Burd et al. (2000)
Mesorhizobium huakuii B3	*Astragalus sinicus*	Cd	Sriprang et al. (2003)
P. putida Flav1-1, *P. putida* PML 2	Arabidopsis	Polychlorinated biphenyls	Narasimhan et al. (2003)
Pseudomonas fluorescens 2–79	*Triticum aestivum*	Trichloroethylene	Yee, Maynard, and Wood (1998)
Pseudomonas sp. M6, *Pseudomonas jessenii* M15	*Ricinus communis*	Ni, Cu and Zn	Rajkumar and Freitas (2008)
P. fluorescens F113	*Medicago sativa*	Polychlorinated biphenyls	Villacieros et al. (2005)
Pseudomonas sp. SRA 2, SRA 1, *B. cereus* SRA 10	*B. juncea, Brassica oxyrrhina*	Ni	Ma, Rajkumar, and Freitas (2009b)
Serratia sp. SY5	*Zea mays*	Cd and Cu	Koo and Cho (2009)

Table 4 Plant Growth Promoting Rhizobacteria (PGPR)-Based Commercial Products Used for Combating Various Diseases in Crops

PGPR	Trade Name	Crop
Agrobacterium radiobacter	Diegall, Galltrol-A, Nogall, Norbac 84C	Fruits, nut, ornamental nursery stock and trees
Azospirillum brasilense	Azo-Green	Turf and forage crops
Bacillus cereus	Pixplus	Cotton
Bacillus pumilus	Sonata ASO, Ballard	Oak trees, nurseries and landscapes
Bacillus subtilis	Epic, HiStick N/T, Kodiak, Rhizo-Plus, Serenade, Subtilex, Quantum 4000, Rhapsody, System 3	Barley, beans, cotton, peanut, pea, rice and soya bean
B. subtilis	Companion	Horticultural crops and turf
B. subtilis + *Bacillus amyloliquefaciens*	Bio Yield	Tomato, cucumber, pepper and tobacco
Burlkholderia cepacia	Blue Circle, Deny, Intercept	Alfalfa, barley, beans, cotton, maize, pea, sorghum and wheat
Pseudomonas auerofaciens	BioJect, Spot-less	Turf and other crops
Pseudomonas sp. + *Azospirillum*	BioJect	Turf and other crops
Pseudomonas chloroaphis	AtEze	Ornamentals and vegetables
P. chloroaphis	Cedomon	Barley, oats, wheat and other cereals
Pseudomonas fluorescens	Blight Ban A506, Conquer, Victus	Almond, apple, cherry, mushroom, potato
Pseudomonas syringae	Bio-save 10, 11, 100, 110, 1000	Strawberry, tomato, citrus and pome fruit
Streptomyces griseoviridis K61	Pixplus	Cotton

origin are now used in China to mitigate soilborne diseases, applied to an area of 20 million ha of different crop plants (Chen, Mei, Lu, Liu, & Kloepper, 1996; Kilian et al., 2000), 25 million ha of soya beans inoculated with *B. japonicum* in South America and approximately 500,000 ha of wheat and maize inoculated with commercial *Azospirillum* inoculants in Argentina and Mexico (Fuentes-Ramirez & Caballero-Mellado, 2005). Recently, in India, the PGPR use has increased considerably due to more awareness about the farming practices and now more than 40 stakeholders from different provinces have registered for mass production of PGPR with Central Insecticide Board, Faridabad, Haryana in collaboration with

Tamil Nadu Agricultural University, Coimbatore, India (Ramakrishnan, Nakkeeran, Chandrasekar, & Doraiswamy, 2001). Field studies conducted in the Indian subcontinent (Johri, Sharma, & Virdi, 2003), Vietnam (Cong et al., 2009), in Latin America, mainly in Argentina and Mexico (Fuentes-Ramirez & Caballero-Mellado, 2005; Hartmann & Bashan, 2009) and in Africa (Atieno, Herrmann, Okalebo, & Lesueur, 2012) have further confirmed the potential of PGPR use in agriculture system.

Despite the promising results in lab and greenhouse studies, the field results show lack of consistency and there exist variation in responses from site to site, year to year and for different crops under consideration. The reason for this inconsistency can be attributed to the fact that the experiments are conducted under controlled conditions in the lab and greenhouse settings, whereas the field studies are influenced by a myriad of biotic and abiotic factors. The inherent heterogeneity of the soil is one of the main hindrance, where introduced bacteria sometimes cannot find an empty niche in the soil. These inoculated bacteria have to compete with the often better adapted native microflora for nutrients and space. They cannot withstand this severe competition with the indigenous microflora and as a result shortly after inoculation into the soil, the bacterial population declines rapidly. A threshold number of cells is essential to obtain the intended positive plant response, for example, 10e6—10e7 cells/plant for the *A. brasilense* and this number too varies among different species (Bashan, 1986). Thus in order to exhibit their positive impact on plant growth, these PGPR must survive and multiply in microhabitats associated with the root surface, in competition with other microbiota, at least for the time needed to express their plant promotion/protection activities.

The factors which govern the success of inoculation are (1) the effectiveness of the introduced bacterial strain in terms of colonization, (2) competence with the indigenous microflora, (3) the species and variety of plant can also be a determining factor since different plant species or cultivars produce different types of root exudates, which also influence the activity of the inoculated microorganisms and also serve as substrates for the formation of biologically active substances (Berg et al., 2013). So, more research in PGPR need to be diverted towards colonization, about their ecology, survival and activity in the plant rhizosphere. Besides this, reproducibility of the effects of microbial inoculants needs to be tested across a wide range of soil types and environmental conditions. Also, when used in conventional agriculture, the microorganisms must also be compatible with the chemical fertilizers and also withstand the pesticides which are commonly used on seeds or foliage to deter the diseases. Thus, a thorough understanding of the mechanisms utilized by the PGPR is imperative to utilize their full potential in the agriculture system and the advent of new and powerful technologies for studying co-operative microbial interactions in the rhizosphere guarantees a greater understanding of these processes and will facilitate their successful application in the field.

6. CONCLUSION

According to the United Nations, it is highly likely we will see 9.6 billion Earthlings by 2050 and up to 11 billion or more by 2100. When we consider that we will need 70% more food by 2050 assuming that we do not make any serious progress in reducing waste, minimizing pollution, slowing population growth or halting the increase in consumption of animal products (FAO, 2011), it is hard to feel hopeful about the future. Thus feeding an ever increasing population is the greatest challenges facing the mankind when the resources are limited, decreasing irrigational water supplies and other environmental concerns further exacerbate this problem.

Since the dawn of civilization, we, the human beings are engaged in various activities which directly or indirectly affect our ecosystems, either way positive or negative. The need and greed for increasing the crop yield has already resulted in overexploitation of the soil ecosystem, and we cannot practice the same conventional agriculture approach to meet this goal since the anthropogenic activities like intensive agriculture, monoculture crops and use of agrochemicals have grave concerns and disturbs the ecosystem functioning. We have to minimize these negative interactions with the ecosystem. Stewardship of both natural and human resources are of prime importance. To solve one problem we should not create another problem which is even bigger in magnitude and difficult to handle.

Keeping in mind the beneficial services rendered by the PGPR (in terms of biofertilization, biocontol and bioremediation) which exhibit positive influence on crop productivity and ecosystem functioning, we should encourage their successful implementation in the main agriculture system. With better research and development, these microbial populations use will become a reality and instrumental to fundamental processes that drive stability and productivity of agro-ecosystems, thus leading us towards an ideal agricultural system which is sustainable, maintains and improves human health, benefits environment and produces enough food for the increasing world population.

ACKNOWLEDGEMENTS

The author sincerely acknowledges all the eminent scientists and researchers engaged in the area of PGPR research.

REFERENCES

Abou-Shanab, R. I., Angle, J. S., & Chaney, R. L. (2006). Bacterial inoculants affecting nickel uptake by *Alyssum murale* from low, moderate and high Ni soils. *Soil Biology & Biochemistry, 38*, 2882–2889.

Adesemoye, A., Torbert, H., & Kloepper, J. (2009). Plant growth-promoting rhizobacteria allow reduced application rates of chemical fertilizers. *Microbial Ecology, 58*, 921–929.

Antoun, H., Beauchamp, C. J., Goussard, N., Chabot, R., & Lalande, R. (1998). Potential of *Rhizobium* and *Bradyrhizobium* species as plant growth promoting rhizobacteria on non-legumes: effect on radishes (*Raphanus sativus* L.). *Plant and Soil, 204*, 57–67.

Arkhipova, T. N., Prinsen, E., Veselov, S. U., Martinenko, E. V., Melentiev, A. I., & Kudoyarova, G. R. (2007). Cytokinin producing bacteria enhance plant growth in drying soil. *Plant and Soil, 292*, 305–315.

Atieno, M., Herrmann, L., Okalebo, R., & Lesueur, D. (2012). Efficiency of different formulations of *Bradyrhizobium japonicum* and effect of co-inoculation of *Bacillus subtilis* with two different strains of *Bradyrhizobium japonicum*. *World Journal of Microbiology & Biotechnology, 28*, 2541–2550.

Backman, P. A., Wilson, M., & Murphy, J. F. (1997). Bacteria for biological control of plant diseases. In N. A. Rechcigl, & J. E. Rechcigl (Eds.), *Environmentally safe approaches to crop disease control* (pp. 95–109). Boca Raton, FL: Lewis Publishers.

Badri, D. V., Weir, T. L., van der Lelie, D., & Vivanco, J. M. (2009). Rhizosphere chemical dialogues: plant-microbe interactions. *Current Opinion in Biotechnology, 20*, 642–650.

Bais, H. P., Park, S. W., Weir, T. L., Callaway, R. M., & Vivanco, J. M. (2004). How plants communicate using the underground information superhighway. *Trends in Plant Science, 9*, 26–32.

Bais, H. P., Weir, T. L., Perry, L. G., Gilroy, S., & Vivanco, J. M. (2006). The role of root exudates in rhizosphere interactions with plants and other organisms. *Annual Review of Plant Biology, 57*, 233–266.

Bashan, Y. (1986). Alginate beads as synthetic inoculant carriers for slow release of bacteria that affect plant growth. *Applied and Environmental Microbiology, 51*, 1089–1098.

Bashan, Y., & de-Bashan, L. E. (2010). Chapter two—how the plant growth-promoting bacterium *Azospirillum* promotes plant growth—a critical assessment. *Advances in Agronomy, 108*, 77–136.

Berendsen, R. L., Pieterse, C. M., & Bakker, P. A. (2012). The rhizosphere microbiome and plant health. *Trends in Plant Science, 17*, 478–486.

Berg, G. (2009). Plant microbe interactions promoting plant growth and health: perspectives for controlled use of microorganisms in agriculture. *Applied Microbiology and Biotechnology, 84*, 11–18.

Berg, G., & Smalla, K. (2009). Plant species and soil type cooperatively shape the structure and function of microbial communities in the rhizosphere. *FEMS Microbiology Ecology, 68*, 1–13.

Berg, G., Zachow, C., Müller, H., Philipps, J., & Tilcher, R. (2013). Next-generation bio-products sowing the seeds of success for sustainable agriculture. *Agronomy, 3*, 648–656.

Bhattacharyya, P. N., & Jha, D. K. (2012). Plant growth-promoting rhizobacteria (PGPR): emergence in agriculture. *World Journal of Microbiology & Biotechnology, 28*, 1327–1350.

Bloemberg, G. V., & Lugtenberg, B. J. J. (2004). Bacterial biofilms on plants: relevance and phenotypic aspects. In M. Ghannoum, & G. A. O'Toole (Eds.), *Microbial biofilms* (pp. 141–159). Washington DC: ASM Press.

Boddey, R. M., Urquiaga, S., Alves, B. J. R., & Reis, V. (2003). Endophytic nitrogen fixation in sugarcane: present knowledge and future applications. *Plant and Soil, 252*, 139–149.

Boopathi, E., & Rao, K. S. (1999). A siderophore from *Pseudomonas putida* type A1: structural and biological characterization. *Biochimica et Biophysica Acta-Protein Structure and Molecular Enzymology, 1435*, 30–40.

Bottini, R., Cassán, F., & Piccoli, P. (2004). Gibberellin production by bacteria and its involvement in plant growth promotion and yield increase. *Applied Microbiology and Biotechnology, 65*, 497–503.

Bouillant, M. L., Miché, L., Ouedraogo, O., Alexandre, G., Jacoud, C., Sallé, G., et al. (1997). Inhibition of *Striga* seed germination associated with sorghum growth promotion by soil bacteria. *Comptes Rendus de l'Academie des Sciences Paris, 320*, 159–162.

Brazil, G. M., Kenefick, L., Callanan, M., Haro, A., de Lorenzo, V., Dowling, D. N., et al. (1995). Construction of a rhizosphere pseudomonad with potential to degrade polychlorinated biphenyls and detection of *bph* gene expression in the rhizosphere. *Applied and Environmental Microbiology, 61*, 1946–1952.

Brimecombe, M. J., De Leij, F. A. A. M., & Lynch, J. M. (2007). Rhizodeposition and microbial populations. In R. Pinton, Z. Varanini, & P. Nannipieri (Eds.), *The rhizosphere: Biochemistry and organic substances at the soil-plant interface* (pp. 73–109). Boca Raton, London, New York: CRC Press, Taylor & Francis Group.

Broadbent, P., Baker, K. F., Franks, N., & Holland, J. (1977). Effect of *Bacillus* spp. on increased growth of seedlings in steamed and non treated soil. *Phytopathology, 67*, 1027–1031.

Broeckling, C. D., Broz, A. K., Bergelson, J., Manter, D. K., & Vivanco, J. M. (2008). Root exudates regulate soil fungal community composition and diversity. *Applied and Environmental Microbiology, 74*, 738–744.

Brown, M. E. (1974). Seed and root bacterization. *Annual Review of Phytopathology, 12*, 181–197.

Bulgarelli, D., Schlaeppi, K., Spaepen, S., Ver Loren van Themaat, E., & Schulze-Lefert, P. (2013). Structure and functions of the bacterial microbiota of plants. *Annual Review of Plant Biology, 64*, 807–838.

Burd, G. I., Dixon, D. G., & Glick, B. R. (2000). Plant growth promoting bacteria that decrease heavy metal toxicity in plants. *Canadian Journal of Microbiology, 46*, 237–245.

Burdman, S., Jurkevitch, E., & Okon, Y. (2000). Recent advances in the use of plant growth promoting rhizobacteria (PGPR) in agriculture. In N. S. Subba Rao, & Y. R. Dommergues (Eds.), *Microbial interactions in agriculture and forestry* (Vol. II, pp. 229–250).

Buysens, S., Heungens, K., Poppe, J., & Hofte, M. (1996). Involvement of pyochelin and pyoverdin in suppression of pythium induced damping off of tomato by *Pseudomonas aeruginosa* 7NSK2. *Applied and Environmental Microbiology, 62*, 865–871.

Carrillo-Castañeda, G., Juárez Muños, J., Peralta-Videa, J. R., Gomez, E., Tiemann, K. J., Duarte-Gardea, M., et al. (2002). Alfalfa growth promotion by bacteria grown under iron limiting conditions. *Advances in Environmental Research, 6*, 391–399.

Catroux, G., Hartmann, A., & Revellin, C. (2001). Trends in rhizobial inoculants production and use. *Plant and Soil, 230*, 21–30.

Cattelan, A. J., Hartel, P. G., & Fuhrmann, J. J. (1999). Screening for plant growth promoting rhizobacteria to promote early soybean growth. *Soil Science Society of America Journal, 63*, 1670–1680.

Chabot, R., Beauchamp, C. J., Kloepper, J. W., & Antoun, H. (1998). Effect of phosphorus on root colonization and growth promotion of maize by bioluminescent mutants of P solubilizing *Rhizobium leguminosarum* biovar *phaseoli*. *Soil Biology & Biochemistry, 30*, 1615–1618.

Chandler, D., Davidson, G., Grant, W. P., Greaves, J., & Tatchell, G. M. (2008). Microbial biopesticides for integrated crop management: an assessment of environmental and regulatory sustainability. *Trends in Food Science & Technology, 19*, 275–283.

Chaparro, J. M., Badri, D. V., Bakker, M. G., Sugiyama, A., Manter, D. K., & Vivanco, J. M. (2013). Root exudation of phytochemicals in *Arabidopsis* follows specific patterns that are developmentally programmed and correlate with soil microbial functions. *PLoS One, 8*(2), e55731. http://dx.doi.org/10.1371/journal.pone.0055731.

Chen, Y., Mei, R., Lu, S., Liu, L., & Kloepper, J. W. (1996). The use of yield increasing bacteria (YIB) as plant growth-promoting rhizobacteria in Chinese agriculture. In R. S. Utkhede, & V. Gupta (Eds.), *Management of soil borne diseases* (pp. 165–176). New Delhi, India: Kalyani Publishers.

Chew, K. (2002). *Georgics* (p. 152). Indianapolis, USA: Hackett Publishing Company.

Chin A-Woeng, T. F., Bloemberg, G. V., & Lugtenberg, B. J. (2003). Phenazines and their role in biocontrol by *Pseudomonas* bacteria. *New Phytologist, 157*, 503–523.

Chung, S., Kong, H., Buyer, J. S., Lakshman, D. K., Lydon, J., & Kim, S. D. (2008). Isolation and partial characterization of *Bacillus subtilis* ME488 for suppression of soilborne pathogens of cucumber and pepper. *Applied Microbiology and Biotechnology, 80*, 115–123.

Compant, S., Reiter, B., Sessitsch, A., Nowak, J., Clement, C., & Ait Barka, E. (2005). Endophytic colonization of *Vitis vinifera* L. by a plant growth promoting rhizobacterium *Burkholderia* sp. Strain PsJN. *Applied and Environmental Microbiology, 71*, 1685–1693.

Cong, P. T., Dung, T. D., Hien, T. M., Hien, N. T., Choudhury, A. T. M. A., Kecskés, K. L., et al. (2009). Inoculant plant growth promoting microorganisms enhance utilisation of urea-N and grain yield of paddy rice in southern Vietnam. *European Journal of Soil Biology, 45*, 52–61.

Crowley, D. E. (2006). Microbial siderophores in the plant rhizosphere. In L. L. Barton, & J. Abadía (Eds.), *Iron nutrition in plants and rhizospheric microorganisms* (pp. 169–198). Netherlands: Springer.

Dadon, T., Nun, N. B., & Mayer, A. M. (2004). A factor from *Azospirillum brasilense* inhibits germination and radicle growth of *Orobanche aegyptiaca*. *Israel Journal of Plant Sciences, 52*(2), 83–86.

Damayanti, T. A., Pardede, H., & Mubarik, N. R. (2007). Utilization of root colonizing bacteria to protect hot-pepper against Tobacco Mosaic Tobamovirus. *Hayati Journal of Bioscience, 14*(3), 105–109.

Deaker, R., Roughley, R. J., & Kennedy, I. R. (2004). Legume seed inoculation technology—a review. *Soil Biology & Biochemistry, 36*, 1275–1288.

DeFlaun, M. F., & Gerba, C. P. (1993). Monitoring recombinant DNA microorganisms and viruses in soil. In F. B. J. Metting (Ed.), *Soil microbial ecology: Application in agricultural and environmental management* (pp. 131–150). Washington: Marcel Dekker Inc.

Delvasto, P., Valverde, A., Ballester, A., Igual, J. M., & Munoz, J. A. (2006). Characterization of brushite as a re-crystallization product formed during bacterial solubilization of hydroxyapatite in batch cultures. *Soil Biology & Biochemistry, 38*, 2645–2654.

Dobbelaere, S., Croonenborghs, A., Thys, A., Vande Broek, A., & Vanderleyden, J. (1999). Phytostimulatory effect of *Azospirillum brasilense* wild type and mutant strains altered in IAA production on wheat. *Plant and Soil, 212*, 153–162.

Dobbelaere, S., Venderleyden, J., & Okon, Y. (2003). Plant growth promoting effects of diazotrophs in the rhizosphere. *Critical Reviews in Plant Sciences, 22*, 107–149.

Dodd, I. C., Zinovkina, N. Y., Safronova, V. I., & Belimov, A. A. (2010). Rhizobacterial mediation of plant hormone status. *Annals of Applied Biology, 157*, 361–379.

Doty, S. L. (2008). Enhancing phytoremediation through the use of transgenics and endophytes. *New Phytologist, 179*, 318–333.

Drogue, B., Combes-Meynet, E., Mënne-Loccoz, Y., Wisniewski-Dyé, F., & Prigent-Combaret, C. (2013). Control of the cooperation between plant growth-promoting rhizobacteria and crops by rhizosphere signals. In F. J. de Bruijn (Ed.), *Molecular microbial ecology of the rhizosphere* (Vols. 1 and 2, pp. 281–294). NJ, USA: John Wiley & Sons, Inc.

Duan, J., Miiller, K. M., Charles, T. C., Vesely, S., & Glick, B. R. (2009). 1-Aminocyclopropane-l-carboxylate (ACC) deaminase genes in rhizobia from Southern Saskatchewan. *Microbial Ecology, 57*, 423–436.

Duffy, B. K., & Défago, G. (1999). Environmental factors modulating antibiotic and siderophore biosynthesis by *Pseudomonas fluorescens* biocontrol strains. *Applied and Environmental Microbiology, 65*, 2429–2438.

Duijff, B. J., De Kogel, W. J., Bakker, P. A. H. M., & Schippers, B. (1994). Influence of pseudobactin 358 on the iron nutrition of barley. *Soil Biology & Biochemistry, 26*, 1681–1688.

Dunleavy, J. (1955). Control of damping-off of sugar beet by *Bacillus subtilis*. *Phytopathology, 45*, 252–257.

FAO. (2011) http://www.fao.org/worldfoodsituation/wfs.

Fuentes-Ramirez, L. E., & Caballero-Mellado, S. (2005). Bacterial biofertilizers. In Z. A. Sadiqui (Ed.), *PGPR: Biological control and biofertilization* (pp. 143–172). Dordrecht, The Netherlands: Springer.

Fulchieri, M., Lucangeli, C., & Bottini, R. (1993). Inoculation with *Azospirillum lipoferum* affects growth and gibberellin status of corn seedling roots. *Plant and Cell Physiology, 34*, 1305–1309.

Germaine, K. J., Liu, X., Cabellos, G. G., Hogan, J. P., & Ryan, D. (2006). Bacterial endophyte-enhanced phytoremediation of the organochlorine herbicide 2,4-dichlorophenoxyacetic acid. *FEMS Microbiology Ecology, 57*, 302–310.

Ghosh, S., Penterman, J. N., Little, R. D., Chavez, R., & Glick, B. R. (2003). Three newly isolated plant growth-promoting bacilli facilitate the seedling growth of canola, *Brassica campestris*. *Plant Physiology and Biochemistry, 41*, 277–281.

Glick, B. R. (1995). The enhancement of plant growth by free living bacteria. *Canadian Journal of Microbiology, 41*, 109–117.

Glick, B. R. (2012). *Plant growth-promoting bacteria: Mechanisms and applications*. Hindawi Publishing Corporation, Scientifica.

Glick, B. R., Karaturovic, D. M., & Newell, P. C. (1995). A novel procedure for rapid isolation of plant growth promoting *Pseudomonas*. *Canadian Journal of Microbiology, 41*, 533–536.

Glick, B. R., Penrose, D. M., & Li, J. D. (1998). A model for the lowering of plant ethylene concentrations by plant growth promoting bacteria. *Journal of Theoretical Biology, 190*, 63–68.

Gottschalk, G. (1986). *Bacterial metabolism*. Berlin; Heidelberg, Germany; New York, USA: Springer.

Govindasamy, V., Senthilkumar, M., Gaikwad, K., & Annapurna, K. (2008). Isolation and characterization of ACC deaminase gene from two plant growth-promoting rhizobacteria. *Current Microbiology, 57*, 312–317.

Gray, E. J., & Smith, D. L. (2005). Intracellular and extracellular PGPR: commonalities and distinctions in the plant–bacterium signaling processes. *Soil Biology & Biochemistry, 37*, 395–412.

Grayston, S. J., Wang, S., Campbell, C. D., & Edwards, A. C. (1998). Selective influence of plant species on microbial diversity in the rhizosphere. *Soil Biology & Biochemistry, 30*, 369–378.

Grichko, V. P., & Glick, B. R. (2001). Flooding tolerance of transgenic tomato plants expressing the bacterial enzyme ACC deaminase controlled by the 35S, rolD or PRB-1b promoter. *Plant Physiology & Biochemistry, 39*, 19–25.

Hartmann, A., & Bashan, Y. (2009). Ecology and application of *Azospirillum* and other plant growth-promoting bacteria (PGPB)—special issue. *European Journal of Soil Biology, 45*, 1–2.

Hellriegel, H., & Wilfarth, H. (1888). Untersuchungen uber die Stickstoffnahrung der Gramineen und Leguminosen. *Beilageheft zu der Z des Vereins Rubenzucker-Industrie Deutschen Reichs*, 234.

Henson, B. J., Watson, L. E., & Barnum, S. R. (2004). The evolutionary history of nitrogen fixation, as assessed by NifD. *Journal of Molecular Evolution, 58*, 390–399.

Hernandez-Rodriguez, A., Heydrich-Perez, M., Acebo-Guerrero, Y., Velazquez-del Valle, M. G., & Hernandez-Lauzardo, A. N. (2008). Antagonistic activity of Cuban native rhizobacteria against *Fusarium verticillioides* (Sacc.) Nirenb. in maize (*Zea mays* L.). *Applied Soil Ecology, 39*, 180–186.

Huang, X. D., El Alawi, Y., Gurska, J., Glick, B. R., & Greenberg, B. M. (2005). A multiprocess phytoremediation system for decontamination of persistent total petroleum hydrocarbons (TPHs) from soils. *Microchemical Journal, 81*, 139–147.

Huang, X. D., El Alawi, Y., Penrose, D. M., Glick, B. R., & Greenberg, B. M. (2004). A multiprocess phytoremediation system for removal of polycyclic aromatic hydrocarbons from contaminated soils. *Environmental Pollution, 130*, 465–476.

Hynes, R. K., Leung, G. C., Hirkala, D. L., & Nelson, L. M. (2008). Isolation, selection and characterization of beneficial rhizobacteria from pea, lentil and chickpea grown in Western Canada. *Canadian Journal of Microbiology, 54*, 248–258.

Iavicoli, A., Boutet Buchala, E. A., & Metraux, J. P. (2003). Induced systemic resistance in *Arabidopsis thaliana* in response to root inoculation with *Pseudomonas fluorescens* CHA0. *Molecular Plant-Microbe Interactions, 16*, 851–858.

Jacobsen, C. S. (1997). Plant protection and rhizosphere colonization of barley by seed inoculated herbicide degrading *Burkholderia (Pseudomonas) cepacia* DBO1 (pRO101) in 2, 4-D contaminated soil. *Plant and Soil, 189*, 139–144.

Jeffries, P., Gianinazzi, S., Perotto, S., Turnau, K., & Barea, J. M. (2003). The contribution of arbuscular mycorrhizal fungi in sustainable maintenance of plant health and soil fertility. *Biology and Fertility of Soils, 37*, 1–16.

Jiang, Z.-Q., Guo, Ya-H., Li, S.-M., Qi, H.-Y., & Guo, J.-H. (2006). Evaluation of biocontrol efficiency of different *Bacillus* preparations and field application methods against phytophthora blight of bell pepper. *Biological Control, 36*, 216–223.

Johri, B. N., Sharma, A., & Virdi, J. S. (2003). Rhizobacterial diversity in India and its influence on soil and plant health. *Advances in Biochemical Engineering Biotechnology, 84*, 49–89.

Kamilova, F., Kravchenko, L. V., Shaposhnikov, A. I., Azarova, T., Makarova, N., & Lugtenberg, B. (2006). Organic acids, sugars and L-tryptophane in exudates of vegetables growing on stonewool and their effects on activities of rhizosphere bacteria. *Molecular Plant-Microbe Interactions, 9*, 250–256.

Kamilova, F., Validov, S., Azarova, T., Mulders, I., & Lugtenberg, B. (2005). Enrichment for enhanced competitive plant root tip colonizers selects for a new class of biocontrol bacteria. *Environmental Microbiology, 7*, 1809–1817.

Kavino, M., Harish, S., Kumar, N., Saravanakumar, D., & Samiyappan, R. (2010). Effect of chitinolytic PGPR on growth: yield and physiological attributes of banana (*Musa* spp.) under field conditions. *Applied Soil Ecology, 45*, 71–77.

Kempster, V. N., Scott, E. S., & Davies, K. A. (2002). Evidence for systemic, cross-resistance in white clover (*Trifolium repens*) and annual medic (*Medicago truncatula* var *truncatula*) induced by biological and chemical agents. *Biocontrol Science and Technology, 12*(5), 615–623.

Kilian, M., Steiner, U., Krebs, B., Junge, H., Schmiedeknecht, G., & Hain, R. (2000). FZB24 *Bacillus subtilis*—mode of action of a microbial agent enhancing plant vitality. *Pflanzenschutz-Nachrichten Bayer, 1*, 72—93.

Kim, J., & Rees, D. C. (1994). Nitrogenase and biological nitrogen fixation. *Biochemistry, 33*, 389—397.

Kloepper, J. W. (1993). Plant-growth-promoting rhizobacteria as biological control agents. In F. B. Metting, Jr. (Ed.), *Soil microbial ecology* (pp. 255—273). NewYork: Marcel Dekker Inc.

Kloepper, J. W., Leong, J., Teintze, M., & Schroth, M. N. (1980). Enhanced plant growth by siderophores produced by plant growth-promoting rhizobacteria. *Nature, 286*, 885—886.

Kloepper, J. W., Ryu, C. M., & Zhang, S. (2004). Induced systemic resistance and promotion of plant growth by *Bacillus* spp. *Phytopathology, 94*, 1259—1266.

Kloepper, J. W., & Schroth, M. N. (1978). Plant growth-promoting rhizobacteria on radishes. In *Proceedings of the 4th international conference on plant pathogenic bacteria* (pp. 879—882). Tours: Gilbert-Clarey.

Koo, S. Y., & Cho, K. S. (2009). Isolation and characterization of a plant growth-promoting rhizobacterium, *Serratia* sp. SY5. *Journal of Microbiology and Biotechnology, 19*, 1431—1438.

Kuffner, M., Puschenreiter, M., Wieshammer, G., Gorfer, M., & Sessitsch, A. (2008). Rhizosphere bacteria affect growth and metal uptake of heavy metal accumulating willows. *Plant and Soil, 304*, 35—44.

Lakshmanan, V., Selvaraj, G., & Bais, H. P. (2014). Functional soil microbiome: belowground solutions to an aboveground problem. *Plant Physiology, 166*, 689—700.

Li, Q., Saleh-Lakha, S., & Glick, B. R. (2005). The effect of native and ACC deaminase-containing *Azospirillum brasilense* Cdl843 on the rooting of carnation cuttings. *Canadian Journal of Microbiology, 51*, 511—514.

Long, S. R. (2001). Genes and signals in the *Rhizobium*-legume symbiosis. *Plant Physiology, 125*, 69—72.

van Loon, L. C. (2007). Plant responses to plant growth-promoting bacteria. *European Journal of Plant Pathology, 119*, 243—254.

van Loon, L. C., Bakker, P. A. H. M., & Pieterse, C. M. J. (1998). Systemic resistance induced by rhizosphere bacteria. *Annual Review of Phytopathology, 36*, 453—483.

Loper, J. E., & Henkels, M. D. (1999). Utilization of heterologous siderophores enhances levels of iron available to *Pseudomonas putida* in the rhizosphere. *Applied and Environmental Microbiology, 65*, 5357—5363.

Loper, J. E., Thompson, B. N., Whistler, C. A., Hagen, M. J., Corbell, N. A., Henkels, M. D., et al. (1997). Biological control mediated by antifungal metabolite production and resource competition: an overview. In A. Ogoshi, K. Kobayashi, Y. Homma, F. Kodama, N. Kondo, & S. Akino (Eds.), *Plant growth-promoting rhizobacteria: Present status and future prospects* (pp. 73—79). Paris: OECD.

Lucy, M., Reed, E., & Glick, B. R. (2004). Applications of free living plant growth-promoting rhizobacteria. *Antonie van Leeuwenhoek, 86*, 1—25.

Lugtenberg, B., & Kamilova, F. (2009). Plant-growth-promoting rhizobacteria. *Annual Review of Microbiology, 63*, 541—556.

Lugtenberg, B. J., Chin A-Woeng, T. F., & Bloemberg, G. V. (2002). Microbe—plant interactions: principles and mechanisms. *Antonie van Leeuwenhoek, 81*, 373—383.

Lugtenberg, B. J. J., & Dekkers, L. C. (1999). What makes *Pseudomonas* bacteria rhizosphere competent. *Environmental Microbiology, 1*, 9—13.

Ma, Y., Rajkumar, M., & Freitas, H. (2009a). Inoculation of plant growth promoting bacterium *Achromobacter xylosoxidans* strain Ax10 for the improvement of copper phytoextraction by *Brassica juncea*. *Journal of Environmental Management, 90*, 831—837.

Ma, Y., Rajkumar, M., & Freitas, H. (2009b). Improvement of plant growth and nickel uptake by nickel resistant-plant-growth promoting bacteria. *Journal of Hazardous Materials, 166*, 1154–1161.

Martinez-Viveros, O., Jorquera, M. A., Crowley, D. E., Gajardo, G., & Mora, M. L. (2010). Mechanisms and practical considerations involved in plant growth promotion by rhizobacteria. *Journal of Soil Science and Plant Nutrition, 10*, 293–319.

Mayak, S., Tirosh, T., & Glick, B. R. (2004). Plant growth-promoting bacteria confer resistance in tomato plants to salt stress. *Plant Physiology & Biochemistry, 42*(6), 565–572.

Moore, F. P., Tanja, B., Brigitte, B., Licy, O., Jaco, V., Daniel, V. L., et al. (2006). Endophytic bacterial diversity in poplar trees growing on a BTEX-contaminated site: the characterisation of isolates with potential to enhance phytoremediation. *Systematic and Applied Microbiology, 29*, 539–556.

Muratova, A. Y., Turkovskaya, O. V., Antonyuk, L. P., Makarov, O. E., Pozdnyakova, L. I., & Ignatov, V. V. (2005). Oil-oxidizing potential of associative rhizobacteria of the genus *Azospirillum*. *Microbiology, 74*, 210–215.

Murphy, J. F., Zehnder, G. W., Schuster, D. J., Sikora, E. J., Polston, J. E., & Kloepper, J. W. (2000). Plant growth-promoting rhizobacterial mediated protection in tomato against tomato mottle virus. *Plant Disease, 84*, 779–784.

Nakkeeran, S., Kavitha, K., Chandrasekar, G., Renukadevi, P., & Fernando, W. G. D. (2006). Induction of plant defence compounds by *Pseudomonas chloraphis* PA23 and *Bacillus subtilis* BSCBE4 in controlling damping-off of hot pepper caused by *Pythium aphanidermatum*. *Biocontrol Science and Technology, 16*, 403–416.

Narasimhan, K., Basheer, C., Bajic, V. B., & Swarup, S. (2003). Enhancement of plant-microbe interactions using a rhizosphere metabolomics driven approach and its application in the removal of polychlorinated biphenyls. *Plant Physiology, 132*, 146–153.

Naureen Zakira, P., Hafeez Adam, H., Roberts Fauzia, Y., & Michael, R. (2009). Identification of rice blast disease-suppressing bacterial strains from the rhizosphere of rice grown in Pakistan. *Crop Protection, 28*, 1052–1060.

Niranjan Raj, S., Deepak, S. A., Basavaraju, P., Shetty, H. S., Reddy, M. S., & Kloepper, J. W. (2003). Comparative performance of formulations of plant growth promoting rhizobacteria in growth promotion and suppression of downy mildew in pearl millet. *Crop Protection, 22*, 579–588.

Ongena, M., Adam, A., Jourdan, E., Paquot, M., Brans, A., Joris, B., et al. (2007). Surfactin and fengycin lipopeptides of *Bacillus subtilis* as elicitors of induced systemic resistance in plants. *Environmental Microbiology, 9*, 1084–1090.

Ongena, M., Daayf, F., Jacques, P., Thonart, P., Benhamou, N., Paulitz, T. C., et al. (1999). Protection of cucumber against pythium root rot by fluorescent pseudomonads: predominant role of induced resistance over siderophores and antibiosis. *Plant Pathology, 48*, 66–76.

Patten, C. L., & Glick, B. R. (2002). Role of *Pseudomonas putida* indoleacetic acid in development of the host plant root system. *Applied and Environmental Microbiology, 68*, 3795–3801.

Perrig, D., Boiero, M. L., Masciarelli, O. A., Penna, C., Ruiz, O. A., & Cassán, F. D. (2007). Plant-growth-promoting compounds produced by two agronomically important strains of *Azospirillum brasilense*, and implications for inoculant formulation. *Applied Microbiology and Biotechnology, 75*, 1143–1150.

Pinton, R., Varanini, Z., & Nannipieri, P. (2007). *The rhizosphere. Biochemistry and organic substances at the soil plant interface* (2nd ed.). Boca Raton, FL, USA: CRC, Taylor and Francis.

Podile, A. R., & Kishore, G. K. (2006). Plant growth-promoting rhizobacteria. In S. S. Gnanamanickam (Ed.), *Plant-associated bacteria* (pp. 195–230). Netherlands: Springer.

Rajkumar, M., & Freitas, H. (2008). Influence of metal resistant-plant growth promoting bacteria on the growth of *Ricinus communis* in soil contaminated with heavy metals. *Chemosphere, 71*, 834–842.

Ramakrishnan, G., Nakkeeran, S., Chandrasekar, G., & Doraiswamy, S. (2001). Biocontrol agents-novel tool to combat plant diseases. In *The III Asia Pacific crop protection conference–2001, 6th–7th September, 2001–New Delhi, India* (pp. 20–39).

Richardson, A. E. (2001). Prospects for using soil microorganisms to improve the acquisition of phosphorus by plants. *Australian Journal of Plant Physiology, 28*, 897–906.

Riefler, M., Novak, O., Strnad, M., & Schmülling, T. (2006). *Arabidopsis* cytokinin receptor mutants reveal functions in shoot growth, leaf senescence, seed size, germination, root development, and cytokinin metabolism. *Plant Cell, 18*, 40–54.

Rodríguez, H., Fraga, R., González, T., & Bashan, Y. (2006). Genetics of phosphate solubilization and its potential applications for improving plant growth-promoting bacteria. *Plant and Soil, 287*, 15–21.

Rudrappa, T., Czymmek, K. J., Par, P. W., & Bais, H. P. (2008). Root-secreted malic acid recruits beneficial soil bacteria. *Plant Physiology, 148*, 1547–1556.

Ryu, C. M., Farag, M. A., Hu, C. H., Reddy, M. S., Wie, H. X., Paré, P. W., et al. (2003). Bacterial volatiles promote growth of *Arabidopsis*. *Proceedings of the National Academy of Sciences of the United States of America, 100*, 4927–4932.

Ryu, C. M., Kim, J., Choi, O., Kim, S. H., & Park, C. S. (2006). Improvement of biological control capacity of *Paenibacillus polymyxa* E681 by seed pelleting on sesame. *Biological Control, 39*, 282–289.

Saleh, S., Huang, X. D., Greenberg, B. M., & Glick, B. R. (2004). Phytoremediation of persistent organic contaminants in the environment. In A. Singh, & O. Ward (Eds.), *Applied bioremediation and phytoremediation: Vol. 1. Soil biology* (pp. 115–134). Berlin: Springer.

Santner, A., Calderon-Villalobos, L. I. A., & Estelle, M. (2009). Plant hormones are versatile chemical regulators of plant growth. *Nature Chemical Biology, 5*, 301–307.

Sharma, A., & Johri, B. N. (2003). Growth promoting influence of siderophore-producing *Pseudomonas* strains GRP3A and PRS9 in maize (*Zea mays* L.) under iron depriving conditions. *Microbiological Research, 158*, 243–248.

Shuhegger, R., Ihring, A., Gantner, S., Bahnweg, G., Knappe, C., Vogg, G., et al. (2006). Induction of systemic resistance in tomato by N-acyl-L homoserine lactone-producing rhizosphere bacteria. *Plant Cell and Environment, 29*, 909–918.

Silva, H. S. A., de Silva, R. R., Macagnan, D., de Almeda Halfeld-Viera, B., Pereira, M. C. B., & Mounteer, A. (2004). Rhizobacterial induction of systemic resistance in tomato plants: non specific protection and increase in enzyme activities. *Biological Control, 29*, 288–295.

Somers, E., Vanderleyden, J., & Srinivasan, M. (2004). Rhizosphere bacterial signalling: a love parade beneath our feet. *Critical Reviews in Microbiology, 30*, 205–240.

Spaepen, S., & Vanderleyden, J. (2011). Auxin and plant-microbe interactions. *Cold Spring Harbor Perspectives in Biology, 3*(4).

Spaepen, S., Vanderleyden, J., & Remans, R. (2007). Indole-3-acetic acid in microbial and microorganism-plant signaling. In F. Unden (Ed.), *FEMS microbiology reviews* (pp. 1–24). New York: Blackwell Publishing Ltd.

Sriprang, R., Hayashi, M., Ono, H., Takagi, M., Hirata, K., & Murooka, Y. (2003). Enhanced accumulation of Cd^{2+} by a *Mesorhizobium* sp. transformed with a gene from *Arabidopsis thaliana* coding for phytochelatin synthase. *Applied and Environmental Microbiology, 69*, 1791–1796.

Tao, G. C., Tian, S. J., Cai, M. Y., & Xie, G. H. (2008). Phosphate solubilizing and mineralizing abilities of bacteria isolated from soils. *Pedosphere, 18*, 515–523.

Tien, T. M., Gaskins, M. H., & Hubbell, D. H. (1979). Plant growth substances produced by *Azospirillum brasilense* and their effect on the growth of pearl millet (*Pennisetum americanum* L.). *Applied and Environmental Microbiology, 37*, 1016–1024.

Timmusk, S., Nicander, B., Granhall, U., & Tillberg, E. (1999). Cytokinin production by *Paenibacillus polymyxa*. *Soil Biology & Biochemistry, 31*, 1847–1852.

Tisdale, S. L., & Nelson, W. L. (1975). *Soil fertility and fertilizers* (3rd ed., p. 694). New York: Macmillan Publishing.

Tsavkelova, E. A., Klimova, S. Y., Cherdyntseva, T. A., & Netrusov, A. I. (2006). Microbial producers of plant growth stimulators and their practical use: a review. *Applied Biochemistry and Microbiology, 42*, 117–126.

Vacheron, J., Desbrosses, G., Bouffaud, M.-L., Touraine, B., Moënne-Loccoz, Y., Muller, D., et al. (2013). Plant growth-promoting rhizobacteria and root system functioning. *Frontiers of Plant Science, 4*, 356.

Vance, C. P., Uhde-Stone, C., & Allan, D. L. (2003). Phosphorus acquisition and use: critical adaptations by plants for securing a nonrenewable resource. *New Phytologist, 157*(3), 423–447.

Verma, J. P., Yadav, J., & Tiwari, K. N. (2010). Impact of plant growth promoting rhizobacteria on crop production. *International Journal of Agricultural Research, 5*, 954–983.

Vessey, J. K. (2003). Plant growth promoting rhizobacteria as biofertilizers. *Plant and Soil, 255*, 571–586.

Villacieros, M., Whelan, C., Mackova, M., Molgaard, J., Sanchez-Contreras, M., & Lloret, J. (2005). Polychlorinated biphenyl rhizoremediation by *Pseudomonas fluorescens* F113 derivatives, using a *Sinorhizobium meliloti* nod system to drive *bph* gene expression. *Applied and Environmental Microbiology, 71*, 2687–2694.

Vivas, A., Biro, B., Ruiz-Lozano, J. M., Barea, J. M., & Azcon, R. (2006). Two bacterial strains isolated from a Zn-polluted soil enhance plant growth and mycorrhizal efficiency under Zn toxicity. *Chemosphere, 52*, 1523–1533.

Wani, P. A., & Khan, M. S. (2010). *Bacillus* species enhance growth parameters of chickpea (*Cicer arietinum* L.) in chromium stressed soils. *Food and Chemical Toxicology, 48*, 3262–3267.

Wani, P. A., Khan, M. S., & Zaidi, A. (2007). Effect of metal tolerant plant growth promoting *Bradyrhizobium* sp. (vigna) on growth, symbiosis, seed yield and metal uptake by green gram plants. *Chemosphere, 70*, 36–45.

Weger, L. A., van Arendonk, J. J., Recourt, K., vander Hofstad, G. A., Weisbeek, P. J., & Lugtenberg, B. (1988). Siderophore-mediated uptake of Fe^{3+} by the plant growth-stimulating *Pseudomonas putida* strain WCS358 and by other rhizosphere microorganisms. *Journal of Bacteriology, 170*, 4693–4698.

Weller, D. M., & Thomashow, L. S. (1994). Current challenges in introducing beneficial microorganisms into the rhizosphere. In F. O'Gara, D. N. Dowling, & B. Boesten (Eds.), *Molecular ecology of rhizosphere microorganisms biotechnology and the release of GMOs* (pp. 1–18). Weinheim: VCH Verlagsgesellschaft.

Wenhua, T., & Hetong, Y. (1997). Research and application of biocontrol of plant diseases and PGPR in China. In A. Ogoshi, K. Kobayashi, Y. Homma, F. Kodama, N. Kondo, & S. Akino (Eds.), *Plant growth-promoting rhizobacteria—Present status and future prospects* (pp. 2–9). Sapporo, Japan: OECD-OCDE.

Wu, C. H., Wood, T. K., Mulchandani, A., & Chen, W. (2006). Engineering plant-microbe symbiosis for rhizoremediation of heavy metals. *Applied and Environmental Microbiology, 72*(2), 1129–1134.

Yang, C. H., & Crowley, D. E. (2000). Rhizosphere microbial community structure in relation to root location and plant iron nutritional status. *Applied and Environmental Microbiology, 66*, 345–351.

Yee, D. C., Maynard, J. A., & Wood, T. K. (1998). Rhizoremediation of trichloroethylene by a recombinant, root-colonizing *Pseudomonas fluorescens* strain expressing toluene ortho-monooxygenase constitutively. *Applied and Environmental Microbiology, 64*, 112–118.
Young, J. P. W., Mutch, L. A., Ashford, D. A., Zézé, A., & Mutch, K. E. (2003). The molecular evolution of host specificity in the rhizobium-legume symbiosis. In R. Hails, H. C. J. Godfray, & J. Beringer (Eds.), *Genes in the environment* (pp. 245–257). Oxford: Blackwell Science.
Zahran, H. H. (2001). Rhizobia from wild legumes: diversity, taxonomy, ecology, nitrogen fixation and biotechnology. *Journal of Biotechnology, 91*, 143–153.
Zaidi, A., Khan, M. S., Ahemad, M., & Oves, M. (2009). Plant growth promotion by phosphate solubilizing bacteria. *Acta Microbiologica et Immunologica Hungarica, 56*, 263–284.
Zaidi, S., Usmani, S., Singh, B. R., & Musarrat, J. (2006). Significance of *Bacillus subtilis* strain SJ 101 as a bioinoculant for concurrent plant growth promotion and nickel accumulation in *Brassica juncea. Chemosphere, 64*, 991–997.
Zeder, M. A. (2009). Evolutionary biology and the emergence of agriculture: the value of co-opted models of evolution in the study of culture change. In *Macroevolution in human prehistory: Evolutionary theory and processual archaeology* (pp. 157–210). New York: Springer.
Zehnder, G. W., Murphy, J. F., Sikora, E. J., & Kloepper, J. W. (2001). Application of rhizobacteria for induced resistance. *European Journal of Plant Pathology, 107*, 39–50.
Zehnder, G. W., Yao, C., Murphy, J. F., Sikora, E. J., & Kloepper, J. W. (2000). Induction of resistance in tomato against cucumber mosaic cucumovirus by plant growth promoting rhizobacteria. *Biological Control, 45*, 127–137.
Zhang, S., Moyne, A. L., Redddy, M. S., & Kloepper, J. W. (2002). The role of salicylic acid in induced systemic resistance elicited by plant growth-promoting rhizobacteria against blue mold of tobacco. *Biological Control, 25*, 288–296.

CHAPTER FIVE

Human Pathogen–Plant Interactions: Concerns for Food Safety

Sarah M. Markland, Kalmia E. Kniel[1]
Department of Animal and Food Sciences, University of Delaware, Newark, DE, USA
[1]Corresponding author: E-mail: kniel@udel.edu

Contents

1. Introduction: Merging Plant Science and Food Science to Address Food Safety 116
2. Human Pathogens on Plants 116
3. Human Pathogen Interactions with Phytobacteria and Rhizobacteria 120
4. Relationships Between Plant Stomata, Circadian Rhythm and Human Pathogens 123
5. Control and Prevention Strategies: Biocontrol and Plant 'Probiotics' 125
6. Attachment and Attraction Mechanisms of Human Pathogens to Plants 127
7. Research Needs and Conclusions 129
References 130

Abstract

The Centers for Disease Control and Prevention estimate the incidence of foodborne illness attributed to fruit and vegetable consumption at three million cases in the United States annually. The means by which food crops become contaminated with foodborne pathogens as well as how these organisms persist within the phyllosphere and rhizosphere of these plants is an extremely complex issue. Outbreaks associated with food crops continue to occur, and it is clear that different approaches are critical to enhance the safety of these foods. Within the last few years, the fields of food safety and plant science have begun to merge to more efficiently address some of the knowledge gaps involving the mechanisms by which human pathogens contaminate plants. We currently know a great deal about the mechanisms by which plant pathogens are able to cause disease in plant crops as well as the immune response of these plants to these pathogens. By studying the relationships between plant pathogens and the plant immune response, scientists have been able to successfully develop biocontrol strategies to reduce crop damage attributed to plant disease. Potential control strategies for human pathogens on plants in the preharvest environment include the use of plant growth promoting bacteria and N-acyl-homoserine lactone-producing bacteria. By studying comparable relationships between foodborne pathogens and plants, we may be able to develop similar strategies to reduce the number of foodborne illnesses associated with fresh produce.

Advances in Botanical Research, Volume 75
ISSN 0065-2296
http://dx.doi.org/10.1016/bs.abr.2015.08.002

1. INTRODUCTION: MERGING PLANT SCIENCE AND FOOD SCIENCE TO ADDRESS FOOD SAFETY

The Centers for Disease Control and Prevention (CDC) estimates the incidence of foodborne illness attributed to fruit and vegetable consumption at three million cases in the United States annually (Painter et al., 2013). The means by which food crops become contaminated with foodborne pathogens as well as how these organisms persist within the phyllosphere and rhizosphere of these plants is a complex issue. Outbreaks continue to occur with food crops, and it is clear that different approaches must be taken to enhance the safety of these foods. While the route of contamination may vary and can occur anywhere along the 'seed to fork' continuum, in some cases, the food crops were likely contaminated in the field (Table 1).

Within the last few years, the fields of food safety and plant science have begun to merge to more efficiently address some of the knowledge gaps involving the mechanisms by which human pathogens contaminate plants. We currently know a great deal about the mechanisms by which plant pathogens are able to cause disease in plant crops as well as the immune response of these plants to these pathogens. There is very little information available regarding the way plants respond to human pathogens. By studying the relationships between plant pathogens and the plant immune response, scientists have been able to successfully develop biocontrol strategies to reduce crop damage attributed to plant disease. This chapter addresses the mechanisms by which human pathogens interact with plants in the preharvest environment and the impact these interactions ultimately have on food safety and public health.

2. HUMAN PATHOGENS ON PLANTS

In general, many species of bacteria have close relationships with plants and in some cases promote plant growth and nitrogen uptake. Similar relationships exist with plant viruses; however, these relationships tend to be more detrimental to the plant when plant disease occurs along with large economic losses and threats to food security and public health. Alternatively, crops may become contaminated with bacterial and viral pathogens that are a threat to human health but not necessarily to plant health. Research has shown that plants can become contaminated with human pathogens in the preharvest environment through a variety of outlets including soil, faeces, irrigation water or water used to apply pesticides and fertilizers,

Table 1 Listing of Multistate Outbreaks Associated with Fresh Produce Where Preharvest Contamination was Suspected from 2006 to 2014

Multistate Outbreaks in the United States Associated with Produce Where Preharvest Contamination was Suspected from 2006 to 2014

Year	Commodity	Pathogen
2014	Bean sprouts	*Salmonella enteritidis*
2014	Cilantro	*Cyclospora cayetanensis*
2014	Clover sprouts	*Escherichia coli* O121
2013	Salad mix	*C. cayetanensis*
2013	Cucumbers	*Salmonella* Saintpaul
2012	Spinach and spring mix	*E. coli O157:H7*
2012	Mangoes	*Salmonella* Branderup
2012	Cantaloupe	*Salmonella Typhimurium* and Newport
2011	Romaine lettuce	*E. coli O157:H7*
2011	Papayas	*Salmonella Agona*
2011	Alfalfa and spicy sprouts	*S. enteritidis*
2011	Cantaloupe	*Salmonella Panama*
2010	Alfalfa sprouts	*Salmonella* I 4, [5], 12:i: (PFGE pattern)
2010	Alfalfa sprouts	*Salmonella* Newport
2009	Alfalfa sprouts	*S.* Saintpaul
2008	Jalapeno peppers, serrano peppers, tomatoes	*S.* Saintpaul
2008	Cantaloupe	*Salmonella* Litchfield
2006	Tomatoes	*S. Typhimurium*
2006	Fresh spinach	*E. coli O157:H7*

Source: CDC (2014) (http://www.cdc.gov/foodsafety/outbreaks/multistate-outbreaks/outbreaks-list.html)

dust, insects, land-applied manures and biosolids and directly from wild or domestic animals (Beuchat, 2002). Leafy greens are considered a high-risk food crop because they have been epidemiologically linked to foodborne illness (Tabana & Halkmanb, 2011) and they are commonly consumed in their raw state, where little or no processing takes place to reduce contaminants. Leafy greens are not all the same when it comes to potential contamination. For example, spinach forms a canopy that may serve as a trap for zoonotic pathogens if the plants become contaminated. The outer leaves of lettuce plants may provide a reservoir for pathogens, but this route of contamination may not lead to illness. Addressing safer ways to manage crop growth and harvest is essential to minimizing microbial contamination (Sobsey, Dean, Knuckles, & Wagner, 1980). To address these issues, we must first develop a better understanding of how human pathogenic bacteria and viruses interact with plants.

It has been previously demonstrated that human pathogenic viruses, such as human norovirus, can be internalized into leafy greens through the roots and foliar surfaces of lettuce plants (Hirneisen & Kniel, 2013; Hirneisen, Sharma, & Kniel, 2012; Wang & Kniel, 2015; Wei, Jin, Sims, & Kniel, 2010, 2011) although it is not known how often this naturally occurs in the field and how internalization physically affects plants. It is important to determine the type of relationships that human pathogens have with plants, whether they be symbiotic, endophytic or antagonistic. For example, it was suggested by Schikora, Carreri, Charpentier, and Hirt (2008) that human pathogenic *Salmonella typhimurium* infect and intracellularly proliferate within tissues of *Arabidopsis thaliana* through both the root and shoot of the plant. *Salmonella* infection also yields disease symptoms in the plants including wilting, chlorosis and death of infected plant organs (Schikora et al., 2008). *Arabidopsis* plants responded immunologically similarly to *Salmonella* as they would to the plant pathogen *Pseudomonas syringae* (DC3000) by induction of the mitogen-activated protein kinase (MAPK) cascades as well as by enhanced expression of pathogenesis related (*PR*) genes (Schikora et al., 2008). There are also studies demonstrating that bacteria normally pathogenic to humans and other mammals can infect plants including *Salmonella enterica*, *Pseudomonas aeruginosa*, *Burkholderia cepacia*, *Erwinia* spp., *Staphylococcus aureus*, *Escherichia coli* O157:H7 and *Listeria monocytogenes* (Haapalainen, van Gestel, Pirhonen, & Taira, 2009; Holden, Pritchard, & Toth, 2009; Milillo, Badamo, Boor, & Wiedmann, 2008; Plotnikova, Rahme, & Ausubel, 2000; Prithiviraj, Bais, Jha, & Vivanco, 2005).

Most of the studies described above have been performed on *A. thaliana*, which according to the National Institutes of Health (NIH), is a small flowering plant that is widely used as a model organism in plant biology (NIH). *Arabidopsis* is a member of the mustard (Brassicaceae) family that includes crop plants such as cabbage, cauliflower and broccoli. It is an important plant for genetic and biological research of crop plants although it has no importance in agriculture itself. Approximately 115 Mb of the 125 Mb genome has been sequenced and annotated, and there are extensive genetic and physical maps available of all five chromosomes (Arabidopsis Genome Initiative, 2000). The life cycle of *A. thaliana* is short compared to that of most crop plants (only ~6 weeks) and it can easily be cultivated in small spaces such as the laboratory or within a biochamber. There are a number of genetic mutant lines and genomic resources available to the scientific community, making *A. thaliana* a good candidate for multinational research in academia, industry and government. Because of all the above listed reasons, it can also be concluded that *Arabidopsis* serves as a good model for studying human foodborne pathogen—plant interactions.

Studies of the interactions between plants and plant pathogens, including the plant immune response, have led to the development of control measures for many devastating plant diseases. Through the study of the way in which plants respond to and effectively fight off plant pathogens, scientists have been able to develop plant-breeding programs for disease-resistant varieties. Many of the current control and prevention strategies for human pathogens on plants are at the postharvest level—once the pathogens have already come into contact with or possibly penetrated the plant tissues. Alternative intervention measures target prevention of the initial contamination event (Cooley, Miller, & Mandrell, 2003). For effective measures to be developed for preharvest interventions, a better understanding of the mechanisms by which human pathogens colonize plants as well as how and if the plants are able to recognize these pathogens is needed (Cooley et al., 2003).

The mechanism by which the plant immune system operates is through the detection of conserved microbial components (Jones & Dangl, 2006). The plant physiological responses to plant infection can be due to activation of the induced systemic resistance (ISR) or systemic acquired resistance pathway (SAR). The SAR pathway is induced if the pathogen is able to elicit a hypersensitive reaction (HR) (Durrant & Dong, 2004; Enyedi, Yalpani, Silverman, & Raskin, 1992), which causes apoptosis in the plant in order to rid itself of the infectious agent. These events may enable the plant to become more resistant to future attacks by the pathogen (Madamanchi & Kuc, 1991; Ross, 1961). When a plant is invaded by a pathogen, one of two defence signalling pathways is activated: a salicylic acid (SA)-dependent pathway or an SA-independent pathway involving jasmonic acid (JA) or ethylene (ET)-signalling molecules (Kunkel & Brooks, 2002). JA-dependent and ET-dependent plant defences are generally activated by necrotrophic pathogens and chewing insects, whereas SA-dependent defences are often triggered by biotrophic pathogens (Pandey & Somssich, 2009). JA and SA signalling usually act antagonistically, but synergism between these two molecules has also been observed (Mur, Kenton, Atzorn, Miersch, & Wasternack, 2006). The interactions among SA-, JA- and ET-dependent pathways are extremely complex (Katagiri, Thilmony, & He, 2002). For example, during an immune response of *A. thaliana* to plant pathogen *P. syringae* DC3000, activation of the JA pathway has been observed to suppress the SA pathway and reduce the plant's general resistance to the pathogen *P. syringae*, among which most plants use the SA pathway (Katagiri et al., 2002). It is also hypothesized that some virulent strains of *P. syringae* take advantage of this antagonistic interaction in order to suppress the *Arabidopsis* defensive response (Katagiri et al., 2002).

3. HUMAN PATHOGEN INTERACTIONS WITH PHYTOBACTERIA AND RHIZOBACTERIA

There is a substantial amount of evidence in the literature that shows the ability of human pathogens to survive in the preharvest environment such as in water, on plants, manure or in fertilizers. While human enteric pathogens are not typically considered to be part of the epiphytic microbial population, this is being considered further as the fields of plant science and food safety begin to merge to address important food safety issues (Lim, Lee, & Heu, 2014). The ability of these pathogens to survive on plants is dependent upon many environmental factors such as ultraviolet exposure, humidity, weather and presence of epiphytic organisms. Results from studies investigating the ability of human pathogens to survive on plants vary considerably depending on the experimental design including whether the experiment was carried out in the field or in a greenhouse environment. However, when applied directly to the foliar region of plants, *E. coli* and *Salmonella* were previously reported to have the ability to survive on parsley plants in the field for 177 and 231 days, respectively (Islam et al., 2004). It is also possible for human pathogens to become internalized in plants through different sources of contamination and may involve diverse plant entry mechanisms (Martíez-Vaz, Fink, Diez-Gonzalez, & Sadowsky, 2014). Internalization can occur via uptake through the root system along with water, via entry through natural plant openings such as the stomata or via would sites on the plant (Martíez-Vaz et al., 2014). While it is possible for internalization to occur in leafy greens, this phenomenon most likely does not occur frequently with the population levels of bacteria normally found in soil or on plants (Erickson et al., 2010; Zhang et al., 2009).

Other factors that influence the colonization of human pathogens on plants include the presence of resident phytobacteria. The presence of plant-associated microbiota can either promote or inhibit the establishment of human enteric pathogens on plants (Lim et al., 2014). One supermarket survey demonstrated that 60% of produce showing symptoms of soft rot also tested positive for the presence of *Salmonella* (Wells & Butterfield, 1997).

Other laboratory studies have shown the presence of other pectinolytic plant pathogens including *Erwinia chrysanthemi* and *Pectobacterium carotovorum* that positively influenced the growth of *Salmonella Typhimurium* and *E. coli* O157:H7 to levels 10 times higher than that on healthy plants (Noel et al., 2010; Yamazaki et al., 2011). Similar studies have shown that this sudden increase of pathogens on plants could be directly correlated with the presence of soft rot symptoms (Brandl, 2008; Goudeu et al., 2013). The diverse epiphytic microbial community on plant tissues varies functionally and structurally depending on the species of the plant (Brandl, Cox, & Teplitski, 2013). Plant pathogens including erwinias, xanthomonads and pseudomonads reside closely within the phyllosphere with human pathogens including proteobacteria, firmicutes, bacteroidetes and actinobacteria (Lopez-Velasco, Tydings, Boyer, & Falkinham, 2012; Rastogi et al., 2012). It was also recently observed that colonization of plant pathogen *Xanthomonas perforans* positively influences the persistence of *S. enterica* on tomatoes (Potnis et al., 2014). More recently, downy mildew disease caused by the oomycete *Bremia lactucae* was shown to promote the colonization of *E. coli* O157:H7 and *S. typhimurium* on romaine lettuce (Simko, Zhou, & Brandl, 2015).

These data suggest the complexity of relationships between plants, resident plant microbiota and human pathogens. Perhaps even more complicated is the less studied interactions of plants, human pathogens and resident rhizobacteria. Human bacterial pathogens have been shown to be able to persist in soils for much longer time periods compared to on plants. *Salmonella typhimurium* and *E. coli* O157:H7 have been shown to persist in soil amended with contaminated compost for >200 days (Islam et al., 2004). Franz, van Hoek, Bouw, and Aarts (2011) showed the ability of *E. coli* O157:H7 to survive for 211 days in manure-amended soil. In another study, *S. typhimurium* was shown to be able to survive for up to 332 days in manure-amended soil (You et al., 2006). Discrepancies between these survival studies in soil have been attributed to different environmental conditions, inoculation methods and microbial enumeration methods (Martínez-Vaz et al., 2014). Environmental factors, such as soil topography, moisture and proximity to water sources have also been shown to increase the frequency of isolating enteric bacteria from vegetable farms (Strawn et al., 2013). The presence of protozoa in soils and subsequent encapsulation of pathogens is hypothesized to have a protective effect that allows for prolonged survival of enteric bacteria in contaminated environments such as manure-amended soils (Brandl et al., 2013). Although there are many factors that affect the persistence of human pathogens on plants or within the soil, there is enough evidence to conclude that human pathogens can survive for long-enough periods that allow them to contaminate plant roots, leaves and fruit to allow them to gain access to the human host (Martínez-Vaz et al., 2014). This further supports the theory that plants are used as a vector by human pathogens to ultimately reach their host.

Rhizobacteria are also able to interact with pathogens within the phyllosphere of plants through complex indirect plant–microbe interactions. Studies suggest that upon detection of a pathogen, a plant is able to recruit bacteria to colonize the root surface, triggering ISR in aerial portions of the plant (Lakshmanan et al., 2012; Ryu et al., 2004) and this ultimately results in the closure of plant stomata (Kumar et al., 2012). Specifically in 2008, Rudrappa et al. found that plants can recruit specific *Bacillus subtilis* spp. to their roots via secretion of malic acid when aerial tissues come into contact with *Pst* DC3000. It has been demonstrated that this specific plant growth promoting rhizobacteria (PGPR) are able to induce stomatal closure when the root is colonized (Kumar et al., 2012).

Interestingly, PGPR have been shown to have the ability to induce stomatal closure in romaine lettuce and spinach plants (Markland, Bais, & Kniel, 2014) (Figures 1 and 2). Over a 5-day study, addition of the same PGPR to the soil of lettuce plants was shown to significantly inhibit the persistence of *L. monocytogenes* (Markland et al., 2014). While this study

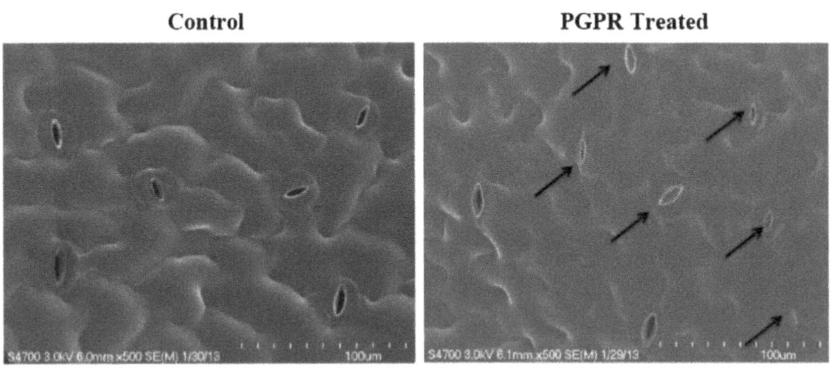

Figure 1 Cryo-scanning electron microscopy images of romaine lettuce leaf sections of PGPR-treated and nontreated plants, showing the effects of PGPR *Bacillus subtilis* UD1022 inoculation on the roots of romaine lettuce plants at 3 h postinoculation. At 3 h postinoculation, more stomata on PGPR-treated plants are closed compared to nontreated plants (controls). Closed stomata are shown by arrows. *Images taken at the University of Delaware Biotechnology Institute Bioimaging Center (Markland et al., 2014).*

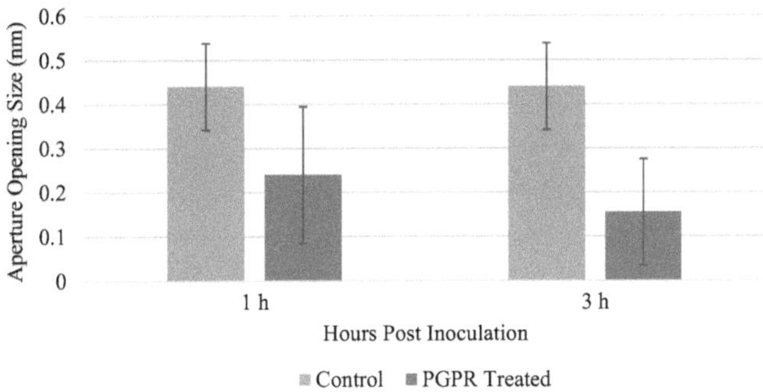

Figure 2 Difference in the aperture opening size of stomata of romaine lettuce where roots were inoculated with water (light) or with PGPR (dark). Results show that stomata are significantly smaller at 3 hpi when roots were treated with PGPR compared to controls. *Data were collected by measuring stomata aperture openings of scanning electron microscopy images using ImageJ Software (National Institutes of Health, Bethesda, MD) (Markland et al., 2014).*

only scratches the surface for the use of PGPR as a biocontrol agent to reduce the risk of human pathogens on plants in the preharvest environment, there is great potential for optimization and use. The use of PGPR has the potential to protect plants from infection by plant pathogens as well as from human pathogens. PGPR can also increase crop yield and may be an economical and sustainable means for protecting public health and crops. Further research in this field may enable the development of new technology that utilizes the natural association between rhizobacteria and plants to minimize contamination by human pathogens in food crops.

4. RELATIONSHIPS BETWEEN PLANT STOMATA, CIRCADIAN RHYTHM AND HUMAN PATHOGENS

A plant stoma is a pore located in the epidermis of the leaf and the stem that functions in gas exchange. A pair of parenchyma cells, also known as guard cells, borders the stoma and controls the size of the pore and the opening and closing of the pore. Carbon dioxide (CO_2) and oxygen (O_2) enter the plant through the stomata and are used in photosynthesis and respiration. The stomata also function in transpiration via the release of water vapour from the plant. Stomata are also known to be important global regulators of the atmospheric environment (Hetherington & Woodward, 2003). Much is known and has been studied involving the anatomy of the stomatal valve, but there is a lack of knowledge as to how exactly stomata sense and respond to their external stimuli (Brodribb & McAdam, 2011), specifically bacterial pathogens. These openings can also serve as a portal of entry for plant pathogens and the opening and closing of the stomata is actually a triggered innate immune response (Melotto, Underwood, Koczan, Nomura, & He, 2006). Although it has been known for quite some time that the phyllosphere is a common place for the colonization of plant and human pathogens (Melotto et al., 2006), there is not enough data to support the mechanism of entry of these pathogens into the plant host and the host immune response involved. Unlike fungal pathogens, bacteria do not have the ability to penetrate the plant epidermis and rely on plant openings or wounds to internalize into plant tissues (Melotto et al., 2006).

For the first time in 2006, Melotto et al. demonstrated that the small pores located on the surface of leaves functioning in gas exchange actually close in response to microbial contact. The authors of this study also show that, in turn, pathogens have the ability to reopen the stomata to gain subsequent entry into the plant. The authors used *Arabidopsis* as a plant model and observed closure of the stomata within 2 h of inoculation with both the plant pathogen *P. syringae* DC3000 and the human pathogen *E. coli* O157:H7. However, after 3 h of incubation with *Pst* DC3000, stomata reopened, whereas stomata on plants inoculated with *E. coli* O157:H7 remained closed throughout the experiment (8 h).

Although human pathogens were not be able to elicit the reopening of plant stomata in this study, stomatal openings are still believed to play a role in contamination of leafy greens. Pathogens have previously been shown to migrate towards stomatal openings under certain conditions and can internalize into plants (Erickson, 2012). A recent study by Markland et al. (2014) also showed that *Salmonella* Newport and *Listeria innocua* were able to reopen stomata on spinach and lettuce plants, respectively, after 3 h post-inoculation on plant leaves. Intrinsic and extrinsic factors including circadian rhythm and photosynthesis (light) are thought to play a role in stomatal opening and closing. Studies have shown that there are direct effects of light on both the defence response in the host plant and on the virulence of the attacking pathogen (Oberpichler et al., 2008). It is believed that plant defences are heightened in response to light; however, some plant defence responses occur independently of light, such as JA production (Zeier, Pink, Mueller, & Berger, 2004). Light has been shown to play a particularly important role in SA-mediated defence responses (Roden & Ingle, 2009). Light is also required for the hypersensitive response (HR), a form of localized programmed cell death at the site of infection, activated during effector-triggered immunity (Roden & Ingle, 2009). Because plant defence is heightened in the light, virulence of pathogens tends to be heightened in the dark. Oberpichler et al. (2008) showed that there is a link between light perception and virulence involving cell motility. The authors of this study showed that bacteria cultured in light had a reduced number of flagella (one or two compared to three to five in the dark) and exhibited reduced motility in colony assays. Virulence was also shown to be directly affected by light where reduced root attachment was observed in tomato and smaller tumour formation in cucumber in the presence of light compared to darkness (Oberpichler et al., 2008). Experiments have demonstrated that having a functional circadian oscillator with the same period as the Earth's rotation gives plants an adaptive advantage and increased fitness (Dodd, Love, & Webb, 2005). It is important to further understand the role of circadian rhythm in plant defences to human enteric pathogens in order to develop better crop management practices to reduce contamination of leafy greens. By understanding how foodborne pathogens utilize plant stomata to internalize and persist on plants, we may also be able to develop biocontrol strategies that control stomatal opening and closing.

5. CONTROL AND PREVENTION STRATEGIES: BIOCONTROL AND PLANT 'PROBIOTICS'

In the preharvest environment, there are many control and prevention strategies currently used by produce growers that help protect plants from infection by plant pathogens and pests that can cause damage to crops. There are other guidelines, including good agricultural practices, which provide information for growers on how to prevent their crops from becoming contaminated with human pathogens (FDA, 1998); however, addressing this issue is much more complex in that we do not know exactly how or why human pathogens interact with plants. It has been hypothesized that the ability for human pathogens to colonize edible plants may be an effective survival strategy that provides a direct route from its excretion in the environment back to its numerous herbivorous and omnivorous hosts (Brandl et al., 2013). The implementation of the Food Safety Modernization Act in 2011 and introduction of the Produce Safety Rules will now require qualified growers to follow more stringent food safety practices that could increase the cost of food production and subsequently raise the cost of food. In addition, food trends including the organic and nongenetically modified organism campaigns have led to a decrease in the use of fertilizers previously used to reduce crop loss as well as higher food prices for these commodities.

It would be ideal to innovate technologies that could simultaneously reduce contamination of crops by plant and human pathogens. One potential solution includes the use of PGPR, which can act as plant 'probiotics'. PGPR are considered part of the natural microflora of plants as well as important contributors to plant health through plant growth promotion or biological disease control. They are often used to induce suppressiveness of plant pathogens by altering the diversity of microorganisms in the rhizosphere (Kloepper et al., 1999). In most cases, biological control by PGPR results from bacterial production of metabolites that directly inhibit the pathogen such as antibiotics, hydrogen cyanide, iron-chelating siderophores and cell-wall degrading enzymes (Kloepper et al., 1999). Plant growth promotion is considered an indirect mechanism of disease control as the time a plant is in a susceptible state is shortened, allowing for the plant to escape infection (Kloepper et al., 1999). PGPR have also been used to reduce plant contamination by foliar pathogens where PGPR control involves induction of plant host defences (Kloepper et al., 1999). As discussed in Section 4 of this chapter, recent studies have demonstrated the ability of PGPR *Bacillus subtilis* to induce stomatal closure in lettuce and spinach as well as reduce the persistence of *L. monocytogenes* on these plants. Through a series of interactions between the human pathogen, the plant and the PGPR, the plant recruits the PGPR by secretion of malic acid to form biofilms on the roots of the plants, which induces plant defences and ultimately stomata closure (Figure 3).

It was also demonstrated that plants have the ability to detect N-acyl-homoserine lactones (AHLs) (Figure 4), used by gram-negative bacteria for quorum sensing behaviour, and are subsequently able to induce system resistance through induction of WRKY and PR-1 genes (Schikora et al., 2011). It has been proposed that N-3-oxo-tetradecanoyl-L-homoserine lactone (oxo-C14-HSL) primes plants for enhanced callose deposition, accumulation of phenolic compounds, lignification of cell walls and increases levels of oxylipins and SA, which favours closure of plant stomata (Schenk et al., 2014). The use of PGPR, AHL-producing bacteria as plant probiotics and biocontrol agents may help provide a sustainable means for growers, including those raising organic crops, to reduce environmental and human health risks associated with growing produce as well as improving productivity and crop yield thus reducing the cost of production and increasing net farm income. Biocontrol and manipulation of pathogens

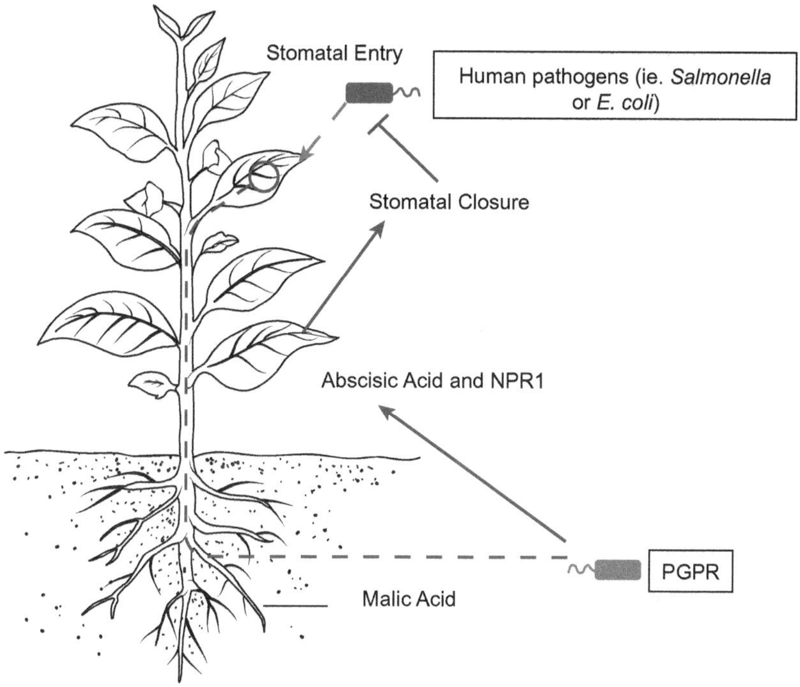

Figure 3 Schematic of the mechanism by which PGPR induces a systemic response (ISR) in plants. (1) Pathogen lands on aerial portions on the leaf of the plant. (2) The plant detects pathogen-associated molecular patterns, such as flagella, which triggers an 'SOS' signal to initiate plant defences and secretes malic acid into the soil. (3) The malic acid signal recruits PGPR to the roots. (4) A systemic response is induced in the plant as well as stomatal closure via SA, NPR1 and ETH to protect the plant from pathogen invasion. *Figure adapted from Lakshmanan et al. (2012).*

on plants can reduce the initial contamination that may be unavoidable, at times, and spread by cross-contamination during washing or during packaging of produce.

6. ATTACHMENT AND ATTRACTION MECHANISMS OF HUMAN PATHOGENS TO PLANTS

The mechanisms by which human pathogens attach to plants are complex and vary greatly between different types of plant–pathogen interactions. There are four main types of interactions that involve attachment of human pathogens to plants including cell surface structures, virulence, motility and biofilm formation (Martínez-Vaz et al., 2014). For example,

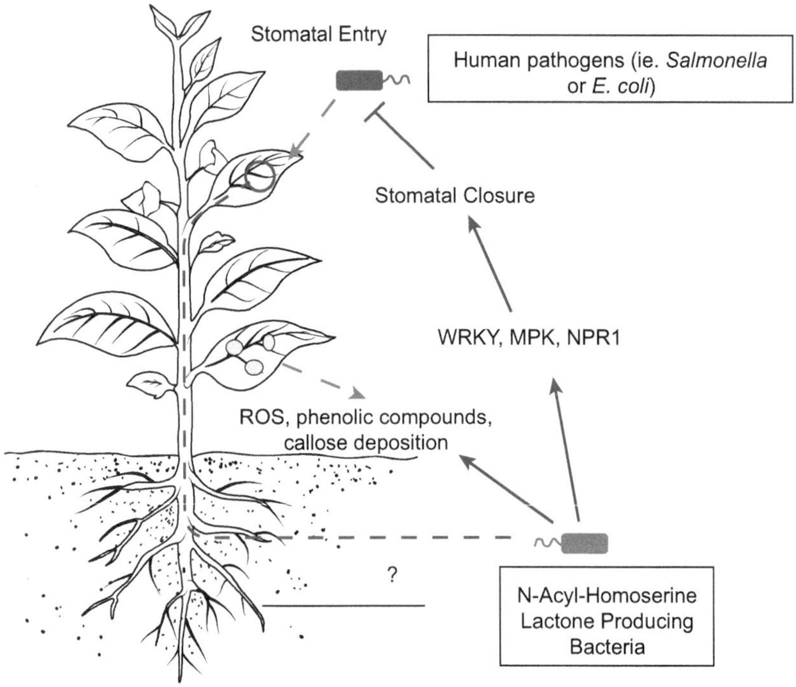

Figure 4 Schematic of the mechanism by which AHL-producing bacteria induces stomata closure in plants. (1) AHL-producing bacteria activates MAPKs and enhances the expression of WRKY transcription factors. (2) Accumulation of reactive oxygen species, phenolic compounds, and callose in the cell walls. (3) Elevated production of phytohormones oxylipin and SA. Induction of callose deposition and enhanced stomatal closure. *Figure adapted from Schikora et al. (2011).*

Salmonella Tennessee was previously found to be able to adhere more efficiently to lettuce and cabbage surfaces compared to other *Salmonella* serotypes likely due to its ability to form biofilms (Patel & Sharma, 2010). The components of the *Salmonella* extracellular matrix are involved in attachment to plant surfaces including cellulose, capsule and fimbriae (Lim et al., 2014). It has also been confirmed that the gene for global stress regulation (*rpoS*) plays an important role in biofilm formation and adhesion regulation (Prigent-Combaret et al., 2001; Raina, Missiakas, Baird, Kumar, & Georgopoulos, 1993) and is required for initial attachment to plant surfaces (Lim et al., 2014). A transcriptional regulator of the LuxR superfamily in *Salmonella* (agfD) regulates curli and cellulose production as well as the O-capsular antigen (Römling, Bian, Hammar, Sierralta, & Normark, 1998). When *S. Typhimurium* presents the rdar morphotype, which is positively regulated by AgfD, it demonstrates improved attachment on tomato plants (Gu et al., 2011). Motility and chemotaxis genes have also been shown to be required for early attachment of *Salmonella* on leafy greens. *Salmonella* utilizes flagellar motility and demonstrates chemotactic behaviour towards leaf exudates for efficient attachment and internalization on lettuce (Kroupitski, Pinto, Brandl, Belausov, & Sela, 2009). However; motility and chemotaxis characteristics among *Salmonella* species may vary depending on bacterial genotypes and environmental conditions (Martínez-Vaz et al., 2014).

Virulence factors that have been shown to play a role in the attachment of *E. coli* O157:H7 to plant surfaces include curli fibres (Macarisin, Patel, Bauchan, Giron, & Sharma, 2012), the type III secretion system (Kenny & Finlay, 1995; Saldana et al., 2009; Xicohtencatl-Cortes, Sanchez Chacon, Saldana, Freer, & Giron, 2009) and capsular production (Hassan & Frank, 2004; Rahn, Beis, Naismith, & Whitfield, 2003). In *E. coli* and *Salmonella*, the yfcR gene, involved in tolerance to multiple stresses as well as biofilm formation, has been shown to play a critical role for attachment to vegetable tissues (Deng, Wang, Rui, Zhang, & Tortorello, 2011; Fink et al., 2012). Transcriptome analysis of human pathogens in close association with plant surfaces, specifically *E. coli* and *Salmonella*, have shown that distinct changes occur in gene expression when pathogens are in association with damaged or intact plant surfaces (Martínez-Vaz et al., 2014). Findings from these studies suggest that mechanisms for attachment of human pathogens in the phyllosphere or the rhizosphere may be species-specific; therefore, caution should be exercised when developing prevention and control strategies for human pathogens on plants in the preharvest environment because strains of the same bacterial species may respond differently to sanitation treatments (Martínez-Vaz et al., 2014). Resident plant bacteria as well as plant pathogens also utilize curli, fimbriae and cellulose to adhere to plant surfaces (Figure 5) (Teplitski, Barak, & Schneider, 2009). This shows that human pathogens use attachment mechanisms similar to those of plant pathogens in order to colonize plants (Martínez-Vaz et al., 2014).

7. RESEARCH NEEDS AND CONCLUSIONS

The study of human pathogens on plants has opened our minds to complicated interactions among bacteria, viruses and plants. There is more to the survival of human pathogens on plant leaf surfaces and in the soil; much more that is yet to be uncovered. Future research in this area will

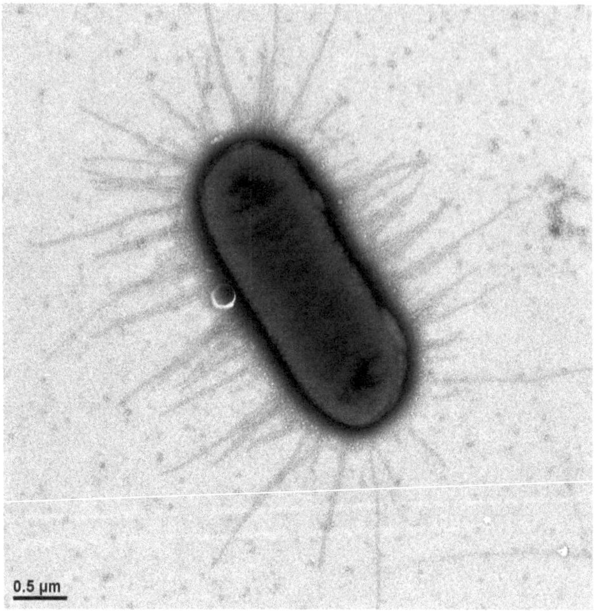

Figure 5 Transmission electron microscopy image of an *Escherichia coli* cell showing long fimbriae appendages extending from the bacterial capsule. *Image taken at the University of Delaware Biotechnology Institute Bioimaging Center.*

generate fundamental information regarding the physical and molecular mechanisms that enable human pathogens to attach, internalize, grow and survive in and on fresh produce, specifically leafy greens. Additionally, this work will generate information on how plants interact with foodborne viruses, specifically norovirus and hepatitis A virus, and if these associations affect the attachment and fate of human pathogens on fresh produce. There are many knowledge gaps involving how plants respond to norovirus contamination, which is the number one cause of foodborne illness in the United States, with a majority of illnesses involving leafy greens. Information concerning viral persistence and survival will impact growing and irrigation practices. Information gained by studying the plant defence response in relation to colonization by noroviruses may also impact preharvest growing practices.

REFERENCES

Arabidopsis Genome Initiative. (2000). Analysis of the genome sequence of the flowering plant *Arabidopsis thaliana*. *Nature, 408*, 796–815.

Beuchat, L. R. (2002). Ecological factors influencing survival and growth of human pathogens on raw fruits and vegetables. *Microbes and Infection, 4*, 413–423.

Brandl, M. T. (2008). Plant lesions promote the rapid multiplication of *Escherichia coli* O157:H7 on postharvest lettuce. *Applied and Environmental Microbiology, 74*, 5285–5289.

Brandl, M. T., Cox, C. E., & Teplitski, M. (2013). *Salmonella* interactions with plants and their associated microbiota. *Phytopathology, 103*, 316–325.

Brodribb, T. J., & McAdam, S. A. (2011). Passive origins of stomatal control in vascular plants. *Science, 331*, 582–585.

Centers for Disease Control and Prevention. (2014). *List of selected multistate foodborne outbreak investigations: 2006–2014*. http://www.cdc.gov/foodsafety/outbreaks/multistate-outbreaks/outbreaks-list.html. Updated February 25, 2015. Accessed 07.03.15.

Cooley, M. B., Miller, W. G., & Mandrell, R. E. (2003). Colonization of *Arabidopsis thaliana* with *Salmonella enterica* and enterohemorrhagic *Escherichia coli* O157:H7 and competition by *Enterobacter asburiae*. *Applied and Environmental Microbiology, 69*, 4915–4926.

Deng, K., Wang, S., Rui, X., Zhang, W., & Tortorello, M. L. (2011). Functional analysis of ycfR and ycfQ in *Escherichia coli* O157:H7 linked to outbreaks of illness associated with fresh produce. *Applied and Environmental Microbiology, 77*, 3952–3959.

Dodd, A. N., Love, J., & Webb, A. A. (2005). The plant clock shows its metal: circadian regulation of cytosolic free Ca(2+). *Trends in Plant Science, 10*, 15–21.

Durrant, W. E., & Dong, X. (2004). Systemic acquired resistance. *Annual Review of Phytopathology, 42*, 185–209.

Enyedi, A. J., Yalpani, N., Silverman, P., & Raskin, I. (1992). Signal molecules in systemic plant resistance to pathogens and pests. *Cell, 70*, 879–886.

Erickson, M. C. (2012). Internalization of fresh produce by foodborne pathogens. *Annual Review of Food Science and Technology, 3*, 283–310.

Erickson, M. C., Webb, C. C., Diaz-Perez, J. C., Phatak, S. C., Silvoy, J. J., Davey, L., et al. (2010). Surface and internalized *Escherichia coli* O157:H7 on field-grown spinach and lettuce treated with spray-contaminated irrigation water. *Journal of Food Protection, 73*, 1023–1029.

Fink, R. C., Black, E. P., Hou, Z., Sugawara, M., Sadowsky, M. J., & Diez-Gonzalez, F. (2012). Transcriptional responses of *Escherichia coli* K-12 and O157:H7 associated with lettuce leaves. *Applied and Environmental Microbiology, 78*, 1752–1764.

Food and Drug Administration. (1998). *Guide to minimize microbial food safety hazards for fresh fruits and vegetables*. Washington, DC: Department of Health and Human Services.

Franz, E., van Hoek, A. H., Bouw, E., & Aarts, H. J. (2011). Variability of *Escherichia coli* O157 strain survival in manure-amended soil in relation to strain origin, virulence profile, and carbon nutrition profile. *Applied and Environmental Microbiology, 77*, 8088–8096.

Goudeau, D. M., Parker, C. T., Zhou, Y., Sela, S., Kroupitski, Y., & Brandl, M. T. (2013). The *Salmonella* transcriptome in lettuce and cilantro soft rot reveals a niche overlap with the animal host intestine. *Applied and Environmental Microbiology, 79*, 250–262.

Gu, G., Hu, J., Cevallos-Cevallos, J. M., Richardson, S. M., Bartz, J. A., & van Bruggen, A.. H. (2011). Internal colonization of *Salmonella enterica* serovar *typhimurium* in tomato plants. *PLoS One, 6*, e27340.

Haapalainen, M., van Gestel, K., Pirhonen, M., & Taira, S. (2009). Soluble plant cell signals induce the expression of the type III secretion system of *Pseudomonas syringae* and upregulate the production of pilus protein HrpA. *Molecular Plant Microbe Interactions, 22*, 282–290.

Hassan, A. N., & Frank, J. F. (2004). Attachment of *Escherichia coli* O157:H7 grown in tryptic soy broth and nutrient broth to apple and lettuce surfaces as related to cell hydrophobicity, surface charge, and capsule production. *International Journal of Food Microbiology, 96*, 103–109.

Hetherington, A. M., & Woodward, F. I. (2003). The role of stomata in sensing and driving environmental change. *Nature, 424*, 901–908.

Hirneisen, K. A., & Kniel, K. E. (2013). Comparative uptake of enteric viruses into spinach and green onions. *Food and Environmental Virology, 5*, 24–34.

Hirneisen, K. A., Sharma, M., & Kniel, K. E. (2012). Human enteric pathogen internalization by root uptake into food crops. *Foodborne Pathogens and Disease, 9*, 396–405.
Holden, N., Pritchard, L., & Toth, I. (2009). Colonization out with the colon: plants as an alternative environmental reservoir for human pathogenic enterobacteria. *FEMS Microbiology Reviews, 33*, 689–703.
Islam, M., Morgan, J., Doyle, M. P., Phatak, S. C., Millner, P., & Jiang, X. (2004). Persistence of *Salmonella enterica* serovar typhimurium on lettuce and parsley and in soils on which they were grown in fields treated with contaminated manure composts or irrigation water. *Foodborne Pathogens and Disease, 1*, 27–35.
Jones, J. D., & Dangl, J. L. (2006). The plant immune system. *Nature, 444*, 323–329.
Katagiri, F., Thilmony, R., & He, S. Y. (2002). The *Arabidopsis thaliana–Pseudomonas syringae* interaction. *The Arabidopsis Book*, 0–35.
Kenny, B., & Finlay, B. B. (1995). Protein secretion by enteropathogenic *Escherichia coli* is essential for transducing signals to epithelial cells. *Proceedings of the National Academy of Sciences of the United States of America, 92*, 7991–7995.
Kloepper, J. W., Rodriguez-Kabana, R., Zehnder, G. W., Murphy, J. F., Sikora, E., & Fernandez, C. (1999). Plant root–bacterial interactions in biological control of soilborne diseases and potential extension to systemic and foliar diseases. *Australasian Plant Pathology, 28*, 21–26.
Kroupitski, Y., Pinto, R., Brandl, M. T., Belausov, E., & Sela, S. (2009). Interactions of *Salmonella enterica* with lettuce leaves. *Journal of Applied Microbiology, 106*, 1876–1885.
Kumar, A. S., Lakshmanan, V., Caplan, J. L., Powell, D., Czymmek, K. J., Levia, D.. F., et al. (2012). Rhizobacteria *Bacillus subtilis* restricts foliar pathogen entry through stomata. *The Plant Journal, 72*, 694–706.
Kunkel, B. N., & Brooks, D. M. (2002). Cross talk between signaling pathways in pathogen defense. *Current Opinion in Plant Biology, 5*, 325–331.
Lakshmanan, V., Kitto, S. L., Caplan, J. L., Hsueh, Y. H., Kearns, D. B., Wu, Y. S., et al. (2012). Microbe-associated molecular patterns-triggered root responses mediate beneficial rhizobacterial recruitment in *Arabidopsis*. *Plant Physiology, 160*, 1642–1661.
Lim, J. A., Lee, D. H., & Heu, S. (2014). The interaction of human enteric pathogens with plants. *The Plant Pathology Journal, 30*, 109–116.
Lopez-Velasco, G., Tydings, H. A., Boyer, R. R., & Falkinham, J. O. (2012). Characterization of interactions between *Escherichia coli* O157:H7 with epiphytic bacteria in vitro and on spinach leaf surfaces. *International Journal of Food Microbiology, 153*, 351–357.
Macarisin, D., Patel, J., Bauchan, G., Giron, J. A., & Sharma, V. K. (2012). Role of curli and cellulose expression in adherence of *Escherichia coli* O157:H7 to spinach leaves. *Foodborne Pathogens and Disease, 9*, 160–167.
Madamanchi, N. R., & Kuc, J. (1991). Induced systemic resistance in plants. In G. T. Cole, & H. C. Hoch (Eds.), *Induced systemic resistance in plants* (pp. 347–362). New York: Plenum Press.
Markland, S. M., Bais, H. P., & Kniel, K. E. (2014). Effects of plant growth promoting rhizobacteria on the immune response of Romaine lettuce and spinach toward human bacterial pathogens. In *International association for food protection annual meeting, Indianapolis, IN* (pp. P1–85).
Martinez-Vaz, B. M., Fink, R. C., Diez-Gonzalez, F., & Sadowsky, M. J. (2014). Enteric pathogen–plant interactions: molecular connections leading to colonization and growth and implications for food safety. *Microbes Environment, 29*, 123–135.
Melotto, M., Underwood, W., Koczan, J., Nomura, K., & He, S. Y. (2006). Plant stomata function in innate immunity against bacterial invasion. *Cell, 126*(5), 969–980.
Milillo, S. R., Badamo, J. M., Boor, K. J., & Wiedmann, M. (2008). Growth and persistence of *Listeria monocytogenes* isolates on the plant model *Arabidopsis thaliana*. *Food Microbiology, 25*, 698–704.

Mur, L. A., Kenton, P., Atzorn, R., Miersch, O., & Wasternack, C. (2006). The outcomes of concentration-specific interactions between salicylate and jasmonate signaling include synergy, antagonism, and oxidative stress leading to cell death. *Plant Physiology, 140*, 249–262.

National Institutes of Health. *Arabidopsis thaliana*. Models for Biomedical Research. http://www.nih.gov/science/models/arabidopsis/. Accessed March 7, 2015.

Noel, J. T., Joy, J., Smith, J. N., Fatica, M., Schneider, K. R., Ahmer, B. M., et al. (2010). *Salmonella* SdiA recognizes N-acyl homoserine lactone signals from *Pectobacterium carotovorum* in vitro, but not in a bacterial soft rot. *Molecular Plant Microbe Interactions, 23*, 273–282.

Oberpichler, I., Rosen, R., Rasouly, A., Vugman, M., Ron, E. Z., & Lamparter, T. (2008). Light affects motility and infectivity of *Agrobacterium tumefaciens*. *Environmental Microbiology, 10*, 2020–2029.

Painter, J. A., Hoekstra, R. M., Ayers, T., Tauxe, R. V., Braden, C. R., Angulo, F.. J., et al. (2013). Attribution of foodborne illnesses, hospitalizations, and deaths to food commodities by using outbreak data, United States, 1998–2008. *Emerging Infectious Diseases, 19*, 407–415.

Pandey, S. P., & Somssich, I. E. (2009). The role of WRKY transcription factors in plant immunity. *Plant Physiology, 150*, 1648–1655.

Patel, J., & Sharma, M. (2010). Differences in attachment of *Salmonella enterica* serovars to cabbage and lettuce leaves. *International Journal of Food Microbiology, 139*, 41–47.

Plotnikova, J. M., Rahme, L. G., & Ausubel, F. M. (2000). Pathogenesis of the human opportunistic pathogen *Pseudomonas aeruginosa* PA14 in *Arabidopsis*. *Plant Physiology, 124*, 1766–1774.

Potnis, N., Soto-Arias, J. P., Cowles, K. N., van Bruggen, A. H., Jones, J. B., & Barak, J. D. (2014). *Xanthomonas perforans* colonization influences *Salmonella enterica* in the tomato phyllosphere. *Applied and Environmental Microbiology, 80*, 3173–3180.

Prigent-Combaret, C., Brombacher, E., Vidal, O., Ambert, A., Lejeune, P., Landini, P., et al. (2001). Complex regulatory network controls initial adhesion and biofilm formation in *Escherichia coli* via regulation of the csgD gene. *Journal of Bacteriology, 183*, 7213–7223.

Prithiviraj, B., Bais, H. P., Jha, A. K., & Vivanco, J. M. (2005). *Staphylococcus aureus* pathogenicity on *Arabidopsis thaliana* is mediated either by a direct effect of salicylic acid on the pathogen or by SA-dependent, NPR1-independent host responses. *The Plant Journal, 42*, 417–432.

Rahn, A., Beis, K., Naismith, J. H., & Whitfield, C. (2003). A novel outer membrane protein, Wzi, is involved in surface assembly of the *Escherichia coli* K30 group 1 capsule. *Journal of Bacteriology, 185*, 5882–5890.

Raina, S., Missiakas, D., Baird, L., Kumar, S., & Georgopoulos, C. (1993). Identification and transcriptional analysis of the *Escherichia coli* htrE operon which is homologous to pap and related pilin operons. *Journal of Bacteriology, 175*, 5009–5021.

Rastogi, G., Sbodio, A., Tech, J. J., Suslow, T. V., Coaker, G. L., & Leveau, J. H. (2012). Leaf microbiota in an agroecosystem: spatiotemporal variation in bacterial community composition on field-grown lettuce. *ISME Journal, 6*, 1812–1822.

Roden, L. C., & Ingle, R. A. (2009). Lights, rhythms, infection: the role of light and the circadian clock in determining the outcome of plant-pathogen interactions. *The Plant Cell, 21*, 2546–2552.

Römling, U., Bian, Z., Hammar, M., Sierralta, W. D., & Normark, S. (1998). Curli fibers are highly conserved between *Salmonella typhimurium* and *Escherichia coli* with respect to operon structure and regulation. *Journal of Bacteriology, 180*, 722–731.

Ross, A. F. (1961). Systemic acquired resistance induced by localized virus infections in plants. *Virology, 14*, 340–358.

Ryu, C. M., Farag, M. A., Hu, C. H., Reddy, M. S., Kloepper, J. W., & Pare, P. W. (2004). Bacterial volatiles induce systemic resistance in *Arabidopsis*. *Plant Physiology, 134*, 1017–1026.

Saldana, Z., Erdem, A. L., Schuller, S., Okeke, I. N., Lucas, M., Sivananthan, A., et al. (2009). The *Escherichia coli* common pilus and the bundle-forming pilus act in concert during the formation of localized adherence by enteropathogenic *E. coli*. *Journal of Bacteriology, 191*, 3451–3461.

Schenk, S. T., Hernandez-Reyes, C., Samans, B., Stein, E., Neumann, C., Schikora, M., et al. (2014). N-Acyl-homoserine lactone primes plants for cell wall reinforcement and induces resistance to bacterial pathogens via the salicylic acid/oxylipin pathway. *The Plant Cell, 26*, 2708–2723.

Schikora, A., Carreri, A., Charpentier, E., & Hirt, H. (2008). The dark side of the salad: *Salmonella typhimurium* overcomes the innate immune response of *Arabidopsis thaliana* and shows an endopathogenic lifestyle. *PLoS One, 3*(5), e2779.

Schikora, A., Schenk, S. T., Stein, E., Molitor, A., Zuccaro, A., & Kogel, K. H. (2011). N-acyl-homoserine lactone confers resistance toward biotrophic and hemibiotrophic pathogens via altered activation of AtMPK6. *Plant Physiology, 157*, 1407–1418.

Simko, I., Zhou, Y., & Brandl, M. T. (2015). Downy mildew disease promotes the colonization of romaine lettuce by *Escherichia coli* O157:H7 and *Salmonella enterica*. *BMC Microbiology, 15*, 19.

Sobsey, M. D., Dean, C. H., Knuckles, M. E., & Wagner, R. A. (1980). Interactions and survival of enteric viruses in soil materials. *Applied and Environmental Microbiology, 40*, 92–101.

Strawn, L. K., Fortes, E. D., Bihn, E. A., Nightingale, K. K., Grohn, Y. T., Worobo, R. W., et al. (2013). Landscape and meteorological factors affecting prevalence of three food-borne pathogens in fruit and vegetable farms. *Applied and Environmental Microbiology, 79*, 588–600.

Tabana, B. M., & Halkmanb, A. K. (2011). Do leafy green vegetables and their ready-to-eat [RTE] salads carry a risk of foodborne pathogens? *Anaerobe, 17*, 286–287.

Teplitski, M., Barak, J. D., & Schneider, K. R. (2009). Human enteric pathogens in produce: un-answered ecological questions with direct implications for food safety. *Current Opinion in Biotechnology, 20*, 166–171.

Wang, Q., & Kniel, K. E. (2015). In *Potential risks and prevention strategies for virus-contaminated seeds, sprouts, and microgreens, abstr microbial systems symposium*. Newark, DE: University of Delaware.

Wei, J., Jin, Y., Sims, T., & Kniel, K. E. (2010). Manure- and biosolids-resident murine norovirus 1 attachment to and internalization by Romaine lettuce. *Applied and Environmental Microbiology, 76*, 578–583.

Wei, J., Jin, Y., Sims, T., & Kniel, K. E. (2011). Internalization of murine norovirus 1 by *Lactuca sativa* during irrigation. *Applied and Environmental Microbiology, 77*, 2508–2512.

Wells, J. M., & Butterfield, J. E. (1997). *Salmonella* contamination associated with bacterial soft rot of fresh fruits and vegetables in the marketplace. *Plant Disease, 81*, 722–726.

Xicohtencatl-Cortes, J., Sanchez Chacon, E., Saldana, Z., Freer, E., & Giron, J. A. (2009). Interaction of *Escherichia coli O157:H7* with leafy green produce. *Journal of Food Protection, 72*, 1531–1537.

Yamazaki, A., Li, J., Hutchins, W. C., Wang, L., Ma, J., Ibekwe, A. M., et al. (2011). Commensal effect of pectate lyases secreted from *Dickeya dadantii* on proliferation of *Escherichia coli* O157:H7 EDL933 on lettuce leaves. *Applied and Environmental Microbiology, 77*, 156–162.

You, Y., Rankin, S. C., Aceto, H. W., Benson, C. E., Toth, J. D., & Dou, Z. (2006). Survival of *Salmonella enterica* serovar Newport in manure and manure-amended soils. *Applied and Environmental Microbiology, 72*, 5777–5783.

Zeier, J., Pink, B., Mueller, M. J., & Berger, S. (2004). Light conditions influence specific defence responses in incompatible plant-pathogen interactions: uncoupling systemic resistance from salicylic acid and PR-1 accumulation. *Planta, 219*, 673–683.

Zhang, G., Ma, L., Beuchat, L. R., Erickson, M. C., Phelan, V. H., & Doyle, M. P. (2009). Lack of internalization of *Escherichia coli* O157:H7 in lettuce (*Lactuca sativa* L.) after leaf surface and soil inoculation. *Journal of Food Protection, 72*, 2028–2037.

CHAPTER SIX

Early Responses in the Root–Rhizobia Interaction

Oswaldo Valdés-López*[,1], Georgina Hernández[§,1]
*Laboratorio de Genómica Funcional de Leguminosas, FES Iztacala, Universidad Nacional Autónoma de México (UNAM), Tlalnepantla, Estado de México, Mexico
[§]Centro de Ciencias Genómicas, Universidad Nacional Autónoma de México (UNAM), Cuernavaca, Morelos, Mexico
[1]Corresponding authors: E-mail: oswaldo_valdesl@yahoo.com.mx; gina@ccg.unam.mx

Contents

1. Innate Immunity Must Be Regulated during the RN Symbiosis 139
2. Searching for Missing Regulators of the Early Responses of the RN Symbiosis 141
3. miRNAs as Regulators of the Early Events of the RN Symbiosis 143
4. Conclusions and Remarks 144
Acknowledgements 144
References 145

Abstract

The symbiotic interaction between legumes and nitrogen-fixing bacteria has a high relevance at ecological and agronomical level. This mutualistic interaction has been extensively studied, over the years. However, recent experimental evidence suggests that there are many gaps that have to be filled-up. For example, large-scale studies indicate the existence of several yet-uncharacterized regulators of the very early events of the legume–rhizobia interaction. In addition, it has been reported that microRNAs play a critical role in the regulation of the early responses. Likewise, different studies suggest that a very critical step to establish this symbiosis is the negative regulation of the plant innate immunity. In this chapter, we will summarize and discuss the most recent discoveries on the control of the early responses of the symbiosis between legumes and nitrogen-fixing bacteria.

Like other plants, legumes can circumvent phosphorus and nitrogen (N_2) deficiencies by establishing mutualistic associations with arbuscular mycorrhiza (AM) fungi. Additionally, legumes can obtain N_2 through mutualistic associations with nitrogen-fixing rhizobia (here after referred as root-nodule symbiosis, RN) (Gutjahr & Parniske, 2013; Venkateshwaran, Volkening, Sussman, & Ané, 2013). These interactions help to reduce soil and water contamination by reducing the need for synthetic fertilizers (Graham & Vance, 2003). To establish AM or RN symbioses, a mutual recognition

Advances in Botanical Research, Volume 75
ISSN 0065-2296
http://dx.doi.org/10.1016/bs.abr.2015.07.003

process between the microorganisms and the legumes is required (Gutjahr & Parniske, 2013; Venkateshwaran et al., 2013). The development of RN is characterized by a high level of host specificity and is regulated by mutual exchange of diffusible signal molecules (Dénarié, Debellé & Promé, 1996; Long, 2001). Flavonoids and isoflavonoids present in the root exudates act as specific inducers of the nodulation genes in the compatible rhizobia (Dénarié, et al., 1996; Hartwig, Maxwell, Joseph, & Philips, 1990). These bacterial genes are responsible for the production and secretion of bacterial lipochitooligosaccharides known as Nod factors (NFs). NFs are perceived by the host plant via the LysM-domain receptor-like kinases Nod Factor Perception (NFP) and LYK3 (Arrighi et al., 2006; Ben Amor et al., 2003). Upon NFs perception, a molecular (metabolic, transcriptional and posttranslational modifications) and developmental reprogramming occurs (Oldroyd, 2013; Venkateshwaran et al., 2013). For example, in the early stages of this interaction, the invading rhizobia trigger root hair deformation, depolarization of the membrane, cytoskeleton reorientation, calcium spiking, cortical cell division, the expression of several genes and posttranslational modification (i.e. phosphorylation and ubiquitination) of a variety of proteins (Ehrhardt, Wais, & Long, 1996; Cardenas et al., 2000; Felle, Kondorosi, Kondorosi, & Schultze, 1998; Hervé, Lefebvre, & Cullimore, 2011; Popp & Ott, 2011; Stacey, Libault, Brechenmacher, Wan, & May, 2006). These modifications are essential to mount a functional RN symbiosis.

Forward and reverse genetics as well as 'omics' (transcriptomics, metabolomics and phosphoproteomics) approaches have significantly contributed to understand RN symbiosis in the model legumes plants *Medicago truncatula* (Medicago) and *Lotus japonicus* (Lotus) and in less degree in the crop legumes common bean (*Phaseolus vulgaris*) and soybean (*Glycine max*). By combining these approaches, different symbiotic genes have been identified and characterized (Limpens et al., 2013; O'Rourke et al., 2014; Takanashi et al., 2012; Zhu, Dahmen, Stacey, & Cheng, 2013). However, most of these studies were performed in relative late stages (e.g. 24 or 48 h after inoculation with the compatible rhizobia or NFs) of the RN symbiosis. This fact can negatively affect the discovery of new players that may control critical early responses of the RN symbiosis. For example, two recent phosphoproteomic studies on Medicago and soybean have revealed that about 90 proteins are phosphorylated after 1 h of NFs treatment or rhizobia inoculation, respectively (Nguyen et al., 2012; Rose et al., 2012). Additionally, a transcriptomic analysis on Medicago roots indicated that 140 genes were differentially regulated after 1 h NFs treatment (Rose et al., 2012). This evidence clearly indicates that there are additional players for the control of critical steps of the RN symbiosis that remain unidentified. This chapter summarizes and discusses the most recent discoveries on the control of the early responses of the RN symbiosis. Events related to nodule development or function, have been reviewed elsewhere.

1. INNATE IMMUNITY MUST BE REGULATED DURING THE RN SYMBIOSIS

Plants detect different potential pathogens through the perception of Microbe-Associated Molecular Patterns (MAMPs), such as flagellin (flg), chitin and lipopolysaccharides, via different receptors located at the membrane surface (Zipfel, 2014). Upon MAMPs perception, a defence response, known as MAMPs-triggered immunity (MTI), is mounted (Newman, Sundelin, Nielsen, & Erbs, 2013). MTI confers plant resistance against different nonadapted pathogens (Jones & Dangl, 2006; Newman et al., 2013). Some of the MTI responses include production of reactive oxygen- or nitrogen species (ROS or NOS, respectively), cytosolic calcium release, activation of mitogen-activated protein kinases, and the expression of different defence-related genes (Bigeard, Colcombet, & Hirt, 2015). To date, no rhizobial MAMPs have been identified as being active in legumes (Gourion, Berrabah, Ratet, & Stacey, 2015). However, different transcriptional and proteomics analyses on Medicago, Lotus and soybean indicate that several defence-related genes/proteins are differentially regulated during the first hours (1–12 h) of the legume–rhizobia interactions (Kouchi et al., 2004; Libault et al., 2010; Lohar et al., 2006). Interestingly, 24 h after rhizobial inoculation, the expression of these genes/proteins returns to basal expression. These evidence suggest that: (1) at the beginning of symbiosis, rhizobia can be recognized as potential pathogen and thus activate the MTI response; and (2) for successful rhizobia colonization, it is necessary to reduce MTI.

There is evidence that supports these two hypotheses. For example, some reports indicate that rhizobia-derived exopolysaccharides (EPS), such as succinoglycans, play some role in the suppression of the MTI at early stages of the RN symbiosis (Aslam et al., 2008; Jones et al., 2008). This is supported by the observation that Medicago plants inoculated with a *Sinorhizobium meliloti exoY* mutant strain, which is deficient for succinoglycan synthesis, showed strong induction of defence-related genes. Additionally, these plants showed less nodule number as compared to plants inoculated with wild-type

S. meliloti. It is proposed that this EPS might suppress the MTI by chelating cytosolic calcium, which is required to mount a functional defence response.

Pathogenic bacteria use the Type III secretion system (T3SS) to deliver protein effectors and block the MTI response (Büttner & He, 2009). By using this strategy, pathogenic bacteria can colonize the plant-host. It has been reported that some rhizobia strains use the T3SS for symbiotic interactions with legume plants (Bartsev et al., 2004; Skorpil et al., 2005; Xin et al., 2012). For example, *Rhizobium* sp. strain NGR234 releases the effectors Nodulation Outer Proteins L (NopL) and NopP (Bartsev et al., 2004; Skorpil et al., 2005) into the cell roots of the legume-host plant. Interestingly, *Tephrosia vogelii*, *Crotalaria juncea* and Lotus plants inoculated with mutant strains of NGR234 unable to release NopL and/or NopP show less nodule number than those plants inoculated with wild-type NGR234 (Bartsev et al., 2004; Skorpil et al., 2005). In addition, it was reported that *Bradyrhizobium elkanii* use the T3SS to establish RN symbiosis with soybean (Okazaki, Kaneko, Sato, & Saeki, 2013). Even a *B. elkanii nodC* mutant strain, which is unable to synthesize NFs, is able to form functional nodules. Interestingly, both wild-type- and *nodC B. elkanii* strains are able to form functional nodule in the *nfr1/lyk3* soybean mutant plant that does not express one component of the NFs receptor (Okazaki et al., 2013). These observations suggest that protein effectors released by rhizobia have a positive role in the establishment of the RN symbiosis. Perhaps one of the roles of these protein effectors is to negatively regulate the MTI in the early stages of the legume–rhizobia interaction. Furthermore, this evidence suggests that the establishment of the RN symbiosis mediated by protein effectors (e.g. NopL and NopP) is independent of the perception of NFs.

Lately, Liang et al. (2013) reported that NFs reduce ROS production triggered by the peptide flg22 in soybean plants. Likewise, it was reported that NFs significantly reduce MTI-related responses (e.g. ROS production and defence-related genes) triggered by flg22 in non-legume plants, such as corn and the model plant *Arabidopsis thaliana* (Liang et al., 2013). These suggest that NFs play an important role in the reduction of the MTI in very early stages of the RN symbiosis. Although the evidence presented in this section clearly indicate that the reduction of the MTI is a critical step to establish functional RN symbiosis, the precise mechanism/role exerted by EPS, protein effectors and NFs remains to be understood. It will be interesting to identify the targets of the protein effectors released by the rhizobia and to determine which modules of the MTI are suppressed by NFs.

2. SEARCHING FOR MISSING REGULATORS OF THE EARLY RESPONSES OF THE RN SYMBIOSIS

As mentioned in previous sections, a molecular- and developmental reprogramming, including transcriptional reprogramming is necessary to establish a functional RN symbiosis. Transcriptomic analyses have led to the identification of hundreds of differentially regulated genes. For example, Czaja et al. (2012) reported that around 900 genes were differentially regulated 24 h after NFs treatment of Medicago roots. In a different study, it was observed that only 158 genes were differentially regulated 24 h after inoculation of Medicago roots with *S. meliloti* (Breakspear et al., 2014). With the transcriptomic information available so far, it has been proposed that the transcriptional reprogramming can be classified in four stages: (1) early signalling events and/or bacterial infection; (2) plant cell differentiation-dependent or (3) independent of bacteroid differentiation and (4) nitrogen fixation (Moreau et al., 2007). However, it is likely that each of these transcriptional stages could have complex subtranscriptional programming or transcriptional modules.

Certainly, transcriptomic analyses have led to identification of new RN symbiosis-related genes, several of them with potential roles as regulator for certain stages of this process. For several years, transcriptional responses that occur after 12- or even 24 h after *S. meliloti* inoculation or NFs treatment were considered as early responses. However, it has been speculated that transcriptional reprogramming might occur immediately after NFs perception. Thus, by doing transcriptional analyses in just "late" stages (e.g. 12 h after rhizobia- or NF treatment) it is likely that several transcriptional responses and potential regulators of the early stages (e.g. 1 h after rhizobia- or NF treatment) remain unknown. To address this, Rose et al. (2012) performed a transcriptional analysis of Medicago roots treated with NFs for 1 h. In this study, 140 differentially regulated genes were identified, the majority of these encoded different transcription factors, protein kinases, cell wall-modifying enzymes, defence-related proteins, transporters and flavonoid biosynthesis proteins. Additionally, the ubiquitin E3 ligase PUB1, which interacts with the Medicago NFs receptor LYK3 and negatively regulates nodulation, was also upregulated after 1 h of NFs treatment (Mbengue et al., 2010; Rose et al., 2012). In an independent study, the transcriptional responses of Medicago roots treated with NFs for 6 h were analyzed thus identifying 600 differentially regulated genes (Czaja et al., 2012).

During the legume–rhizobia interaction, a variety of proteins undergo posttranslational modifications, such as phosphorylation (Oldroyd, 2013; Venkateshwaran et al., 2013). The study on different mutant plants that do not express certain protein kinases (e.g. the NFs receptor LYK3 and the kinase DMI2/SYMRK) clearly shows that protein phosphorylation is a critical step for the establishment of the RN symbiosis (Oldroyd, 2013; Rose et al., 2012; Venkateshwaran et al., 2013). Likewise, different phosphoproteomic analyses have demonstrated that protein phosphorylation occurs in almost the entire RN symbiotic process (Oldroyd, 2013). Interestingly, there are two reports that demonstrate that protein phosphorylation occurs as early as 1 h after the NFs signal is perceived (Nguyen et al., 2012; Rose et al., 2012). Both transcriptomic and phosphoproteomic approaches have clearly shown that molecular reprogramming occurs right after the NFs perception and also suggest the existence of different players that might control in some way early stages of the RN symbiosis.

Certainly both transcriptomic and phosphoproteomic analyses have helped to identify several new potential players of the early stages of the RN symbiosis. However, the role of these new players remains unknown. There are two main limitations for deciphering the function of these potential regulators. The first one is related to the identification of the biological relevance of the candidate gene, and the second one is the lack of criteria to select the right candidate. To solve these two limitations, some researchers have suggested to integrate both transcriptomic and phosphoproteomic data and build regulatory networks (Zhu et al., 2013). Applying regulatory network analysis in combination with functional enrichment analysis, it will be possible to identify which gene could have a potential regulatory role and which biological function it might control. Additionally, it has been demonstrated that several members of the order Brassicales have lost several genes involved in the establishment of symbiosis with rhizobia (Delaux et al., 2014). Thus, it has been also proposed that absent genes in Brassicales (e.g. *A. thaliana*) genome and whose expression is regulated in response to rhizobia or by purified NFs might play a role in the RN symbiosis. For example, the *VAPIRYN* gene, which is absent in the *Arabidopsis* genome, play a critical role in the establishment of the legume–rhizobia symbiosis. Thus, combining regulatory network- and synteny (to determinate whether a gene is absent in a certain genome) analyses could help to select and identify new regulators of the RN symbiosis.

3. miRNAs AS REGULATORS OF THE EARLY EVENTS OF THE RN SYMBIOSIS

Like other physiological processes, the RN symbiosis has to be finely controlled at different levels. This symbiosis is controlled at the transcriptional level by different transcription factors, such as Nodule Inception (NIN), Nodulation Signaling Pathway1 (NSP1)/NSP2 and NF-YA/B (formerly called as HAP2) (Hirsch et al., 2009; Kaló et al., 2005; Smit et al., 2005). Recently, it has been demonstrated that RN symbiosis is also regulated at posttranscriptional level by microRNAs (miRNAs) (Simon, Meyers, & Sherrier, 2009). For example, miR166 and miR169 control the root and nodule development in Medicago (Boualem et al., 2008; Combier et al., 2006). Since the legume–rhizobia interaction is a multilayer process, it is likely that additional miRNAs are involved. To address this, different research groups have performed high-throughput sequencing analysis of small RNAs in different stages of the RN symbiosis (De Luis et al., 2012; Formey et al., 2014; Reynoso, Blanco, Bailey-Serres, Crespi, & Zanetti, 2013; Subramanian et al., 2008). These studies indicated that most of the miRNAs that are regulated during this symbiotic interaction belong to miRNA families conserved across different plant species (Subramanian et al., 2008). However, these have also led to the identification of novel miRNA families, including some that are only present in legumes (Subramanian et al., 2008). Some of these legume-specific miRNA families play a role in nodule development. For instance, the miRNAs miR482, miR1512 and miR1515 are positive regulators of nodule development in soybean (Li, Deng, Wu, Subramanian, & Yu, 2010).

In an attempt to identify miRNAs that could have some role in the control of early responses of the RN symbiosis, De Luis et al. (2012) massively sequenced miRNAs to identify differentially regulated miRNAs in Lotus roots inoculated with rhizobia for 3 h.

Unfortunately, no significant changes were found in this study. However, preliminary observations indicate that the abundance of the miRNAs miR2873c, miR2665, miR5207 and miR396c is significantly modified in Medicago roots treated for 1 h with NFs (Ané, J.M., personal communication). Interestingly, putative targets of these miRNAs are E3 ligases. Two of these E3 ligases are SINA2 and SINA5, which play a role in nodule development in Medicago (Den Herder et al., 2008). Additionally, it has been reported that the transcription factor NSP2, which acts immediately after the decoding of the calcium spiking, is the target gene of the miRNA miR171h (Hofferek, Medrinna, Gaude, Krajinski, & Devers, 2014). By overexpressing miR171h, it was demonstrated this miRNA is a negative regulator of the RN symbiosis (Hofferek et al., 2014). Most recently, we demonstrated that the miR172c, which has the transcription factor AP2-1 as target gene, indirectly regulate the expression of the transcription factors NF-YA1, NSP2 and CYCLOPS as well as the gene *FLOT2*, all of them acting immediately after the decoding of the calcium spiking (Nova-Franco et al., 2015). This observation suggests that miRNAs can regulate some of the earliest responses of the RN symbiosis. It is likely that another miRNAs, or even another type of noncoding small RNAs, participate in the regulation of early responses of the RN symbiosis. However, the experimental strategy to identify them must be changed. For instance, a strategy using only root hairs that are single cell and where the NFs are perceived has been developed to avoid "signal dilution" that is highly frequently in those studies performed in whole roots (Libault et al., 2010). Likewise, different studies have demonstrated that purified NFs induce quick symbiotic responses than the whole rhizobia. Thus working with purified NFs and root hairs is a promising strategy to identify different yet-unidentified regulators, including miRNAs, of the early responses of the RN symbiosis.

4. CONCLUSIONS AND REMARKS

Although that transcriptomic (for both coding and noncoding RNAs) and phosphoproteomics analyses have demonstrated that the molecular reprogramming in response to rhizobia initiates as early as 1 h, there are gaps that have to be filled-up to better understand this symbiosis. For instance, it is important to implement strategies to identify potential new regulators. Once identified and characterized the new regulators of the early response of this symbiosis, it will be interesting to see whether this knowledge can be harnessed to make possible that important crop plant, such as corn, can interact with rhizobia and reduce the dependency or artificial N_2 fertilizer.

ACKNOWLEDGEMENTS

O.V.L research is supported by grant from CONACyT (grant number: 219759) and PAPCA-FES Iztacala UNAM (grant number: FESI-DIP-PAPCA-2014-3). O.V.L. and G.H. research projects are supported by PAPIIT-UNAM (grant numbers: IA203815 and IN202213, respectively).

REFERENCES

Arrighi, J. F., Barre, A., Ben Amor, B., Bersoult, A., Soriano, L. C., Mirabella, R., et al. (2006). The *Medicago truncatula* lysine [corrected] motif-receptor-like kinase gene family includes NFP and new nodule-expressed genes. *Plant Physiology, 142*, 265–279.

Aslam, S. N., Newman, M. A., Erbs, G., Morrissey, K. L., Chinchilla, D., Boll, T., et al. (2008). Bacterial polysaccharides suppress induced innate immunity by calcium chelation. *Current Biology*, 1078–1083.

Bartsev, A. V., Deakin, W. J., Boukli, N. M., McAlvin, C. B., Stacey, G., Malnoë, P., et al. (2004). NopL, an effector protein of *Rhizobium* sp. NGR234, thwarts activation of plant defense reactions. *Plant Physiology, 134*, 871–879.

Ben Amor, B., Shaw, S. L., Oldroyd, G. E., Maillet, F., Penmetsa, R. V., Cook, D., et al. (2003). The NFP locus of *Medicago truncatula* controls an early step of nod factor signal transduction upstream of a rapid calcium influx and root hair deformation. *Plant Journal, 34*, 495–506.

Bigeard, F., Colcombet, J., & Hirt, H. (2015). Signaling mechanism in pattern-triggered immunity (PTI). *Molecular Plant, 8*, 521–539.

Boualem, A., Laporte, P., Jovanovic, M., Laffont, C., Plet, J., Combier, J. P., et al. (2008). MicroRNA166 controls root and nodule development in *Medicago truncatula. Plant Journal, 54*, 876–887.

Breakspear, A., Liu, C., Roy, S., Stacey, N., Rogers, C., Trick, M., et al. (2014). The root hair "infectome" of *Medicago truncatula* uncovers changes in cell cycle genes and reveals a requirement for auxin signaling in rhizobial infection. *The Plant Cell, 26*, 4680–4701.

Büttner, D., & He, S. Y. (2009). Type III protein secretion in plant pathogenic bacteria. *Plant Physiology, 150*, 1656–1664.

Cardenas, L., Holdaway-Clarke, T. L., Sanchez, F., Quinto, C., Feijo, J. A., Kunkel, J. G., et al. (2000). Ion changes in legume root hairs responding to Nod factors. *Plant Physiology, 123*, 443–452.

Combier, J. P., Frugier, F., de Billy, F., Boualem, A., El-Yahyaoui, F., Moreau, S., et al. (2006). *MtHAP2-1* is a key transcriptional regulator of symbiotic nodule development regulated by microRNA169 in *Medicago truncatula. Genes Development, 20*, 3084–3088.

Czaja, L. F., Hogekamp, C., Lamm, P., Maillet, F., Martinez, E. A., Samin, E., et al. (2012). Transcriptional responses toward diffusible signals from symbiotic microbes reveal *MtNFP*- and *MtDMI3*-dependent reprogramming of host gene expression by arbuscular mycorrhizal fungal lipochitooligosaccharides. *Plant Physiology, 159*, 1671–1685.

De Luis, A., Markmann, K., Cognat, V., Holt, D. B., Charpentier, M., Parniske, M., et al. (2012). Two MicroRNAs linked to nodule infection and nitrogen-fixing ability in the legume *Lotus japonicum. Plant Physiology, 160*, 2137–2154.

Delaux, P. M., Varala, K., Edger, P. P., Coruzzi, G. M., Pires, J. C., & Ané, J. M. (2014). Comparative phylogenomics uncovers the impact of symbiotic associations on host genome evolution. *PLoS Genetics, 10*, e1004487.

Den Herder, G., De Keyser, A., De Rycke, R., Rombauts, S., Van de Velde, W., Clemente, M. R., et al. (2008). Seven in absentia proteins affects plant growth and nodulation in *Medicago truncatula. Plant Physiology, 148*, 369–382.

Dénarié, J., Debellé, F., & Promé, J. C. (1996). *Rhizobium* lipo-chitooligosaccharide nodulation factors: signaling molecules mediating recognition and morphogenesis. *Annual Review of Biochemistry, 65*, 503–535.

Ehrhard, D. W., Wais, R., & Long, S. R. (1996). Calcium spiking in plant root hairs responding to *Rhizobium* nodulation signals. *Cell, 85*, 673–681.

Felle, H. H., Kondorosi, E., Kondorosi, A., & Schultze, M. (1998). The role of ion fluxes in Nod factor signalling in *Medicago sativa. Plant Journal, 13*, 455–463.

Formey, D., Sallet, E., LelandaisBrière, C., Ben, C., Bustos-Sanmamed, P., Niebel, A., et al. (2014). The small RNA diversity from *Medicago truncatula* roots under biotic interactions evidences the environmental plasticity of the miRNAome. *Genome Biology, 15*, 457.
Gourion, B., Berrabah, F., Ratet, P., & Stacey, G. (2015). *Rhizobium*-legume symbioses: the crucial role of plant immunity. *Trends in Plant Sciences, 20*, 186–194.
Graham, P. H., & Vance, C. P. (2003). Legumes: importance and constrains to greater use. *Plant Physiology, 131*, 872–877.
Gutjahr, C., & Parniske, M. (2013). Cell and development biology of arbuscular mycorrhiza symbiosis. *Annual Review of Cell and Developmental Biology, 29*, 593–617.
Hartwig, U. A., Maxwell, C. A., Joseph, C. M., & Philips, D. A. (1990). Chrysoeriol and luteolin released from alfalfa seeds induce *nod* genes in *Rhizobium meliloti*. *Plant Physiology, 92*, 116–122.
Hervé, C., Lefebvre, B., & Cullimore, J. (2011). How many E3 ubiquitin ligase are involved in the regulation of nodulation? *Plant Signaling and Behavior, 6*, 660–664.
Hirsch, S., Kim, J., Muñoz, A., Heckmann, A. B., Downie, J. A., & Oldroyd, G. E. (2009). GRAS proteins form a DNA binding complex to induce gene expression during nodulation signaling in *Medicago truncatula*. *The Plant Cell, 21*, 545–557.
Hofferek, V., Medrinna, A., Gaude, N., Krajinski, F., & Devers, E. A. (2014). Mir171h restricts root symbioses and shows like its target *NSP2* a complex transcriptional regulation in *Medicago truncatula*. *BMC Plant Biology, 14*, 199.
Jones, D. G., & Dangl, J. L. (2006). The plant immune system. *Nature, 444*, 323–329.
Jones, K. M., Sharapova, N., Lohar, D. P., Zhang, J. Q., VandenBosch, K. A., & Walker, G. (2008). Differential response of the plant *Medicago truncatula* to its symbiont *Sinorhizobium meliloti* or an exopolysaccharide-deficient mutant. *Proceedings of the National Academy of Sciences of the United States of America, 105*, 704–709.
Kaló, P., Gleason, C., Edwards, A., Marsh, J., Mitra, R. M., Hirsch, S., et al. (2005). Nodulation signaling in legumes requires NSP2, a member of the GRAS family of transcriptional regulators. *Science, 308*, 1786–1789.
Kouchi, H., Shimomura, K., Hata, S., Hirota, A., Wu, G. J., & Kumagai, H. (2004). Large-scale analysis of gene expression profiles during early stages of root nodule formation in a model legume, *Lotus japonicum*. *DNA Research, 11*, 253–274.
Li, H., Deng, Y., Wu, T., Subramanian, S., & Yu, O. (2010). Misexpression of miR482, miR1512, and miR1515 increases soybean nodulation. *Plant Physiology, 153*, 1759–1770.
Liang, Y., Cao, Y., Tanaka, K., Thibivilliers, S., Wan, J., Choi, J., et al. (2013). Nonlegumes respond to rhizobia Nod factors by suppressing the innate immune response. *Science, 341*, 1384–1387.
Libault, M., Farmer, A., Brechenmacher, L., Drnevich, J., Langley, R. J., Bilgin, D. D., et al. (2010). Complete transcriptome of the soybean root hair cell, a single-cell model, and its alteration in response to *Bradyrhizobium japonicum* infection. *Plant Physiology, 152*, 541–552.
Limpens, E., Moling, S., Hooiveld, G., Pereira, P. A., Bisseling, T., Becker, J. D., et al. (2013). Cell- and tissue-specific transcriptome analyses of *Medicago truncatula* root nodule. *PLoS One, 8*, 64377.
Lohar, D. P., Sharapova, N., Endre, G., Peñuela, S., Samac, D., Town, D., et al. (2006). Transcript analysis of early nodulation events in *Medicago truncatula*. *Plant Physioloy, 140*, 221–234.
Long, S. R. (2001). Genes and signals in the *Rhizobium*-legume symbiosis. *Plant Physiology, 125*, 69–72.
Mbengue, M., Camut, S., de Carvalho-Nieble, F., Deslandes, L., Froidure, S., Klaus-Heisen, D., et al. (2010). The *Medicago truncatula* E3 ubiquitin ligase PUB1 interacts with the LYK3 symbiotic receptor and negatively regulates infection and nodulation. *The Plant Cell, 22*, 3474–3488.

Moreau, S., Verdenaud, M., Ott, T., Letort, S., de Billy, F., Niebel, A., et al. (2007). Transcription reprogramming during root nodule development in *Medicago truncatula*. *PLoS One, 6*, e16463.

Newman, M. A., Sundelin, T., Nielsen, J. T., & Erbs, E. (2013). MAMP (microbe-associated molecular pattern) triggered immunity in plants. *Frontiers in Plant Sciences, 4*, 139.

Nguyen, T. H. N., Brechenmacher, L., Aldrich, J. T., Clauss, T. R., Gritsenko, M. A., Hixson, K. K., et al. (2012). Quantitative phosphoproteomic analysis of soybean root hairs inoculated with *Bradyrhizobium japonicum*. *Molecular and Cellular Proteomics, 11*, 1140–1155.

Nova-Franco, B., Iñiguez, L. P., Valdés-López, O., Alvarado-Affantranger, X., Leija, A., Fuentes, S. I., et al. (2015). The micro-RNA172c - APETALA2-1 node as a key regulator of the common bean-rhizobia nitrogen fixation symbiosis. *Plant Physiology, 168*, 273–291.

Okazaki, S., Kaneko, T., Sato, S., & Saeki, K. (2013). Hijacking of leguminous nodulation signaling by the rhizobial type III secretion system. *Proceedings of the National Academy of Sciences of the United States of America, 10*, 17131–17136.

Oldroyd, G. E. D. (2013). Speak, friend and enter: signalling systems that promote beneficial symbiotic associations in plants. *Nature Reviews Microbiology, 11*, 252–263.

O'Rourke, J. A., Iñiguez, L. P., Fu, F., Bucciarelli, B., Miller, S. S., Jackson, S. A., et al. (2014). An RNA-seq based gene expression atlas of the common bean. *BMC Genomics, 15*, 866.

Popp, C., & Ott, T. (2011). Regulation of signal transduction and bacterial infection during root nodule symbiosis. *Current Opinion in Plant Biology, 14*, 458–467.

Reynoso, M. A., Blanco, F. A., Bailey-Serres, J., Crespi, M., & Zanetti, M. E. (2013). Selective recruitment of mRNAs and miRNAs to polyribosomes in response to rhizobia infection in *Medicago truncatula*. *Plant Journal, 73*, 289–301.

Rose, C. M., Venkateshwaran, M., Volkening, J. D., Grimsrud, P. A., Maeda, J., Bailey, D. J., et al. (2012). Rapid phosphoproteomic and transcriptomic changes in the rhizobia-legume symbiosis. *Molecular and Cellular Proteomics, 11*, 724–744.

Simon, S. A., Meyers, B. C., & Sherrier, D. J. (2009). MicroRNAs in the rhizobia legume symbiosis. *Plant Physiology, 151*, 1002–1008.

Skorpil, P., Saad, M. M., Boukli, N. M., Kobayashi, H., Ares-Orpel, F., Broughton, W. J., et al. (2005). NopP, a phosphorylated effector of *Rhizobium* sp. strain NGR234, is a major determinant of nodulation of the tropical legumes *Flemingia congesta* and *Tephrosia vogelii*. *Molecular Microbiology, 57*, 1304–1317.

Smit, P., Raedts, J., Portyanko, V., Debellé, F., Gough, C., Bisseling, T., et al. (2005). NSP1 of the GRAS protein family is essential for rhizobial Nod factor-induced transcription. *Science, 308*, 1789–1791.

Stacey, G., Libault, M., Brechenmacher, L., Wan, J., & May, G. D. (2006). Genetics and functional genomics of legume nodulation. *Current Opinion in Plant Biology, 9*(2), 110–121.

Subramanian, S., Fu, Y., Sunkar, R., Barbazuk, W. B., Zhu, J. K., & Yu, O. (2008). Novel and nodulation-regulated microRNAs in soybean roots. *BMC Genomics, 9*, 160.

Takanashi, K., Takanashi, H., Sakurai, N., Sugiyama, A., Suzuki, H., Shibata, D., et al. (2012). Tissue-specific transcriptome analysis in nodules of *Lotus japonicus*. *Molecular Plant-Microbe Interactions, 25*, 869–876.

Venkateshwaran, M., Volkening, J. D., Sussman, M. R., & Ané, J. M. (2013). Symbiosis and the social network of higher plants. *Current Opinion in Plant Biology, 16*, 118–127.

Xin, D. W., Liao, S., Xie, Z. P., Hann, D. R., Steinle, L., Boller, T., et al. (2012). Functional analysis of NopM, a nobel E3 ubiquitin ligase (NEL) domain effector of *Rhizobium* sp. strain NGR234. *PLoS Pathogens, 8*, e1002707.

Zhu, M., Dahmen, J. L., Stacey, G., & Cheng, J. (2013). Predicting gene regulatory networks of soybean nodulation from RNA-Seq transcriptomic data. *BMC Bioinformatics, 14*, 278.

Zipfel, C. (2014). Plant pattern-recognition receptors. *Trends in Immonology, 35*, 345–351.

CHAPTER SEVEN

The Implications of Polyploidy for the Evolution of Signalling in Rhizobial Nodulation Symbiosis

Adrian F. Powell*, Jeff J. Doyle†,1
*Section of Plant Biology, School of Integrative Plant Science, Cornell University, Ithaca, NY, USA
†Section of Plant Breeding and Genetics and Section of Plant Biology, School of Integrative Plant Science, Cornell University, Ithaca, NY, USA
[1]Corresponding author: E-mail: jjd5@cornell.edu

Contents

1. Introduction 150
2. Nodulation Signalling: Determinants of Symbiotic Interactions 152
3. WGD as a Process Altering Plant Signalling, Biotic Interactions and Ecological and Geographical Ranges 158
4. Potential Implications of Polyploidy for Nodulation Signalling 162
   4.1 Flavonoids 163
   4.2 Nod Factor Receptors 167
   4.3 Resistance Genes 167
   4.4 Additional Avenues for Effects of Polyploidy on Signalling Mechanisms via Plant Growth and Cytokinins 168
5. Evidence of Roles for Polyploidy in the Evolution of Nodulation and Rhizobial Signalling 170
6. Current and Future Studies of Polyploidy and Nodulation: Challenges and Opportunities 175
   6.1 Studying Effects at Several Timescales: Natural Paleopolyploids, Neopolyploids and Synthetics 176
   6.2 Challenges of Genotype-By-Genotype Interactions 178
7. Conclusion 179
References 180

Abstract

Nodulation is a symbiotic interaction between soil bacteria and plant hosts, most notably between rhizobia and legumes. This interaction is important for plant hosts, since it enables them to access atmospheric nitrogen made available by the bacteria. The establishment of this symbiosis requires several levels of communication and signalling between the plant host and the bacterial symbiont. The evolution of nodulation, and of the signalling mechanisms involved, appears to have involved recruitment of

Advances in Botanical Research, Volume 75
ISSN 0065-2296
http://dx.doi.org/10.1016/bs.abr.2015.09.007

genes from pre-existing processes. Polyploidy, or whole genome duplication, has the capacity to provide duplicate genes that can be recruited for novel functions, and genomic analyses have recently given rise to speculation that polyploidy has played an important role in the evolution of nodulation. Assessing the implications of polyploidy for nodulation enables predictions about possible outcomes and, while several challenges exist for studying the effects of polyploidy on nodulation signalling, testing of these predictions in a variety of study systems will yield insights into the evolutionary consequences of polyploidy for nodulation symbioses.

1. INTRODUCTION

Polyploidy, or whole genome duplication (WGD), and nitrogen-fixing symbioses have both been instrumental in contributing to the evolution and diversity of plants. Nitrogen-fixing symbioses involve interactions between host plants and bacterial symbionts, generally enabling the plants to acquire nitrogen that would be inaccessible otherwise (Vance, 2001; White, Prell, James, & Poole, 2007). Such symbioses, particularly those termed 'rhizobial' interactions that involve nodule-forming interactions with bacteria ('rhizobia') from numerous genera of α- and β-proteobacteria, constitute an important aspect of the diversity of the legume family (Leguminosae) (Sprent, 2009). WGD events, in turn, are also believed to have had an important role in the evolution and diversification of plants (Madlung, 2013; Weiss-Schneeweis et al., 2013). Given the importance of both polyploidy and nodulation, interest in the intersection between the two phenomena has increased in recent years.

Rhizobial nodulation symbiosis provides a model for understanding the implications of polyploidy for biotic signalling and interactions, given that many aspects of the signalling between plants and rhizobia have been elucidated. Similarly, a basic understanding of the potential evolutionary effects of polyploidy has been developed. Thus, it is possible to make predictions about the effects of polyploidy on interactions between plants and rhizobia by examining particular elements that relate to signalling. Genomic studies have also provided perspectives on the possible role of polyploidy in the evolution and refinement of nodulation symbiosis and signalling. Along with these recent advances and current opportunities, there are also challenges to the study of connections between these two phenomena.

Much of the seminal, critical work in understanding the signalling mechanisms underlying interactions between plants and rhizobia was conducted using limited sets of plant and rhizobial genotypes, often under laboratory conditions. Such experimental systems are necessary, and they have yielded important results. However, we are also interested in rhizobial interactions at the level of *species*; in our case, our focus is primarily on the plant species that serve as hosts. Ultimately, in considering the interactions for a plant species, we would like to understand the changes in interactions that can occur at the level of species over time, as well as over spatial variability, which includes environmental interactions.

Researchers have elucidated numerous positive and negative determinants of rhizobial symbioses (e.g. Broughton, Jabbouri, & Perret, 2000; Keen & Staskawicz, 1988; Oldroyd, Murray, Poole, & Downie, 2011). Many of these involve signalling components, such as secondary metabolite signals in the rhizosphere (see below). The status of each determinant will yield a proximal outcome and, while some sets of determinants act in a sequential manner where the outcome of one determinant is not possible without a prior outcome, each determinant and its outcomes exist in a complex network of interacting determinants. The fitness consequences, benefits and costs of the interactions will also interact with the other variables to modulate probabilities of outcomes and, ultimately, to determine the range of symbiotic interactions. Nevertheless, each signalling determinant or variable can be visualized as defining a 'signalling space' (Figure 1). This space represents the total hypothetical set of rhizobial genotypes for which the given variable under consideration in a particular legume host *genotype* will yield an outcome that permits symbiotic nodulation. This can also be conceptualized for a host plant *species*, where the species signalling space will comprise the rhizobial genotype by plant genotype ($G_{plant} \times G_{rhizobium}$) combinations that will yield outcomes conducive to symbiosis. However, each variable, considered singly, may enable conducive outcomes with different sets of rhizobial genotypes, so as additional variables are considered and the signalling spaces for each variable are overlain, the possible symbiotic interactions will generally become reduced and, as additional non-signalling variable spaces are included, the realized or actual symbiotic space, for a host plant species or particular genotype, is reached. The signalling components that serve as determinants in rhizobial interactions, and the potential implications of polyploidy for these determinants, can be understood in the context of the signalling spaces conceived in this manner. These spaces (notably the overall symbiotic space) will also be variable over space (environment) and time (species evolution).

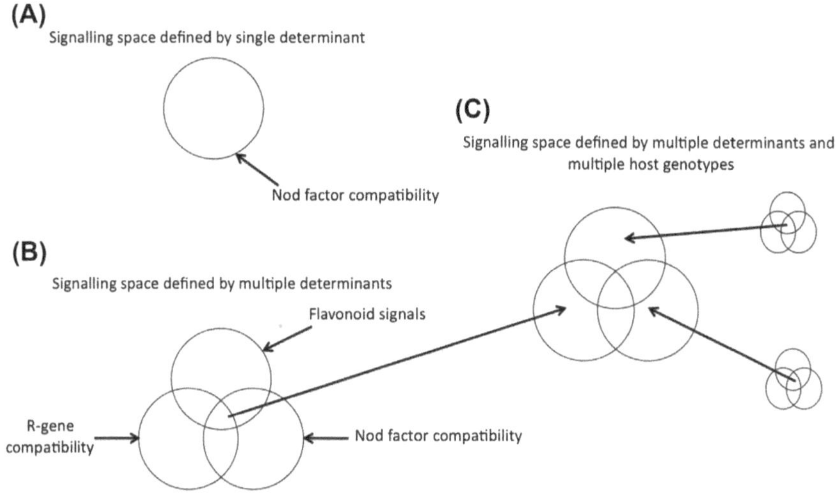

Figure 1 Visualization of the signalling spaces defined by determinants of symbiotic nodulation. Each determinant of nodulation can be understood as having a direct outcome that is needed for symbiosis. The direct outcomes of these determinants act to define the host–symbiont combinations that are effective (functional). (A) For any single host plant genotype, any single signalling determinant (e.g. flavonoid stimulation of Nod factor production, Nod factor binding to NFR genes, etc.) delineates a 'signalling space', which is the subset of all possible rhizobial genotypes for which the determinant will lead to a direct outcome that is necessary or conducive to nodulation. (B) As the signalling spaces of sequentially acting determinants are overlain (intersect), the composite signalling space will either remain the same or become more restricted. (C) This conception of the signalling space can be applied not only to a single plant host genotype, but also to a host species, where the species signalling space is defined as the set (union) of rhizobial genotypes with which the sum of host plant genotypes that make up the species can interact. As with the signalling space of a single plant genotype, one or multiple determinants can be considered.

2. NODULATION SIGNALLING: DETERMINANTS OF SYMBIOTIC INTERACTIONS

Many aspects of the evolution of nodulation and nodulation signalling appear to be marked by the recruitment of pre-existing genetic capacities, including those related to developmental programs and symbiotic interactions, notably signalling components involved in the evolutionarily older fungal mycorrhizal mutualism and in microbial defence responses (Mathesius, 2003). The recruitment or co-option of these programs ('exaptation' as defined by Gould & Vrba, 1982) also indicate that the concept of 'deep homology' is relevant to understanding the evolution of signalling in rhizobial symbiosis. This term describes cases where, even when an organ or trait (i.e. character) may have evolved independently in several lineages and is therefore considered nonhomologous, the independent origins have involved recruitment of homologous programs or components (Shubin, Tabin, & Carroll, 2009). Such deep homology and recruitment have been proposed for several aspects of rhizobial nodulation-related signalling (Doyle, 2011; Soyano & Hayashi, 2014).

Successful establishment of rhizobial nitrogen-fixing nodule-forming symbioses requires numerous steps, in which signalling plays a critical role in various forms and at various stages, including initial communication between the plant host and rhizobia, signal transduction and root hair responses, bacterial invasion, infection thread formation and growth, and nodule organogenesis (Oldroyd & Downie, 2008). Many of the mechanisms and key steps have been elucidated, particularly in the root hair-mediated rhizobial symbioses that are our primary focus (Figure 2). Variations in plant metabolites and molecules employed in rhizosphere signalling between plant and bacteria, the plant receptors for perception of bacterial signals,

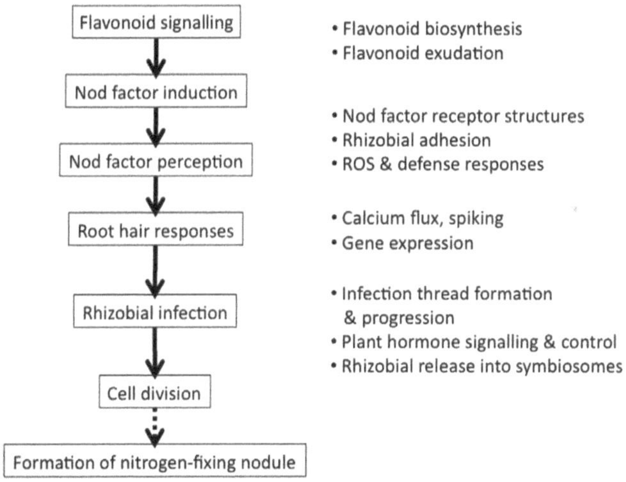

Figure 2 Summary schematic of steps required in the formation of rhizobial nodulation symbioses via root hair infection. Additional components involved at each stage are noted on the right. The interaction is initiated through the exudation of plant secondary metabolites, which lead to Nod factor induction in rhizobia. The Nod factors are perceived by the plant via Nod factor receptors. This is followed by signal transduction and a variety of responses, including root hair curling and formation of the bacterial infection thread. The infection proceeds to reach the site of cortical cell division, where rhizobia are released into symbiosomes. ROS, Reactive oxygen species.

and additional signalling components required for nodulation can serve to enhance or lessen symbiotic interactions.

The plant metabolites involved in establishing successful symbiotic relationships have been well characterized in various model papilionoid legume systems, including *Lotus japonicus, Medicago truncatula* and *Glycine max* (soybean), along with corresponding rhizobial symbionts (Oldroyd & Downie, 2008; Oldroyd, 2013). For nodulation symbiosis to be established, the initial requirement is for plants to exude compounds that are able to trigger the induction of *nod* genes in the rhizobia. Compatible compounds, notably flavonoids, interact with the transcriptional regulator NodD in the bacteria, enhancing the DNA-binding affinity of NodD to a consensus sequence in the promoter region of *nod* genes (termed the *nod* box), activating transcription of the *nod* genes (Peck, Fisher, & Long, 2006). Noninducing flavonoids are competitive inhibitors, indicating that they also bind to NodD and can enhance *nod* box binding affinity of NodD, but without leading to *nod* gene transcription (Peck et al., 2006). While evidence has been documented of *nod* gene induction by compounds such as aldonic acids (Gagnon & Ibrahim, 1998), jasmonates (Mabood, Souleimanov, Khan, & Smith, 2006) and betaines (Phillips, Joseph, & Maxwell, 1992), flavonoids constitute the primary class of compounds implicated in *nod* gene induction in legumes (Cesco, Neumann, Tomasi, Pinton, & Weisskopf, 2010; Oldroyd, 2013).

Induction of *nod* genes, in turn, leads to the production of Nod factors (lipochitooligosaccharides) that are secreted by rhizobial bacteria and perceived by the plant, triggering subsequent symbiotic responses. Perception of the compatible Nod factors leads to signal transduction, transcriptional activation of the early nodulation (ENOD) genes and physiological responses including root hair curling and infection thread formation (Ferguson et al., 2010). The backbone of Nod factors is composed of *N*-acetylglucosamines, with an *N*-linked acyl chain on the terminal nonreducing residue (Cullimore & Gough, 2013; D'Haeze & Holsters, 2002). Substantial diversity is possible in the structure of Nod factors synthesized and secreted by rhizobia, and this is largely due to variability in the acyl group, as well as substituents at one or both of the nonreducing- and reducing-terminal residues (D'Haeze and Holsters, 2002). Different species and strains of rhizobia have varying capacities for synthesizing Nod factors with distinctive structures (D'Haeze and Holsters, 2002). In general, broad host range rhizobia produce a correspondingly broad array of Nod factors; narrow host range rhizobia produce a more restricted and specific set

(Cullimore & Gough, 2013). For example, the promiscuous strain NGR234 synthesizes a diverse family of Nod factors with various combinations of substitutions, encoded by a series of *nod* genes (Price et al., 1992). Following the induction and release of Nod factors by rhizobia in response to host plant compounds, the host plants then perceive the Nod factors, primarily through Nod factor receptors (NFRs). Here, again, there is a requirement for compatibility between the Nod factor and the plant receptors. Diversity of Nod factor structures, with their decorations, and NFRs have been implicated in the specificity of interactions in a variety of legume genera (Cullimore & Gough, 2013).

NFRs were identified and characterized in *L. japonicus* by Madsen et al. (2003), and Radutoiu et al. (2003). Radutoiu et al. (2003) cloned *NFR1* in *L. japonicus* and showed that it was necessary for the earliest responses to Nod factors. NFR5 was also identified as essential for nodulation in *L. japonicus* (Madsen et al., 2003). It has been proposed that the NFR5 and NFR1 proteins form a heterodimer and function together in Nod factor perception (Arrighi et al., 2006; Oldroyd, 2013). The extracellular domains of NFR1 and NFR5 were found to be similar to the LysM (lysin motif) domains of previously identified chitinases and peptidoglycan-binding proteins (Gust, Willmann, Desaki, Grabherr, & Nürnberger, 2012; Madsen et al., 2003; Radutoiu et al., 2003). Diverse LysM proteins are thus involved in binding a variety of molecules that are structurally similar to Nod factors, including the microbe-associated molecular patterns such as fungal chitin and bacterial peptidoglycan (Gust et al., 2012). Op den Camp, Streng, et al. (2011) also found that, in the only known instance of a nonlegume that nodulates with rhizobial bacteria, *Parasponia andersonii*, a LysM Nod factor receptor protein involved in mycorrhizal association has been recruited for nodulation symbiosis. Their data suggest that the *PaNFP* gene identified serves a dual function in both mycorrhizal (Myc factor) signal reception and Nod factor reception, which is plausible given the apparent structural similarity of Nod factors to Myc factors as well (Gust et al., 2012).

The importance of NFRs in determining specificity between legume hosts and rhizobial partners has been demonstrated through transformation studies. The expression of the *NFR1* and *NFR5* genes from *L. japonicus* in *M. truncatula* and *Lotus filicaulis* enabled nodulation of these latter species with *Mesorhizobium loti*, which commonly nodulates *L. japonicus* but not the other species (Radutoiu et al., 2007). While nodulation capability was transferred, the developmental activation in *M. truncatula* resulted in the indeterminate nodules typical for that species, as opposed to the determinate nodules formed by *L. japonicus*. Using domain swap and mutant experiments, the critical role of the LysM domains of the NFRs in determining specificity was also demonstrated (Radutoiu et al., 2007). Overexpression of receptors can lead to increased nodule numbers (Indrasumunar et al., 2011), suggesting that the amount or density of receptors can also lead to variation in the amount of interaction with Nod factor molecules, and the consequent response, that is possible. While most evidence for the role of Nod factor receptors in nodulation was obtained through studies of mutants, more recently, Broghammer et al. (2012) obtained direct evidence showing binding of NFR1 and NFR5 to Nod factors. In addition to Nod factor receptors, lectin nucleotide phosphohydrolases have also been identified as proteins that bind to Nod factors at the plant surface and are thus also thought to mediate symbiotic interactions with rhizobia (De Hoff, Brill, & Hirsch, 2009; Etzler et al., 1999).

Toll-interleukin receptor/nucleotide-binding site/leucine-rich repeat (TIR-NBS-LRR) resistance genes (*R* genes) have also been implicated in determining rhizobial interactions in cultivars of *G. max* (Yang, Tang, Gao, Krishnan, & Zhu, 2010). *Rfg1* and *Rj2* alleles are known to confer the specificity for nodulation with particular strains of *Bradyrhizobium japonicum* and *Sinorhizobium fredii* (e.g. Nakano, Yamakawa, Ikeda, & Ishizuka, 1997; see summary in Hayashi et al., 2012), the characterization of these alleles as *R* genes is a more recent development (Yang et al., 2010). The particular mechanisms of action or the roles of these *R* genes in inhibiting nodulation remain unresolved, although it appears that failed infection thread growth is involved, which may be due to plant defence responses by perception of bacterial effectors, leading to recognition of incompatible rhizobia as parasites rather than mutualists (Yang et al., 2010). Another gene, *Rj4*, which is known to restrict nodulation of *G. max* with specific incompatible strains has recently been found to encode a thaumatin-like protein; it is a member of the pathogenesis-related protein family 5, which is implicated plant host resistance (Hayashi et al., 2014). Proteins, including nodulation outer proteins, are also secreted by rhizobia through type III and type IV secretion systems, and these proteins have additional potential for both positive and negative effects on symbiosis, which further emphasizes the importance of plant defence in determining rhizobial interactions and also highlights the likely evolutionary connections between pathogenic and mutualistic symbioses (Deakin & Broughton, 2009).

In addition to root hair responses, symbiotic responses are also triggered in cortical cells, leading to the expression of additional genes required for nodule organogenesis (Oldroyd & Downie, 2008). Plant hormones have been implicated in several aspects of signalling and regulation of nodulation. Cytokinins and cytokinin-related signalling, in particular, appear to have important roles in the development and regulation of nodule formation. A histidine kinase cytokinin receptor in cortical cells is necessary for signalling subsequent to the root hair cell responses following initial contact with rhizobia. RNA interference of the cytokinin receptor homologue in *M. truncatula* (*MtCRE1*) resulted in strongly reduced nodulation, including decreased progression of rhizobial infection and formation of nodule primordia (Gonzalez-Rizzo, Crespi, & Frugier, 2006). Gain-of-function mutation of the orthologous gene in *L. japonicus* led to spontaneous nodule organogenesis (Tirichine et al., 2007). Experiments in *L. japonicus* over-expressing cytokinin oxidase, the primary enzyme for degradation of cytokinins, also led to both a decrease in the number of nodules formed per plant and an increase in the number of lateral roots (Lohar et al., 2004). In contrast to this positive role in nodule formation, recent work also suggests that cytokinins may be implicated in the shoot-deived inhibition of nodulation that is part of the autoregulation of nodulation (AON) pathway in legumes (Sasaki et al., 2014). AON is another process that provides feedback regulation of nodule number, and genes in this pathway have been identified from mutants of several species, including soybean, possessing increased nodule numbers (Miyahara et al., 2008; Searle et al., 2003). Other hormones have also been found to have roles in the control of nodule development and nodule number, including ethylene, which is a negative regulator of nodule number (Gresshoff et al., 2009; Lohar, Stiller, Kam, Stacey, & Gresshoff, 2009; Penmetsa et al., 2008).

In the context of the evolution of nodulation, the genes involved in many of the processes underlying nodulation, including signalling, often appear to have additional, related functions within a given plant species (Yokota & Hayashi, 2011). This confirms the long-held hypothesis that, in many cases, genes with pre-existing functions have been recruited to play roles in nodulation (Doyle, 1994). A notable example of this type of recruitment is apparent in the early stages of nodulation signalling, where the genes involved in the common symbiosis pathway were recruited from the older arbuscular mycorrhizal (AM) symbiosis (Soyano & Hayashi, 2014). These genes include those encoding nuclear cation channels (*POLLUX, CASTOR/DMI1*), a plasma membrane localized LRR receptor kinase (*SYMRK/DMI2*), and a calcium-dependent and calmodulin-dependent protein kinase (*CCaMK/DMI3*), among others (Ané et al.,

2004; Charpentier et al., 2008; Endre et al., 2002; Lévy et al., 2004; Mitra et al., 2004; Stracke et al., 2002).

3. WGD AS A PROCESS ALTERING PLANT SIGNALLING, BIOTIC INTERACTIONS AND ECOLOGICAL AND GEOGRAPHICAL RANGES

Polyploidy is a prominent process in the evolutionary history of flowering plants, having been suggested to underlie their origin and diversification (De Bodt, Maere, & Van de Peer, 2005; Vanneste, Maere, & Van de Peer, 2014). Wood et al. (2009) estimated that 15% of all speciation events in flowering plants have involved polyploidy. Moreover, it is now recognized that all flowering plants have a polyploid past (Jiao et al., 2011), and that the genomes of most modern species show evidence of more than one WGD — for example, the soybean genome shows evidence of the two polyploidy events shared with all flowering plants, a whole genome triplication shared with eudicots (Jiao et al., 2011), a WGD shared with many legumes (Cannon et al., 2015) and a 5–10 MYA polyploidy event unique to the genus *Glycine* (Egan & Doyle, 2010). Thus, it is indisputable that polyploidy has produced successful lineages and has played a significant role in plant evolution.

Recruitment of duplicated genes can provide a means of evolving novel or distinct functions. For example, recent work on nodulation suggests that gene duplication has given rise to the GRAS transcription factors NSP1 and RAM1, which function downstream of the common symbiosis pathway (Gobbato et al., 2012). RAM1 has a role primarily in the mycorrhizal symbiosis, while NSP1 has a role in nodulation and does not appear to have any function in plant response to mycorrhizal signalling (Gobbato et al., 2012). In the case of the gene family of *MtLYK3-LjNFR1a*, several duplication events are believed to have preceded the evolution of nodulation, the origin of legumes and thus any polyploidy events in the legumes, nevertheless contributing to the evolutionary trajectory towards nodulation (De Mita, Streng, Bisseling, & Geurts, 2014).

If the duplication and subsequent evolution of single genes can contribute to the evolution of novel traits, including diversification of symbiotic interactions and components of nodulation signalling, duplication of the entire set of genes as well as regulatory sequences by polyploidy has the potential to greatly amplify these processes, providing duplicates across entire biosynthetic, signalling and regulatory networks. Evolution is driven by mutation, and polyploidy can be thought of as a super-mutation, leading to evolutionary novelty; Freeling and Thomas (2006) have suggested that the retention and recruitment of duplicated genes, specifically by polyploidy, is a major driver of morphological novelty and complexity.

While gene duplicates are often lost following a WGD event, several models suggest ways in which natural selection can lead to retention of both paralogous copies, such as neofunctionalization and subfunctionalization (Innan & Kondrashov, 2010). In the case of nodulation signalling, if a gene with a pre-existing function were duplicated, one copy could be recruited for a novel function in nodulation, while the other copy could maintain the original function. This provides an example of neofunctionalization. Alternately, if a gene developed both nodulation-related and non-nodulation-related functions and was subsequently duplicated, the two functions could be divided between the two resultant copies, and subsequently each paralogue could evolve for specialization in each function. This process is referred to as subfunctionalization.

Quantitative dosage effects will also play a role in determining the initial retention of certain types of gene duplicates; WGD can provide an opportunity for retention of genes involved in pathways where selection for specific stoichiometry means that single gene duplications are less likely to be favoured, while the maintenance of stoichiometric relationships occurring through a WGD would be favoured (Birchler & Veitia, 2012). The classes of genes that are subject to dosage balance in this manner often include transcription factors and the components of signalling pathways (Blanc & Wolfe, 2004; Maere et al., 2005; Paterson et al., 2006). In the context of nodulation, retention of duplicates in both of these cases could contribute to the evolution of novel signalling or the enhancement and modulation of existing signalling mechanisms.

Autopolyploidy and allopolyploidy are two endpoints on a spectrum of genetic and taxonomic possibilities for WGD. Autopolyploidy is typically defined as a WGD event occurring involving genetically similar genomes, often from within the same species. In contrast, allopolyploidy involves a WGD event coupled with hybridization between two or more progenitors, often from distinct species. Both result in doubled genomes and the concomitant effects specifically related to an increase in DNA content (nucleotypic effects, e.g. Bennett, 1972), while allopolyploidy also involves potential effects of genome merger, such as heterosis (Washburn & Birchler, 2014). In an allopolyploid's complex genome, genetic loci that were orthologous in the diploid progenitors are termed homoeologues (Wendel & Doyle,

2005). Allopolyploids thus at least initially have the opportunity to deploy the homoeologous gene copies from both progenitors, often with variation in homoeologue expression bias (Chen, 2010; Grover et al., 2012; Yoo, Szadkowski, & Wendel, 2013). WGD can result in transgressive, novel expression patterns and homoeologue silencing in natural polyploids (Yoo, Liu, Pires, Soltis, & Soltis, 2014). The changes to patterns of epigenetic regulation and gene expression are referred to as 'genomic shock' (McClintock, 1984). Furthermore, unequal expression of homoeologous copies is thought to lead to fractionation (loss of one copy in each homoeologue gene pair); bias in expression of homoeologous copies can be a legacy of diploid progenitors, with one transcriptome dominant over the other (Buggs et al., 2014). It is hypothesized that in allopolyploids this process can lead to biased fractionation across the duplicated genome (Garsmeur et al., 2014; Schnable, Springer, & Freeling, 2011). In addition to fractionation, biases in gene conversion can also alter the expected additive pattern of homoeologous copies in allopolyploids (Flagel, Wendel, & Udall, 2012).

Polyploidy can thus lead to substantial chromosomal rearrangement or loss, genetic and epigenetic changes, and gene expression changes that, while in many cases deleterious, have the capacity to generate increased variability and potential for adaptation. Allopolyploidy, in particular, can confer advantages in changing environments, including variable climate and habitat disturbance (Combes, Cenci, Baraille, Bertrand, & Lashermes, 2012; Fawcett & Van de Peer, 2010; Stebbins, 1985). Furthermore, polyploids are often found to show an increased ability to colonize new habitats and alter their ranges. Evidence indicates that polyploidy can broaden environmental distributions and that polyploids can have larger ranges than their diploid progenitors (Green, Ramsey, & Ramsey, 2011, 2013; Hijmans et al., 2007; Lowry & Lester, 2006), as would be expected of plants with a greater capacity for colonization, although it may not be a general tendency (Harbert, Brown, & Doyle, 2014; Martin & Husband, 2009). Pandit, Pocock, and Kunin (2011) conducted a comprehensive study and meta-analysis of invasive species in the world and found that polyploidy correlated strongly with invasiveness. This could be due to alterations that affect polyploids in numerous ways and incorporate a variety of effects of polyploidy including phenotypic plasticity, ecological tolerance and niche differentiation (te Beest et al., 2012). Of particular interest here, te Beest et al. (2012) highlighted the importance of biotic interactions and, in particular, changes to biotic signalling for apparent increased invasiveness and colonization ability of polyploids. Connecting invasiveness of polyploids in this manner suggests that polyploids can possess broadened signalling spaces, compared to related diploids. Given the emerging awareness of the significance of the microbiome in plant and animal health and ecology (Berendsen, Pieterse, & Bakker, 2012; Cho & Blaser, 2012; Ezenwa, Gerardo, Inouye, Medina, & Xavier, 2012), it is possible that a polyploid's ability to occupy and colonize new habitats or geographic ranges may be partly due to its ability to interact with diverse microbes. In the case of legumes, this could include compatible rhizobial signalling interactions available to the plant, coupled with the supply and demand for available soil nitrogen.

Given the many genetic, physiological and ecological changes that can result from polyploidy, researchers have begun examining the effects of polyploidy on biotic interactions. While the effects of polyploidy on biotic interactions are not yet fully elucidated (Soltis, Buggs, Doyle, & Soltis, 2010), recent research has yielded progress. Several studies have established that polyploidy has the potential to generate differences in, and diversification of, biotic interactions. In pollination interactions, *Heuchera grossulariifolia* polyploids have suites of pollinators that differ in the relative visitation rates and contributions to seed set of particular pollinators between diploid and autopolyploid plants (Segraves & Thompson, 1999; Thompson & Merg, 2008; Thompson, Nuismer, & Merg, 2004). Thompson and Merg (2008) note that these differential visitation rates of pollinators on diploids and polyploids are likely due to chemical signalling cues employed by the pollinators, in addition to any morphological differences, since certain diploid and polyploid populations studied were morphologically indistinguishable; explanations in terms of chemical signalling cues suggest alterations to the signalling space in the polyploids. Damage from herbivory also varies between diploids and polyploids in a number of study systems (Hull-Sanders, Johnson, Owen, & Meyer, 2009a; Münzbergová, 2006; Nuismer & Thompson, 2001). Differences in performance of the insect herbivore *Spodoptera exigua* on *Solidago gigantea* (giant goldenrod) plants of differing cytotypes were also attributed to changes in secondary chemistry (Hull-Sanders et al., 2009a; Hull-Sanders, Johnson, Owen, Meyer, 2009b), indicating further modulation of the interaction space.

That symbiotic interactions with soil microorganisms can also be affected by polyploidy has been shown in relation to AM associations. Studies of the *Gymnadenia conopsea* group (Orchidaceae) found that plants of different ploidy levels formed associations with distinct suites of mycorrhizal fungi (Těšitelová et al., 2013). In this work, the segregation of mycorrhizal symbionts by ploidy level was most strongly observed in the comparison of diploids and derived autopolyploids growing in close proximity at sites containing mixtures of plants of several ploidy levels. This research suggests the potential for differences in signalling between plants of varying ploidy and mycorrhizal symbionts in that the interactions with different suites of symbionts may be indicative of signalling specificity and differentiation of signalling spaces. Subsequent work has confirmed interactive effects of ploidy and AM inoculation in other genera, affecting plant growth and mycorrhizal parameters (Sudová, Pánková, Rydlová, Münzbergová, & Suda, 2014), which also suggests effects of polyploidy on signalling and interaction spaces. Differing growth responses to particular symbionts between autopolyploids and diploids also suggest that divergent selection can lead to differentiation of signalling subsequent to polyploid formation. However, the effects of polyploidy can vary with taxon. In a study by Sudová, Rydlová, Münzbergová, and Suda (2010), differences in mycorrhizal growth responses were not observed between diploids and polyploids for *Campanula gentilis* and *Pimpinella saxifraga*, though the growth of *Aster amellus* diploids was positively responsive to mycorrhizal inoculation, while hexaploids were either not responsive or negatively responsive. In these experiments, ploidy did not have any effect on mycorrhizal parameters, including extraradical mycelial growth and the extent of mycorrhizal colonization (Sudová et al., 2010).

This research indicates the potential for plant polyploidy to alter interactions between plants and various types of symbionts, notably soil microorganisms, and suggests that alterations in signalling may be connected to the differences observed. The next section explores the potential implications of polyploidy for nodulation signalling and symbiosis, with attention devoted to effects on particular components of signalling.

4. POTENTIAL IMPLICATIONS OF POLYPLOIDY FOR NODULATION SIGNALLING

The visualization of symbiotic and signalling spaces introduced earlier is useful in considering effects of polyploidy, since polyploidy can alter these spaces by affecting diverse processes in plants, leading to a variety of possible outcomes. If we consider a scenario with two diploid progenitors with distinct but overlapping hypothetical symbiotic spaces, we can predict a variety of outcomes in the symbiotic space of an allopolyploid, including additive, parental and transgressive patterns (Figure 3). The following sections will explore a number of effects through which polyploidy can alter these hypothetical symbiotic spaces. We will consider alterations to metabolite

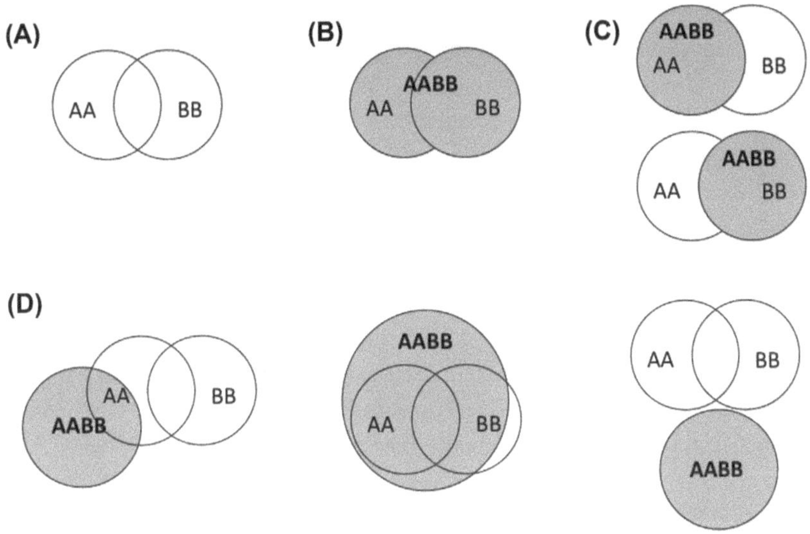

Figure 3 Visualizations of species-level signalling spaces (defined by multiple determinants) and outcomes of allopolyploidy. (A) The composite signalling space illustrated in Figure 1(C) is shown for two diploid species (genome designations AA and BB), representing the rhizobial genotypes for which interactions are possible with plant genotypes (A_1A_1, A_1A_2, A_2A_2, …; B_1B_1, B_1B_2, B_2B_2, …) that make up each host species. In this case, the two species have partly overlapping, partly nonoverlapping signalling spaces. When an allopolyploid is formed from the two diploid progenitors, a variety of outcomes are possible for the signalling space of the allopolyploid, including (B) a signalling space that comprises the summed signalling spaces of both diploids; (C) a signalling space that is the same as one diploid or the other (possibly due to dominance of one homoeologous genome over the other) or (D) some combination of the signalling spaces of the diploid progenitors, including potential transgressive (nonparental) signalling spaces.

signalling in exudates, signal reception, as well as additional signalling mechanisms and effects on signalling.

4.1 Flavonoids

Polyploidy can alter the biosynthesis of secondary metabolites. These effects include increased accumulation and concentration of secondary metabolites (e.g. Hull-Sanders et al., 2009a; Lavania et al., 2012), including changes in flavonoid pathways (Griesbach & Kamo, 1996). Qualitative variations in flavonoid profiles between ploidy levels, including novel compounds in polyploids that are not found in progenitor diploids, have also been documented (Levy & Levin, 1971, 1974). Historically, such qualitative differences in flavonoid profiles, particularly when additive, were often used as evidence for determining the parentage of allopolyploids (Mears, 1980).

The synthesis of novel, more diverse flavonoids, and novel combinations of flavonoids, can occur as a result of diversification of genes affecting the biosynthetic processes by neofunctionalization or through the effects of hybridity (including differential regulation or novel regulatory patterns) in an allopolyploid. Qualitative differences in exuded flavonoids can alter symbiotic outcomes because, as noted above, there is specificity between particular flavonoid compounds and the response of *nod* gene induction in particular strains of rhizobial bacteria (Aoki, Akashi, & Ayabe, 2000; Long, 1996). A given flavonoid can have distinct inducing and inhibiting effects on Nod factor synthesis across various species and strains of rhizobia (Kape, Parniske, & Werner, 1991; Kosslak, Bookland, Barkei, Paaren, & Appelbaum, 1987, Kosslak, Joshi, Bowen, Paaren, Appelbaum, 1990; Yokoyama, 2008). Similarly, polyploidy-induced alterations to gene expression that affect enzymes involved in the synthesis of signalling compounds could have quantitative effects on symbiotic interactions, increasing or decreasing the magnitude and intensity of the signalling. It is also conceivable that, in terms of absolute exudation on a per-plant basis, changes to the biomass of biosynthetic tissues in polyploids could also increase or decrease the total amount of flavonoids produced. In vitro, increases in flavonoid concentration can lead to increased Nod factor synthesis (e.g. Yokoyama, 2008), which in turn can lead to different calcium responses (Shaw & Long, 2003). In optimized hydroponic systems, exogenous application of inducing flavonoids stimulates nodulation (Novák et al., 2002). Such induction responses can increase with flavonoid concentration over a certain range, beyond which, in certain cases, the response decreases, potentially due to toxicity effects (Bolaños-Vásquez & Werner, 1997; Novák et al., 2002; Pueppke et al., 1998). In general, any change in biosynthesis of particular flavonoids that affects the profile of exuded compounds would be expected to lead to alterations in the magnitude of induction or inhibition of Nod factor synthesis, which would alter the probabilities of outcomes within the signalling space defined by flavonoid signalling.

In the context of signalling spaces, one can propose a broad, hypothetical signalling space delineated solely by the exudation of flavonoids and their capacity for inducing or inhibiting Nod factor production. The size of the signalling space can initially be taken as all rhizobial genotypes for which a plant's exudate yields a nonzero probability of Nod factor induction. The relative size of the signalling space under consideration can be increased or decreased depending on setting a biologically relevant Nod factor induction threshold. The exclusion or inclusion of a rhizobial strain from a particular signalling space established by a given Nod factor induction threshold will be determined by the total Nod factor inducing capability of the mixture of compounds in the exudate of a plant. This could be calculated by summing the products of the total amount of each compound released by the plant and taken in by a given bacterium and the Nod factor inducing or inhibiting effect of each compound. Additional terms are also required to account for interactive effects between the compounds that will change the total induction or inhibition capacity of the exudate for any particular rhizobial genotype. In this framework, the effect of quantitative or qualitative changes in flavonoid profiles due to polyploidy could be approximated by comparing the concentrations of compounds and inductive capabilities of the compounds in the polyploids with those of the diploids.

While changes in the biosynthesis of flavonoids due to polyploidy are critical to consider, in the context of nodulation, it is the exudation of flavonoids from the root into the rhizosphere that affects the potential for signalling with rhizobial bacteria. This requires consideration of ATP-binding cassette (ABC) and multidrug and toxic compound extrusion (MATE) transporters. ABC-type transporters have been implicated in the exudation of flavonoids from legume root cells, and evidence suggests they could have a variable affinity for different flavonoid compounds (Sugiyama, Shitan, & Yazaki, 2007). Competitive inhibition assays using soybean plasma membrane vesicles found that the isoflavonoid daidzein, as well as biochanin A and formononetin, inhibited genistein transport, while the flavanone naringenin and the flavonol kaempferol did not, suggesting the latter were not readily transported (Sugiyama et al., 2007). Genistin, the glucoside of genistein, also inhibited genistein transport, but appeared to do so to a lesser degree than the daidzein, biochanin A and formononetin (Sugiyama et al., 2007). MATE transporters have not been reported as being involved in exudation from the root or transport across the plasma membrane, but several MATE transporters have been found that are involved in transport of flavonoids into the vacuole, leading to suggestions that MATE transporters may also play a role in flavonoid exudation (Gomez et al., 2009; Marinova et al., 2007; Zhao & Dixon, 2009). Given the importance of transporters in exudation and varying capacities for transporting different flavonoids, simply synthesizing novel compounds alone would not necessarily translate into their exudation from the plant root tissues; different transporters may be required for effective exudation of particular flavonoids. The observation that mutations to individual transporters can lead to changes throughout the biosynthetic networks involved in flavonoid production (e.g. Zhao et al., 2011) suggests that changes to the number and specificities of transporters could have a profound influence on rhizosphere signalling. Thus, for synthesis of novel or more diverse flavonoids in polyploids to affect nodulation signalling, the capacity to transport the novel array of compounds must also be present among the transport proteins that participate in the exudation of flavonoid compounds from the root tissues to the rhizosphere. In a polyploid, the genes encoding the necessary transporters may be available due to their presence in the diploid progenitors; alternately, novel transport capacities could theoretically be developed by mutation (creation of new alleles) or by neofunctionalization of duplicated genes.

Similarly, quantitative increases in the biosynthesis of flavonoids would not necessarily lead to greater exudation, but would depend on the density of transporters available for exudation. This would, in turn, depend in a complex manner on the number of transporters per cell and the external cell surface area. This is not expected to be constant between polyploids and diploids, since cell size and, therefore, surface-to-volume ratios are often altered by WGD (Coate et al., 2012; Lavania et al., 2012; Levin, 1983). Alterations to transcriptome size and differential translational regulation will also have effects on the number of flavonoid transporters on the root surface; both types of changes have been documented in polyploids (Coate & Doyle, 2010, 2015; Coate, Bar, & Doyle, 2014). These aspects will also be relevant in the following section on genetic diversity and gene expression.

Although flavonoids can be important in the attraction of rhizobia, they are not necessarily the primary compounds that serve this purpose. Early studies on the chemotaxis of rhizobia found that with some rhizobial species, such as *Rhizobium leguminosarum* and *Sinorhizobium meliloti*, chemotaxis occurred in response not only to flavonoids but also to nutrients, specifically organic acids (Aguilar et al., 1988; Armitage, Gallagher, & Johnston, 1988; Caetano-Anollés, Crist-Estes, & Bauer, 1988). In these species, flavonoids that were *nod* gene inducers were also found to be chemotactic attractants. However, in *B. japonicum*, chemotaxis occurred in response to nutrients but not flavonoids (Barbour, Hattermann, & Stacey, 1991; Kape et al., 1991). In this case, it appeared that chemotactic molecules were distinct from *nod* gene inducers. More recent work has examined varying chemotaxis of *B. japonicum* strains to nutrient attractants (Althabegoiti et al., 2008) and also explores the bacterial genes and receptors involved in perception of chemotactic signals in various rhizobial species (Miller, Yost, Hynes, & Alexandre, 2007;

Webb, Hildreth, Helm, & Scharf, 2014; Yost, Rochepeau, & Hynes, 1998). Given the chemotactic importance of non-flavonoid compounds, the impacts of polyploidy on rhizobial interactions may extend beyond the effects on flavonoid biosynthesis and exudation to the synthesis and exudation of organic acids and other nutrients as well, when considering the contribution of chemotaxis as a variable defining the signalling space.

4.2 Nod Factor Receptors

In the context of nodulation signalling, polyploids have the potential to possess increased diversity of Nod factor receptors, thereby enabling perception of a broader array of rhizobial signals. As fixed hybrids, allopolyploid species can, at least initially, combine Nod factor receptors from different diploid species. Neofunctionalization and further diversification of Nod factor receptors could increase the capacity to perceive a broader array of bacterial Nod factors having diverse structural properties. A general prediction is that greater Nod factor receptor diversity should be associated with a greater range of bacterial symbiont partners with which root hair curling is elicited (Indrasumunar et al., 2010; Radutoiu et al., 2007). Thus, one hypothesis is that the allopolyploids, as fixed hybrids, will have equal or greater symbiont ranges (interaction spaces) relative to their diploid progenitors. However, tests of selection and differences in expression may reveal early fractionation, neutral evolution and loss of particular gene copies, in which case the allopolyploid may not have an increased diversity of receptor genes and the expected concomitant increased symbiotic range. Further, polyploidy-related modulation of the expression of plant receptors could also have quantitative effects on symbiotic interactions, increasing or decreasing the magnitude and intensity of the signalling.

4.3 Resistance Genes

R genes encode R proteins, which play an important role in plant defence against pathogens. This contribution to resistance can be achieved by several potential mechanisms, including through the R protein binding directly to pathogen avirulence molecules (effectors) or by monitoring cellular targets of effectors (according to the 'guard hypothesis'; Jones & Dangl, 2006). Tandem duplications and segmental genomic duplications have contributed to *R* gene family expansion, though paralogous genes may also be lost following duplication in some cases, and duplicated genes have subsequently diverged due additional accumulated mutations (Meyers, Kaushik, & Nandety, 2005). Thus, both genome duplication and fixed hybridity (in allopolyploids) have the capacity to increase diversity and create novel combinations of *R* genes, which in turn would affect potential biotic interactions, including nodulation. In particular, variations on the allelic combinations of *Rfg1* and *Rj2* in *G. max*, mentioned earlier, could enhance or reduce the ability of a polyploid to form nodules with particular bacterial species or strains. The general predictions for *R* genes in polyploids stand in opposition to those for the Nod factor receptors, because greater diversity of *R* genes would be expected to lead to a more restricted symbiont range, given that particular alleles at the locus serve to prevent nodulation with certain symbionts. Thus, when considering interaction outcomes defined by *R* genes, one could predict that allopolyploid fixed hybrids, for example, would have an equal or more restricted signalling space than their diploid progenitors, unless the allopolyploids silence *R* genes from one progenitor. That *R* genes may belong to the general class of 'duplication resistant' genes (Paterson et al., 2006) is plausible given preferential NB-LRR gene loss in *G. max* following polyploidy (Ashfield et al., 2012). The general lability of *R* genes is suggested by extreme copy number variation in several plant taxa (Zhang et al., 2010).

4.4 Additional Avenues for Effects of Polyploidy on Signalling Mechanisms via Plant Growth and Cytokinins

Polyploidy often has effects on plant growth. In general, wild autopolyploids, along with induced neopolyploids, tend to have larger organs than related diploids (Ramsey & Ramsey, 2014; Ramsey & Schemske, 2002;). Such changes to mature plant size and plant growth rates can be related to alterations to nutrient demand and uptake. Early studies found greater tolerance of *Nicotiana* polyploids for growth in nutrient-poor calcareous soils (Noguti, Oka, & Otuka, 1940); more recent work also found greater tolerance to calcium treatments in polyploid *S. gigantea* (Schlaepfer, Edwards, & Billeter, 2010) and greater salinity tolerance in tetraploid *Brassica rapa* (Meng et al., 2011). Effects of ploidy resulting in increased or decreased nutrient uptake efficiency, however, appear to be species-specific, in the cases of potassium and sulphate (Cacco, Ferrari, & Lucci, 1976). If polyploidy results in alterations to nitrogen uptake and demand, such effects would be likely to yield changes in rhizosphere signalling. Flavonoid and metabolite exudation into the rhizosphere is changed by high nitrogen treatments: nitrogen deficiency can trigger the biosynthesis of flavonoids and nitrate application decreases flavonoid accumulation, including the nodulation signalling isoflavonoids daidzein and genistein in soybean (Cho & Harper, 1991; Lea, Slimestad, Smedvig, & Lillo, 2007). In addition to effects on secondary metabolite accumulation, changes in plant growth and nutrient regulation are also connected to differences in hormonal concentrations and signalling (Depuydt & Hardtke, 2011; Kiba, Kudo, Kojima, & Sakakibara, 2011; Krouk et al., 2011). Cytokinin levels, as noted above, have the potential to affect the number of nodules formed (Lohar et al., 2004; Sasaki et al., 2014); thus certain changes to hormonal signalling in polyploid plants can have additional effects on symbiotic interactions. Furthermore, cytokinins have the capacity to repress a number of transporters of nutrients, including nitrate, ammonium, sulphate and phosphate (Sakakibara, Takei, & Hirose, 2006), while cytokinin transport and biosynthesis can also be regulated by nitrate availability (Takei, Sakakibara, Taniguchi, & Sugiyama, 2001, Takei et al., 2004). Nitrogen availability relative to plant demand, then, has the potential for interacting effects on cytokinin signalling and flavonoid biosynthesis, both of which have implications for signalling and interactions between legumes and rhizobia. If there were changes in the response of a polyploid to the relative quantities of signalling molecules, such as cytokinins, the densities of receptors for these signals or the sensitivity of receptors to the signals would lead to alterations, as noted above, in nodule formation, in addition to altered plant growth. Altered cytokinin signalling in the polyploid would also have the potential to interact with nitrogen transport, thereby affecting flavonoid biosynthesis and possibly flavonoid signalling into the rhizosphere.

Op den Camp, De Mita, et al. (2011) examined the duplication of genes in the cytokinin signalling pathway through a WGD event in the papilionoid legume lineage to determine if this duplication could lead to evolution of nodulation-related functions. The authors identified a gene pair of cytokinin response regulators that were duplicated at the time of the WGD event, and their data appear to indicate that the two genes have differing expression patterns and responses to Nod factors, with MtRR9 having the strongest induction in response to symbiotic signals, while MtRR11 showed nonsymbiotic expression. They suggested that the WGD event in the papilionoids had an effect on nodulation signalling through the duplication of genes in the cytokinin pathway.

Although the potential impacts of polyploidy on nodulation symbioses and signalling can be explored to make predictions, extensive studies to match those conducted with pollinators, herbivores and AM fungi remain necessary. Nevertheless, recent genomic studies have generated interest in the interface between polyploidy and rhizobial interactions. These studies have yielded insights into the evolutionary connections between polyploidy and nodulation, with genetic data relating specifically to signalling processes.

5. EVIDENCE OF ROLES FOR POLYPLOIDY IN THE EVOLUTION OF NODULATION AND RHIZOBIAL SIGNALLING

Nitrogen-fixing symbioses have evolved in four plant orders (Cucurbitales, Rosales, Fagales and Fabales) that comprise the nitrogen-fixing clade (NFC) of the fabid group of rosid eudicots; the instances of nodulation in this clade are thought to be nonhomologous (Doyle, 2011). Prominent within this group is the legume family, where symbioses with rhizobial bacteria evolved (Figure 4); with the exception of *Parasponia* in the order Rosales, the other nitrogen-fixing symbioses in the NFC involve plants that are nodulated by Gram-positive soil bacteria in the genus *Frankia* (Pawlowski & Sprent, 2008). The Cucurbitales, Rosales and Fagales each contain a relatively small number of taxa that form nodulating associations. In contrast, rhizobial nodulation symbioses in legumes are widespread and extremely diverse, such that Gram-negative bacteria from 11 genera interact with a wide taxonomic range of legumes (Pawlowski & Sprent, 2008). There is also ample diversity of structure and growth in legumes (e.g. indeterminate nodules that maintain a persistent meristem or determinate nodules that do not), method of infection (e.g. rhizobia enter the plant tissues through root hairs, or by epidermal or crack entry) and nutrient transport (e.g. nitrogen is exported from the nodule as amides or ureides) among the symbioses between legumes and rhizobia (Sprent, 2007). Here, we review recent genomic studies that have suggested connections between the evolution of nodulation in legumes and WGD events and have led to speculation about the contribution of polyploidy to the development and refinement of nodulation interactions (Li, Zhang, Li, Dunwell, & Zhang, 2013; Young et al., 2011).

The exact role of WGD events in the evolution of nodulation has been difficult to ascertain. In part, this is due to the difficulty of determining the timing of WGD events and in pinpointing instances of the novel evolution of nodulation. The ability to distinguish between ancestral and derived states is necessary to understand the evolution of any trait. In the case of many legume genera, the states are not known for nodulation (Doyle, 2011). For example, the genus *Senna* does not nodulate (Sprent, 2009). If none of its ancestors were able to nodulate, lack of nodulation would be an

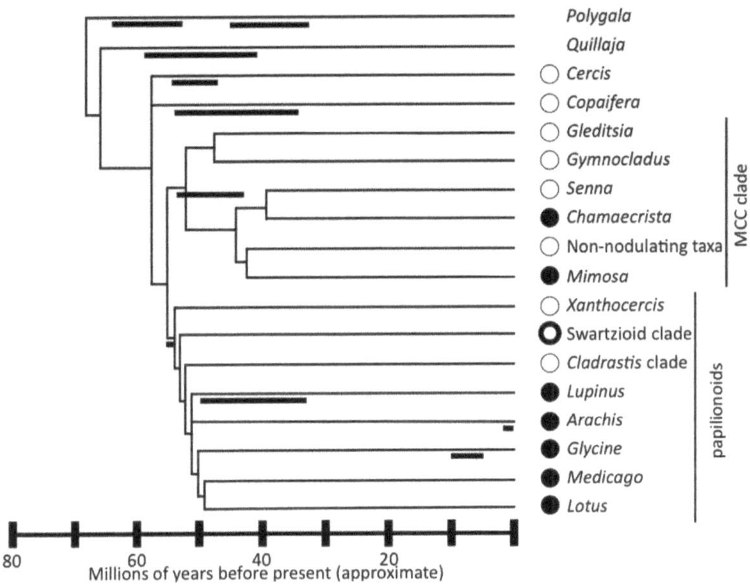

Figure 4 Summary legume phylogeny *(Adapted from Cannon et al. (2015).)* showing selected taxa (for more detailed figure, see Doyle, 2011). Filled circles indicate that many species in the clade nodulate, while empty circles indicate that no members of the clade are known to nodulate. The circle for the Swartzioid clade is meant to indicate that some members of the clade nodulate. Timing of hypothesized genome duplication events is indicated using red bars. *Polygala* and *Quillaja* are nonlegume outgroups, neither of which nodulates; these genera and legumes are members of the order Fabales, one of several orders of the nitrogen-fixing clade, where nodulation occurs sporadically. Nodulation in legumes occurs only in the Mimosoid–Cassiinae–Caesalpinieae (MCC) clade and in the papilionoids, and may have originated in their common ancestor. However, this hypothesis is not parsimonious given the absence of nodulation from most genera in the MCC clade outside of the core mimosoids (represented here by *Mimosa*, but including over 3000 species, nearly all of which nodulate), and from some of the 'basal' lineages of papilionoids (here represented by *Xanthocercis*, the *Cladrastis* clade, and some members of the Swartzioid clade). It is more parsimonious to hypothesize multiple independent origins of nodulation, one or more times each in the MCC and papilionoid clades (Doyle, 2011). Independent origins of nodulation are known.

ancestral ('primitive') character; but if its ancestors were able to nodulate, but this ability has been lost, the absence of nodulation would be a derived state. Thus, it is at present unknown how many independent origins of nodulation there have been in the legume family (Doyle, 2011). Similarly, the number and timing of polyploidy events in the legumes has not been known until recently (Cannon et al., 2010, 2015). Moreover, when WGD events are older than a few million years, the ancestor or ancestors of the polyploid typically are not known, because diploid progenitors are extinct. Finally, it is difficult to reconstruct ancestral states for gene families due to processes including gene birth and death, concerted evolution and the evolution of gene expression.

Despite these difficulties, progress has been made in assessing the role of polyploidy in the evolution of nodulation. Researchers have proposed the unique evolution of a still-unknown 'predisposition' or 'precursor' for nodulation in a common ancestor of the NFC (Soltis et al., 1995). A recent modelling study updates this hypothesis, reaffirming that this predisposition evolved a single time, over 100 MYA, and suggesting that subsequent evolution led to 'actualization' that enabled formation of symbiotic associations in nodule structures (Werner, Cornwell, Sprent, Kattge, & Kiers, 2014). In turn, this actualization was followed by additional processes of refinement, elaboration and diversification in different lineages. Elements of signalling cascades are clearly candidates for the evolution of the precursor and steps in the process of actualization leading to effective nodulation symbioses.

Constraining the analysis to rhizobial symbiosis in legumes, establishing timing of WGD events and instances of actualization, or evolution of nodulation proper, has also been difficult. While fossil evidence exists for AM symbioses (Remy, Taylor, Hass, & Kerp, 1994; Taylor, Remy, Hass, & Kerp, 1995), there are no known fossil nodules (Sprent, 2007). The most recent common ancestor of legumes existed around 60 MYA, and the early evolution of the family was rapid, with the major lineages all established within 10–15 MY (Bruneau, Mercure, Lewis, & Herendeen, 2008; Lavin, Herendeen, & Wojciechowski, 2005). Nodulation has been rarely lost in two major lineages – over 95% of papilionoids nodulate, and a comparable percentage of mimosoids do as well – with several other legume groups also able to nodulate. Given uncertainties in legume phylogeny, incomplete data on nodulation, and an inability to discriminate loss of nodulation from ancestral inability to nodulate (discussed above), we still do not know how many times nodulation was gained or lost in the family. However, focusing only on papilionoids, we know that the common papilionoid ancestor experienced a WGD event (Cannon et al., 2015), and given our current understanding of legume phylogeny (Cardoso et al., 2012, 2013), it seems most likely that this ancestor did not nodulate. Thus, the timing and potential coincidence of this WGD event with the advent or refinement of rhizobial symbioses has led to speculation that the polyploidy event in the papilionoid lineage had a critical role in the evolution of legume nodulation.

Recent evidence, primarily derived from genomic and transcriptomic data on gene retention and expression, has been used to support the proposition that the WGD event shared by the papilionoid legumes enabled the refinement and specialization of various nodulation-related processes (e.g. Kim, Parupalli, Azam, Lee, & Varshney, 2013; Li et al., 2013; Young et al., 2011). The evidence supporting the role of this WGD in the development of nodulation centres on the retention of gene duplicates originating from the papilionoid WGD where one member of each pair is currently involved in nodulation-related processes. This, coupled with differential expression across tissues and functions, suggests that the WGD contributed genetic material that served in the development and/or refinement of nodulation.

For example, in their analysis of the *M. truncatula* genome, Young et al. (2011) found that the nodulation-related genes *NFP*, the Nod factor receptor involved in rhizobial signal perception, and *ERN1* both have duplicates (*LYR1* and *ERN2*, respectively) that date to the papilionoid WGD, and have contrasting expression patterns, suggesting functional specialization. While *NFP* and *ERN1* are implicated in and expressed during nodulation, *LYR1* and *ERN2* are associated with mycorrhizal infection. This is in contrast to *P. andersonii*, the nonlegume nodulating with rhizobia: *P. andersonii* possesses a single *NFP* orthologue that functions in both rhizobial and mycorrhizal interactions (Op den Camp, Streng, et al., 2011). This suggests that the signalling receptors *NFP* and *LYR1* developed their specialized functions following duplication via the papilionoid WGD.

Li et al. (2013), working with the sequenced genomes of four papilionoid legumes, found further examples of apparent preferential gene retention following duplication during the papilionoid WGD, where a member of a pair of duplicate genes had a role in nodulation, indicating the polyploidy event 'might have induced the emergence of critical symbiotic genes and increased the complexity of the symbiotic signalling pathway'. They found genes in the flavonoid biosynthesis pathway that were retained in duplicate and suggest that 'more abundant and diverse flavonoids would be synthesized as a result of the polyploidy event and that the enrichment of flavonoids might be adaptive for the complex signalling required for legume—Rhizobium symbiosis' (Li et al., 2013). Some caution is warranted in interpreting this evidence; while possession of duplicated genes encoding an enzyme may be found to lead to increased enzyme concentrations and biosynthetic activity, functional divergence of the homoeologous gene copies would be required for synthesizing a greater diversity of flavonoids. In addition to retention of duplicates involved in flavonoid biosynthesis, this work also highlighted retention of duplicates in other ontology terms that include processes involved in nodulation. These included signalling receptors; nodule organogenesis (particularly gene ontology terms including hormone-mediated signalling pathways and cytokinin-mediated signalling pathways); infection thread formation and growth; and nutrient transport, exchange and metabolism. These hypotheses provide fertile ground for further, more detailed studies on individual gene families and genes.

As noted above, however, any simple account of a connection between polyploidy and the evolution of nodulation based solely on the relationship of nodulation to the papilionoid WGD event is complicated by several additional considerations. The presence of nodulation in the genus *Chamaecrista*, a legume that does not share the papilionoid WGD, also indicates that this particular WGD event was not necessary for the appearance of nodulation, even in legumes (Cannon et al., 2010). On this basis, Cannon et al. (2010) suggested that polyploidy did not necessarily predate nodulation in all legumes and, consequently, there can be no necessary role ascribed to the papilionoid-specific WGD. A recent, more comprehensive phylogenomic study, however, found evidence of an additional polyploidy event at the base of the Mimosoid—Cassiinae—Caesalpinieae (MCC) clade, which includes the genus *Chamaecrista* (Cannon et al., 2015). Thus, nodulation in the MCC clade may yet be connected to an MCC-specific WGD event. However, just as there are non-nodulating taxa in the papilionoid subfamily that experienced the papilionoid-specific WGD event, there are also non-nodulating taxa that shared the MCC-specific event (e.g. *Gleditsia*, *Gymnocladus*, *Senna*); indeed, outside the mimosoid clade, most members of the MCC do not nodulate. Moreover, evidence was also found for separate polyploidies in the history of the Cercideae and Detarieae (Cannon et al., 2015), which are groups in which there is no known nodulation (Sprent, 2009). The presence of these non-nodulating lineages that have undergone WGD events in their evolutionary history reinforces the observation that polyploidy does not inevitably lead to nodulation in legumes, despite the presence of the nodulation precursor. It remains unclear, though, whether these non-nodulating taxa represent multiple losses (preceded by a single origin of nodulation) or whether there were, conversely, multiple independent gains of nodulation. According to the models of Werner et al. (2014), the 'actualization' of nodulation occurred independently on numerous occasions in the MCC clade; however, this modelling study used a very different tree than that which has been used in other legume phylogenies and does not provide definitive information on where nodulation originated in the clade. Even assuming the evolution of a shared predisposition for nodulation in all members of the Fabales, it appears that subsequent polyploidy events do not necessarily lead to nodulation, given evidence of independent polyploidy events in the non-nodulating *Polygala* and *Quillaja* lineages (Cannon et al., 2015).

The current evidence indicates that no single WGD event appears to have been necessary for the evolution of rhizobial signalling mechanisms and symbiosis throughout the legume family. If there were a necessary connection between the two events, the interpretation would require acknowledgement of the distinct, lineage-specific WGD events in nodulating legumes and the apparent absence of a single WGD event in the most recent common ancestor of all nodulating lineages. While it is plausible that polyploidy can, and likely did, provide raw materials for the elaboration or refinement of nodulation signalling and symbioses (as it does for all other plant processes), it is also still uncertain that WGD is sufficient for the evolution of rhizobial nodulation in all legumes or even in papilionoids. There are, nevertheless, several examples of critical genes involved in papilionoid nodulation, including some encoding signalling functions, that arose as duplicates due to WGD in the evolutionary history of the legume family. Additional tests of the effects of more recent polyploidy events in nodulating legume lineages could help to clarify the potential of WGD events to enhance or yield further refinements.

6. CURRENT AND FUTURE STUDIES OF POLYPLOIDY AND NODULATION: CHALLENGES AND OPPORTUNITIES

Studies using genomic data to examine the connection between polyploidy and the origin or refinement of nodulation have highlighted a need for additional evidence and direct experimentation. Several early inoculation studies attempted to examine the relationship between polyploidy and nodulation in a restricted number of study systems (Leps, Brill, & Bingham, 1980; Weir, 1961). These studies, however, were primarily conducted in synthetically generated autopolyploids and preceded the substantial advances in understanding of the signalling mechanisms underlying nodulation, many of which have occurred since the 1980s (e.g. reviewed in Broughton et al., 2000; Ferguson et al., 2010; Long, 1996; Oldroyd, 2013). Moreover, substantial progress was also made over the course of these decades in advancing the study of polyploidy through the use of molecular techniques (Soltis et al., 2010). Thus, there is now a great opportunity to engage again in direct study of the interaction between these phenomena, given both our enhanced understanding of the molecular mechanisms underlying the evolutionary context and relevance of polyploidy for nodulation indicated by recent genomic studies. To this end, our research has turned to experiments testing relationships between polyploidy and nodulation, flavonoid exudation, receptor diversity and transcriptome-level responses, primarily using *Glycine* subgenus *Glycine* as our study system, which includes a complex of recently formed allopolyploids (Sherman-Broyles et al., 2014).

Along with the opportunities for advancing this area of research, several challenges exist for any future work on the connection between polyploidy and nodulation. These include issues related to variability of the effects of polyploidy at different timescales and the importance of assessing the progenitor contributions to the formation of polyploids, given the sensitivity of rhizobial interactions to genotype-by-genotype-by-environment (G × G × E) interactions. Each of these issues is broadly relevant to the study of polyploids and their ecological interactions, but here we consider particular aspects that will be important in assessing the implications for nodulation symbioses and signalling.

6.1 Studying Effects at Several Timescales: Natural Paleopolyploids, Neopolyploids and Synthetics

Polyploidy can have variable effects at different timescales, and this is a challenge for developing any general understanding of the effects polyploidy can have on nodulation. While ancient polyploidy events and the genomes of ancient paleopolyploids have been the source of data pertaining to the evolution of nodulation, direct evaluations are generally not possible, since the progenitors of these polyploids are not known or available (e.g. Young et al., 2011). Natural, recently formed polyploids, though, often present opportunities to study both the extant progenitors and the polyploids to which they gave rise (Hegarty et al., 2013). Even here, however, caution is required in making inferences, since the formation of polyploids may involve radical genetic and epigenetic changes (e.g. 'genomic shock' of McClintock, 1984) and selection has had the capacity and time to alter rhizobial symbioses in the polyploid plants, such that the effects of selection would be combined with any effects due to polyploidy (Hegarty et al., 2013). In addition, the diploid progenitors and their symbiotic interactions will also have evolved and, in their current ecological responses, may not be accurate representatives of the plants that originally gave rise to the polyploid. In general, even when comparison of natural polyploids and diploid progenitors is possible, such a comparison necessarily confounds the near-term effects of polyploidy, which include effects of changes to cell size and division, development and gene dosage, with longer-term effects, including altered responses to natural selection (Ramsey & Ramsey, 2014). These issues complicate any direct comparisons that could be attempted between extant polyploids and presumed diploid progenitors. Not coincidentally, studies of natural polyploids must necessarily involve consideration of species rather than individual genotypes, since the relevant genotypes are unknown and unsampled if indeed they still exist.

When available, synthetic polyploids present an excellent opportunity to study the immediate effects of polyploidy. The effect of genome doubling itself can be studied, removing from consideration subsequent effects of natural selection in an ecologically relevant context, and can also be used to separate effects of genome duplication from those of genome merger by examining both autopolyploids and allopolyploids (Hegarty et al., 2013). With nodulation-related plant signalling mechanisms, as outlined above, enhanced exudation of a broader diversity of signalling compounds or the expression of Nod factor receptors from distinct diploid progenitors might be such immediate outcomes of polyploidy. However, constraints to the inferences that can be made from the study of synthetic polyploids are precisely due to the isolation from effects at longer timescales noted above, when natural selection has shaped the symbiotic interactions. Interactions between host plant and rhizobial symbiont can exist along a continuum from mutualism to parasitism (Denison & Kiers, 2004); with respect to changes in nodulation-related signalling, interactions that are initially enabled or made possible by polyploidy may later be selected against if they adversely impact fitness. While the expression and silencing of one member of some classes of progenitor genes (duplication-resistant genes) appears to be predictable, stochastic alterations are also observed over the early generations of synthetically formed allopolyploids (e.g. Gaeta, Pires, Iniguez-Luy, Leon, & Osborn, 2007; Gaeta et al., 2009; Madlung & Wendel, 2013; Wang et al., 2004). Thus, a polyploid's biosynthesis of signalling molecules, receptors and other elements of rhizobial signal transduction pathways could also change rapidly in the generations following its formation. These issues, coupled with the fact that the environmental range (and thus the adaptive context for rhizobial symbiosis) is often altered over time for polyploids relative to their diploid progenitors (te Beest et al., 2012), suggest that a combination of studies examining both synthetic and natural (i.e. non-synthetic) polyploids, including established and very recently formed neopolyploids, will be needed in order to develop a complete picture of the implications of polyploidy for nodulation.

6.2 Challenges of Genotype-By-Genotype Interactions

The definition of 'species' continues to be debated (e.g. De Queiroz, 2007), but for our present purposes it can be defined as a collection of genotypes, each with different potentials for genetic interaction (in addition to gene flow, the ability to independently assort and recombine to produce diversity). Consideration of rhizobial nodulation in a species necessarily requires consideration of interactions between genotypes of the host plants (G_{plant}) by the genotypes of the rhizobia ($G_{rhizobia}$) by the environment (E). The variability of $G_{plant} \times G_{rhizobia}$ and $G_{plant} \times G_{rhizobia} \times E$ interactions has been observed in a number of cases (e.g. Heath, Burke, & Stinchcombe, 2012; Heath & Tiffin, 2007). This variability is relevant when studying polyploids for several reasons. Polyploids may be formed multiple times by various combinations of diploid progenitor genotypes and may incorporate further variation from its progenitors by gene flow after formation (Soltis & Soltis, 2009). Even so, formation will involve only a subset of genotypes from within the diploid progenitor population. Thus, the variability of these interactions in a set of plant genotypes within a progenitor species has the ability to affect the symbiotic outcomes in the polyploid produced from them. In a natural context, the symbiotic capacity of a polyploid will depend on the genotypes it has 'sampled' and incorporated from the progenitor population. When a polyploid arises multiple times, the resulting polyploids may interbreed to form a single polyploid species; however, if polyploids from different origins are separated geographically or genetically, the result can be multiple reproductively isolated polyploid species (Soltis & Soltis, 2009). Isolated polyploid species arising from multiple origins would produce low diversity in each individual polyploid species; multiple origins with sampling of a significant part of the diversity of the diploid progenitors, followed by recombination (including independent assortment) could produce a single, much more diverse, polyploid species that would be more likely both to sample more of the diploid diversity (though the sum of diversity across all of the polyploid taxa might be the same in both cases) and to produce much more recombinational diversity. In turn, these differences in diversity will affect the resultant symbiotic signalling capabilities and interaction spaces of the polyploid species. In such cases, comparisons of symbiotic interactions should ideally be made between the polyploids and the diploid progenitor genotypes that contributed to its formation, rather than other genotypes of the diploid species, though this would be difficult in all but the most recently formed natural polyploids and would depend on sampling.

Allopolyploids thus involve the complications of interactive effects due to hybridity and the combination of two diploid progenitor genotypes. Allopolyploid complexes nevertheless provide opportunities to understand the emergent effects of polyploidy on nodulation. While repeated formation of allopolyploids, with different combinations of diploid progenitors, can provide additional complexity, it also provides opportunities to examine a naturally replicated system, where emergent properties can be observed across independent instances of polyploid formation among closely related species. An additional challenge arises, however, given the combination of the effects of hybridity and genome doubling in allopolyploids. Nevertheless, this challenge can be addressed in natural systems by relating levels of hybridity with particular effects to determine whether a possible effect of allopolyploidy correlates strongly with the level of hybridity (i.e. the genetic distance or divergence between progenitor species). Effects that are strongly associated with allopolyploids, but do not correlate well with hybridity levels are thus less likely to be due to hybridity and may be attributable to the phenomenon of genome doubling in the allopolyploids (and subsequent selection acting on them). In the context of rhizobial symbiosis, untangling such effects is important, since there is evidence that both hybridity and polyploidy can, for example, affect biosynthesis of flavonoid compounds (e.g. Levy & Levin, 1971, 1974; Orians, 2000; Schwarze, 1959). Thus, changes in rhizobial interactions observed between an allopolyploid and its diploid progenitors may be attributable either to hybridity, genome doubling or both.

7. CONCLUSION

In recent years, researchers have expressed interest in the possibility of engineering biological nitrogen fixation in cereals, due to the potential value of reducing nitrogen fertilizer application (Beatty & Good, 2011; Oldroyd & Dixon, 2014). Such potential applications have stimulated additional interest in understanding the evolution of nodulation in legumes and other NFC species. Analyses of genomic data from nodulating legumes known to have experienced WGD have suggested an important role for polyploidy in the evolution of nodulation in such species. While several challenges exist to the direct evaluation of the connections between polyploidy and nodulation, the opportunities for increased understanding are great. Many of the signalling mechanisms underlying nodulation have now been elucidated. The methods available for accurately identifying and assessing polyploids have improved. Thus, future research involving the combined study of diverse nodulating systems will yield insights into the emergent implications of polyploidy for rhizobial signalling and symbiotic interactions with plants.

REFERENCES

Aguilar, J. M. M., Ashby, A. M., Richards, A. J., Loake, G. J., Watson, M. D., & Shaw, C. H. (1988). Chemotaxis of *Rhizobium leguminosarum* biovar *phaseoli* towards flavonoid inducers of the symbiotic nodulation genes. *Journal of General Microbiology, 134*(10), 2741–2746.

Althabegoiti, M. J., López-García, S. L., Piccinetti, C., Mongiardini, E. J., Pérez-Giménez, J., Quelas, J. I., et al. (2008). Strain selection for improvement of *Bradyrhizobium japonicum* competitiveness for nodulation of soybean. *FEMS Microbiology Letters, 282*(1), 115–123.

Ané, J. M., Kiss, G. B., Riely, B. K., Penmetsa, R. V., Oldroyd, G. E., Ayax, C., et al. (2004). *Medicago truncatula DMI1* required for bacterial and fungal symbioses in legumes. *Science (New York, N.Y.), 303*(5662), 1364–1367.

Aoki, T., Akashi, T., & Ayabe, S. (2000). Flavonoids of leguminous plants: structure, biological activity, and biosynthesis. *Journal of Plant Research, 113*(4), 475–488.

Armitage, J., Gallagher, A., & Johnston, A. (1988). Comparison of the chemotactic behaviour of *Rhizobium leguminosarum* with and without the nodulation plasmid. *Molecular Microbiology, 2*(6), 743–748.

Arrighi, J. F., Barre, A., Ben Amor, B., Bersoult, A., Soriano, L. C., Mirabella, R., et al. (2006). The *Medicago truncatula* lysin [corrected] motif-receptor-like kinase gene family includes *NFP* and new nodule-expressed genes. *Plant Physiology, 142*(1), 265–279.

Ashfield, T., Egan, A. N., Pfeil, B. E., Chen, N. W., Podicheti, R., Ratnaparkhe, M. B., et al. (2012). Evolution of a complex disease resistance gene cluster in diploid *Phaseolus* and tetraploid *Glycine*. *Plant Physiology, 159*(1), 336–354.

Barbour, W. M., Hattermann, D. R., & Stacey, G. (1991). Chemotaxis of *Bradyrhizobium japonicum* to soybean exudates. *Applied and Environmental Microbiology, 57*(9), 2635–2639.

Beatty, P. H., & Good, A. G. (2011). Future prospects for cereals that fix nitrogen. *Science (New York, N.Y.), 333*(6041), 416–417.

te Beest, M., Le Roux, J. J., Richardson, D. M., Brysting, A. K., Suda, J., Kubesova, M., et al. (2012). The more the better? The role of polyploidy in facilitating plant invasions. *Annals of Botany, 109*(1), 19–45.

Bennett, M. D. (1972). Nuclear DNA content and minimum generation time in herbaceous plants. *Proceedings of the Royal Society of London. Series B, Biological Sciences, 181*(1063), 109–135.

Berendsen, R. L., Pieterse, C. M., & Bakker, P. A. (2012). The rhizosphere microbiome and plant health. *Trends in Plant Science, 17*(8), 478–486.

Birchler, J. A., & Veitia, R. A. (2012). Gene balance hypothesis: connecting issues of dosage sensitivity across biological disciplines. *Proceedings of the National Academy of Sciences of the United States of America, 109*(37), 14746–14753.

Blanc, G., & Wolfe, K. H. (2004). Functional divergence of duplicated genes formed by polyploidy during *Arabidopsis* evolution. *The Plant Cell, 16*(7), 1679–1691.

Bolaños-Vásquez, M., & Werner, D. (1997). Effects of *Rhizobium tropici, R. etli,* and *R. leguminosarum* bv. *phaseoli* on *nod* gene-inducing flavonoids in root exudate of *Phaseolus vulgaris. Molecular Plant-Microbe Interactions, 10*(3), 339–346.

Broghammer, A., Krusell, L., Blaise, M., Sauer, J., Sullivan, J. T., Maolanon, N., et al. (2012). Legume receptors perceive the rhizobial lipochitin oligosaccharide signal molecules by direct binding. *Proceedings of the National Academy of Sciences of the United States of America, 109*(34), 13859–13864.

Broughton, W. J., Jabbouri, S., & Perret, X. (2000). Keys to symbiotic harmony. *Journal of Bacteriology, 182*(20), 5641–5652.

Bruneau, A., Mercure, M., Lewis, G. P., & Herendeen, P. S. (2008). Phylogenetic patterns and diversification in the caesalpinioid legumes this paper is one of a selection of papers published in the special issue on systematics research. *Botany, 86*(7), 697–718.

Buggs, R. J., Wendel, J. F., Doyle, J. J., Soltis, D. E., Soltis, P. S., & Coate, J. E. (2014). The legacy of diploid progenitors in allopolyploid gene expression patterns. *Philosophical Transactions of the Royal Society of London. Series B, Biological Sciences, 369*(1648). http://dx.doi.org/10.1098/rstb.2013.0354.

Cacco, G., Ferrari, G., & Lucci, G. (1976). Uptake efficiency of roots in plants at different ploidy levels. *The Journal of Agricultural Science, 87*(03), 585–589.

Caetano-Anollés, G., Crist-Estes, D. K., & Bauer, W. D. (1988). Chemotaxis of *Rhizobium meliloti* to the plant flavone luteolin requires functional nodulation genes. *Journal of Bacteriology, 170*(7), 3164–3169.

Cannon, S. B., Ilut, D., Farmer, A. D., Maki, S. L., May, G. D., Singer, S. R., et al. (2010). Polyploidy did not predate the evolution of nodulation in all legumes. *PLoS One, 5*(7), e11630.

Cannon, S. B., McKain, M. R., Harkess, A., Nelson, M. N., Dash, S., Deyholos, M. K., et al. (2015). Multiple polyploidy events in the early radiation of nodulating and nonnodulating legumes. *Molecular Biology and Evolution, 32*(1), 193–210.

Cardoso, D., Pennington, R., de Queiroz, L., Boatwright, J., Van Wyk, B., Wojciechowski, M., et al. (2013). Reconstructing the deep-branching relationships of the papilionoid legumes. *South African Journal of Botany, 89*, 58–75.

Cardoso, D., de Queiroz, L. P., Pennington, R. T., de Lima, H. C., Fonty, E., Wojciechowski, M. F., et al. (2012). Revisiting the phylogeny of papilionoid legumes: new insights from comprehensively sampled early-branching lineages. *American Journal of Botany, 99*(12), 1991–2013.

Cesco, S., Neumann, G., Tomasi, N., Pinton, R., & Weisskopf, L. (2010). Release of plant-borne flavonoids into the rhizosphere and their role in plant nutrition. *Plant and Soil, 329*(1–2), 1–25.

Charpentier, M., Bredemeier, R., Wanner, G., Takeda, N., Schleiff, E., & Parniske, M. (2008). *Lotus japonicus* CASTOR and POLLUX are ion channels essential for perinuclear calcium spiking in legume root endosymbiosis. *The Plant Cell, 20*(12), 3467–3479.

Chen, Z. J. (2010). Molecular mechanisms of polyploidy and hybrid vigor. *Trends in Plant Science, 15*(2), 57–71.

Cho, I., & Blaser, M. J. (2012). The human microbiome: at the interface of health and disease. *Nature Reviews Genetics, 13*(4), 260–270.

Cho, M. J., & Harper, J. E. (1991). Effect of inoculation and nitrogen on isoflavonoid concentration in wild-type and nodulation-mutant soybean roots. *Plant Physiology, 95*(2), 435–442.

Coate, J. E., Bar, H., & Doyle, J. J. (2014). Extensive translational regulation of gene expression in an allopolyploid (*Glycine dolichocarpa*). *The Plant Cell, 26*(1), 136–150.
Coate, J. E., & Doyle, J. J. (2010). Quantifying whole transcriptome size, a prerequisite for understanding transcriptome evolution across species: an example from a plant allopolyploid. *Genome Biology and Evolution, 2*, 534–546.
Coate, J. E., & Doyle, J. J. (2015). Variation in transcriptome size: are we getting the message? *Chromosoma, 124*(1), 27–43.
Coate, J. E., Luciano, A. K., Seralathan, V., Minchew, K. J., Owens, T. G., & Doyle, J. J. (2012). Anatomical, biochemical, and photosynthetic responses to recent allopolyploidy in *Glycine dolichocarpa* (Fabaceae). *American Journal of Botany, 99*(1), 55–67.
Combes, M. C., Cenci, A., Baraille, H., Bertrand, B., & Lashermes, P. (2012). Homeologous gene expression in response to growing temperature in a recent allopolyploid (*Coffea arabica* L.). *The Journal of Heredity, 103*(1), 36–46.
Cullimore, J., & Gough, C. (2013). Lipochitooligosaccharide perception and the basis of partner recognition in root endosymbioses. In F. J. de Bruijn (Ed.), *Molecular microbial ecology of the rhizosphere* (Vols 1 and 2, pp. 483–494). Hoboken: Wiley.
D'Haeze, W., & Holsters, M. (2002). Nod factor structures, responses, and perception during initiation of nodule development. *Glycobiology, 12*(6), 79R–105R.
De Bodt, S., Maere, S., & Van de Peer, Y. (2005). Genome duplication and the origin of angiosperms. *Trends in Ecology and Evolution, 20*(11), 591–597.
De Hoff, P. L., Brill, L. M., & Hirsch, A. M. (2009). Plant lectins: the ties that bind in root symbiosis and plant defense. *Molecular Genetics and Genomics, 282*(1), 1–15.
De Mita, S., Streng, A., Bisseling, T., & Geurts, R. (2014). Evolution of a symbiotic receptor through gene duplications in the legume–rhizobium mutualism. *New Phytologist, 201*(3), 961–972.
De Queiroz, K. (2007). Species concepts and species delimitation. *Systematic Biology, 56*(6), 879–886.
Deakin, W. J., & Broughton, W. J. (2009). Symbiotic use of pathogenic strategies: rhizobial protein secretion systems. *Nature Reviews Microbiology, 7*(4), 312–320.
Denison, R. F., & Kiers, E. T. (2004). Lifestyle alternatives for rhizobia: mutualism, parasitism, and forgoing symbiosis. *FEMS Microbiology Letters, 237*(2), 187–193.
Depuydt, S., & Hardtke, C. S. (2011). Hormone signalling crosstalk in plant growth regulation. *Current Biology, 21*(9), R365–R373.
Doyle, J. J. (1994). Phylogeny of the legume family: an approach to understanding the origins of nodulation. *Annual Review of Ecology and Systematics, 25*, 325–349.
Doyle, J. J. (2011). Phylogenetic perspectives on the origins of nodulation. *Molecular Plant-Microbe Interactions, 24*(11), 1289–1295.
Egan, A. N., & Doyle, J. (2010). A comparison of global, gene-specific, and relaxed clock methods in a comparative genomics framework: dating the polyploid history of soybean (*Glycine max*). *Systematic Biology, 59*(5), 534–547.
Endre, G., Kereszt, A., Kevei, Z., Mihacea, S., Kaló, P., & Kiss, G. B. (2002). A receptor kinase gene regulating symbiotic nodule development. *Nature, 417*(6892), 962–966.
Etzler, M. E., Kalsi, G., Ewing, N. N., Roberts, N. J., Day, R. B., & Murphy, J. B. (1999). A nod factor binding lectin with apyrase activity from legume roots. *Proceedings of the National Academy of Sciences of the United States of America, 96*(10), 5856–5861.
Ezenwa, V. O., Gerardo, N. M., Inouye, D. W., Medina, M., & Xavier, J. B. (2012). Animal behavior and the microbiome. *Science (New York, N.Y.), 338*(6104), 198–199.
Fawcett, J. A., & Van de Peer, Y. (2010). Angiosperm polyploids and their road to evolutionary success. *Trends in Evolutionary Biology, 2*(1), e3.
Ferguson, B. J., Indrasumunar, A., Hayashi, S., Lin, M., Lin, Y., Reid, D. E., et al. (2010). Molecular analysis of legume nodule development and autoregulation. *Journal of Integrative Plant Biology, 52*(1), 61–76.

Flagel, L. E., Wendel, J. F., & Udall, J. A. (2012). Duplicate gene evolution, homoeologous recombination, and transcriptome characterization in allopolyploid cotton. *BMC Genomics, 13*, 302. http://dx.doi.org/10.1186/1471-2164-13-302.

Freeling, M., & Thomas, B. C. (2006). Gene-balanced duplications, like tetraploidy, provide predictable drive to increase morphological complexity. *Genome Research, 16*(7), 805–814.

Gaeta, R. T., Pires, J. C., Iniguez-Luy, F., Leon, E., & Osborn, T. C. (2007). Genomic changes in resynthesized *Brassica napus* and their effect on gene expression and phenotype. *The Plant Cell, 19*(11), 3403–3417.

Gaeta, R. T., Yoo, S., Pires, J., Doerge, R., Chen, Z. J., & Osborn, T. C. (2009). Analysis of gene expression in resynthesized *Brassica napus* allopolyploids using Arabidopsis 70mer oligo microarrays. *PLoS One, 4*(3), e4760.

Gagnon, H., & Ibrahim, R. K. (1998). Aldonic acids: a novel family of nod gene inducers of *Mesorhizobium loti, Rhizobium lupini*, and *Sinorhizobium meliloti*. *Molecular Plant-Microbe Interactions, 11*(10), 988–998.

Garsmeur, O., Schnable, J. C., Almeida, A., Jourda, C., D'Hont, A., & Freeling, M. (2014). Two evolutionarily distinct classes of paleopolyploidy. *Molecular Biology and Evolution, 31*(2), 448–454.

Gobbato, E., Marsh, J. F., Vernié, T., Wang, E., Maillet, F., Kim, J., et al. (2012). A GRAS-type transcription factor with a specific function in mycorrhizal signaling. *Current Biology, 22*(23), 2236–2241.

Gomez, C., Terrier, N., Torregrosa, L., Vialet, S., Fournier-Level, A., Verries, C., et al. (2009). Grapevine MATE-type proteins act as vacuolar H+-dependent acylated anthocyanin transporters. *Plant Physiology, 150*(1), 402–415.

Gonzalez-Rizzo, S., Crespi, M., & Frugier, F. (2006). The *Medicago truncatula* CRE1 cytokinin receptor regulates lateral root development and early symbiotic interaction with *Sinorhizobium meliloti*. *The Plant Cell, 18*(10), 2680–2693.

Gould, S. J., & Vrba, E. S. (1982). Exaptation-a missing term in the science of form. *Paleobiology, 8*(1), 4–15.

Green, A. F., Ramsey, T. S., & Ramsey, J. (2011). Phylogeny and biogeography of ivies (*Hedera* spp., Araliaceae), a polyploid complex of woody vines. *Systematic Botany, 36*(4), 1114–1127.

Green, A. F., Ramsey, T. S., & Ramsey, J. (2013). Polyploidy and invasion of English ivy (*Hedera* spp., Araliaceae) in North American forests. *Biological Invasions, 15*(10), 2219–2241.

Gresshoff, P. M., Lohar, D., Chan, P., Biswas, B., Jiang, Q., Reid, D., et al. (2009). Genetic analysis of ethylene regulation of legume nodulation. *Plant Signaling and Behavior, 4*(9), 818–823.

Griesbach, R., & Kamo, K. (1996). The effect of induced polyploidy on the flavonols of petunia 'Mitchell'. *Phytochemistry, 42*(2), 361–363.

Grover, C., Gallagher, J., Szadkowski, E., Yoo, M., Flagel, L., & Wendel, J. (2012). Homoeolog expression bias and expression level dominance in allopolyploids. *New Phytologist, 196*(4), 966–971.

Gust, A. A., Willmann, R., Desaki, Y., Grabherr, H. M., & Nürnberger, T. (2012). Plant LysM proteins: modules mediating symbiosis and immunity. *Trends in Plant Science, 17*(8), 495–502.

Harbert, R. S., Brown, A. H., & Doyle, J. J. (2014). Climate niche modeling in the perennial *Glycine* (Leguminosae) allopolyploid complex. *American Journal of Botany, 101*(4), 710–721.

Hayashi, M., Saeki, Y., Haga, M., Harada, K., Kouchi, H., & Umehara, Y. (2012). *Rj (rj)* genes involved in nitrogen-fixing root nodule formation in soybean. *Breeding Science, 61*(5), 544.

Hayashi, M., Shiro, S., Kanamori, H., Mori-Hosokawa, S., Sasaki-Yamagata, H., Sayama, T., et al. (2014). A thaumatin-like protein, Rj4, controls nodule symbiotic specificity in soybean. *Plant and Cell Physiology, 55*(9), 1679–1689.

Heath, K. D., Burke, P. V., & Stinchcombe, J. R. (2012). Coevolutionary genetic variation in the legume-rhizobium transcriptome. *Molecular Ecology, 21*(19), 4735–4747.

Heath, K. D., & Tiffin, P. (2007). Context dependence in the coevolution of plant and rhizobial mutualists. *Proceedings of the Royal Society of London. Series B, Biological Sciences, 274*(1620), 1905–1912.

Hegarty, M., Coate, J., Sherman-Broyles, S., Abbott, R., Hiscock, S., & Doyle, J. (2013). Lessons from natural and artificial polyploids in higher plants. *Cytogenetic and Genome Research, 140*(2–4), 204–225.

Hijmans, R. J., Gavrilenko, T., Stephenson, S., Bamberg, J., Salas, A., & Spooner, D. M. (2007). Geographical and environmental range expansion through polyploidy in wild potatoes (*Solanum* section *Petota*). *Global Ecology and Biogeography, 16*(4), 485–495.

Hull-Sanders, H. M., Johnson, R. H., Owen, H. A., & Meyer, G. A. (2009a). Effects of polyploidy on secondary chemistry, physiology, and performance of native and invasive genotypes of *Solidago gigantea* (Asteraceae). *American Journal of Botany, 96*(4), 762–770.

Hull-Sanders, H. M., Johnson, R. H., Owen, H. A., & Meyer, G. A. (2009b). Influence of polyploidy on insect herbivores of native and invasive genotypes of *Solidago gigantea* (Asteraceae). *Plant Signaling and Behavior, 4*(9), 893–895.

Indrasumunar, A., Kereszt, A., Searle, I., Miyagi, M., Li, D., Nguyen, C. D., et al. (2010). Inactivation of duplicated nod factor receptor 5 (NFR5) genes in recessive loss-of-function non-nodulation mutants of allotetraploid soybean (*Glycine max* L. merr.). *Plant and Cell Physiology, 51*(2), 201–214.

Indrasumunar, A., Searle, I., Lin, M., Kereszt, A., Men, A., Carroll, B. J., et al. (2011). Nodulation factor receptor kinase 1α controls nodule organ number in soybean (*Glycine max* L. merr). *The Plant Journal, 65*(1), 39–50.

Innan, H., & Kondrashov, F. (2010). The evolution of gene duplications: classifying and distinguishing between models. *Nature Reviews Genetics, 11*(2), 97–108.

Jiao, Y., Wickett, N. J., Ayyampalayam, S., Chanderbali, A. S., Landherr, L., Ralph, P. E., et al. (2011). Ancestral polyploidy in seed plants and angiosperms. *Nature, 473*(7345), 97–100.

Jones, J. D., & Dangl, J. L. (2006). The plant immune system. *Nature, 444*(7117), 323–329.

Kape, R., Parniske, M., & Werner, D. (1991). Chemotaxis and nod gene activity of *Bradyrhizobium japonicum* in response to hydroxycinnamic acids and isoflavonoids. *Applied and Environmental Microbiology, 57*(1), 316–319.

Keen, N., & Staskawicz, B. (1988). Host range determinants in plant pathogens and symbionts. *Annual Reviews in Microbiology, 42*(1), 421–440.

Kiba, T., Kudo, T., Kojima, M., & Sakakibara, H. (2011). Hormonal control of nitrogen acquisition: roles of auxin, abscisic acid, and cytokinin. *Journal of Experimental Botany, 62*(4), 1399–1409.

Kim, D. H., Parupalli, S., Azam, S., Lee, S., & Varshney, R. K. (2013). Comparative sequence analysis of nitrogen fixation-related genes in six legumes. *Frontiers in Plant Science, 4*, 300.

Kosslak, R. M., Bookland, R., Barkei, J., Paaren, H. E., & Appelbaum, E. R. (1987). Induction of *Bradyrhizobium japonicum* common *nod* genes by isoflavones isolated from *Glycine max*. *Proceedings of the National Academy of Sciences of the United States of America, 84*(21), 7428–7432.

Kosslak, R. M., Joshi, R. S., Bowen, B. A., Paaren, H. E., & Appelbaum, E. R. (1990). Strain-specific inhibition of *nod* gene induction in *Bradyrhizobium japonicum* by flavonoid compounds. *Applied and Environmental Microbiology, 56*(5), 1333–1341.

Krouk, G., Ruffel, S., Gutiérrez, R. A., Gojon, A., Crawford, N. M., Coruzzi, G. M., et al. (2011). A framework integrating plant growth with hormones and nutrients. *Trends in Plant Science, 16*(4), 178–182.
Lavin, M., Herendeen, P. S., & Wojciechowski, M. F. (2005). Evolutionary rates analysis of Leguminosae implicates a rapid diversification of lineages during the tertiary. *Systematic Biology, 54*(4), 575–594.
Lavania, U. C., Srivastava, S., Lavania, S., Basu, S., Misra, N. K., & Mukai, Y. (2012). Autopolyploidy differentially influences body size in plants, but facilitates enhanced accumulation of secondary metabolites, causing increased cytosine methylation. *The Plant Journal, 71*(4), 539–549.
Lea, U. S., Slimestad, R., Smedvig, P., & Lillo, C. (2007). Nitrogen deficiency enhances expression of specific MYB and bHLH transcription factors and accumulation of end products in the flavonoid pathway. *Planta, 225*(5), 1245–1253.
Leps, W. T., Brill, W. J., & Bingham, E. (1980). Effect of alfalfa ploidy on nitrogen fixation. *Crop Science, 20*(4), 427–430.
Levin, D. A. (1983). Polyploidy and novelty in flowering plants. *American Naturalist, 122*(1), 1–25.
Lévy, J., Bres, C., Geurts, R., Chalhoub, B., Kulikova, O., Duc, G., et al. (2004). A putative Ca2+ and calmodulin-dependent protein kinase required for bacterial and fungal symbioses. *Science (New York, N.Y.), 303*(5662), 1361–1364.
Levy, M., & Levin, D. A. (1971). The origin of novel flavonoids in *Phlox* allotetraploids. *Proceedings of the National Academy of Sciences of the United States of America, 68*(7), 1627–1630.
Levy, M., & Levin, D. A. (1974). Novel flavonoids and reticulate evolution in the *Phlox pilosa-P. drummondii* complex. *American Journal of Botany, 61*(2), 156–167.
Li, Q. G., Zhang, L., Li, C., Dunwell, J. M., & Zhang, Y. M. (2013). Comparative genomics suggests that an ancestral polyploidy event leads to enhanced root nodule symbiosis in the papilionoideae. *Molecular Biology and Evolution, 30*(12), 2602–2611.
Lohar, D. P., Schaff, J. E., Laskey, J. G., Kieber, J. J., Bilyeu, K. D., & Bird, D. M. (2004). Cytokinins play opposite roles in lateral root formation, and nematode and rhizobial symbioses. *The Plant Journal, 38*(2), 203–214.
Lohar, D., Stiller, J., Kam, J., Stacey, G., & Gresshoff, P. M. (2009). Ethylene insensitivity conferred by a mutated *Arabidopsis* ethylene receptor gene alters nodulation in transgenic *Lotus japonicus*. *Annals of Botany, 104*(2), 277–285.
Long, S. R. (1996). Rhizobium symbiosis: nod factors in perspective. *The Plant Cell, 8*(10), 1885–1898.
Lowry, E., & Lester, S. (2006). The biogeography of plant reproduction: potential determinants of species' range sizes. *Journal of Biogeography, 33*(11), 1975–1982.
Mabood, F., Souleimanov, A., Khan, W., & Smith, D. (2006). Jasmonates induce nod factor production by *Bradyrhizobium japonicum*. *Plant Physiology and Biochemistry, 44*(11), 759–765.
Madlung, A. (2013). Polyploidy and its effect on evolutionary success: old questions revisited with new tools. *Heredity, 110*(2), 99–104.
Madlung, A., & Wendel, J. F. (2013). Genetic and epigenetic aspects of polyploid evolution in plants. *Cytogenetic and Genome Research, 140*(2–4), 270–285.
Madsen, E. B., Madsen, L. H., Radutoiu, S., Olbryt, M., Rakwalska, M., Szczyglowski, K., et al. (2003). A receptor kinase gene of the LysM type is involved in legume perception of rhizobial signals. *Nature, 425*(6958), 637–640.
Maere, S., De Bodt, S., Raes, J., Casneuf, T., Van Montagu, M., Kuiper, M., et al. (2005). Modeling gene and genome duplications in eukaryotes. *Proceedings of the National Academy of Sciences of the United States of America, 102*(15), 5454–5459.
Marinova, K., Pourcel, L., Weder, B., Schwarz, M., Barron, D., Routaboul, J. M., et al. (2007). The *Arabidopsis* MATE transporter TT12 acts as a vacuolar flavonoid/H+ -antiporter

active in proanthocyanidin-accumulating cells of the seed coat. *The Plant Cell, 19*(6), 2023–2038.

Martin, S. L., & Husband, B. C. (2009). Influence of phylogeny and ploidy on species ranges of North American angiosperms. *Journal of Ecology, 97*(5), 913–922.

Mathesius, U. (2003). Conservation and divergence of signalling pathways between roots and soil microbes—the rhizobium-legume symbiosis compared to the development of lateral roots, mycorrhizal interactions and nematode-induced galls. In J. J. Abe (Ed.), *Roots: The dynamic interface between plants and the earth* (pp. 105–119). Dordrecht: Kluwer.

McClintock, B. (1984). The significance of responses of the genome to challenge. *Science, 226*, 792–801.

Mears, J. A. (1980). Chemistry of polyploids: a summary with comments on *Parthenium* (Asteraceae-Ambrosiinae). In W. H. Lewis (Ed.), *Polyploidy* (pp. 77–101). New York: Springer.

Meng, H., Jiang, S., Hua, S., Lin, X., Li, Y., Guo, W., et al. (2011). Comparison between a tetraploid turnip and its diploid progenitor (*Brassica rapa* L.): the adaptation to salinity stress. *Agricultural Sciences in China, 10*(3), 363–375.

Meyers, B. C., Kaushik, S., & Nandety, R. S. (2005). Evolving disease resistance genes. *Current Opinion in Plant Biology, 8*(2), 129–134.

Miller, L. D., Yost, C. K., Hynes, M. F., & Alexandre, G. (2007). The major chemotaxis gene cluster of *Rhizobium leguminosarum* bv. *viciae* is essential for competitive nodulation. *Molecular Microbiology, 63*(2), 348–362.

Mitra, R. M., Gleason, C. A., Edwards, A., Hadfield, J., Downie, J. A., Oldroyd, G. E., et al. (2004). A Ca2+/calmodulin-dependent protein kinase required for symbiotic nodule development: gene identification by transcript-based cloning. *Proceedings of the National Academy of Sciences of the United States of America, 101*(13), 4701–4705.

Miyahara, A., Hirani, T. A., Oakes, M., Kereszt, A., Kobe, B., Djordjevic, M. A., et al. (2008). Soybean nodule autoregulation receptor kinase phosphorylates two kinase-associated protein phosphatases in vitro. *The Journal of Biological Chemistry, 283*(37), 25381–25391.

Münzbergová, Z. (2006). Ploidy level interacts with population size and habitat conditions to determine the degree of herbivory damage in plant populations. *OIKOS, 115*(3), 443–452.

Nakano, Y., Yamakawa, T., Ikeda, M., & Ishizuka, J. (1997). Nodulation of *Rj*-soybean varieties with *Rhizobium fredii* USDA 193 under limited supply of nutrients. *Soil Science and Plant Nutrition, 43*(4), 929–932.

Noguti, Y., Oka, H., & Otuka, T. (1940). Studies on the polyploidy of *Nicotiana* induced by the treatment with colchicine. II. growth rate and chemical analysis of diploid and its autotetraploids in *Nicotiana rustica* and *N. tabacum*. *Japanese Journal of Botany, 10*, 343–364.

Novák, K., Chovanec, P., Skrdleta, V., Kropacova, M., Lisa, L., & Nemcova, M. (2002). Effect of exogenous flavonoids on nodulation of pea (*Pisum sativum* L.). *Journal of Experimental Botany, 53*(375), 1735–1745.

Nuismer, S. L., & Thompson, J. N. (2001). Plant polyploidy and non-uniform effects on insect herbivores. *Proceedings of the Royal Society of London. Series B, Biological Sciences, 268*(1479), 1937–1940.

Oldroyd, G. E. (2013). Speak, friend, and enter: signalling systems that promote beneficial symbiotic associations in plants. *Nature Reviews Microbiology, 11*(4), 252–263.

Oldroyd, G. E., & Dixon, R. (2014). Biotechnological solutions to the nitrogen problem. *Current Opinion in Biotechnology, 26*, 19–24.

Oldroyd, G. E., & Downie, J. A. (2008). Coordinating nodule morphogenesis with rhizobial infection in legumes. *Annual Review of Plant Biology, 59*, 519–546.

Oldroyd, G. E., Murray, J. D., Poole, P. S., & Downie, J. A. (2011). The rules of engagement in the legume-rhizobial symbiosis. *Annual Review of Genetics, 45*, 119–144.

Op den Camp, R. H., De Mita, S., Lillo, A., Cao, Q., Limpens, E., Bisseling, T., et al. (2011). A phylogenetic strategy based on a legume-specific whole genome duplication yields symbiotic cytokinin type-A response regulators. *Plant Physiology, 157*(4), 2013–2022.

Op den Camp, R., Streng, A., De Mita, S., Cao, Q., Polone, E., Liu, W., et al. (2011). LysM-type mycorrhizal receptor recruited for rhizobium symbiosis in nonlegume *Parasponia*. *Science (New York, N.Y.), 331*(6019), 909–912.

Orians, C. M. (2000). The effects of hybridization in plants on secondary chemistry: implications for the ecology and evolution of plant-herbivore interactions. *American Journal of Botany, 87*(12), 1749–1756.

Pandit, M. K., Pocock, M. J., & Kunin, W. E. (2011). Ploidy influences rarity and invasiveness in plants. *Journal of Ecology, 99*(5), 1108–1115.

Paterson, A. H., Chapman, B. A., Kissinger, J. C., Bowers, J. E., Feltus, F. A., & Estill, J. C. (2006). Many gene and domain families have convergent fates following independent whole-genome duplication events in *Arabidopsis, Oryza, Saccharomyces* and *Tetraodon*. *Trends in Genetics, 22*(11), 597–602.

Pawlowski, K., & Sprent, J. I. (2008). Comparison between actinorhizal and legume symbiosis. In K. Pawlowski, & W. E. Newton (Eds.), *Nitrogen-fixing actinorhizal symbioses* (pp. 261–288). Dordrecht: Springer.

Peck, M. C., Fisher, R. F., & Long, S. R. (2006). Diverse flavonoids stimulate NodD1 binding to *nod* gene promoters in *Sinorhizobium meliloti*. *Journal of Bacteriology, 188*(15), 5417–5427.

Penmetsa, R. V., Uribe, P., Anderson, J., Lichtenzveig, J., Gish, J., Nam, Y. W., et al. (2008). The *Medicago truncatula* ortholog of *Arabidopsis EIN2*, sickle, is a negative regulator of symbiotic and pathogenic microbial associations. *The Plant Journal, 55*(4), 580–595.

Phillips, D. A., Joseph, C. M., & Maxwell, C. A. (1992). Trigonelline and stachydrine released from alfalfa seeds activate NodD2 protein in *Rhizobium meliloti*. *Plant Physiology, 99*(4), 1526–1531.

Price, N., Relíc, B., Talmont, F., Lewin, A., Promé, D., Pueppke, S., et al. (1992). Broad-host-range rhizobium species strain NGR234 secretes a family of carbamoylated, and fucosylated, nodulation signals that are O-acetylated or sulphated. *Molecular Microbiology, 6*(23), 3575–3584.

Pueppke, S. G., Bolaños-Vásquez, M. C., Werner, D., Bec-Ferte, M. P., Prome, J. C., & Krishnan, H. B. (1998). Release of flavonoids by the soybean cultivars McCall and peking and their perception as signals by the nitrogen-fixing symbiont *Sinorhizobium fredii*. *Plant Physiology, 117*(2), 599–606.

Radutoiu, S., Madsen, L. H., Madsen, E. B., Felle, H. H., Umehara, Y., Grønlund, M., et al. (2003). Plant recognition of symbiotic bacteria requires two LysM receptor-like kinases. *Nature, 425*(6958), 585–592.

Radutoiu, S., Madsen, L. H., Madsen, E. B., Jurkiewicz, A., Fukai, E., Quistgaard, E. M., et al. (2007). LysM domains mediate lipochitin–oligosaccharide recognition and *NFR* genes extend the symbiotic host range. *The EMBO Journal, 26*(17), 3923–3935.

Ramsey, J., & Ramsey, T. S. (2014). Ecological studies of polyploidy in the 100 years following its discovery. *Philosophical Transactions of the Royal Society of London. Series B, Biological Sciences, 369*(1648). http://dx.doi.org/10.1098/rstb.2013.0352.

Ramsey, J., & Schemske, D. W. (2002). Neopolyploidy in flowering plants. *Annual Review of Ecology and Systematics, 33*, 589–639.

Remy, W., Taylor, T. N., Hass, H., & Kerp, H. (1994). Four hundred-million-year-old vesicular arbuscular mycorrhizae. *Proceedings of the National Academy of Sciences of the United States of America, 91*(25), 11841–11843.

Sakakibara, H., Takei, K., & Hirose, N. (2006). Interactions between nitrogen and cytokinin in the regulation of metabolism and development. *Trends in Plant Science, 11*(9), 440–448.

Sasaki, T., Suzaki, T., Soyano, T., Kojima, M., Sakakibara, H., & Kawaguchi, M. (2014). Shoot-derived cytokinins systemically regulate root nodulation. *Nature Communications, 5*, 4983.
Schlaepfer, D. R., Edwards, P. J., & Billeter, R. (2010). Why only tetraploid *Solidago gigantea* (Asteraceae) became invasive: a common garden comparison of ploidy levels. *Oecologia, 163*(3), 661–673.
Schnable, J. C., Springer, N. M., & Freeling, M. (2011). Differentiation of the maize subgenomes by genome dominance and both ancient and ongoing gene loss. *Proceedings of the National Academy of Sciences of the United States of America, 108*(10), 4069–4074.
Schwarze, P. (1959). Untersuchungen über die gesteigerte flavonoidproduktion in *Phaseolus*-artbastarden (*Phaseolus vulgaris* × *Phaseolus coccineus*). *Planta, 54*(2), 152–161.
Searle, I. R., Men, A. E., Laniya, T. S., Buzas, D. M., Iturbe-Ormaetxe, I., Carroll, B. J., et al. (2003). Long-distance signaling in nodulation directed by a CLAVATA1-like receptor kinase. *Science (New York, N.Y.), 299*(5603), 109–112.
Segraves, K., & Thompson, J. (1999). Plant polyploidy and pollination: floral traits and insect visits to diploid and tetraploid *Heuchera grossulariifolia*. *Evolution, 53*(4), 1114–1127.
Shaw, S. L., & Long, S. R. (2003). Nod factor elicits two separable calcium responses in *Medicago truncatula* root hair cells. *Plant Physiology, 131*(3), 976–984.
Sherman-Broyles, S., Bombarely, A., Powell, A. F., Doyle, J. L., Egan, A. N., Coate, J. E., et al. (2014). The wild side of a major crop: Soybean's perennial cousins from down under. *American Journal of Botany, 101*(10), 1651–1665.
Shubin, N., Tabin, C., & Carroll, S. (2009). Deep homology and the origins of evolutionary novelty. *Nature, 457*(7231), 818–823.
Soltis, D. E., Buggs, R. J., Doyle, J. J., & Soltis, P. S. (2010). What we still don't know about polyploidy. *Taxon, 59*(5), 1387–1403.
Soltis, P. S., & Soltis, D. E. (2009). The role of hybridization in plant speciation. *Annual Review of Plant Biology, 60*, 561–588.
Soltis, D. E., Soltis, P. S., Morgan, D. R., Swensen, S. M., Mullin, B. C., Dowd, J. M., et al. (1995). Chloroplast gene sequence data suggest a single origin of the predisposition for symbiotic nitrogen fixation in angiosperms. *Proceedings of the National Academy of Sciences of the United States of America, 92*(7), 2647–2651.
Soyano, T., & Hayashi, M. (2014). Transcriptional networks leading to symbiotic nodule organogenesis. *Current Opinion in Plant Biology, 20*, 146–154.
Sprent, J. I. (2007). Evolving ideas of legume evolution and diversity: a taxonomic perspective on the occurrence of nodulation. *New Phytologist, 174*(1), 11–25.
Sprent, J. I. (2009). *Legume nodulation: A global perspective*. Chichester: Wiley-Blackwell.
Stebbins, G. L. (1985). Polyploidy, hybridization, and the invasion of new habitats. *Annals of the Missouri Botanical Garden, 72*(4), 824–832.
Stracke, S., Kistner, C., Yoshida, S., Mulder, L., Sato, S., Kaneko, T., et al. (2002). A plant receptor-like kinase required for both bacterial and fungal symbiosis. *Nature, 417*(6892), 959–962.
Sudová, R., Pánková, H., Rydlová, J., Münzbergová, Z., & Suda, J. (2014). Intraspecific ploidy variation: a hidden, minor player in plant-soil-mycorrhizal fungi interactions. *American Journal of Botany, 101*(1), 26–33.
Sudová, R., Rydlová, J., Münzbergová, Z., & Suda, J. (2010). Ploidy-specific interactions of three host plants with arbuscular mycorrhizal fungi: does genome copy number matter? *American Journal of Botany, 97*(11), 1798–1807.
Sugiyama, A., Shitan, N., & Yazaki, K. (2007). Involvement of a soybean ATP-binding cassette-type transporter in the secretion of genistein, a signal flavonoid in legume-rhizobium symbiosis. *Plant Physiology, 144*(4), 2000–2008.
Takei, K., Sakakibara, H., Taniguchi, M., & Sugiyama, T. (2001). Nitrogen-dependent accumulation of cytokinins in root and the translocation to leaf: Implication of cytokinin

species that induces gene expression of maize response regulator. *Plant and Cell Physiology, 42*(1), 85–93.

Takei, K., Ueda, N., Aoki, K., Kuromori, T., Hirayama, T., Shinozaki, K., et al. (2004). *AtIPT3* is a key determinant of nitrate-dependent cytokinin biosynthesis in *Arabidopsis*. *Plant and Cell Physiology, 45*(8), 1053–1062.

Taylor, T., Remy, W., Hass, H., & Kerp, H. (1995). Fossil arbuscular mycorrhizae from the Early Devonian. *Mycologia, 87*(4), 560–573.

Thompson, J. N., & Merg, K. F. (2008). Evolution of polyploidy and the diversification of plant-pollinator interactions. *Ecology, 89*(8), 2197–2206.

Thompson, J. N., Nuismer, S. L., & Merg, K. (2004). Plant polyploidy and the evolutionary ecology of plant/animal interactions. *Biological Journal of the Linnean Society, 82*(4), 511–519.

Tirichine, L., Sandal, N., Madsen, L. H., Radutoiu, S., Albrektsen, A. S., Sato, S., et al. (2007). A gain-of-function mutation in a cytokinin receptor triggers spontaneous root nodule organogenesis. *Science (New York, N.Y.), 315*(5808), 104–107.

Těšitelová, T., Jersáková, J., Roy, M., Kubátová, B., Těšitel, J., Urfus, T., et al. (2013). Ploidy-specific symbiotic interactions: divergence of mycorrhizal fungi between cytotypes of the *Gymnadenia conopsea* group (Orchidaceae). *New Phytologist, 199*(4), 1022–1033.

Vance, C. P. (2001). Symbiotic nitrogen fixation and phosphorus acquisition. Plant nutrition in a world of declining renewable resources. *Plant Physiology, 127*(2), 390–397.

Vanneste, K., Maere, S., & Van de Peer, Y. (2014). Tangled up in two: a burst of genome duplications at the end of the cretaceous and the consequences for plant evolution. *Philosophical Transactions of the Royal Society of London. Series B, Biological Sciences, 369*(1648). http://dx.doi.org/10.1098/rstb.2013.0353.

Wang, J., Tian, L., Madlung, A., Lee, H. S., Chen, M., Lee, J. J., et al. (2004). Stochastic and epigenetic changes of gene expression in *Arabidopsis* polyploids. *Genetics, 167*(4), 1961–1973.

Washburn, J. D., & Birchler, J. A. (2014). Polyploids as a "model system" for the study of heterosis. *Plant Reproduction, 27*(1), 1–5.

Webb, B. A., Hildreth, S., Helm, R. F., & Scharf, B. E. (2014). *Sinorhizobium meliloti* chemoreceptor McpU mediates chemotaxis toward host plant exudates through direct proline sensing. *Applied and Environmental Microbiology, 80*(11), 3404–3415.

Weir, J. (1961). A comparison of the nodulation of diploid and tetraploid varieties of red clover inoculated with different rhizobial strains. *Plant and Soil, 14*(1), 85–89.

Weiss-Schneeweiss, H., Emadzade, K., Jang, T. S., & Schneeweiss, G. M. (2013). Evolutionary consequences, constraints and potential of polyploidy in plants. *Cytogenetic and Genome Research, 140*(2–4), 137–150.

Wendel, J., & Doyle, J. (2005). Polyploidy and evolution in plants. In R. J. Henry (Ed.), *Plant diversity and evolution: Genotypic and phenotypic variation in higher plants* (pp. 97–117). Wallingford: CABI.

Werner, G. D., Cornwell, W. K., Sprent, J. I., Kattge, J., & Kiers, E. T. (2014). A single evolutionary innovation drives the deep evolution of symbiotic N_2-fixation in angiosperms. *Nature Communications, 5*, 4087.

White, J., Prell, J., James, E. K., & Poole, P. (2007). Nutrient sharing between symbionts. *Plant Physiology, 144*(2), 604–614.

Wood, T. E., Takebayashi, N., Barker, M. S., Mayrose, I., Greenspoon, P. B., & Rieseberg, L. H. (2009). The frequency of polyploid speciation in vascular plants. *Proceedings of the National Academy of Sciences of the United States of America, 106*(33), 13875–13879.

Yang, S., Tang, F., Gao, M., Krishnan, H. B., & Zhu, H. (2010). *R* gene-controlled host specificity in the legume-rhizobia symbiosis. *Proceedings of the National Academy of Sciences of the United States of America, 107*(43), 18735–18740.

Yokota, K., & Hayashi, M. (2011). Function and evolution of nodulation genes in legumes. *Cellular and Molecular Life Sciences, 68*(8), 1341–1351.

Yokoyama, T. (2008). Flavonoid-responsive *nodY-lacZ* expression in three phylogenetically different bradyrhizobium groups. *Canadian Journal of Microbiology, 54*(5), 401–410.

Yoo, M., Liu, X., Pires, J. C., Soltis, P. S., & Soltis, D. E. (2014). Nonadditive gene expression in polyploids. *Annual Review of Genetics, 48*, 485–517.

Yoo, M., Szadkowski, E., & Wendel, J. (2013). Homoeolog expression bias and expression level dominance in allopolyploid cotton. *Heredity, 110*(2), 171–180.

Yost, C. K., Rochepeau, P., & Hynes, M. F. (1998). *Rhizobium leguminosarum* contains a group of genes that appear to code for methyl-accepting chemotaxis proteins. *Microbiology (Reading, England), 144*(7), 1945–1956.

Young, N. D., Debellé, F., Oldroyd, G. E., Geurts, R., Cannon, S. B., Udvardi, M. K., et al. (2011). The *Medicago* genome provides insight into the evolution of rhizobial symbioses. *Nature, 480*(7378), 520–524.

Zhang, M., Wu, Y. H., Lee, M. K., Liu, Y. H., Rong, Y., Santos, T. S., et al. (2010). Numbers of genes in the NBS and RLK families vary by more than four-fold within a plant species and are regulated by multiple factors. *Nucleic Acids Research, 38*(19), 6513–6525.

Zhao, J., & Dixon, R. A. (2009). MATE transporters facilitate vacuolar uptake of epicatechin 3′-O-glucoside for proanthocyanidin biosynthesis in *Medicago truncatula* and *Arabidopsis*. *The Plant Cell, 21*(8), 2323–2340.

Zhao, J., Huhman, D., Shadle, G., He, X. Z., Sumner, L. W., Tang, Y., et al. (2011). MATE2 mediates vacuolar sequestration of flavonoid glycosides and glycoside malonates in *Medicago truncatula*. *The Plant Cell, 23*(4), 1536–1555.

CHAPTER EIGHT

Plant Cues and Factors Influencing the Behaviour of Beneficial Nematodes as a Belowground Indirect Defense

Jared Gregory Ali*,§,1, Elizabeth Davidson-Lowe§

*Department of Entomology, Pennsylvania State University, State College, PA, USA

§Department of Entomology, Michigan State University, East Lansing, MI, USA

[1]Corresponding author: E-mail: jga8@psu.edu

Contents

1. Nematode Diversity 193
2. A Nematode's Sensory Tool Kit 194
3. Entomopathogenic Nematodes 195
4. Indirect Defences and Herbivore-Induced Plant Cues Belowground 196
5. Sampling Belowground Herbivore-Induced Volatiles 200
6. Manipulation of Root Volatiles for Agricultural Improvement 203
7. Multitrophic Interactions and Paradox Surrounding Belowground Indirect Defences 206

Acknowledgements 209

References 209

Abstract

Plant signals play diverse roles to the many organisms that surround them. One facet of this is their ability to manipulate organisms in a manner that protects them or harms herbivores that feed on them. This relationship has more recently been recognized to occur belowground. Here, we discuss these belowground interactions, techniques and findings, focussing on entomopathogenic nematodes and soil nematode chemotaxis in response to plant root cues and potential implications for agroecosystems and fundamental concepts in ecological trophic cascades.

Plants influence the behaviour of and modify community composition of soil-dwelling organisms through the exudation of organic molecules (Bais, Weir, Perry, Gilroy, & Vivanco, 2006). Given the chemical complexity of the soil matrix, soil-dwelling organisms have evolved the ability to detect and respond to these cues for successful foraging due to their dependency on green plants, either directly or indirectly, as their primary source of

Advances in Botanical Research, Volume 75
ISSN 0065-2296
http://dx.doi.org/10.1016/bs.abr.2015.08.003

energy. The second trophic level, or the primary consumers, needs plants for nourishment, while the third trophic level, or secondary consumers such as predators or parasites, might utilize plant cues to locate their herbivore host. Several decades of work have shown that animals use plant-derived physical (visual, tactile), and/or chemical (olfactory, gustatory) cues to locate and accept their food source (Schoonhoven, van Loon, & Dicke, 2005). Although most multitrophic interaction studies have been done using predominantly aboveground communities (Price et al., 1980; Tscharntke & Hawkins, 2002; Vet & Dicke, 1992), the last decade has shown a marked increase in exploring soil—plant—animal interactions (Bonkowski, Villenave, & Griffiths, 2009; De Deyn et al., 2003; van der Putten, 2003; Rasmann & Agrawal, 2008; Soler, Erb, & Kaplan, 2013), and how plants mediate interactions between aboveground and belowground communities (van Dam, 2009; Erb, Ton, Degenhardt, & Turlings, 2008; Kaplan, Halitschke, Kessler, Sardanelli, & Denno, 2008; Soler et al., 2013; Wardle et al. 2004). Roots are frequently the storage site for nutrients and metabolites, and they are utilized as food and shelter for numerous soil-dwelling organisms (van Dam, 2009; Hunter, 2001). This belowground facet of the plant is no doubt an epicentre of ecological interactions for all groups of soil inhabitants.

Whether at the roots or shoots of a plant, the intense pressure on the first trophic level has led to evolutionary mechanisms for both direct and indirect plant defence. Chemical toxins and physical barriers make up much of what is designated as direct defence. Indirect defence is described by mechanisms that either attract natural enemies or improve their foraging in search of herbivores, thus facilitating control of herbivore populations (Karban & Baldwin, 1997). General patterns and theories for chemically mediated interactions structuring aboveground communities (Karban & Baldwin, 1997; Price et al., 1980; Vet & Dicke, 1992) can also be applied belowground (Ali, Campos-Herrera, Alborn, Duncan, & Stelinski, 2013; van der Putten, 2009; Rasmann, Bauerle, Poveda, & Vannette, 2011; van Tol et al., 2001). However, striking differences exist between aboveground and belowground subsystems. The soil matrix is composed of all three phases (gas, liquid and solid), which can impact mobility, behaviour, signalling and interactions between organisms in a different manner than when living aboveground. This will influence the mobility of organisms belowground, which often do not surpass >1 m^2 in their lifetime, whereas aboveground, vertebrates can explore >1 ha per day (Rasmann, Ali, Helder, & van der Putten, 2012). Moreover, the physiochemical legacy of the soil, shaped by high metabolic retention capacity, slow turnover of chemical metabolites, slower diffusion and higher heterogeneity, suggests longer times for ecological interactions to occur (Coleman & Crossley, 2004). Cumulatively, these factors should impose selection for particular stimuli to be perceived by soil organisms. This makes sense because soil inhabitants are strongly limited in using visual information, but can use chemical and tactile cues to communicate and behave (Johnson & Nielsen, 2012; Jones, 2002). The group of soil organisms we will focus on in this review belongs to the phylum Nematoda.

1. NEMATODE DIVERSITY

This chapter is particularly focussed on plant–soil–nematode interactions and how root exudates can impact searching and foraging behaviour of beneficial entomopathogenic nematodes. However, we provide here a brief review of nematodes, which are among the most diverse groups of soil organisms. Each square metre of soil may contain millions of individual nematodes belonging to >400 species (Gaugler & Bilgrami, 2004), and to at least seven functional types: plant feeding nematodes (e.g. *Pratylenchus, Heterodera, Meloidogyne, Helicotylenchus* and *Aphelenchoides*), plant-associated nematodes (e.g. *Tylenchus* and *Dorylaimellus*), fungal hyphae-feeding nematodes (e.g. *Aphelenchus, Aphelenchoides, Leptonchus* and *Diphtherophora*), bacterial-feeding nematodes (e.g. *Rhabditis, Plectus, Cephalobus* and *Caenorhabditis*), nematodes that feed on unicellular eukaryotes, animal parasitic nematodes (e.g. *Heterorhabditis* and *Steinernema*), and omnivorous nematodes (e.g. *Dorylaimida*) (Yeates, 1999). However, it is worth noting that more than one feeding habit can occur within a genus (e.g. the genus *Aphelenchoides* harbours fungivorous and plant parasitic species) or even within a single species (Yeates, 1999). Feeding habits may thus be the result of convergent evolution (hence, taxonomically heterogeneous groups).

Because of their abundance, systematic and functional diversities, and their representation in multiple trophic levels in the soil food web, nematodes have a strong influence on ecosystem dynamics and functioning (Rasmann et al., 2012). They have been shown to influence soil nutrient cycling, growth rate, health and yield of plants, as well as populations of other soil inhabitants. For example, herbivores have been shown to influence plant yield (Chitwood, 2002), plant community composition (De Deyn et al., 2003) and successional dynamics (De Deyn et al., 2003).

Microbial feeders can stimulate nutrient cycling and plant yield (Fu, Ferris, Brown, & Plant, 2005; Ingham, Trofymow, Ingham, & Coleman, 1985), detritivorous nematodes can contribute up to 40% of total organic matter mineralization (De Ruiter et al., 1993), and predators and parasites can reduce arthropod populations (Kaya & Gaugler, 1993), which, in turn, can have cascading effects on plant performance (Rasmann et al., 2012).

2. A NEMATODE'S SENSORY TOOL KIT

Although it is hard to imagine complex sensory machinery in such a small creature, as in all animals, nematode behaviour is a sequence of integrating and synchronizing several external stimuli leading to responses (e.g. locomotion, movement, feeding, mating and penetration) (Gaugler & Bilgrami, 2004). A nematode's sensory apparatus allows it to use chemical, electrical, light, mechanical and temperature stimuli (Jones, 2002) to orientate, move and locate a sexual partner or energy sources (food) in the soil (Lee, 2002).

Although there is variation in their detailed structure, nematode sense organs can essentially be subdivided into cuticular and internal sense organs. Cuticular sense organs are generally composed of a sheath cell, a socket cell and a variable number of dendritic processes, and are responsible for detecting chemical, mechanical and temperature-related stimuli. Internal sense organs are more diverse, and are mainly responsible for detecting mechanical stimuli as well as light stimuli (Jones, 2002). Invariably, the largest and most complex of the nematode sense organs are the amphids, which are exposed to the external environment by a pore in the cuticle, and which primarily function as chemoreceptors. In *Caenorhabditis elegans*, the tail's bilateral sensory organs, called phasmids, are also shown to function as chemoreceptors, and help the nematode to orient towards or away from the chemical stimuli (Hilliard, Bargmann, & Bazzicalupo, 2002). It should be noted that there are some nematode clades that function without phasmids according to Holterman et al. (2006). As all feeding types are represented in these clades, it is concluded that phasmids are not essential for vital chemoreception functions. Because of the complexity and abundance of soil chemicals compared to other physical stimuli, chemoreception by nematodes is undoubtedly the most important source of nuanced stimulus (Jones, 2002).

Responses by nematodes to chemical stimuli have been extensively studied in the bacteriophagous nematode, *C. elegans*, some plant-parasitic nematodes (e.g. *Meloidogyne* and *Globodera* spp.), and in an increasing number of animal-parasitic nematodes (e.g. *Heterorhabditis* and *Steinernema* spp.). Our review will focus on beneficial natural enemies, an animal-parasitic group of entomopathogenic nematodes.

3. ENTOMOPATHOGENIC NEMATODES

Entomopathogenic nematodes comprise two families with one genus each: Steinernematidae (genus *Steinernema*) and Heterorhabditidae (genus *Heterorhabditis*). These nematode families exclusively grow and reproduce inside arthropod hosts, and third instar infective juveniles leave the cadaver. The infective juvenile (dauer juvenile) is the only stage that can survive without food for long periods while foraging for alternative hosts in the soil (Gaugler, 2002). All members of these families are actually bacteriophagous, having evolved the ability to carry and introduce symbiotic bacteria into the body cavities of insects (Chaston et al., 2011). Bacteria then reproduce in the insect, thus providing the food for the nematodes to complete their life cycle (Poinar, 1990). Because of their ability to kill the majority of insect orders and families in the soil and the relative ease of large scale culturing in artificial solid or liquid media, they have been promoted as exceptionally good candidates for the biological control of insect pests of roots in crop fields (Gaugler & Kaya, 1990). In general, entomopathogenic nematode foraging strategies can be divided into two broad categories: a cruiser form, which is highly mobile, and an ambusher form (sit-and-wait) (Gaugler, 2002). Cruising foragers have a higher probability of finding sedentary and cryptic resources than ambushers, and ambush foragers are more effective at finding resources with high mobility (Lewis, Campbell, Griffin, Kaya, & Peters, 2006). However, direct evidence suggests that foraging strategies used by different infective juvenile species to find a host vary along a continuum between ambush and cruise foragers, and may be plastic depending on contexts (Ennis, Dillon, & Griffin, 2010; Lewis et al., 2006).

It was suggested that entomopathogenic nematode attraction to a suitable host can integrate different possible cues such as temperature, electric potential, carbon dioxide and various organic and inorganic substances, although, no specific compound was found to be responsible for entomopathogenic nematode attraction towards the insect host (Boff, Zoon, & Smits, 2001; Kaya & Gaugler, 1993), and it was generally assumed that nematode orientation and aggregation were due to unspecific signalling such as carbon dioxide emissions. For example, Lewis, Gaugler, and Harrison (1993) found that *Steinernema glaseri* responded positively to volatile cues from an insect host and that this response was eliminated if CO_2 was removed. A similar response was later found by Grewal, Lewis, Gaugler and Campbell (1994) for other cruiser *Steinernema* spp. and for two species of *Heterorhabditis*. This general response to unspecified volatile cues was later extended to many other *Steinernema* spp. (Campbell, Lewis, Stock, Nadler, & Kaya, 2003). On the other hand, it was argued that CO_2 should mainly function as a short-range attractant, playing a role in host penetration through the spiracles (Ishibashi & Kondo, 1990). It seems also very unlikely that such a general signal can be unambiguously exploited by foraging nematodes looking for a specific arthropod host feeding on roots, as demonstrated by Bilgrami, Kondo, and Yoshiga (2001), who found that *S. glaseri* was attracted to plant tissue from roots and leaves from *Asparagus officinalis* and *Tagetes erecta*, but not to nitrogenous insect products (Bilgrami et al., 2001). It was simultaneously proposed that entomopathogenic nematodes can use arthropod herbivore-induced plant cues to locate the site of wounding, and thus the host (Boff et al., 2001; van Tol et al., 2001).

4. INDIRECT DEFENCES AND HERBIVORE-INDUCED PLANT CUES BELOWGROUND

The manipulation of beneficial natural enemies is not a new subject, yet the notion has received renewed interest with the discovery of herbivore-induced plant volatiles (HIPVs) as attractants for natural enemies (Hare, 2011; Kaplan, 2011; Turlings & Wäckers, 2004). Despite being an important concept, the study of indirect defence cues as a driver of ecological interactions has largely (i) been restricted to only half of the plant that is aboveground tissues (Hunter, 2001) and (ii) focussed largely on the cue's ability to attract natural enemies of herbivores and/or improve their foraging success (Kaplan, 2011; Turlings, Hiltpold, & Rasmann, 2012). As documented within the plant canopy, belowground herbivory also imparts significant natural selection favouring the evolution of root defences and traits that not only mediate associations with the third trophic level, but an array of trophic interactions (Erb & Lu, 2013). Understanding the cascading consequences of HIPVs among competing species can reveal information that is crucial to interpreting the potential multiple roles of information-based induced plant defence.

Indirect plant defence is described as a phenomenon of plant defence strategy where plants attract enemies of the herbivore when 'attacked' (Dicke & Sabelis, 1988; Turlings, Tumlinson, & Lewis, 1990). Plant volatile organic compounds (VOCs), which are mainly comprised of terpenoids, fatty acid derivatives, phenyl propanoids and benzenoids (Dudareva, Pichersky, & Gershenzon, 2004), have particularly been the centre of intensive studies of plant—herbivore—predator interactions for more than two decades. VOC blends can be complex, comprising hundreds of compounds, some of which are not produced by intact or mechanically damaged plants but are synthesized de novo in response to herbivore attack (Mumm & Dicke, 2010; Turlings & Wäckers, 2004).

Primarily due to methodological constraints, most research on plant VOCs released after insect herbivory has been conducted mainly among interactions playing out on the surfaces of leaf tissues, that is, aboveground. However, an increasing number of studies are showing that herbivore-induced belowground volatiles might also trigger predator attraction in the soil. For example, *Neoseiulus cucumeris*, female predatory mites of rust mites (*Aceria tulipae*), responded to belowground volatile signals of tulip bulbs infested by *A. tulipae*, but not to volatiles of untreated or wounded bulbs (Aratchige, Lesna, & Sabelis, 2004). Two inspiring papers demonstrated for the first time that unknown emissions of odorous cues were responsible for attracting entomopathogenic nematodes to insect damaged roots (Boff et al., 2001; van Tol et al., 2001). To date, few additional tritrophic interactions implying belowground HIPV signalling have been described both in agricultural systems (Ali, Alborn, & Stelinski, 2010, 2011; Rasmann et al., 2005) and in natural environments (Rasmann et al., 2011). Ali et al. (2010) have demonstrated that citrus roots, upon feeding by the root weevil, *Diaprepes abbreviates*, emit several terpenes in the surrounding soil. Using belowground olfactometers, Ali et al. (2010) could show that the entomopathogenic nematode, *Steinernema diaprepesi*, was significantly more attracted by citrus roots induced by the insect pest larvae than by mechanically damaged roots or by empty control pots. However, Ali et al. (2011) recently pointed out that insect-induced roots of citrus trees not only had variable responses to nematodes with separate foraging strategies but could also attract the phytopathogenic nematode, *Tylenchulus semipenetrans*. Consequently, this may reduce the exploitation of citrus induced VOC emission in biological control strategies targeting *Diaprepes abbreviates* where rootstocks are not resistant to *T. semipenetrans*.

Outside of agricultural settings, Rasmann et al. (2011) showed that the common milkweed, *Asclepias syriaca*, which is generally fed on by the specialist root herbivore larvae of the cerambycid beetle, *Tetraopes tetraophthalmus*, can release volatiles in the soil. Increased emissions of VOCs after damage were correlated with increased entomopathogenic nematodes, *Heterorhabditis bacteriophora*, in laboratory experiments. Subsequent field trials demonstrated that soil inoculation of nematodes benefitted the plants by restoring the biomass lost due to herbivory to that of control levels (Rasmann et al., 2011). Whether this was correlated with higher levels of emission in damaged plants was not, however, assessed. Root emissions of *A. syriaca* plants are a very complex mixture of >30 compounds, of which only few are described as being in the terpene family (Rasmann et al., 2011). Such a complex blend by itself impedes the assessment of which particular compounds are really responsible for the attraction. This is a problem that is well known for aboveground systems, where the emerging picture is that VOC production in plants is the result of diffuse selection due to multiple players interacting with the plant (Kessler & Heil, 2011). As aboveground, the functional role of belowground compounds, individual components or complex blends, would benefit our knowledge of organisms' intimate relationship with the plant, advanced metabolomics, and multivariate statistical tools (van Dam & Poppy, 2008).

Another example of highly complex volatile blends comes from the roots of cotton (*Gossypium herbaceum*). After feeding by the generalist root feeder larvae of the chrysomelid beetle, *Diabrotica balteata*, cotton plants were scored to emit >10 compounds, among which at least seven terpenoid volatiles were observed (Rasmann & Turlings, 2008). Among all cotton VOCs induced by the chrysomelid larva, the sesquiterpenoid, aristolene, was discussed as being a good candidate for playing a major role in *Heterorhabditis megidis* nematode attraction. This, however, remains to be confirmed in future studies (Rasmann & Turlings, 2008). In the same study, among corn and cotton, nematode preference was also tested against damaged roots of cowpea (*Vigna unguiculata*) plants. In contrast to corn and cotton, cowpea plants emitted almost undetectable amounts of volatiles, which also resulted in lower nematode attraction (Rasmann & Turlings, 2008).

The corn system, first described by Rasmann et al. (2005), is still the best known belowground tritrophic interaction. Upon attack by the voracious larvae of the western corn rootworm, *Diabrotica virgifera virgifera*, European maize varieties emit in soil the sesquiterpene, (*E*)-β-caryophyllene (EβC) (Kollner et al., 2008; Rasmann et al., 2005), a highly attractive HIPV to entomopathogenic nematodes, *H. megidis*, in the laboratory as well as in the field (Hiltpold, Baroni, Toepfer, Kuhlmann, & Turlings, 2010a; Kollner et al., 2008; Rasmann et al., 2005). A series of experiments with various corn lines and synthetic compounds have shown that EβC is an ideal compound to diffuse through the complex belowground soil compartment, that it is among the least costly terpenoids capable of travelling within the soil matrix (Hiltpold & Turlings, 2008) and that it is subject to selection (Kollner et al., 2008). Its production within the root system appears to be systemic even though the root area upon feeding emits more of the volatile and is more attractive to *H. megidis* than distal regions (Hiltpold, Erb, Robert, & Turlings, 2011).

More recently, Hallem et al. (2011) reported positive chemotaxis of the two entomopathogenic nematodes, *H. bacteriophora* and *Steinernema carpocapsae*, to several VOCs, such as methyl salicylate, hexanol, heptanol, undecyl acetate or 4,5-dimethylthiazole. Interestingly, they showed that several volatiles repelled the nematodes. Similar effects of VOCs on the behaviour of the entomopathogenic nematodes were observed by Hiltpold, Baroni, Toepfer, Kuhlmann, & Turlings (2010b) and Rasmann and Turlings (2008). However, no volatiles were identified. Most of the VOCs involved in belowground tritrophic interactions remain unknown, but an increasing effort is being invested into this field of research. Understanding more of these complex interactions would not only allow a better understanding of the rhizosphere, but could also offer ecologically sound alternatives in pest management in agricultural systems.

There still remains a gap in our interpretation of how insect parasitic nematodes would have become sensitive to indirect cues of host location, such as HIPVs. Answers may lie in the life histories of closely related nematode taxa, along with their associated bacteria. For example, the insect parasitic nematode genus, *Heterorhabditis*, most closely resembles a genus of marine nematodes, *Pellioditis* (Dougherty & Nigon, 1949). Species from *Pellioditis* are selective bacterial feeders that occur in the intertidal and coastal regions (Poinar, 1993). There is evidence that the heterorhabditids evolved in a coastal habitat from free-living microbiotrophic marine nematodes (Hara, Gaugler, Kaya, & Lebeck, 1991; Poinar, 1993). The bioluminescent bacteria that are responsible for the pathogenic effects of *Heterorhabditis* on invertebrates are *Photorhabdus luminescens*. *Photorhabdus luminescens* are believed to originate from a marine shore habitat, where there are many reports of living and dead marine invertebrates containing luminescent bacteria (Harvey, 1952). *Pellioditis marina*, a candidate for a pelloiditid that could have evolved into an insect parasitic heterorhabditid, can survive on a luminescent bacterium that is similar to *P. luminescens* (Tietjen, Lee, Rullman, Greengart, & Trompeter, 1970). It becomes easy to imagine a scenario that could permit a free-living bacterial feeder like *P. marina* to evolve into an insect parasitic nematode, where an injective juvenile came into contact with and retained bacteria lethal to invertebrates. This early heterorhabditid need only to parasitize a littoral and beach dwelling crustacean, and a shift from crustacean to an insect would not have been difficult. Possible hosts would be root-feeding weevils (*Otiorhynchus* spp., Curculionidae), which are found along seacoasts on the roots of beachgrasses (e.g. *Ammophila arenaria* or seashore wormwood, *Artemisia maritime*), or scarabid larvae of *Aegialia arenaria* (Scarabidae) along coastal dwelling and beachgrass root feeders. Interestingly, scarabs and curculionids are currently known to be among the most susceptible soil insects to *Heterorhabditis* nematodes (Poinar & Georgis, 1990). If sensitivity to a volatile signal is as inheritable as demonstrated by studies of Hiltpold et al. (2010b), it is not hard to see how entomopathogenic nematode sensitivity to HIPVs is not only possible but also probable. Future work could evaluate this potential relationship by examining herbivore-induced compounds released by plant roots in sandy coastal regions, along with entomopathogenic nematode bioassays.

5. SAMPLING BELOWGROUND HERBIVORE-INDUCED VOLATILES

The technical difficulties associated with dynamics of the soil ecosystem have been a major limitation in studying belowground multitrophic interactions. Soil is an opaque, triphasic medium, making the analysis of individual factors and their interactions difficult (Rasmann et al., 2012). Most research has been based on in vitro analysis of individual factors. Researchers hoping to study factors associated with root signals are often directed to the study of root diffusates, root leachates and/or root exudates. Diffusate is used to convey nonvolatile substances diffusing through the soil and establishing a gradient. Leachate refers to a method of obtaining an extract from the roots, more than it does to the solution itself. Exudate is most often restricted to liquids that gradually 'ooze' from their source, but can be applied to volatiles as well. Approaches to evaluate root volatiles have only recently been developed and applied in contexts of chemical and evolutionary ecology. As mentioned earlier, all of these can serve as signals and cues for herbivores and their natural enemies. However, in this review, we focus on HIPVs as they are often shown to be involved in multitrophic interactions that include natural enemies and have been shown to be detectable from intact plants. Although methods exist for the evaluation of additional exudates from intact plant roots (i.e. polydimethylsiloxane for solid-phase root zone extraction, see Mohney et al., 2009), thus far they have not been used to evaluate induction related to root herbivores.

Rasmann et al. (2005) provided for the first time the evaluation and identification of indirect volatile defences of maize roots using solid-phase microextraction (SPME). SPME is a method of sampling volatiles without the use of solvents. In short, an adsorbent-coated fused silica fibre with properties similar to a gas chromatography (GC) column can collect volatile compounds from the headspace of a sample. The volatile compounds, once fixed to the SPME fibre, can then be thermally desorbed in an injection port of a gas chromatograph and further analysed and/or identified when coupled with known standards or libraries of mass spectroscopy (MS). In order to sample the effects of root herbivory on plant produced VOCs, Rasmann et al. (2005) crushed flash-frozen roots, either fed-upon or not fed-upon, into a fine powder. This powder was then exposed to the SPME fibre. This allowed for the volatiles that had accumulated in either treatment to be sampled and compared with GC—MS. SPME is a rapid and simple extraction method that does not require the use of solvents, and its detection limits can reach parts per trillion (ppt) levels for certain compounds (Pawliszyn, 2009). However, one of the limitations of this technique is that it is a destructive method of sampling root material. The plant and herbivore must be separated and volatiles from this interaction can only be examined after harvesting and crushing the plant tissues.

Ali et al. (2010) were able to nondestructively sample belowground herbivore-induced volatiles from citrus roots using a flow-through dynamic sampling technique coupled with adsorbent traps. In this case, volatiles can be collected and extracted by elution of an adsorbent with low boiling point solvents. Adsorbent traps are typically made of glass tubes filled with the granulated adsorbent, held in place by stainless steel mesh, glass wool plugs or Teflon-fitted rings. By connecting the adsorbent trap to a vacuum pump and pulling air through glass chambers containing intact citrus plants, either with or without feeding larvae, Ali et al. (2010) was able to sample volatiles associated with belowground herbivory nondestructively and in situ. The volatiles collected on this trap are rinsed using a nonpolar (e.g. methanol, hexane and dichloromethane) solvent and analysed with GC—MS. This method allows for the sample to be retained in a solvent, which can be analysed more than once. The solvents containing root volatiles were also tested in sand-filled two-choice bioassay chambers. By using this approach, Ali et al. (2010, 2011) found evidence for entomopathogenic nematode attraction to volatiles from infested citrus roots. Additionally, a soil probe was recently used to sample soil volatiles in Florida citrus groves at a depth of 20 cm (Ali et al., 2012), which constitutes an important advance in the study of volatiles in situ. This soil probe was developed to collect the compound on adsorbent traps by using a vacuum pump to pull air from the soil. Recently, a nondestructive method similar to the one used by Ali et al. (2010) to collect root volatiles was also successfully used to detect (*E*)-β-caryophyllene from maize roots (Robert et al., 2013) and work by Abraham, Giacomuzzi, & Angeli (2015) isolated induced volatiles from apple trees after damage by hoot herbivores.

Proton-transfer-reaction mass spectrometry (PTR-MS) is an additional nondestructive analysis that has recently been applied to the detection of belowground trace HIPVs in real-time (Danner, Samudrala, Cristescu, & van Dam, 2012). Briefly, the PTR-MS technology continuously pumps air to be analysed through a drift tube reactor and a fraction of the VOCs is ionized in proton-transfer reactions with hydronium ions (H_3O^+). The ionized molecules typically form a protonated molecular ion $[M + H]^+$, in which M is the molecular mass of the parent molecule. Organic compounds, such as aldehydes, ketones, alcohols, oxygenated aromatic and aliphatic compounds, will be readily protonated (Danner et al., 2012). The major advantage of using proton transfer is that it utilizes a soft ionization method; therefore, it generally leads to less overall fragmentation of the product ions, and the mass of the product ion equals the VOC mass plus one (Danner et al., 2012). At the end of the drift tube, the reagent and product ions are measured by a quadrupole mass spectrometer, and the product ion signal is proportional to the VOC mixing ratio. PTR-MS allows numerous VOCs to be monitored with a high sensitivity (10−100 ppt) and rapid response time (1−10 s) (de Gouw & Warneke, 2007).

There are many advantages of the PTR-MS method. The more common methods of collecting plant VOCs onto filters with polymer adsorbents make it necessary to preconcentrate the sample before analysis by collecting volatiles during a range of minutes to hours; this reduces the ability to resolve the timing of VOC emission measurements (Danner et al., 2012). Furthermore, solvents used to elute VOCs from the adsorbents may introduce contamination before GC analysis. A major disadvantage is that PTR-MS only determines the mass of product ions, which is, of course, not a unique indicator of the VOC identity. It is clear that isomers cannot be distinguished, and the interpretation of mass spectra is further complicated by the formation of cluster ions and the fragmentation of product ions. A significant experimental effort has therefore been made to characterize the specificity of PTR-MS for different VOCs.

PTR-MS already has been successfully applied to analyse VOCs produced by aboveground (infested) plant parts. Two recent examples show its promise as a technique to evaluate belowground interactions in preliminary PTR-MS results on herbivore-induced root responses in *Brassica* species. Danner et al. (2012) monitored VOCs emanating from roots of potted turnip plants (*Brassica rapa* subsp. *rapa* var. Nancy) during infestation with a belowground herbivore, the larvae of the cabbage root fly (*Delia radicum*). The resulting mass scans found that the intensities of several molecular masses are enhanced in root fly-infested *B. rapa* roots (Danner et al., 2012). In a second example, Danner et al. (2012) monitored the induction of VOCs in *Brassica juncea* roots after infestation with *D. radicum* in real-time and compared it to a control treatment. Initially, they detected only a low emission rate of compounds, which steadily increased with longer feeding times of the root flies. In control plants, the VOC emissions remained at a very low level, allowing a clear distinction between control and infested plants within a few hours after infestation (Danner et al., 2012).

In general, all of these techniques are effective and informative in different ways. The nondestructive sampling techniques are useful in evaluating belowground interactions in situ and may potentially prove useful in additional contexts. However, the properties of the surrounding soil may interfere and make resolution difficult with the potential for significant background. In this way, SPME eliminates such background, but can introduce complications from tissue maceration, where enzymes or oxidation can rapidly change the chemical profile and may not accurately represent the blend released from intact living organisms (Heath & Manukian, 1992; Tollsten & Bergström, 1988). Perhaps a combination of techniques and refinements of approaches will produce the best resolution for the dynamics of an individual system and the factors of concern.

6. MANIPULATION OF ROOT VOLATILES FOR AGRICULTURAL IMPROVEMENT

The substantial advances in research on molecular mechanisms and ecological signalling of insect herbivore-induced VOCs open promising prospects of manipulating the release of these compounds to enhance crop protection. Encouraging examples from laboratory and field experiments support this approach to develop novel ecologically sound crop protection strategies.

Manipulating the plant emitting VOCs appears as a first straightforward approach. Aboveground, for instance, Thaler (1999) applied jasmonic acid on tomato plants in an experimental field. This treatment resulted in the emission of typical VOC blends that the plant would have produced upon herbivore attack. Additionally, the rate of lepidopteran parasitism increased on plants that were treated with the phytohormone compared to control plants (Thaler, 1999). Supporting this plant manipulation approach, Rostàs and Eggert (2008) obtained a significant systemic resistance by treating maize plants with the salicylic acid-mimic, benzo-(1,2,3)-thiadiazole-7-carbothioic acid (S)-methyl ester, resulting in an increased resistance against the fungal phytopathogen, *Setosphaeria turcica*. Upon caterpillar attack, the treated plants were more attractive for the parasitoid wasp, *Microplitis rufiventris*, than plants damaged by the insect larvae only. Control experiments showed that these results were due to the plant-meditated effect rather than to the chemical treatment itself (Rostas & Eggert, 2008). So far, there is no published example of such manipulation in the belowground compartment. Ali et al. (2012) have recently increased mortality of root pests in the field by enhancing host location of naturally occurring entomopathogenic nematodes in citrus and blueberry crops with the application of the citrus root volatile, 1,5-dimethylcyclodeca-1,5,7-triene (pregeijerene).

When the induced VOC blend is clearly identified, it could be considered to genetically manipulate the plant to either (i) make the plant more attractive for beneficial predators or parasitoids or (ii) restore a new phenotype that was lost due to natural or human selection, thus re-establishing predator or parasitoid attraction. (i) Again, the first examples of this approach are aboveground. Indeed, it has been first demonstrated in the model plant, *Arabidopsis thaliana*, in which a linalool/nerolidol synthase gene, FaNES1, from strawberry was introduced. This resulted in the constitutive release of (3S)-(*E*)-nerolidol, rendering the plants attractive to predatory mites, *Phytoseiulus persimilis* (Kappers et al., 2005). Later, *A. thaliana* was transformed with a TPS10 gene, coding for sesquiterpenes typically present in the blends emitted by plants upon caterpillar attack (Schnee et al., 2006). The engineered plants were attractive to the parasitoid wasp, *Cotesia marginiventris*, but only after they had learnt to associate the HIPV blend with the presence of their hosts. (ii) More recently, the terpene synthase gene,

TPS23, has been identified in maize (Kollner et al., 2008). TPS23 is responsible for the synthesis of EβC, a key attractant for some entomopathogenic nematodes (Rasmann et al., 2005). Most of the European maize varieties and Teosinte produce this sesquiterpene, whereas American varieties do not (Kollner et al., 2008; Rasmann et al., 2005). This indicates a shift in the gene activity through breeding selection (Kollner et al., 2008). In 2009, Degenhardt and colleagues were able to restore the ability of maize to recruit entomopathogenic nematodes by inserting a TPS23 gene from *Origanum vulgare* into a nonproducing maize line (Degenhardt, 2009). In the field, the transformed maize line was significantly more attractive for the entomopathogenic nematode, *H. megidis*, compared to the wild type, leading to a better protection for transformed plants. Transformed plants received far less damage by the root pest, *D. v. virgifera*, and the transformation significantly reduced beetle emergence. Overall, this restored indirect defence resulted from the constitutive emission of EβC in the soil (Degenhardt, 2009). This was the first demonstration in the field that plant genotype engineering could enhance biological control. It has to be noticed that because genetic modifications are still controversial, it should also be possible to incorporate highly attractive volatile compounds into new varieties using classical breeding programmes.

In an inundative biological control strategy, the manipulation of the biological control agent can be considered as an option to enhance pest management (Hoy, 1982) – a classic idea that was put into practice with some relative success (Beckendorf & Hoy, 1985; Hoy, 2000). In that perspective, entomopathogenic nematodes appear as good candidates. Several studies have succeeded in selecting beneficial traits such as host finding (Gaugler & Campbell, 1991; Gaugler, Campbell, & McGuire 1989), virulence (Peters & Ehlers, 1998; Tomalak, 1994) and tolerance to temperature (Grewal, Gaugler, & Wang, 1996) or desiccation (Strauch, Oestergaard, Hollmer, & Ehlers, 2004).

For specific entomopathogenic nematodes, the knowledge of key attractants is now available (Hiltpold et al., 2010a; Rasmann et al., 2005). This prompted Hiltpold et al. (2010a) to evaluate whether selection for enhanced responsiveness to the crucial root signal, EβC, could improve the efficiency of nematodes in controlling the larvae of the chrysomelid beetle, *D. v. virgifera*. Using belowground six-arm olfactometers, a strain of the nematode, *H. bacteriophora*, was successfully selected. Originally, this nematode did not respond to EβC (Hiltpold et al., 2010b) even though its effectiveness in controlling western corn rootworm larvae is high (Kurzt, Hiltpold, Turlings, Kuhlmann, & Toepfer, 2009). The selected strain responded much better to EβC in laboratory experiments and was also able to significantly control the pest in the field when the belowground signal was present (Hiltpold et al., 2010a). Because of the strong selective pressure, there were minor trade-offs in the infectiousness of the selected strain. Yet, the higher responsiveness to the HIPV overbalanced these weaknesses (Hiltpold et al., 2010a). The establishment and the persistence in the field were not influenced by the selection process (Hiltpold et al., 2010a). These results reflect the great potential of selecting beneficial organisms for a better and faster response, resulting in higher infection rates. Although there are certain constraints, such as knowledge of key compound/blends and the laborious selection process, selecting for specific nematode strains could be coupled with selection of more attractive plant genotypes, making biological control of insect pests a success.

7. MULTITROPHIC INTERACTIONS AND PARADOX SURROUNDING BELOWGROUND INDIRECT DEFENCES

There is a problem similar to the one found in aboveground systems (Hare, 2011), where the emerging picture is that volatile production in plants is the result of diffuse selection due to multiple players interacting with the plant. In effect, organic molecule emissions can have unintended effects on nontarget nematodes, and we believe that nematode-induced changes in soil chemical characteristics do contribute to the structuring of specific nematode communities around roots and might impede optimal soil management practices. For example, the insect herbivore-induced emissions of terpenes in citrus plants, which were shown to attract entomopathogenic nematodes (Ali et al., 2010, 2011), were also shown to attract plant parasitic nematodes, *T. semipenetrans* (Ali et al., 2011). This may reduce the exploitation of citrus-induced volatile emission in biological control strategies targeting the root weevil, *Diaprepes abbreviates*, where rootstocks are not otherwise resistant to *T. semipenetrans*. Undoubtedly, future work is needed to complement the relatively meagre literature on the exact nature of compounds driving nematode foraging behaviour. However, as for aboveground insect communities (Poelman, Galiart, Raaijmakers, Van Loon, & van Dam, 2008), we predict that plants have the potential to mediate indirect interactions between different nematode species and shape nematode community structure belowground.

The best example of this was demonstrated by Ali et al. (2013), who reported evidence that an HIPV (pregeijerene) attracts free-living bacteria-feeding nematode 'hyperparasites' living within the rhizosphere, in addition to previously reported attraction of entomopathogenic nematodes (Ali et al., 2012). In previous work, it was demonstrated that citrus roots attract beneficial nematodes when they are fed upon by root herbivores (Ali et al., 2010). Later, it was found that plant parasites were also attracted by this same HIPV (Ali et al., 2011). However, these initial investigations consisted of laboratory bioassays and did not allow for analysis of multitrophic interactions in a natural field setting.

A semiochemical basis for multitrophic relationships (citrus plant—herbivore—biological control agent and natural enemies) was studied in the field (Ali et al., 2013). Molecular probes were used together with in situ recovery of HIPVs to demonstrate that naturally occurring entomopathogenic nematode as well as free-living bactivorous nematodes that sometimes compete with entomopathogenic nematodes were attracted by the volatile compound, pregeijerene (1,5-dimethylcyclodeca-1,5,7-triene), which emanates from citrus roots when fed upon by weevils (Ali et al., 2012, 2013). The net effect of the free-living bactivorous nematodes on entomopathogenic nematodes efficacy in this system is unknown, and the effects of these interactions on plant production/yield need to be measured to understand whether manipulating indirect plant defences is practical for sustainable pest control.

The influence of plant cues on their surrounding community is a dynamic area of current research (Kaplan, 2012). Although much effort has been focussed on induced plant volatiles as a result of plant damage, fewer studies have extensively examined the diverse impact of competing species occupying different trophic levels, particularly in studies of indirect defence (Kaplan, 2012). Many studies have shown that HIPVs attract beneficial natural enemies of herbivores, yet quantification of plant fitness benefits as a result of such altered (induced) volatile profiles is rare (Allison & Daniel Hare, 2009; Hare, 2011; Kost & Heil, 2007). Two mechanisms have been proposed to explain why fitness benefits are difficult to detect in information-based (e.g. HIPV) defences. Firstly, it may be that the attracted natural enemies do not immediately kill their hosts (e.g. parasitoids), which results in continued damage to the plant (Kessler & Heil, 2011). Secondly, complex interactions within the community may result in no net effect because of the lack of specificity of the cue or signal perception and/or attractiveness (Kessler & Heil, 2011; Robert et al., 2013). Both aboveground and belowground chemical ecology studies rarely take such competing factors into consideration simultaneously. Ali et al. (2013) findings indicate that a belowground HIPV can have diverse functions, and although plant fitness was not measured, we found evidence for complex community interactions.

Predators and parasitoids can have significant effects on herbivore performance and consequences on population density (De Moraes, Lewis, Pare, Alborn, & Tumlinson, 1998; Kessler & Baldwin, 2001, 2004; Thaler, 1999). However, there is less evidence that specialist natural enemies use more specific search cues than those used by generalists (Steidle & Van Loon, 2003). Investigating belowground, there are even fewer studies examining cues that influence foraging strategies of natural enemies. Recent investigations have found evidence for differences in entomopathogenic nematode response to HIPVs as a function of natural enemy foraging strategy (Ali et al., 2011), specificity of belowground herbivore-induced volatile production and associated natural enemy response (Rasmann & Turlings, 2008), and influences of plant genotype on indirect defences (Hiltpold et al., 2010a; Rasmann et al., 2005). The behaviour of entomopathogenic nematodes to direct cues emitted from hosts demonstrates that parasitic nematodes have evolved specialized olfactory systems that likely contribute to appropriate host selection (Dillman et al., 2012). Furthermore, Hiltpold et al. (2010b) observed 'learning' behaviour of an entomopathogenic nematode in response to an HIPV. However, the free-living bactivorous nematode 'hyperparasites', which were attracted to the HIPV in our system, have not been examined in this manner previously. Results suggest that higher trophic levels may 'eavesdrop' on the same cue used by entomopathogenic nematodes to enhance host location.

One important aspect of the currently described results is the possibly reduced benefit of an HIPV to the plant given that this cue apparently attracts competitors of the beneficial natural enemies of the soil-borne herbivores. Although we found evidence for competition between herbivore natural enemies and other guilds, such as hyperparasites, the direct consequences on plant defence were not measured in this investigation. It is possible that these competitors may reduce entomopathogenic nematode population growth; however, there is little evidence that they reduce the insecticidal efficacy of entomopathogenic nematodes (Campos-Herrera, El-Borai, & Duncan, 2012; Duncan, Dunn, Bague, & Nguyen, 2003). It may be inconsequential whether a cue increases competition between competing species occupying different trophic levels (parasites and their hyperparasites) if the result is reduced overall density of the herbivore species, despite this competition. However, if this competition reduces parasite species rapidly, due to the action of their hyperparasites, the herbivore population will likely soon rebound, leading to questions about the possible application of such HIPV cues for pest control in agricultural settings. In the currently described belowground system, we observed species interactions that may result in outcomes similar to those postulated by Poelman et al. (2012) aboveground, where the effect of HIPVs on natural enemy response may be attenuated due to simultaneous attraction of competitors or hyperparasites, possibly resulting in overall negative consequences for plant survival. However, augmentative releases of entomopathogenic nematodes for biological control might counterbalance the negative effects of their natural enemies on pest mortality. Future work on how HIPVs influence hyperparasite—parasite—prey interactions within the subterranean environment are needed to fully explore induced plant defence, which may lead to practical applications in pest management, and will increase our basic understanding of the organismal networks that are influenced by these induced cues.

ACKNOWLEDGEMENTS

The authors thank the USDA—NIFA—AFRI for the fellowship awarded to JGA (2012-67012-19821).

REFERENCES

Abraham, J., Giacomuzzi, V., & Angeli, S. (2015). Root damage to apple plants by cockchafer larvae induces a change in volatile signals below- and above-ground. *Entomologia Experimentalis Et Applicata, 156*(3), 279—289.

Ali, J. G., Alborn, H. T., Campos-Herrera, R., Kaplan, F., Duncan, L. W., Rodriguez-Saona, C., et al. (2012). Subterranean, herbivore-induced plant volatile increases biological control activity of multiple beneficial nematode species in distinct habitats. *PLoS One,* 7, e38146.

Ali, J. G., Alborn, H. T., & Stelinski, L. L. (2010). Subterranean herbivore-induced volatiles released by citrus roots upon feeding by *Diaprepes abbreviatus* recruit entomopathogenic nematodes. *Journal of Chemical Ecology, 36*, 361—368.

Ali, J. G., Alborn, H. T., & Stelinski, L. L. (2011). Constitutive and induced subterranean plant volatiles attract both entomopathogenic and plant parasitic nematodes. *Journal of Ecology, 99*, 26—35.

Ali, J. G., Campos-Herrera, R., Alborn, H., Duncan, L., & Stelinski, L. (2013). Sending mixed messages: a trophic cascade produced by a belowground herbivore-induced cue. *Journal of Chemical Ecology, 39*, 1140—1147.

Allison, J. D., & Daniel Hare, J. (2009). Learned and naïve natural enemy responses and the interpretation of volatile organic compounds as cues or signals. *New Phytologist, 184*, 768—782.

Aratchige, N. S., Lesna, I., & Sabelis, M. W. (2004). Below-ground plant parts emit herbivore-induced volatiles: olfactory responses of a predatory mite to tulip bulbs infested by rust mites. *Experimental and Applied Acarology, 33*, 21—30.

Bais, H. P., Weir, T. L., Perry, L. G., Gilroy, S., & Vivanco, J. M. (2006). The role of root exudates in rhizosphere interactions with plants and other organisms. *Annual Review of Plant Biology, 57*(1), 233–266. http://dx.doi.org/10.1146/annurev.arplant.57.032905.105159.

Beckendorf, S. K., & Hoy, M. A. (1985). Genetic improvement of arthropod natural enemies through selection, hybridization or genetic engineering techniques. In M. A. Hoy, & D. C. Herzog (Eds.), *Biological control in agricultural IPM systems*. Orlando, FL: Academic.

Bilgrami, A. L., Kondo, E., & Yoshiga, T. (2001). Attraction response of *Steinernema glaseri* in presence of plant root and leaves. *Annals of Plant Protection Sciences, 9*, 258–263.

Boff, M. I. C., Zoon, F. C., & Smits, P. H. (2001). Orientation of *Heterorhabditis megidis* to insect hosts and plant roots in a Y-tube sand olfactometer. *Entomologia Experimentalis et Applicata, 98*, 329–337.

Bonkowski, M., Villenave, C., & Griffiths, B. (2009). Rhizosphere fauna: the functional and structural diversity of intimate interactions of soil fauna with plant roots. *Plant and Soil, 321*, 213–233.

Campbell, J. F., Lewis, E. E., Stock, S. P., Nadler, S., & Kaya, H. K. (2003). Evolution of host search strategies in entomopathogenic nematodes. *Journal of Nematology, 35*, 142–145.

Campos-Herrera, R., El-Borai, F. E., & Duncan, L. W. (2012). Wide interguild relationships among entomopathogenic and free-living nematodes in soil as measured by real time qPCR. *Journal of Invertebrate Pathology, 111*, 126–135.

Chaston, J. M., Suen, G., Tucker, S. L., Andersen, A. W., Bhasin, A., Bode, E., et al. (2011). The entomopathogenic bacterial endosymbionts *Xenorhabdus* and *Photorhabdus*: convergent lifestyles from divergent genomes. *PLoS One, 6*, e27909.

Chitwood, D. J. (2002). Phytochemical based strategies for nematode control. *Annual Review of Phytopathology, 40*, 221–249.

Coleman, D. C., & Crossley, D. A., Jr. (2004). *Fundamentals of soil ecology*.

van Dam, N. M. (2009). Belowground herbivory and plant defenses. *Annual Review of Ecology, Evolution and Systematics, 40*, 373–391.

van Dam, N. M., & Poppy, G. M. (2008). Why plant volatile analysis needs bioinformatics – detecting signal from noise in increasingly complex profiles. *Plant Biology, 10*, 29–37.

Danner, H., Samudrala, D., Cristescu, S. M., & van Dam, N. M. (2012). Tracing hidden herbivores: time-resolved non-invasive analysis of belowground volatiles by proton-transfer-reaction mass spectrometry (PTR-MS). *Journal of Chemical Ecology, 38*, 785–794.

De Deyn, G. B., Raaijmakers, C. E., Zoomer, H. R., Berg, M. P., de Ruiter, P. C., Verhoef, H. A., et al. (2003). Soil invertebrate fauna enhances grassland succession and diversity. *Nature, 422*, 711–713.

De Moraes, C. M., Lewis, W. J., Pare, P. W., Alborn, H. T., & Tumlinson, J. H. (1998). Herbivore-infested plants selectively attract parasitoids. *Nature, 393*, 570–573.

De Ruiter, P. C., Moore, J. C., Zwart, K. B., Bouwman, L. A., Hassink, J., Bloem, J., et al. (1993). Simulation of nitrogen mineralization in the below-ground food webs of two winter wheat fields. *Journal of Applied Ecology, 30*, 95–106.

Degenhardt, J. (2009). Indirect defense responses to herbivory in grasses. *Plant Physiology, 149*, 96–102.

Dicke, M., & Sabelis, M. W. (1988). How plants obtain predatory mites as bodyguards. *Netherlands Journal of Zoology, 38*, 148–165.

Dillman, A. R., Guillermin, M. L., Lee, J. H., Kim, B., Sternberg, P. W., & Hallem, E. A. (2012). Olfaction shapes host–parasite interactions in parasitic nematodes. *Proceedings of the National Academy of Sciences of the United States of America, 109*, E2324–E2333.

Dougherty, E. C., & Nigon, V. (1949). A new species of the free-living nematode genus *Rhabdilis* of interest in comparative physiology and genetics. *Journal of Parasitology, 35*(Suppl.), 11.

Dudareva, N., Pichersky, E., & Gershenzon, J. (2004). Biochemistry of plant volatiles. *Plant Physiology, 135*, 1893–1902.

Duncan, L. W., Dunn, D. C., Bague, G., & Nguyen, K. (2003). Competition between entomopathogenic and free-living bactivorous nematodes in larvae of the weevil *Diaprepes abbreviatus*. *Journal of Nematology, 35*, 187–193.

Ennis, D. E., Dillon, A. B., & Griffin, C. T. (2010). Simulated roots and host feeding enhance infection of subterranean insects by the entomopathogenic nematode *Steinernema carpocapsae*. *Journal of Invertebrate Pathology, 103*, 140–143.

Erb, M., & Lu, J. (2013). Soil abiotic factors influence interactions between belowground herbivores and plant roots. *Journal of Experimental Botany, 64*, 1295–1303.

Erb, M., Ton, J., Degenhardt, J., & Turlings, T. C. J. (2008). Interactions between arthropod-induced aboveground and belowground defenses in plants. *Plant Physiology, 146*, 867–874.

Fu, S., Ferris, H., Brown, D., & Plant, R. (2005). Does the positive feedback effect of nematodes on the biomass and activity of their bacteria prey vary with nematode species and population size? *Soil Biology and Biochemistry, 37*, 979–1987.

Gaugler, R. (2002). *Entomopathogenic nematology*. CABI.

Gaugler, R., & Bilgrami, A. L. (Eds.). (2004). *Nematode behaviour*. New York, USA: CABI.

Gaugler, R., Campbell, J. F., & McGuire, T. R. (1989). Selection for host-finding in *Steinernema feltiae*. *Journal of Invertebrate Pathology, 54*(3), 363–372.

Gaugler, R., & Campbell, J. F. (1991). Selection for enhanced host-finding of scarab larvae (Coleoptera, Scarabaeidae) in an entomopathogenic nematode. *Environmental Entomology, 20*(2), 700–706.

Gaugler, R., & Kaya, H. K. (1990). *Entomopathogenic Nematodes in Biological Control*. Boca Raton, Florida, USA: CRC Press.

de Gouw, J., & Warneke, C. (2007). Measurements of volatile organic compounds in the earth's atmosphere using proton-transfer-reaction mass spectrometry. *Mass Spectrometry Reviews, 26*, 223–257.

Grewal, P. S., Lewis, E. E., Gaugler, R., & Campbell, J. F. (1994). Host finding behaviour as a predictor of foraging strategy in entomopathogenic nematodes. *Parasitology, 108*, 207–215.

Grewal, P. S., Gaugler, R., & Wang, Y. (1996). Enhanced cold tolerance of the entomopathogenic nematode *Steinernema feltiae* through genetic selection. *Annals of Applied Biology, 129*, 335–341.

Hallem, E. A., Dillman, A. R., Hong, A. V., Zhang, Y. J., Yano, J. M., DeMarco, S. F., et al. (2011). A sensory code for host seeking in parasitic nematodes. *Current Biology, 21*, 377–383.

Hara, A. H., Gaugler, R., Kaya, H. K., & Lebeck, L. M. (1991). Natural populations of entomopathogenic nematodes (Rhabditida: Heterorhabditidae, Steinernematidae) from the Hawaiian Islands. *Environmental Entomology, 20*(1), 211–216.

Hare, J. D. (2011). Ecological role of volatiles produced by plants in response to damage by herbivorous insects. *Annual Review of Entomology, 56*, 161–180.

Harvey, E. N. (1952). *Bioluminescence*. New York: Academic.

Heath, R. R., & Manukian, A. (1992). Development and evaluation of systems to collect volatile semiochemicals from insects and plants using a charcoal-infested medium for air purification. *Journal of Chemical Ecology, 20*, 593–608.

Hilliard, M. A., Bargmann, C. I., & Bazzicalupo, P. (2002). *C. elegans* responds to chemical repellents by integrating sensory inputs from the head and the tail. *Current Biology, 12*, 730–734.

Hiltpold, I., Baroni, M., Toepfer, S., Kuhlmann, U., & Turlings, T. C. J. (2010a). Selective breeding of entomopathogenic nematodes for enhanced attraction to a root signal did not reduce their establishment or persistence after field release. *Plant Signaling and Behavior, 5*, 1450–1452.

Hiltpold, I., Baroni, M., Toepfer, S., Kuhlmann, U., & Turlings, T. C. J. (2010b). Selection of entomopathogenic nematodes for enhanced responsiveness to a volatile root signal helps to control a major root pest. *Journal of Experimental Biology, 213*(14), 2417–2423.
Hiltpold, I., Erb, M., Robert, C. A. M., & Turlings, T. C. J. (2011). Systemic root signalling in a belowground, volatile-mediated tritrophic interaction. *Plant Cell and Environment, 34*, 1267–1275.
Hiltpold, I., & Turlings, T. C. J. (2008). Belowground chemical signaling in maize: when simplicity rhymes with efficiency. *Journal of Chemical Ecology, 34*, 628–635.
Holterman, M., van der Wurff, A., van den Elsen, S., van Megen, H., Bongers, T., Holovachov, O., et al. (2006). Phylum-wide analysis of SSU rDNA reveals deep phylogenetic relationships among nematodes and accelerated evolution toward crown Clades. *Molecular Biology and Evolution, 23*, 1792–1800.
Hoy, M. A. (1982). Aerial dispersal and field efficacy of a genetically improved strain of the spider-mite predator metaseiulus-occidentalis. *Entomologia Experimentalis Et Applicata, 32*(3), 205–212.
Hoy, M. A. (2000). Transgenic arthropods for pest management programs: Risks and realities. *Experimental and Applied Acarology, 24*(5–6), 463–495.
Hunter, M. D. (2001). Out of sight, out of mind: the impacts of root-feeding insects in natural and managed systems. *Agricultural and Forest Entomology, 3*, 3–9.
Ingham, R. E., Trofymow, J. A., Ingham, E. R., & Coleman, D. C. (1985). Interactions of bacteria, fungi and their nematode grazers on nutrient cycling and plant growth. *Ecological Monographs, 55*, 119–140.
Ishibashi, N., & Kondo, E. (1990). Behavior of infective juveniles. In R. Gaugler, & H. K. Kaya (Eds.), *Entomopathogenic nematodes in biological control.* Boca Raton: CRC.
Johnson, S. N., & Nielsen, U. N. (2012). Foraging in the dark—chemically mediated host plant location by belowground insect herbivores. *Journal of Chemical Ecology, 38*(6), 604–614.
Jones, J. (2002). *Nematode sense organs. The biology of nematodes* (pp. 369–387). New York, NY, USA: Taylor and Francis Inc.
Kaplan, I. (2011). Attracting carnivorous arthropods with plant volatiles: the future of biocontrol or playing with fire? *Biological Control, 60*, 77–89.
Kaplan, I. (2012). Trophic complexity and the adaptive value of damage-induced plant volatiles. *PLoS Biology, 10*, e1001437.
Kaplan, I., Halitschke, R., Kessler, A., Sardanelli, S., & Denno, R. F. (2008). Constitutive and induced defenses to herbivory in above- and belowground plant tissues. *Ecology, 89*, 392–406.
Kappers, I. F., Aharoni, A., van Herpen, T., Luckerhoff, L. L. P., Dicke, M., & Bouwmeester, H. J. (2005). Genetic engineering of terpenoid metabolism attracts bodyguards to *Arabidopsis*. *Science, 309*, 2070–2072.
Karban, R., & Baldwin, I. T. (1997). *Induced responses to herbivory* (1st ed.). Chicago: The University of Chicago Press.
Kaya, H., & Gaugler, R. (1993). Entomopathogenic nematodes. *Annual Review of Entomology, 38*, 181–206.
Kessler, A., & Baldwin, I. T. (2001). Defensive function of herbivore-induced plant volatile emissions in nature. *Science, 291*, 2141–2144.
Kessler, A., & Baldwin, I. T. (2004). Herbivore-induced plant vaccination. Part I. The orchestration of plant defenses in nature and their fitness consequences in the wild tobacco *Nicotiana attenuata*. *The Plant Journal, 38*, 639–649.
Kessler, A., & Heil, M. (2011). The multiple faces of indirect defences and their agents of natural selection. *Functional Ecology, 25*, 348–357.
Kollner, T. G., Held, M., Lenk, C., Hiltpold, I., Turlings, T. C. J., Gershenzon, J., et al. (2008). A maize (E)-beta-caryophyllene synthase implicated in indirect defense responses against herbivores is not expressed in most American maize varieties. *The Plant Cell, 20*, 482–494.

Kost, C., & Heil, M. (2007). The defensive role of volatile emission and extrafloral nectar secretion for lima bean in nature. *Journal of Chemical Ecology, 34*, 2–13.

Kurzt, B., Hiltpold, I., Turlings, T. C. J., Kuhlmann, U., & Toepfer, S. (2009). Comparative susceptibility of larval instars and pupae of the western corn rootworm to infection by three entomopathogenic nematodes. *Biocontrol, 54*, 255–262.

Lee, D. L. (2002). *The biology of nematodes.*

Lewis, E., Campbell, J., Griffin, C., Kaya, H., & Peters, A. (2006). Behavioral ecology of entomopathogenic nematodes. *Biological Control, 38*, 66–79.

Lewis, E. E., Gaugler, R., & Harrison, R. (1993). Response of cruiser and ambusher entomopathogenic nematodes (Steinernematidae) to host volatile cues. *Canadian Journal of Zoology, 71*, 765–769.

Mohney, B. K., Matz, T., LaMoreaux, J., Wilcox, D. S., Gimsing, A. L., Mayer, P., et al. (2009). In situ silicone tube microextraction: a new method for undisturbed sampling of root-exuded thiophenes from marigold (*Tagetes erecta* L.) in soil. *Journal of Chemical Ecology, 35*, 1279–1287.

Mumm, R., & Dicke, M. (2010). Variation in natural plant products and the attraction of bodyguards involved in indirect plant defense. *Canadian Journal of Zoology-Revue Canadienne De Zoologie, 88*, 628–667.

Pawliszyn, J. (2009). *Handbook of solid phase microextraction.* Chemical Industry Press.

Peters, A., & Ehlers, R. U. (1998). Evaluation and selection for enhanced nematode pathogenicity against *Tipula* spp. In N. Simoes, N. Boemare, & R. U. Ehlers (Eds.), *Pathogenicity of entomopathogenic nematodes versus insect defense mechanisms: Impact on selection of virulent strains.* Brussels: European Commission Publication COST819.

Poelman, E. H., Bruinsma, M., Zhu, F., Weldegergis, B. T., Boursault, A. E., Jongema, Y., et al. (2012). Hyperparasitoids use herbivore-induced plant volatiles to locate their parasitoid host. *PLoS Biology, 10*, e1001435.

Poelman, E. H., Galiart, R. J. F. H., Raaijmakers, C. E., Van Loon, J. J. A., & van Dam, N. M. (2008). Performance of specialist and generalist herbivores feeding on cabbage cultivars is not explained by glucosinolate profiles. *Entomologia Experimentalis et Applicata, 127*, 218–228.

Poinar, G. O., Jr. (1990). Biology and taxonomy of Steinernematidae and Heterorhabditidae. *Entomopathogenic nematodes in biological control* (pp. 23–62). Boca Raton Fla: CRC Press.

Poinar, G. O. (1993). Origins and phylogenetic relationships of the entomophilic rhabditids, *Heterorhabditis* and *Steinernema. Fundamental and Applied Nematology, 16*, 333–338.

Poinar, G. O., & Georgis, R. (1990). Description and field application of the HP88 strain of *Heterorhabditis* bacteriophora. *Revue de Nematologie, 13*, 387–393.

Price, P. W., Bouton, C. E., Gross, P., McPheron, B. A., Thompson, J. N., & Weis, A. E. (1980). Interaction between three trophic levels: Influence of plants on interactions between insect herbivores and natural enemies. *Annual Review of Ecology and Systematics, 11*, 41–65.

van der Putten, W. H. (2003). Plant defense belowground and spatiotemporal processes in natural vegetation. *Ecology, 84*(9), 2269–2280.

van der Putten, W. H. (2009). A multitrophic perspective on functioning and evolution of facilitation in plant communities. *Journal of Ecology, 97*, 1131–1138.

Rasmann, S., & Agrawal, A. A. (2008). In defense of roots: a research agenda for studying plant resistance to belowground herbivory. *Plant Physiology, 146*, 875–880.

Rasmann, S., Ali, J. G., Helder, J., & van der Putten, W. H. (2012). Ecology and evolution of soil nematode chemotaxis. *Journal of Chemical Ecology, 38*, 615–628.

Rasmann, S., Bauerle, T. L., Poveda, K., & Vannette, R. (2011). Predicting root defence against herbivores during succession. *Functional Ecology, 25*, 368–379.

Rasmann, S., Kollner, T. G., Degenhardt, J., Hiltpold, I., Toepfer, S., Kuhlmann, U., et al. (2005). Recruitment of entomopathogenic nematodes by insect-damaged maize roots. *Nature, 434*, 732–737.

Rasmann, S., & Turlings, T. C. J. (2008). First insights into specificity of belowground tritrophic interactions. *Oikos, 117*, 362–369.

Robert, C. A. M., Erb, M., Hiltpold, I., Hibbard, B. E., Gaillard, M. D. P., Bilat, J., et al. (2013). Genetically engineered maize plants reveal distinct costs and benefits of constitutive volatile emissions in the field. *Plant Biotechnology Journal, 11*, 628–639.

Rostàs, M., & Eggert, K. (2008). Ontogenetic and spatio-temporal patterns of induced volatiles in *Glycine max* in the light of the optimal defence hypothesis. *Chemoecology, 18*, 29–38.

Schnee, C., Kollner, T. G., Held, M., Turlings, T. C. J., Gershenzon, J., & Degenhardt, J. (2006). The products of a single maize sesquiterpene synthase form a volatile defense signal that attracts natural enemies of maize herbivores. *Proceedings of the National Academy of Sciences of the United States of America, 103*, 1129–1134.

Schoonhoven, L., van Loon, J., & Dicke, M. (2005). *Insect–plant biology* (2nd ed.). New York: Oxford University Press.

Soler, R., Erb, M., & Kaplan, I. (2013). Long distance root–shoot signalling in plant–insect community interactions. *Trends in Plant Science, 18*, 149–156.

Steidle, J. L. M., & Van Loon, J. J. A. (2003). Dietary specialization and infochemical use in carnivorous arthropods: testing a concept. *Entomologia Experimentalis et Applicata, 108*, 133–148.

Strauch, O., Oestergaard, J., Hollmer, S., & Ehlers, R. U. (2004). Genetic improvement of the desiccation tolerance of the entomopathogenic nematode *Heterorhabditis bacteriophora* through selective breeding. *Biological Control, 31*, 218–226.

Thaler, J. S. (1999). Jasmonate-inducible plant defences cause increased parasitism of herbivores. *Nature, 399*, 686–688.

Tietjen, J. H., Lee, J. J., Rullman, J., Greengart, A., & Trompeter, J. (1970). Gnotobiotic culture and physiological ecology of the marine nematode Rhabditis marina Bastian1. *Limnology and Oceanography, 15*(4), 535–543. http://dx.doi.org/10.4319/lo.1970.15.4.0535.

van Tol, R. W. H. M., van der Sommen, A. T. C., Boff, M. I. C., van Bezooijen, J., Sabelis, M. W., & Smits, P. H. (2001). Plants protect their roots by alerting the enemies of grubs. *Ecology Letters, 4*, 292–294.

Tollsten, L., & Bergström, G. (1988). Headspace volatiles of whole plants and macerated plant parts of *Brassica* and *Sinapis*. *Phytochemistry, 27*, 4013–4018.

Tomalak, M. (1994). Selective breeding of *Steinernema feltiae* Filipjev (Nematoda, Steinernematidae) for improved efficacy in control of a mushroom fly, *Lycoriella solani* Winnertz (Diptera, Sciaridae). *Biocontrol Science and Technology, 4*(2), 187–198.

Tscharntke, T., & Hawkins, B. A. (Eds.). (2002). *Multitrophic level interactions*. Cambridge University Press.

Turlings, T. C. J., Hiltpold, I., & Rasmann, S. (2012). The importance of root-produced volatiles as foraging cues for entomopathogenic nematodes. *Plant and Soil, 358*, 51–60.

Turlings, T. C. J., Tumlinson, J. H., & Lewis, W. J. (1990). Exploitation of herbivore-induced plant odors by host-seeking parasitic wasps. *Science, 250*, 1251–1253.

Turlings, T. C. J., & Wäckers, F. (2004). Recruitment of predators and parasitoids by herbivore-injured plants. In R. T. Cardé, & J. G. Millar (Eds.), *Advances in insect chemical ecology* (pp. 21–75). Cambridge University Press.

Vet, L. E. M., & Dicke, M. (1992). Ecology of infochemical use by natural enemies in a tritrophic context. *Annual Review of Entomology, 37*, 141–172.

Wardle, D. A., Bardgett, R. D., Klironomos, J. N., Setälä, H., van der Putten, W. H., & Wall, D. H. (2004). Ecological linkages between aboveground and belowground biota. *Science, 304*(5677), 1629–1633.

Yeates, G. W. (1999). Effects of plants on nematode community structure. *Annual Review of Phytopathology, 37*, 127–149.

CHAPTER NINE

Forest Canopy Precipitation Partitioning: An Important Plant Trait Influencing the Spatial Structure of the Symbiotic Soil Microbial Community

Carl L. Rosier[*,§,1], **Leslie Dean Moore**[¶], **Tiehang Wu**[¶], **John T. Van Stan**[||]

[*]Boone Pickens School of Geology, Oklahoma State University, Stillwater, OK, USA
[§]College of Agriculture and Natural Resources, University of Delaware, Newark, DE, USA
[¶]Department of Biology, Georgia Southern University, Statesboro, GA, USA
[||]Department of Geology and Geography, Georgia Southern University, Statesboro, GA, USA
[1]Corresponding author: E-mail: crosier@udel.edu

Contents

1. Introduction: Canopy Precipitation Partitioning and Its Potential Influence on Symbiotic Root Microbial Communities 216
2. Potential Throughfall and Stemflow Impacts on N-fixing Nodule-Forming Bacterial Signalling Pathways 219
3. Potential Throughfall and Stemflow Impacts on Endo- and Ectomycorrhizal Fungi Host Root Signalling 225
4. Future Directions for Evaluating Throughfall and Stemflow Influences on Symbiotic Microbes 230

Acknowledgement 231
References 232

Abstract

Diversity and structure of the symbiotic soil microbial community significantly influences essential soil ecosystem services including nutrient cycling, C-mineralization and stabilization/formation of soil structure. The overall structure and function of the symbiotic microbial community can also impact trees' ability to acquire nutrients, potentially influencing its overall fitness. Interestingly, factors affecting below-ground signalling pathways between tree hosts and their symbiotic microbes (e.g. soil moisture and chemistry) can be altered by the tree's canopy structure partitioning precipitation. In fact, canopy precipitation partitioning produces persistent patterns in the receipt of meteoric water (and the solutes it carries) to soils via two pathways: throughfall (drips from canopy surfaces) and stemflow (the funnelling of droplets to an infiltration area around the trunk). If these above-ground, tree-directed mechanisms (throughfall and

Advances in Botanical Research, Volume 75
ISSN 0065-2296
http://dx.doi.org/10.1016/bs.abr.2015.09.005

stemflow) altered edaphic conditions in the active zone surrounding the roots, it may be possible for canopy-derived hydrologic fluxes to influence host–symbiont signalling communication. In this chapter, we discuss and evaluate past research that indicates canopy precipitation partitioning may alter how microbial symbionts (specifically N-fixing bacteria and mycorrhizal fungi) interact with tree roots by changing soil characteristics (pH, micro–macronutrient status and concentration of allelopathic compounds). The potential for throughfall and stemflow to influence tree-symbiotic microbe communication has not been previously explored, but past literature indicate it could significantly alter the profile, and perhaps even the function, of root microbial symbiont communities. Our literature synthesis and evaluation concludes with suggested directions for future efforts to evaluate the existence (and extent of biogeochemical implications) of this process.

1. INTRODUCTION: CANOPY PRECIPITATION PARTITIONING AND ITS POTENTIAL INFLUENCE ON SYMBIOTIC ROOT MICROBIAL COMMUNITIES

Precipitation across forest canopies reaches the soil surface by two major pathways: (1) via drip from canopy surfaces and through gaps (called 'throughfall') or (2) through the funnelling of entrained water droplets along outlying crown components to the tree stem (i.e. 'stemflow'). Throughfall represents the vast majority of precipitation reaching soils for all forest ecosystems – 60–90% (Levia, Keim, Carlyle-Moses, & Frost, 2011). Spatial patterns in throughfall are temporally persistent, resulting in soil locations that can be characterized as 'hot' spots receiving abundant throughfall or 'cold' spots receiving limited throughfall (Guswa, 2012; Keim, Skaugset, Weiler, et al., 2005; Zimmermann, Zimmermann, Elsenbeer, et al., 2009). The patterning of throughfall to soils results from interactions between storms and a forest's canopy structure, making these patterns ecosystem-dependent, and to some degree species-dependent (Fathizadeh et al., 2014; Keim, Skaugset, Weiler, et al., 2005; Nadkarni & Sumera, 2004). Stemflow rarely accounts for >10% of rainfall, yet its infiltration area is highly concentrated about the stem and, if canopy structural and meteorological conditions are appropriate, stemflow can input large quantities of water (>100 L) to an area typically no larger than 0.5 m^2 on a consistent basis (Levia, Van Stan, et al., 2011; Van Stan, Van Stan, Levia, et al., 2014). However, if species-dependent canopy structural conditions are inefficient at capturing and funnelling rainfall to the stem, then stemflow inputs to near-stem soils can be low in comparison to even throughfall cold spots (e.g. Bryant, Bhat, Jacobs, et al., 2005; Moore, Winkler, Carlyle Moses, et al., 2008; Valente, David, Gash, et al., 1997).

Combined, these canopy-derived hydrologic fluxes create intricate, but persistent, spatial patterns in water receipt to forest soils that depend, in large part, on species-dependent canopy structural characteristics. It is, therefore, well known that spatial patterns in throughfall and stemflow inputs can impact edaphic properties, e.g. storm-based soil moisture responses (Coenders-Gerrits, Hopp, Savenije, Pfister, et al., 2013), salt concentrations (Rosier et al., 2015) and soil solution chemistry (Chang & Matzner, 2000). However, the influence of canopy-derived water flux patterns on soil microbial community structure has been minimally investigated (Bollen, Chen, Lu, Tarrant, et al., 1968; Ceccherini et al., 2008; Rosier et al., 2015; Wilkinson & Anderson, 2001). Moreover, no studies known to the authors have assessed the potential for canopy-derived water flux patterns to impact root symbionts despite the evidence that fine root patterns can match throughfall patterns (e.g. Ford & Deans, 1978) and that stemflow can preferentially travel along root pathways (e.g. Li, Yang, Li, Lin, et al., 2009; Martinez-Meza & Whitford, 1996; Schwärzel, Ebermann, Schalling, et al., 2012).

As throughfall and stemflow drain across canopy surfaces (leaves, bark, epiphytes, etc.), they become differentially enriched with a variety of elements, including ionic micro- and macronutrients (Macinnis-Ng, Flores, Muller, Schwendenmann, et al., 2012; Rosier et al., 2015; Staelens, De Schrijver, Verheyen, et al., 2007), heavy metals (Aubert et al., 2002; Avila & Rodrigo, 2004; Rodrigo, Avila, Gomez-Bolea, et al., 1999), dissolved organic matter (Coxson, McIntyre, Vogel, et al., 1992; Schulze, Borken, Matzner, et al., 2011; Van Stan et al., 2015) and particulate organic matter (Le Mellec, Meesenburg, Michalzik, et al., 2010; Levia et al., 2013). Some organic matter components identified in canopy-derived hydrologic fluxes contain phenolic compounds (Bischoff et al., 2014) commonly associated with allelochemicals (Wardle, Nilsson, Gallet, Zackrisson, et al., 1998), as well as biolabile exogenous sugars (Coxson et al., 1992) and proteinlike fluorophores (Van Stan et al., 2015). These elements differentially enrich throughfall and stemflow depending on species structural traits. For example, greater bark roughness increases the scavenging of dry-deposited aerosols (Figure 1) available to throughfall and stemflow, while also enhancing leaching of internal elements (e.g. K^+, Ca^{2+} and Mg^{2+}) due to the greater hydrologic residence times required to drain off rougher surfaces (Levia, Keim, et al., 2011). Branch inclination angle also affects the residence time of entrained rain droplets (i.e. lower angle equates to a slower drainage, Figure 1), increasing greater chemical enrichment of stemflow or throughfall (Levia, Keim, et al., 2011). A species which hosts substantial epiphyte cover

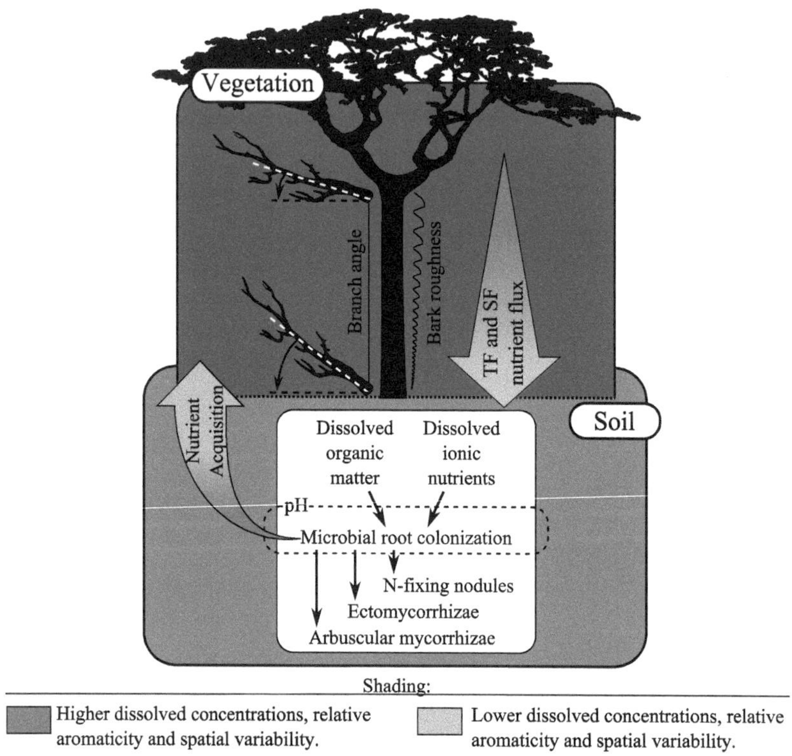

Figure 1 The potential changes in edaphic soil conditions produced by contrasting net precipitation (i.e. throughfall and stemflow) effects on symbiotic microbial community. (See colour plate)

(*Quercus virginiana* and *Tillandsia usneoides* in the southeastern United States, for example) can further alter materials leached or washed off by precipitation drainage through the canopy (Van Stan et al., 2015). Therefore, the elemental character, concentration and flux of canopy-derived hydrologic fluxes are also connected to species-dependent canopy structural traits. Because of this, patterns in throughfall and stemflow chemistry have been linked to spatial variability in soil pH, soil chemistry and soil solution chemistry (Chang & Matzner, 2000; Matschonat & Falkengren-Grerup, 2000; Rosier et al., 2015; Van Stan, 2012). Even patterns in physiological below-ground structures/processes have been observed to correlate to throughfall patterns, including fine root distribution and plant water uptake (Bouten, Heimovaara, Tiktak, et al., 1992; Ford & Deans, 1978). Moreover, a recent study connected canopy structural alterations to throughfall amount and chemistry to soil microbial community structure (Rosier et al., 2015).

As precipitation partitioning into throughfall and stemflow patterns at the soil surface can be linked to changes in soil ecosystem chemistry (e.g. pH, CEC and OM) and general microbial community structure, could this plant trait significantly impact the spatial structure of the symbiotic soil microbial community by altering chemical signalling pathways?

Using root microbial symbionts as an example (nodule-forming bacteria and mycorrhizal fungi), our chapter will examine this question via review, discussion and evaluation of literature selections. Figure 1 represents the influence of tree species with differing canopy structural characteristics on the chemical speciation, concentration and flux of micro- and macronutrient ions (as this is the most-researched aspect of throughfall and stemflow chemistry). Microbial clades responsible for generating symbiotic relationships with host plants will respond uniquely to chemically contrasting canopy-derived hydrologic flux and the changes in edaphic conditions produced by contrasting net precipitation (Figure 1). In this response, each of the soil conditions altered by throughfall and stemflow patterns can significantly influence the capacity of beneficial microorganisms to establish a symbiotic relationship with a plant host via either enhancement or interruption of signalling pathways. In turn, this may impact a tree species' ability to acquire nutrients from root microbial symbionts (Figure 1), potentially influencing overall fitness. To the best of our knowledge no previous work has suggested, much less assessed, the potential of tree canopy manipulation of precipitation as a possible mechanism influencing plant symbiotic microbe signalling. Therefore, this chapter will discuss findings from the extant literature to formulate and support the hypothesis that canopy precipitation partitioning (into throughfall and stemflow) may affect (1) root symbiotic bacteria (root nodule N-fixing bacteria) and (2) mycorrhizal fungi (specifically, ecto- and endomycorrhiza). The chapter concludes with suggested future directions for the evaluation of this hypothesis.

2. POTENTIAL THROUGHFALL AND STEMFLOW IMPACTS ON N-FIXING NODULE-FORMING BACTERIAL SIGNALLING PATHWAYS

N-fixing bacteria are particularly well known for their symbiotic relationships with a variety of plant species, accounting for up to 20% of the N acquired by that plant per year (van der Heijden, Bardgett, Van Straalen, et al, 2008). Due to nitrogen limitation of most environments, several tree species form symbiont relationships with N-fixing bacteria (e.g. Rhizobia,

Frankia and cyanobacteria), most of which undergo nodulation as a result of their symbiosis (Mergaert et al., 2006; Santi, Bogusz, Franche, et al., 2013). Rhizobia form symbiosis with the majority of leguminous (*Fabaceae*) trees (Franche, Lindström, Elmerich, et al., 2009) as well as trees in the nonlegume genus *Parasponia* (*Ulmaceae*) (Sprent & Parsons, 2000). *Frankia* are a group of Actinobacteria that form relationships with a broader spectrum of trees from eight different families, all of which are woody trees or shrubs with the exception of the *Datisca* family (Santi, Bogusz, Franche, et al., 2013). While many rhizobia (Masson-Boivin, Giraud, Perret, Batut, et al., 2009) and *Frankia* (Sprent & Parsons, 2000) can sustain non-symbiotic growth, most favour nodule formation and function as endophytic symbionts. This is predominately a result of the ideal conditions that nodules provide as a location for the anaerobic fixation process. The formation of nodules can occur via a number of different pathways most of which involve the use of various nodule factors (NF) to initiate nodule formation (Cohn, Bradley Day, Stacey, et al., 1998; Geurts, Fedorova, Bisseling, et al., 2005).

Since fine root distribution has been linked to throughfall patterns (Ford & Deans, 1978) and stemflow preferentially supplies water and nutrients along root pathways (Johnson & Lehmann, 2006; Backnäs, Laine-Kaulio, Kløve, et al., 2012), there appears to be potential for both processes to influence microbial density along root pathways. The potential for throughfall and stemflow to affect the capacity of bacteria and plant to initiate symbiosis is more likely than its influence on fine root distribution (Figure 2). Is it possible that throughfall and stemflow from favourable species-dependent canopy structural characteristics supplies roots with water and nutrients, potentially promoting or preventing nodulation. For both hydrological

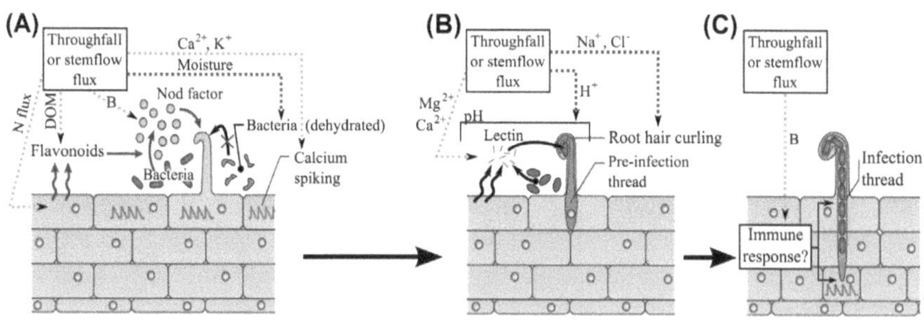

Figure 2 The potential mechanisms for throughfall and stemflow to influence the signalling pathways of bacteria and host root during (A) initiation of symbiosis, (B) attachment of nodule-forming bacteria to root hairs, (C) termination of nodule development. *Figure adapted from Oldroyd (2013).* (See colour plate)

processes, throughfall and stemflow influences on root nodulation will most likely be associated with several inputs including H^+, dissolved salts (e.g. Na, Cl), variable moisture regimes and species-specific nutrient flux/constituency, all of which have been shown to influence a variety of N-fixing symbiosis (Zahran, 1999), especially during initiation of the nodulation process (Figure 2(A)). Furthermore, since nodulation is initiated by root-secreted nodulation factors, the leaching of allelopathic compounds from canopy surfaces by stemflow and throughfall could also play an influential role in nodulation.

Throughfall and stemflow can alter soil pH by (1) increased H^+ flux resulting from bark leaching during the leafless season, (2) decreased H^+ flux during leafed-season proton buffering/neutralization or (3) washing dry and wet acid deposition from canopy surfaces to the soil (André, Jonard, Ponette, et al., 2008a, 2008b; Nakanishi et al., 2001). For N-fixing bacteria, pH tolerance and adaptations to extreme acidity or alkalinity depends on the symbiont. Throughfall or stemflow initiated increase in soil pH can interfere with the attachment of nodule-forming bacteria to root hairs (Figure 2(B)). The capacity for N-fixing bacteria to form nodules under extreme pHs has been observed at both ends of the pH spectrum: *Frankia* are known to nodulate red alder (*Alnus rubra* Bong.) in acidic soils with pH lower than 4.5 (Knowlton & Dawson, 1983) as well as black alder (*Alnus glutinosa* Gaertn.) at pH as high 12 (Oliveira, Castro, Dodd, Vosátka, et al., 2005). However, in both cases, a return to an optimal pH was favoured as nodulation was low at these extremes. While optimal pH varies across systems, a decrease in nodulation as deviation from optimal increases is consistent (Martin, Posavatz, Myrold, et al., 2003; Rejili et al., 2012; Surange, Wollum, Kumar, Nautiyal, et al., 1997). Along with the direct effects of pH, there are a number of confounding effects including an increase in toxicity of Al (Flis, Glenn, & Dilworth, 1993) and manganese (Coventry & Evans, 1989) at low pH and a decrease in PO_4 solubility (Coventry & Evans, 1989; Lindsay, 1979). Previous research has shown throughfall's and stemflow's ability to concentrate heavy metals (including Al) depends on the scavenging efficiency of species-dependent canopy structural traits (Eisalou, Sengonul, Gokbulak, Serengil, Uygur, et al., 2013; Levia et al., 2015; Nikodem et al., 2010). Soil salinity can also be altered by throughfall and stemflow transporting dry-deposited sea salts (Na and Cl, in particular) from branch and leaf surfaces to the soil (e.g. Levia, Keim, et al., 2011; Rosier et al., 2015) and has a measurably negative effect on root hair curling (Figure 2(B)). Though many N-fixing relationships are salt tolerant, high salt concentrations can prevent initiation of nodulation (Zahran, 1999) with several *Rhizobium*-legume symbiosis (Rejili et al., 2012; Shamseldin & Werner, 2005) and *Frankia* symbiosis (Dawson & Gibson, 1987; Sande & Young, 1992) having nodulation depressed at high soil NaCl concentrations. Thus, increased soil salinity caused by the wash-off of dry deposition in, for example, maritime settings with substantial sea salt aerosol deposition (as observed in Rosier et al., 2015) could lead to the suppression of the initial symbiotic process. Where throughfall or stemflow inputs to soils are low, soil moisture may be highly influential for N-fixing symbiosis (Figure 2(A)). Many N-fixing bacteria demonstrate irregular morphology when exposed to low water potential (Zahran, 1999). This lowers the infection rate of the bacteria and therefore decreases nodulation (Figure 2(A)). Drought conditions can also cause senescence of nodules, severing the symbiotic ties that were already formed (Dudeja, Giri, Saini, Suneja-Madan, Kothe, et al., 2012; Porcel, Barea, & Ruiz-Lozano, 2003).

Throughfall and stemflow processes have also been shown to influence soil micronutrient fluxes, including those necessary for N-fixation: Fe, B, Mn and Mo. While all of these elements are important plant nutrients, Fe and Mo are the primary constituents of the nitrogenase enzyme known as FeMoCo (Fe—Mo cofactor; Franche, Lindström, Elmerich, et al., 2009). Návar, González, Gonzalez, et al. (2009) found that when air masses pass over land uses emitting Fe, forest canopy-derived fluxes can become significantly enriched with Fe. As the presence of Fe has been shown to increase nodulation in legumes (Rojas, Perry, Li, Ganio, et al., 2002) and Fe deficiency decreases the initiation and development of nodules (Abdelmajid, Karim, & Chedly, 2008), seasonal Fe pulses like those observed by Návar, González, Gonzalez, et al. (2009) may prove important. Mo is also deposited onto canopies by passing air masses and can enrich net precipitation if lichens, or other epiphytes, are not actively acquiring Mo along the throughfall and stemflow pathways (Campbell, Bradfield, Prescott, Fredeen, et al., 2010). Mo is important to N-fixation (Bellenger, Xu, Zhang, Morel, & Kraepiel, 2014) though, outside of its function in the FeMoCo enzyme, limited information is available regarding its role in nodulation. Mo's relationship to precipitation has recently been characterized, indicating that the majority of the Mo within the soil environment is derived from bulk precipitation (Gardner, Lyons, Litt, Ogden, et al., 2014), further solidifying the potential importance of Mo in throughfall and stemflow. Limited research exists regarding canopy-derived B fluxes, or its flux dynamics, yet two studies known to the authors have found very high B concentrations in both throughfall (Cividini,

Lemarchand, Chabaux, Boutin, Pierret, et al., 2010) and stemflow (Frost & Levia, 2013) compared to other hydrologic fluxes in forested watersheds. B deficiency has been shown to diminish the quality of root exudates (Figure 2(A)), causing a reduction in the secretion of nod factors (signalling molecules which stimulate root curling) limiting root hair curling (Redondo-Nieto, Bonilla, Bolanos, Rivilla, El-Hamdaoui, et al., 2001). Plants in limited B environments have been shown to terminate nodule development, identifying symbiotic bacterium partners as potential pathogens (Reguera, Bonilla, Bolaños, et al., 2010; Figure 2(C)). Finally, Mn assists in root colonization by mediating lectin binding to the root hair tips (Gonzalez-Guerrero, Matthiadis, Saez, Long, et al, 2014). Studies suggest that Mn is leached into the soil environment as a result of stemflow interaction with senescing canopies, and the loss of Mn can be exacerbated by acidic precipitation (Helmisaari & Mälkönen, 1989; Norden, 1991). These conditions are fairly common in tropical areas where throughfall and stemflow effects on soil conditions may explain the increased presence of many legumes and actinorizal plants (Kopacek, Turek, Hejzlar, Santruckova, et al, 2009).

Throughfall and stemflow are highly enriched in several macronutrients including K, S, Mg, yet their presence in the soil environment supports a minimal role in symbiotic N-fixation (Divito & Sadras, 2014; Kumar, 2011; Miller & Sirois, 1983). K, one of the most abundant macronutrients in throughfall and stemflow (Levia & Frost, 2003; Van Stan, Levia, Inamdar, Lepori-Bui, Mitchell, et al., 2012), potentially serves as a counterion to Ca (Gonzalez-Guerrero, Matthiadis, Saez, Long, et al, 2014) where it may nominally influence Ca spiking during nodule initiation (Figure 2(A)). S is a component of the FeMoCo enzyme (Furnholm & Tisa, 2014); since S is rarely a limiting nutrient in most ecosystems, it's abundance in throughfall or stemflow is unlikely to influence N-fixation. Ca, N and P, on the other hand, are important to symbiotic nodule-forming associations (Figure 2) and are present in throughfall and stemflow (Van Stan, Levia, Inamdar, Lepori-Bui, Mitchell, et al., 2012; Levia & Germer, 2015). Ca spiking is the primary reaction associated with host recognition of nodulation factors (Limpens & Bisseling, 2003; Masson-Boivin, Giraud, Perret, Batut, et al., 2009; Santi, Bogusz, Franche, et al., 2013). Ca also affects root hair attachment (Gonzalez-Guerrero, Matthiadis, Saez, Long, et al, 2014) (Figure 2(A)) and its influence may be inversely related to soil pH (Crannell, Tanaka, Myrold, et al., 1994; Zahran, 1999) (Figure 2(B)). This is especially important since Ca is widely reported as one of the most abundant canopy-derived cation fluxes for tree species (including legume species) (Návar, González, Gonzalez, et al., 2009). In the case of N and P both appear to play an integral role in regulating N-fixation. There is evidence that fixed N initiates downregulation of N-fixation (Batterman, Wurzburger, Hedin, et al., 2013; Isaac, Harmand, Drevon, et al., 2011; Limpens & Bisseling, 2003). This suggests that the leaching (Kopacek, Turek, Hejzlar, Santruckova, et al, 2009; Muoghalu & Oakhumen, 2000) or lack thereof (Dezzeo & Chacón, 2006) of NO_3, NH_3 and dissolved organic nitrogen in throughfall and stemflow can directly influence nodulation for a system that is not phosphorus-dependent (Figure 2(A)). The addition of P to the soil environment has been shown to increase nodule number in several studies (Ares, Burner, Brauer, et al., 2009; Isaac, Harmand, Drevon, et al., 2011; Kumar, 2011) as the FeMoCo enzyme requires a high energy input (16 mol of ATP per mole of N_2 reduction) and is irreversibly inactivated by the presence of O_2 despite the aerobic nature of these bacteria (Masson-Boivin, Giraud, Perret, Batut, et al., 2009). Due to the importance of P in N-fixation (Dezzeo & Chacón, 2006; Kopacek, Turek, Hejzlar, Santruckova, et al, 2009; Muoghalu & Oakhumen, 2000), throughfall and stemflow may have a direct influence on symbiotic communication. This may be of particular importance to stemflow, as Backnäs, Laine-Kaulio, Kløve, et al. (2012) suggest that stemflow transport along root preferential flow pathways may not only supply roots with PO_4, but also contains organic acids which enhance weathering of PO_4 from rock surfaces.

The presence of allelopathic compounds within throughfall and stemflow may also play an integral role influencing the signalling between plant roots and N-fixing bacteria. Allelopathic compounds involved in N-fixing symbiosis are most often phenolic compounds (i.e. flavonoids) (Figure 2(A)) released by plants influencing the expression of *nod* genes (Brencic & Winans, 2005; Dudeja, Giri, Saini, Suneja-Madan, Kothe, et al., 2012). Additionally, several allelopathic phenolic compounds have also been identified as microbial inhibitors (Brencic & Winans, 2005; John & Sarada, 2012) directly affecting specific clades of the microbial community. Bischoff et al. (2014) identified several highly concentrated phenolic compounds within throughfall and stemflow which have been shown to inhibit plant germination and growth (Hane, Hamburg, Barber, & Plaut, 2003; Muscolo, Panuccio, Sidari, et al., 2001). Furthermore, several species of *Acacia*, a genus of the Fabaceae family, have been used to study the allelopathic effects of throughfall on soil processes; however, no study to the knowledge of the authors has directly linked the effect of allelopathic compounds within throughfall and stemflow on N-fixing symbionts. This same trend holds for leguminous trees—*rhizobia* and *Frankia*–actinorhiza relationships. We hypothesize that there is a clear connection between allelopathic compounds within throughfall and stemflow and the potential to influence N-fixation symbiosis. Further study of these systems is paramount when considering the necessity of understanding how anthropogenic activity impacts the intricately intertwined processes that fix N, the primary source for new nitrogen to natural systems.

3. POTENTIAL THROUGHFALL AND STEMFLOW IMPACTS ON ENDO- AND ECTOMYCORRHIZAL FUNGI HOST ROOT SIGNALLING

The majority of fine roots (<0.1 mm) in forest ecosystems host mycorrhizal symbionts, and most temperate forest species host the two common types: arbuscular mycorrhizal (AM) and ectomycorrhizal (ECM) (Newman & Rydell, 1987). Distinct morphological, physiological and biochemical differences exist between AM and ECM. ECM roots are enclosed by a dense mantle with thick-walled hyphae extending away from the root surface. In the case of AM roots, no mantle is present and thin-walled hyphae extend from the intracellular regions of the host root into the soil environment (Smith & Read, 2008). Although more than 90% of terrestrial plant species host mycorrhizas (representing ~8000 plant and ~10,000 fungal species; Taylor & Alexander, 2005), only about 3–10% are reported as ECM, making the majority of these plant–fungal symbioses AM (Churchland & Grayston, 2014; Meyer, 1973; Smith & Read, 2008). Despite this small percentage, trees in forest ecosystems are widely reported as forming ECM relationships, especially gymnosperms that dominate temperate forests (Courty, Franc, Garbaye, et al., 2010; Newman & Rydell, 1987). The most common temperate ECM species are in the *Pinaceae, Salicaceae, Betulaceae, Fagaceae and Myrtaceae* families (Meyer, 1973), yet AM species are hosted by over 85% of plant species (Wang & Qiu, 2006). More interestingly, some genera of plant species form both ECM and AM on the same plant, such as *Populus, Salix, Eucalyptus* (Chilvers, Lapeyrie, Horan, et al., 1987; Lapeyrie & Chilvers, 1985; Lodge & Wentworth, 1990) and leguminous tree species (*Fabaceae*) (Frioni, Minasian, Volfovicz, et al., 1999). Perhaps, when roots from the same plant host both ECM and AM, patterns may be attributable to altered root–microbe signals resulting from stemflow preferential flow along particular roots or throughfall patterns?

The specific plant-derived signalling molecule instigating mycorrhizal symbiosis has yet to be identified (Harrision, 2005). Mycorrhizal spores in close vicinity to host roots respond by increasing respiratory activity and upregulation of hyphal branching, precursory activities leading to the initiation of symbiosis (Pinior, Wyss, Piché, Vierheilig, et al., 1999; Vierheilig, Lerat, & Piche, 2003). Application of host root exudates also elicits a similar spore response (Akiyama, Matsuzaki, Hayashi, et al., 2005; Buee, Rossignol, Jauneau, Ranjeva, Bécard, et al., 2000). Further research suggests that specific components of host root exudates such as flavonoids and hormones potentially induce metabolic or other symbiotic initiation process in mycorrhiza spores (Morandi, 1996; Vierheilig, Bago, Albrecht, Poulin, & Piché, 1998). There is evidence that, besides specific signalling compounds released from plant roots, mycorrhizal spores may also respond to environmental cues including nutrient enrichment (Hepper, 1979; Miranda & Harris, 1994), changes in soil chemistry (Bartolome-Esteban & Schenck, 1994; Clark, 1997) and addition of allelopathic compounds (Cantor, Hale, Aaron, Traw, Kalisz, et al., 2011; Roberts & Anderson, 2001). Throughfall and stemflow may provide these environmental cues. For example, the only study known to the authors which identified symbiotic fungal genera in stemflow-influenced soils (Rosier et al., in review) identified several unique ectomycorrhizal communities within soils experiencing contrasting stemflow generation (American beech vs yellow poplar). Rosier et al. (in review) attributed differences in symbiotic fungal community structure as a result of water stress, nutrient availability and changes in soil chemistry generated by tree species-dependent seasonal stemflow dynamics.

Soil pH is an important factor influencing plant nutrient availability (Marschner, 1995) and under extreme pH conditions often induce mineral stress. Differential effects of soil pH on several mycorrhizal fungal taxa have been widely investigated (Cairney & Meharg, 1999). For example, ECM such as *Cenococcum* spp. have been reported more abundant or unaffected by increasing levels of soil acidification; however, E-strain ascomycetes and *Thelephora terrestris* ECM growth are negatively affected at lower soil pH (Danielson & Visser, 1989). The general consensus suggests that ECM maintain a pH optimum between 4 and 6, experiencing decreased functionality and occurrence outside of this range (Abuzinadah & Read, 1986; Chalot & Brun, 1998; Colpaert & van Laere, 1996; Mayerhofer, Fraser, Kernaghan, et al., 2015). Additionally, ECM proteolytic activity (enzymes responsible for protein metabolism facilitating N availability) is highly sensitive to changes in soil pH (Abuzinadah & Read, 1986; Nygren, Edqvist, Elfstrand, Heller, Taylor, et al., 2007). The alteration of ECM enzyme functionality due to extreme soil pH levels suggests reduced symbiotic benefit for the host which in turn could alter the ECM community structure. AM isolates have also been observed in soils across an extensive pH range (pH 4.0–7.5) (Clark & Zeto, 1996). Under extreme soil pH conditions (<5.0) AM colonization has been found to enhance plant biomass and survival in acidic soils via increases in nutrient uptake (Clark, Zeto, & Zobel, 1999; Heijne, van Dam, Heil, & Bobbink, 1996) as well as reducing the toxic effects of Al uptake (Cuenca, Andrade, & Meneses, 2001; Ouimet, Camire, & Furlan, 1996) improving overall plant fitness. van Aarle, Olsson, Söderström et al. (2002) observed that the growth potential of AM mycelia is directly influenced by soil pH and AM species' response to changes in soil pH may vary between species. While differing effects of soil pH levels may not affect certain species of the AM community, or the potential root colonization rate, experimental evidence suggests that AM species' richness and community structure may be significantly reduced in soil experiencing shifts in soil pH (Wang, Stribley, Tinker, Walker, et al., 1993). As mentioned earlier, throughfall and stemflow can alter soil pH via seasonal dynamics in leachable surface features (bark vs leaf surfaces), proton buffering/neutralization and wash-off of acid deposition (André, Jonard, Ponette, et al., 2008a, 2008b; Nakanishi et al., 2001).

Soils exposed to differing throughfall and stemflow regimes also differ significantly in soil P (e.g. Backnäs, Laine-Kaulio, Kløve, et al., 2012; Rosier et al., 2015). AM symbiosis greatly benefits plant fitness by improving P uptake as AM provides a very effective pathway of scavenging P from soil environment and rapidly delivering it to cortical cells within the host root, bypassing direct uptake. However, high concentrations of soil P content greatly decrease the percentage of root length of plant infected by AM fungi (Abbott & Robson, 1979; Smith, Jakobsen, Grønlund, Smith, et al., 2011). Such a response has been correlated to reductions in concentrations of soluble carbohydrates (Jasper, Robson, Abbott, et al., 1979; Same, Robson, Abbott, et al., 1983) and free amino-nitrogen compounds in root exudates (Graham, Leonard, Menge, et al., 1981; Ratnayake, Leonard, & Menge, 1978). Therefore it stands to reason that the preferential transport and leaching of PO_4 by throughfall and stemflow observed by Backnäs, Laine-Kaulio, Kløve, et al. (2012) may hamper AM root colonization. In contrast, low available soil P while limiting plant growth increases root membrane permeability, leading to a net loss of root metabolites at sufficient levels to elicit germination and growth of AM fungi during pre- and postinfection (Graham, Leonard, Menge, et al., 1981). As such, roots in soils where throughfall and stemflow inputs contain moderate-to-low P, as a result of primarily bark contact or due to low P deposition (for example), may support increased AM association.

Throughfall- or stemflow-mediated AM colonization would result in improving root P nutrition and reduce membrane-mediated loss of root metabolites (a similar process was noted by Graham, Leonard, Menge, et al. (1981) without the consideration of throughfall or stemflow as a mediator). ECM colonization also benefits the host plant under P-poor conditions through enhanced soil exploration and P acquisition by a wider range of extrametrical mycelium hyphal patterns and enzymes (as described by Agerer, 2001). Subsequent concentrations of P within the soil environment, in turn, influence ECM development, as P fertilization affects physical root development and metabolite content on which ECM formation relies (Rousseau & Reid, 1991). In this way, throughfall and stemflow P flux would have similar effect on ECM root colonization as described for AM.

A substantial and growing body of work shows the ability of throughfall and stemflow to transport heavy metals to soils and even accumulate them in soils (Aubert et al., 2002; Avila & Rodrigo, 2004; Eisalou, Sengonul, Gokbulak, Serengil, Uygur, et al., 2013; Levia et al., 2015; Nikodem et al., 2010; Rodrigo et al., 1999). High concentrations of toxic metals adversely affect both ECM and AM root colonization potential functions as well as diversity (Cairney & Meharg, 1999). ECM and AM symbiosis provides a direct link between soil and host roots and can therefore be of significant importance increasing the heavy metal availability and toxicity to plants. Increased metal concentrations including Cd, Cu, Ni, Pb and/or Zn have been reported to reduce percentage of ECM host root colonization (Chappelka, Kush, Runion, Meier, Kelley, et al., 1991; Dixon & Buschena, 1988). Hartley, Cairney, Freestone, Woods, and Meharg (1999) observed that increasing concentrations of Cd and Zn decreased ECM diversity. However, increased soil Pb concentration suggested differential effects regarding different ECM morphotypes. Greater concentrations of Pb resulted in a linear decrease in the diversity of morphotypes colonizing *Pinus taeda*; however, ECM species *Cenococcum geophilum* increased 300% in occurrence when compared to uncontaminated soil (Chappelka, Kush, Runion, Meier, Kelley, et al., 1991). Similar to the findings presented for ECM, soils exposed to heavy metal contamination have been found to reduce or completely eliminate AM host root colonization (Gildon & Tinker, 1981). Certain metals maintain greater toxicity than others as negative correlations between Zn concentrations and AM colonization have been reported (Boyle & Paul, 1988). Additionally, AM species' richness and diversity is similarly affected by the presence of heavy metals. Del Val, Barea, and Azcon-Aguilar (1999) measured an increase in

AM species' richness and diversity at moderate levels of soil contamination; however, this trend was reversed when soils were amended with greater rates of heavy metal contamination. Several investigations have observed a similar trend, decrease in AM diversity as a result of increasing metal contamination (Pawlowska & Charvat, 2004; Ouziad, 1999; Regvar, Groznik, Goljevscek, & Gogala, 2001; Tonin, Vandenkoornhuyse, Joner, Straczek, & Leyval, 2001). The reasons underlying stress-related changes in the diversity of both ECM and AM populations due to the presence of heavy metals are not completely understood. Initial hypothesis suggests that a genetic predisposition must exist between some species of ECM and AM allowing certain species to thrive under increasing levels of metal contamination. However, a secondary hypothesis would suggest that metal contamination is interfering with root—symbiont signalling reducing the potential of ECM and AM species to colonize host roots. Understanding the effects of heavy metal pollution on physiological, ecological and applied aspects of mycorrhizal colonization is important as increasing levels of ECM and AM species diversity positively influences land rehabilitation and restoration success (Haselwandter & Bowen, 1996; Leyval, Turnau, Haselwandter, et al., 1997).

Allelopathic compounds are biochemicals produced by organisms that influence growth, survival and reproduction success of neighbouring/competing organisms. During mycorrhizal colonization, both the host and symbiont are modified through the release of several metabolites and signalling molecules establishing symbiosis. Allelopathic compounds released from the canopy during throughfall and stemflow could potentially influence host—symbiont communication. For example, allelopathic compounds liberated during throughfall or stemflow could potentially affect several processes important to the initial establishment of mycorrhiza symbiosis including early contact, root-cell loosening and structure development/formation of mycorrhizal symbiont. Phytohormones have been identified as effective in ECM development. Phytohormones are metabolites produced in low quantities influencing the development of plant organs. Well-identified phytohormones include auxin, ethylene, gibberellins, abscisic acid and cytokinins, and further molecules such as salicylic acid, brassinosteriod and jasmonic acid. During ECM colonization of host roots, phytohormones are upregulated, mediating ECM development. Current research efforts suggest that primary fungal signals perceived by host roots may be volatile phytohormones eliminating the physical contact requirement between host and ECM (Felten, Martin, Legué, et al., 2012). Case in point, auxin has been identified as an initial signalling molecule, modifying root development (i.e. cell loosening and Hartig net development) during ECM establishment. AM symbiosis also occurs via complex exchange of signals between the host root and AM, beginning with pre-colonization phase, formation of appressoria and penetration of the host root (Smith & Read, 2008). Several research efforts suggest that flavonoids are critical signalling molecules stimulating spore germination, hyphal growth and root colonization (Hassan & Mathesius, 2012; Larose et al., 2002; Vierheilig, Bago, Albrecht, Poulin, & Piché, 1998). Harrison (2005) has extensively reviewed several of the signalling pathways present in the AM host root symbiosis. To the knowledge of the authors no study has quantified the presence of either phytohormones or flavonoids within throughfall or stemflow. However, Gonzalezy, Souto, and Pastoriza (2012) suggests that stemflow is a significant mechanism facilitating the transfer of canopy derived phytohoromones to the soil environment.

4. FUTURE DIRECTIONS FOR EVALUATING THROUGHFALL AND STEMFLOW INFLUENCES ON SYMBIOTIC MICROBES

These connections between symbiotic N-fixing microbes, mycorrhiza fungi and throughfall and stemflow are currently theoretical, based on the literature reviewed, and have never been directly tested. However, the processes we have discussed are directly testable in the field, greenhouse and laboratory. In the field, for example, throughfall water and nutrient supply patterns can be easily quantified through inexpensive installations using irregular grids of HDPE funnels and bottles, as has been performed across a variety of forest ecosystems (e.g. Fathizadeh et al., 2014; Gerrits, Pfister, & Savenije, 2010; Keim, Skaugset, Weiler, et al., 2005; Nanko, Onda, Ito, & Moriwaki, 2011; Zimmermann, Wilcke, & Elsenbeer, 2007). Once these patterns are determined (and throughfall spatial patterns have often been found to be temporally persistent; Guswa, 2012), root sampling can be performed and the degree of nodulation and mycorrhizal infection determined between areas of high and low throughfall water or nutrient input. As for the effect of stemflow water or chemistry on root nodulation and mycorrhiza, dye tracer studies in the field have been efficacious in identifying roots situated within the macropores through which stemflow was preferentially transported (Li, Yang, Li, Lin, et al., 2009; Martinez-Meza & Whitford, 1996; Schwärzel, Ebermann, Schalling, et al., 2012). After exhumation of the roots to identify those roots contacted by the dye tracer, the roots can be sampled and their degree of colonization by nodules and mycorrhiza determined. Stemflow-contacted roots may then be compared with dye-free roots that were similarly analyzed for their degree of symbiosis. Even more interesting would be to compare the community structures of these (dyed and non-dyed) root microbial symbionts to see whether plant species colonized by both AM and ECM, for example, are correlated to differing stemflow (or even throughfall) exposures. The authors are unaware of any studies showing that tracer dyes commonly used for subsurface hydrologic studies have a measureable impact on microbial community structure or function and, thus, would not anticipate that the use of these dyes would alter the symbiosis significantly prior to sampling and analysis.

To control confounding variables present at most field sites — spatially inconsistent soil texture, chemistry, moisture, competition with other species, etc. — greenhouses may be used with tree saplings. A rainfall simulator (e.g. Keim, Skaugset, & Weiler, 2006) within the greenhouse would allow for controlled throughfall generation under consistent rainfall conditions. This arrangement has been used consistently by the ecohydrology community for improved understanding of precipitation partitioning (see review of Friesen, Lundquist, and Van Stan (2015) for example) and would provide comparable, repeatable generation of throughfall and stemflow patterns for extended periods. As such, the long-term exposure of root symbionts to controlled canopy-derived hydrological and nutrient flux patterns could be determined. Under such conditions, dye tracers for stemflow tracking would be easier to apply, stemflow preferential flow pathways may be better controlled and the roots will be easier to exhume for analysis. After a satisfactory period of time under consistent simulated rainfall conditions and canopy structural conditions in the greenhouse, both the structure and simulated rainfall conditions could then be altered to observe whether (or to what degree) nodule-forming and mycorrhizal microbes respond. Canopy manipulations could represent a number of disturbances predicted to increase with global climate change — from ice storm and hurricane damage (Houlton & Driscoll, 2011; McDowell, 2011), to fires (Onodera & Van Stan, 2011), or insect infestations (Michalzik, 2011), and even lightning (Romps, Seeley, Vollaro, Molinari, et al., 2014).

ACKNOWLEDGEMENT

C.L.R. acknowledges that ideas and concepts generated for this chapter were supported via funding from the NSF Christina River Basin Critical Zone Observatory National Science Foundation grant (0724971).

REFERENCES

van Aarle, I. M., Olsson, P. A., Söderström, B., et al. (2002). Arbuscular mycorrhizal fungi respond to the substrate pH of their extraradical mycelium by altered growth and root colonization. *New Phytologist, 155*, 173–182.

Abbott, L. K., & Robson, A. D. (1979). A quantitative study on the spores and anatomy of mycorrhizas formed by a species of Glomus, with special reference to its taxonomy. *Australian Journal of Botany, 27*, 363–375.

Abdelmajid, K., Karim, B. H., & Chedly, A. (2008). Symbiotic response of common bean (phaseolus vulgaris L.) to iron deficiency. *Acta Physiologiae Plantarum, 30*, 27–34.

Abuzinadah, R. A., & Read, D. J. (1986). The role of proteins in the nitrogen nutrition of ectomycorrhizal plants. I. Utilization of peptides and proteins by ectomycorrhizal fungi. *New Phytologist, 103*, 481–493.

Agerer, R. (2001). Exploration types of ectomycorrhizae. A proposal to classify ectomycorrhizal mycelial systems according to their patterns of differentiation and putative ecological importance. *Mycorrhiza, 11*, 107–114.

Akiyama, K., Matsuzaki, K. I., Hayashi, H., et al. (2005). Plant sesquiterpenes induce hyphal branching in arbuscular mycorrhizal fungi. *Nature, 435*, 824–827.

André, F., Jonard, M., Ponette, Q., et al. (2008a). Effects of biological and meteorological factors on stemflow chemistry within a temperate mixed oak-beech stand. *Science of the Total Environment, 393*, 72–83.

André, F., Jonard, M., Ponette, Q., et al. (2008b). Spatial and temporal patterns of throughfall chemistry within a temperate mixed oak-beech stand. *Science of the Total Environment, 397*, 215–228.

Ares, A., Burner, D. M., Brauer, D. K., et al. (2009). Soil phosphorus and water effects on growth, nutrient and carbohydrate concentrations, d13C, and nodulation of mimosa (*Albizia julibrissin* Durz.) on a highly weathered soil. *Agroforestry Systems, 76*, 317–325.

Aubert, D., Stille, P., Probst, A., Gauthier-Lafaye, F., Pourcelot, L., Del Nero, M., et al. (2002). Characterization and migration of atmospheric REE in soils and surface waters. *Geochimica et Cosmochimica Acta, 66*, 3339–3350.

Avila, A., & Rodrigo, A. (2004). Trace metal fluxes in bulk deposition, throughfall and stemflow at two evergreen oak stands in NE Spain subject to different exposure to the industrial environment. *Atmospheric Environment, 38*, 171–180.

Backnäs, S., Laine-Kaulio, H., Kløve, B., et al. (2012). Phosphorous forms and related soil chemistry in preferential flowpaths and the soil matrix of a forested podzolic till soil profile. *Geoderma, 189*, 50–64.

Bartolome-Esteban, H., & Schenck, N. C. (1994). Spore germination and hyphal growth of arbuscular mycorrhizal fungi in relation to soil aluminum saturation. *Mycologia*, 217–226.

Batterman, S. A., Wurzburger, N., Hedin, L. O., et al. (2013). Nitrogen and phosphorus interact to control tropical symbiotic N_2 fixation: a test in *Inga punctata*. *Journal of Ecology, 101*, 1400–1408.

Bellenger, J. P., Xu, Y., Zhang, X., Morel, F. M. M., & Kraepiel, A. M. L. (2014). Possible contribution of alternative nitrogenases to nitrogen fixation by asymbiotic N_2-fixing bacteria in soils. *Soil Biology and Biochemistry, 69*, 413–420.

Bischoff, S., Schwarz, M. T., Siemens, J., Thieme, L., Wilcke, W., Michalzik, B., et al. (2014). Properties of dissolved and total organic matter in throughfall, stemflow and forest floor leachate of Central European forests. *Biogeosciences, 12*, 2695–2706.

Bollen, W. B., Chen, C. S., Lu, K. C., Tarrant, R. F., et al. (1968). Effect of stemflow precipitation on chemical and microbiological soil properties beneath a single alder tree. In *Biology of alder: Proceedings of a symposium, 14* pp. 149–156.

Bouten, W., Heimovaara, T. J., Tiktak, A., et al. (1992). Spatial patterns of throughfall and soil water dynamics in a Douglas fir stand. *Water Resources Research, 28*, 3227–3233.

Boyle, M., & Paul, E. A. (1988). Vesicular-arbuscular mycorrhizal associations with barley on sewage-amended plots. *Soil Biology and Biochemistry, 20*, 945–948.

Brencic, A., & Winans, S. C. (2005). Detection of and response to signals involved in host-microbe interactions by plant-associated bacteria. *Microbiology and Molecular Biology Reviews, 69*, 155–194.

Bryant, M. L., Bhat, S., Jacobs, J. M., et al. (2005). Measurement and modeling of throughfall variability for five forest communities in the southeastern US. *Journal of Hydrology, 312*, 95–108.

Buee, M., Rossignol, M., Jauneau, A., Ranjeva, R., Bécard, G., et al. (2000). The pre-symbiotic growth of arbuscular mycorrhizal fungi is induced by a branching factor partially purified from plant root exudates. *Molecular Plant-Microbe Interactions, 13*, 693–698.

Cairney, J. W. G., & Meharg, A. A. (1999). Influences of anthropogenic pollution on mycorrhizal fungal communities. *Environmental Pollution, 106*, 169–182.

Campbell, J., Bradfield, G. E., Prescott, C. E., Fredeen, A. L., et al. (2010). The influence of overstory *Populus* on epiphytic lichens in subboreal spruce forests of British Columbia. *Canadian Journal of Forest Research, 40*, 143–154.

Cantor, A., Hale, A., Aaron, J., Traw, M. B., Kalisz, S., et al. (2011). Low allelochemical concentrations detected in garlic mustard-invaded forest soils inhibit fungal growth and AMF spore germination. *Biological invasions, 13*, 3015–3025.

Ceccherini, M. T., Ascher, J., Angelli, A., Certini, G., Pietramellara, G., Piovanelli, C., et al. (2008). Tree bark and soil ammonia oxidizers: a molecular study on a historical forest of Central Italy. *Fresenius Environmental Bulletin, 17*, 882–889.

Chalot, M., & Brun, A. (1998). Physiology of organic nitrogen acquisition by ectomycorrhizal fungi and extomycorrhizas. *FEMS Microbiology Reviews, 22*, 21–44.

Chang, S. C., & Matzner, E. (2000). The effect of beech stemflow on spatial patterns of soil solution chemistry and seepage fluxes in a mixed beech/oak stand. *Hydrological Processes, 14*, 135–144.

Chappelka, A. H., Kush, J. S., Runion, G. B., Meier, S., Kelley, W. D., et al. (1991). Effects of soil-applied lead on seedling growth and ectomycorrhizal colonization of loblolly-pine. *Environmental Pollution, 72*, 307–316.

Chilvers, G. A., Lapeyrie, F. F., Horan, D. P., et al. (1987). Ectomycorrhizal vs endomycorrhizal fungi within the same root-system. *New Phytologist, 107*, 441–448.

Churchland, C., & Grayston, S. (2014). Specificity of plant-microbe interactions in the tree mycorrhizosphere biome and consequences for soil C cycling. *Frontiers in Microbiology, 5*, 1–20.

Cividini, D., Lemarchand, D., Chabaux, F., Boutin, R., Pierret, M. C., et al. (2010). From biological to lithological control of the B geochemical cycle in a forest watershed (Strengbach, Vosges). *Geochimica et Cosmochimica Acta, 74*, 3143–3163.

Clark, R. B. (1997). Arbuscular mycorrhizal adaptation, spore germination, root colonization, and host plant growth and mineral acquisition at low pH. *Plant and Soil, 192*, 15–22.

Clark, R. B., & Zeto, S. K. (1996). Mineral acquisition by mycorrhizal maize grown on acid and alkaline soil. *Soil Biology and Biochemistry, 28*, 1495–1503.

Clark, R. B., Zeto, S. K., & Zobel, R. W. (1999). Arbuscular mycorrhizal fungal isolate effectiveness on growth and root colonization of *Panicum virgatum* in acidic soil. *Soil Biology and Biochemistry, 31*, 1757–1763.

Coenders-Gerrits, A. M. J., Hopp, L., Savenije, H. H. G., Pfister, L., et al. (2013). The effect of spatial throughfall patterns on soil moisture patterns at the hillslope scale. *Hydrology Earth System Science, 17*, 1749–1763.

Cohn, J., Bradley Day, R., Stacey, G., et al. (1998). Legume nodule organogenesis. *Trends in Plant Science, 3*, 105–110.

Colpaert, J. V., & van Laere, A. (1996). A comparison of the extracellular enzyme activities of two ectomycorrhizal and leaf-saprotrophic basidiomycete colonizing beech leaf litter. *New Phytologist, 134*, 133–141.
Courty, P. E., Franc, A., Garbaye, J., et al. (2010). Temporal and functional pattern of secreted enzyme activities in an ectomycorrhizal community. *Soil Biology and Biochemistry, 42*, 2022–2025.
Coventry, D. R., & Evans, J. (1989). In A. D. Robson (Ed.), *Symbiotic nitrogen fixation and soil acidity*. Marrickville; Australia: Academic Press.
Coxson, D. S., McIntyre, D. D., Vogel, H. J., et al. (1992). Pulse release of sugars and polyols from canopy bryophytes in tropical montane rain forest (Guadeloupe, French West Indies). *Biotropica, 24*, 121–133.
Crannell, W. K., Tanaka, Y., Myrold, D. D., et al. (1994). Calcium and pH interaction on root nodulation of nursery-grown red alder (*Alnus rubra* Bong.) seedlings by Frankia. *Soil Biology and Biochemistry, 26*, 607–614.
Cuenca, G., Andrade, Z. D., & Meneses, E. (2001). The presence of aluminium in arbuscular mycorrhizas of *Clusia multiflora* exposed to increased acidity. *Plant and Soil, 231*, 233–241.
Danielson, R. M., & Visser, S. (1989). Effects of forest soil acidification on ectomycorrhizal and vesicular arbuscular mycorrhizal development. *New Phytologist, 112*, 41–47.
Dawson, J. O., & Gibson, A. H. (1987). Sensitivity of selected Frankia isolates from Casuarina, Allocasuarina and North American host plants to sodium chloride. *Physiologia Plantarum, 70*, 272–278.
Del Val, C., Barea, J. M., & Azcon-Aguilar, C. (1999). Diversity of arbuscular mycorrhizal fungus populations in heavy-metal-contaminated soils. *Applied and Environmental Microbiology, 65*, 718–723.
Dezzeo, N., & Chacón, N. (2006). Nutrient fluxes in incident rainfall, throughfall, and stemflow in adjacent primary and secondary forests of the Gran Sabana, southern Venezuela. *Forest Ecology and Management, 234*, 218–226.
Divito, G. A., & Sadras, V. O. (2014). How do phosphorus, potassium and sulphur affect plant growth and biological nitrogen fixation in crop and pasture legumes? A meta-analysis. *Field Crops Research, 156*, 161–171.
Dixon, R. K., & Buschena, C. A. (1988). Response of ectomycorrhizal *Pinus-Banksiana* and *Picea-Glauca* to heavy-metals in soil. *Plant and Soil, 105*, 265–271.
Dudeja, S. S., Giri, R., Saini, R., Suneja-Madan, P., Kothe, E., et al. (2012). Interaction of endophytic microbes with legumes. *Journal of Basic Microbiology, 52*, 248–260.
Eisalou, H. K., Sengonul, K., Gokbulak, F., Serengil, Y., Uygur, B., et al. (2013). Effects of forest canopy cover and floor on chemical quality of water in broad leaved and coniferous forests of Istanbul, Turkey. *Forest Ecology and Management, 289*, 371–377.
Fathizadeh, O., Attarod, P., Keim, R. F., Stein, A., Amiri, G. Z., Darvishsefat, A. A., et al. (2014). Spatial heterogeneity and temporal stability of throughfall under individual *Quercus brantii* trees. *Hydrological Processes, 28*, 1124–1136.
Felten, J., Martin, F., Legué, V., et al. (2012). Signaling in ectomycorrhizal symbiosis. In *Signaling and communication in plant symbiosis* (pp. 123–142). Heidelberg: Springer Berlin.
Flis, S. E., Glenn, A. R., & Dilworth, M. J. (1993). The interaction between aluminium and root nodule bacteria. *Soil Biology and Biochemistry, 25*, 403–417.
Ford, E. D., & Deans, J. D. (1978). The effects of canopy structure on stemflow, throughfall and interception loss in a young Sitka spruce plantation. *Journal of Applied Ecology, 15*, 905–917.
Franche, C., Lindström, K., Elmerich, C., et al. (2009). Nitrogen-fixing bacteria associated with leguminous and non-leguminous plants. *Plant and Soil, 321*, 35–59.
Friesen, J., Lundquist, J., & Van Stan, J. T. (2015). Evolution of forest precipitation water storage measurement methods. *Hydrological Processes, 29*, 2504–2520.

Frioni, L., Minasian, H., Volfovicz, R., et al. (1999). Arbuscular mycorrhizae and ectomycorrhizae in native tree legumes in Uruguay. *Forest Ecology and Management, 115*, 41–47.
Frost, E. E., & Levia, D. F. (2013). Boron stemflow chemistry in relation to species and season in a temperate deciduous forest. In *American Geophysical Union Fall Meeting* (pp. B31A–B0361).
Furnholm, T. R., & Tisa, L. S. (2014). The ins and outs of metal homeostasis by the root nodule actinobacterium Frankia. *BMC Genomics, 15*, 964–1003.
Gardner, C. B., Lyons, W. B., Litt, G., Ogden, F. L., et al. (2014). Rock-derived micronutrient transport in the tropics: molybdenum cycling in deeply-weathered Panama soils. *Procedia Earth and Planetary Science, 10*, 266–270.
Gerrits, A. M. J., Pfister, L., & Savenije, H. H. G. (2010). Spatial and temporal variability of canopy and forest floor interception in a beech forest. *Hydrological Processes, 24*, 3011–3025.
Geurts, R., Fedorova, E., Bisseling, T., et al. (2005). Nod factor signaling genes and their function in the early stages of rhizobium infection. *Current Opinion in Plant Biology, 8*, 346–352.
Gildon, A., & Tinker, P. B. (1981). A heavy metal tolerant strain of mycorrhizal fungus. *Transactions British Mycological Society, 77*, 648–649.
Gonzalezy, L., Souto, X., & Pastoriza, J. (2012, December). Allelopathy in forest ecosystems. In Allelopathy in Ecological Agriculture and Forestry: Proceedings of the III International Congress on Allelopathy in Ecological Agriculture and Forestry, Dharwad, India, 18–21 August 1998 (p. 183). Springer Science & Business Media.
Gonzalez-Guerrero, M., Matthiadis, A., Saez, A., Long, T. A., et al. (2014). Fixating on metals: new insights into the role of metals in nodulation and symbiotic nitrogen fixation. *Frontiers in Plant Science, 5*, 45.
Graham, J. H., Leonard, R. T., Menge, J. A., et al. (1981). Membrane-mediated decrease in root exudation responsible for phorphorus inhibition of vesicular-arbuscular mycorrhiza formation. *Plant Physiology, 68*, 548–552.
Guswa, A. J. (2012). Canopy vs roots: production and destruction of variability in soil moisture and hydrologic fluxes. *Vadose Zone Journal, 11*, 1–13.
Hane, E. N., Hamburg, S. P., Barber, A. L., & Plaut, J. A. (2003). Phytotoxicity of American beech leaf leachate to sugar maple seedlings in a greenhouse experiment. *Canadian Journal of Forest Research, 33*, 814–821.
Harrison, M. J. (2005). Signaling in the arbuscular mycorrhizal symbiosis. *Annual Review Microbiology, 59*, 19–42.
Hartley, J., Cairney, J. W. G., Freestone, P., Woods, C., & Meharg, A. A. (1999). The effects of multiple metal contamination on ectomycorrhizal Scots pine (*Pinus sylvestris*) seedlings. *Environmental Pollution, 106*, 413–424.
Haselwandter, K., & Bowen, G. D. (1996). Mycorrhizal relations in trees for agroforestry and land rehabilitation. *Forest Ecology and Management, 81*, 1–17.
Hassan, S., & Mathesius, U. (2012). The role of flavonoids in root-rhizosphere signaling: opportunities and challenges for improving plant-microbe interactions. *Journal of Experimental Botany, 63*, 3429–3444.
van der Heijden, M. G. A., Bardgett, R. D., Van Straalen, N. M., et al. (2008). The unseen majority: soil microbes as drivers of plant diversity and productivity in terrestrial ecosystems. *Ecology Letters, 11*, 296–310.
Heijne, B., van Dam, D., Heil, G. W., & Bobbink, R. (1996). Acidification effects on vesicular-arbuscular mycorrhizal (VAM) infection, growth and nutrient uptake of established heathland herb species. *Plant and Soil, 179*, 197–206.
Helmisaari, H. S., & Mälkönen, E. (1989). Acidity and nutrient content of throughfall and soil leachate in three *Pinus sylvestris* stands. *Scandinavian Journal of Forest Research, 4*, 13–28.
Hepper, C. M. (1979). Germination and growth of *Glomus caledonius* spores: the effects of inhibitors and nutrients. *Soil Biology and Biochemistry, 11*, 269–277.

Houlton, B. Z., & Driscoll, C. T. (2011). The effects of ice storms on the hydrology and biogeochemistry of forests. In D. Carlyle-Moses, D. Levia, & T. Tanaka (Eds.), *Forest hydrology and biogeochemistry: Synthesis of past research and future directions: Vol. 216. Ecological studies series* (pp. 623–642). Heidelberg: Springer-Verlag.

Isaac, M. E., Harmand, J. M., Drevon, J. J., et al. (2011). Growth and nitrogen acquisition strategies of *Acacia senegal* seedlings under exponential phosphorus additions. *Journal of Plant Physiology, 168*, 776–781.

Jasper, D. A., Robson, A. D., Abbott, L. K., et al. (1979). Phosphorus and the formation of vesicular-arbuscular mycorrhizas. *Soil Biology and Biochemistry, 11*, 501–505.

John, J., & Sarada, S. (2012). Role of phenolics in allelopathic interactions. *Allelopathy Journal, 29*, 215–229.

Johnson, M. S., & Lehmann, J. (2006). Double-funneling of trees: stemflow and root-induced preferential flow. *Ecoscience, 13*, 324–333.

Keim, R. F., Skaugset, A. E., Weiler, M., et al. (2005). Temporal persistence of spatial patterns in throughfall. *Journal of Hydrology, 314*, 263–274.

Keim, R. F., Skaugset, A. E., & Weiler, M. (2006). Storage of water on vegetation under simulated rainfall of varying intensity. *Advances in Water Resources, 29*, 974–986.

Knowlton, S., & Dawson, J. (1983). Effects of pseudomonas-cepacia and cultural-factors on the nodulation of *Alnus-rubra* roots by Frankia. *Canadian Journal of Botany-Revue Canadienne de Botanique, 61*, 2877–2882.

Kopacek, J., Turek, J., Hejzlar, J., Santruckova, H., et al. (2009). Canopy leaching of nutrients and metals in a mountain spruce forest. *Atmospheric Environment, 43*, 5443–5453.

Kumar, J. (2011). Effect of phosphorus and sulphur application on performance of vegetable pea (*Pisum sativum* L.) cv. pant matar-2. *Legume Research, 34*, 292–295.

Lapeyrie, F. F., & Chilvers, G. A. (1985). An endomycorrhiza succession associated with enhanced growth of *Eucalyptus-dumosa* seedlings planted in a Calcareous soil. *New Phytologist, 100*, 93–104.

Larose, G., Chenevert, R., Moutoglis, P., Gagne, S., Piche, Y., & Vierheilig, H. (2002). Flavonoid levels in roots of *Medicago sativa* are modulated by the developmental stage of the symbiosis and the root colonizing arbuscular mycorrhizal fungus. *Journal of Plant Physiology, 159*, 1329–1339.

Le Mellec, A., Meesenburg, H., Michalzik, B., et al. (2010). The importance of canopy-derived dissolved and particulate organic matter (DOM and POM) — comparing throughfall solution from broadleaved and coniferous forests. *Annals of Forest Science, 67*, 411.

Levia, D. F., & Frost, E. E. (2003). A review and evaluation of stemflow literature in the hydrologic and biogeochemical cycles of forested and agricultural ecosystems. *Journal of Hydrology, 274*, 1–29.

Levia, D. F., & Germer, S. (2015). A review of stemflow generation dynamics and stemflow-environment interactions in forests and shrublands. *Reviews in Geophysics, 53*, 673–714.

Levia, D. F., Keim, R. F., Carlyle-Moses, D. E., & Frost, E. E. (2011). Throughfall and stemflow in wooded ecosystems. In D. Carlyle-Moses, D. Levia, & T. Tanaka (Eds.), *Forest hydrology and biogeochemistry: synthesis of past research and future directions: Vol. 216. Ecological studies series* (pp. 425–444). Heidelberg: Springer-Verlag.

Levia, D. F., Van Stan, J. T., Siegert, C. M., Inamdar, S. P., Mitchell, M. J., Mage, S. M., et al. (2011). Atmospheric deposition and corresponding variability of stemflow chemistry across temporal scales in a mid-Atlantic broadleaved deciduous forest. *Atmospheric Environment, 45*, 3046–3054.

Levia, D. F., Michalzik, B., Bischoff, S., Näthe, K., Legates, D. R., Grusell, M.-C., et al. (2013). Measurement and modeling of diameter distributions of particulate matter in terrestrial solutions. *Geophysical Research Letters, 40*, 1317–1321.

Levia, D. F., Shiklomanov, A. N., Van Stan, J. T., Scheick, C. E., Inamdar, S. P., Mitchell, M. J., et al. (2015). Calcium and aluminum cycling in a temperate broadleaved deciduous forest of the eastern United States: relative impacts of tree species, canopy state, and flux type. *Environmental Monitoring and Assessment, 187*, 458.

Leyval, C., Turnau, K., Haselwandter, K., et al. (1997). Effect of heavy metal pollution on mycorrhizal colonization and function: physiological, ecological and applied aspects. *Mycorrhiza, 7*, 139–153.

Li, X. Y., Yang, Z. P., Li, Y. T., Lin, H., et al. (2009). Connecting ecohydrology and hydropedology in desert shrubs: stemflow as a source of preferential flow in soils. *Hydrology and Earth System Science, 13*, 1133–1144.

Limpens, E., & Bisseling, T. (2003). Signaling in symbiosis. *Current Opinion in Plant Biology, 6*, 343–350.

Lindsay, W. L. (1979). *Chemical equilibria in soils*. New York: Wiley.

Lodge, D. J., & Wentworth, T. R. (1990). Negative associations among VA-mycorrhizal fungi and some ectomycorrhizal fungi inhabiting the same root system. *Oikos, 57*, 347–356.

Macinnis-Ng, C. M. O., Flores, E. E., Muller, H., Schwendenmann, L., et al. (2012). Rainfall partitioning into throughfall and stemflow and associated nutrient fluxes: land use impacts in a lower montane tropical region of Panama. *Biogeochemistry, 111*, 661–676.

Marschner, H. (1995). In *Mineral nutrition of higher plants* (2nd ed.). Elsevier, 889 p.

Martin, K., Posavatz, N., Myrold, D., et al. (2003). Nodulation potential of soils from red alder stands covering a wide age range. *Plant and Soil, 254*, 187–192.

Martinez-Meza, E., & Whitford, W. G. (1996). Stemflow, throughfall and channelization of stemflow by roots in three Chihuahuan desert shrubs. *Journal of Arid Environments, 32*, 271–287.

Masson-Boivin, C., Giraud, E., Perret, X., Batut, J., et al. (2009). Establishing nitrogen-fixing symbiosis with legumes: how many rhizobium recipes? *Trends in Microbiology, 17*, 458–466.

Matschonat, G., & Falkengren-Grerup, U. (2000). Recovery of soil pH, cation-exchange capacity and the saturation of exchange sites from stemflow-induced soil acidification in three Swedish beech (*Fagus sylvatica* L.) forests. *Scandinavian Journal of Forest Research, 15*, 39–48.

Mayerhofer, M. S., Fraser, E., Kernaghan, G., et al. (2015). Acid protease production in fungal root endophytes. *Mycologia, 107*, 1–11.

McDowell, W. H. (2011). Impacts of hurricanes on forest hydrology and biogeochemistry. In D. Carlyle-Moses, D. Levia, & T. Tanaka (Eds.), *Forest hydrology and biogeochemistry: synthesis of past research and future directions: Vol. 216. Ecological studies series* (pp. 643–658). Heidelberg: Springer-Verlag.

Mergaert, P., Uchiumi, T., Alunni, B., Evanno, G., Cheron, A., Catrice, O., et al. (2006). Eukaryotic control on bacterial cell cycle and differentiation in the rhizobium-legume symbiosis. *Proceedings of the National Academy of Sciences of the United States of America, 103*, 5230–5235.

Meyer, F. H. (1973). Distribution of ectomycorrhizae in native and man-made forests. In G. C. Mark, & T. T. Kozlowski (Eds.), *Ectomycorrhizae: Their ecology and physiology* (pp. 79–105). New York and London: Academic Press.

Michalzik, B. (2011). Insects, infestation, and nutrient fluxes. In D. Carlyle-Moses, D. Levia, & T. Tanaka (Eds.), *Forest hydrology and biogeochemistry: Synthesis of past research and future directions: Vol. 216. Ecological studies series* (pp. 557–580). Heidelberg: Springer-Verlag.

Miller, R. W., & Sirois, J. C. (1983). Calcium and magnesium effects on symbiotic nitrogen fixation in the alfalfa (*M. sativa*) – *Rhizobium meliloti* system. *Physiologia Plantarum, 58*, 464–470.

Miranda, J. D., & Harris, P. J. (1994). Effects of soil phosphorus on spore germination and hyphal growth of arbuscular mycorrhizal fungi. *New Phytologist, 128*(1), 103–108.

Moore, R. D., Winkler, R. D., Carlyle-Moses, D. E., et al. (2008). Watershed response to the McLure forest fire: presentation summaries from the Fishtrap Creek workshop, March 2008. *Streamline Watershed Management Bulletin, 12*, 1–11.

Morandi, D. (1996). Occurrence of phytoalexins and phenolic compounds on endomycorrhizal interactions, and their potential role in biological control. *Plant and Soil, 185*, 241–251.

Muoghalu, J. I., & Oakhumen, A. (2000). Nutrient content of incident rainfall, throughfall and stemflow in a Nigerian secondary lowland rainforest. *Applied Vegetation Science, 3*, 181–188.

Muscolo, A., Panuccio, M., Sidari, M., et al. (2001). The effect of phenols on respiratory enzymes in seed germination. *Plant Growth Regulation, 35*, 31–35.

Nadkarni, N. M., & Sumera, M. M. (2004). Old-growth forest canopy structure and its relationship to throughfall interception. *Forest Science, 50*, 290–298.

Nakanishi, A., Shibata, H., Inokura, Y., Nakao, T., Toda, H., Sato, F., et al. (2001). Chemical characteristics in stemflow of Japanese cedar in Japan. *Water Air and Soil Pollution, 130*, 709–714.

Nanko, K., Onda, Y., Ito, A., & Moriwaki, H. (2011). Spatial variability of throughfall under a single tree: experimental study of rainfall amount, raindrops, and kinetic energy. *Agricultural and Forest Meteorology, 151*, 1173–1182.

Návar, J., González, J. M., Gonzalez, H., et al. (2009). Gross precipitation and throughfall chemistry in legume species planted in Northeastern México. *Plant and Soil, 318*, 15–26.

Newman, E. I., & Rydell, P. (1987). The distribution of mycorrhizas among families of vascular plants. *New Phytologist, 106*, 745–751.

Nikodem, A., Kodesova, R., Drabek, O., Bubenickova, L., Boruvka, L., Pavlu, L., et al. (2010). A numerical study of the impact of precipitation redistribution in a beech forest canopy on water and aluminum transport in a Podzol. *Vadose Zone Journal, 9*, 238–251.

Norden, U. (1991). Acid deposition and throughfall fluxes of elements as related to tree species in deciduous forests of South Sweden. *Water Air and Soil Pollution, 60*, 209–230.

Nygren, C. M. R., Edqvist, J., Elfstrand, M., Heller, G., Taylor, A. F. S., et al. (2007). Detection of extracellular protease activity in different species and genera of ectomycorrhizal fungi. *Mycorrhiza, 17*, 241–248.

Oldroyd, G. E. (2013). Speak, friend, and enter: signalling systems that promote beneficial symbiotic associations in plants. *Nature Reviews Microbiology, 11*, 252–263.

Oliveira, R. S., Castro, P. M. L., Dodd, J. C., Vosátka, M., et al. (2005). Synergistic effect of *Glomus intraradices* and Frankia spp. on the growth and stress recovery of *Alnus glutinosa* in an alkaline anthropogenic sediment. *Chemosphere, 60*, 1462–1470.

Onodera, S.-I., & Van Stan, J. T. (2011). Effect of forest fires on hydrology and biogeochemistry of watersheds. In D. Carlyle-Moses, D. Levia, & T. Tanaka (Eds.), *Forest hydrology and biogeochemistry: Synthesis of past research and future directions: Vol. 216. Ecological studies series* (pp. 599–621). Heidelberg: Springer-Verlag.

Ouimet, R., Camire, C., & Furlan, V. (1996). Effect of soil base saturation and endomycorrhization on growth and nutrient status of sugar maple seedlings. *Canadian Journal of Soil Science, 76*, 109–115.

Ouziad, F. (1999). *Charakterisierun der arbuskulaeren Mykorrhiza Pilz-Flora an ausgewaehlten Standorten mit molekularbiologischen Methoden* (Diploma thesis). Germany: University of Cologne.

Pawlowska, T. E., & Charvat, I. (2004). Heavy-metal stress and development patterns of arbuscular mycorrhizal fungi. *Applied and Environmental Microbiology, 70*, 6643–6649.

Pinior, A., Wyss, U., Piché, Y., Vierheilig, H., et al. (1999). Plants colonized by AM fungi regulate further root colonization by AM fungi through altered root exudation. *Canadian Journal of Botany, 77*, 891–897.

Porcel, R., Barea, J., & Ruiz-Lozano, J. (2003). Antioxidant activities in mycorrhizal soybean plants under drought stress and their possible relationship to the process of nodule senescence. *New Phytologist, 157*, 135–143.

Ratnayake, M., Leonard, R. T., & Menge, J. A. (1978). Root exudation in relation to supply of phosphorus and its possible relevance to mycorrhizal formation. *New Phytologist, 81*, 543–552.

Redondo-Nieto, M., Bonilla, I., Bolanos, L., Rivilla, R., El-Hamdaoui, A., et al. (2001). Boron deficiency affects early infection events in the pea-rhizobium symbiotic interaction. *Australian Journal of Plant Physiology, 28*, 819–823.

Reguera, M., Bonilla, I., Bolaños, L., et al. (2010). Boron deficiency results in induction of pathogenesis-related proteins from the PR-10 family during the legume–rhizobia interaction. *Journal of Plant Physiology, 167*, 625–632.

Regvar, M., Groznik, N., Goljevscek, K., & Gogala, N. (2001). Diversity of arbuscular mycorrhizal fungi from various disturbed ecosystems in Slovenia. *Acta Biologica Slovenica, 44*, 27–34.

Rejili, M., Mahdhi, M., Fterich, A., Dhaoui, S., Guefrachi, I., Abdeddayem, R., et al. (2012). Symbiotic nitrogen fixation of wild legumes in tunisia: soil fertility dynamics, field nodulation and nodules effectiveness. *Agriculture, Ecosystems and Environment, 157*, 60–69.

Roberts, K. J., & Anderson, R. C. (2001). Effect of garlic mustard [*Alliaria petiolata* (Beib. Cavara & Grande)] extracts on plants and arbuscular mycorrhizal fungi. *The American Midland Naturalist, 146*, 146–152.

Rodrigo, A., Avila, A., Gomez-Bolea, A., et al. (1999). Trace metal contents in *Parmelia caperata* (L.) Ach. compared to bulk deposition, throughfall and leaf-wash fluxes in two holm oak forests in Montseny (NE Spain). *Atmospheric Environment, 33*, 359–367.

Rojas, N., Perry, D., Li, C., Ganio, L., et al. (2002). Interactions among soil biology, nutrition, and performance of actinorhizal plant species in the H.J. Andrews' experimental forest of Oregon. *Applied Soil Ecology, 19*, 13–26.

Romps, D. M., Seeley, J. T., Vollaro, D., Molinari, J., et al. (2014). Projected increase in lightning strikes in the United States due to global warming. *Science, 346*, 851–854.

Rosier, C. L., Van Stan, J. T., Moore, D. M., Schrom, J. O. S., Wu, T., Reichard, J., et al. (2015). Forest canopy structural controls over throughfall affect soil microbial community structure in an epiphyte-laden maritime oak stand. *Ecohydrology* (in press).

Rosier, C.L. Levia, D.F. VanStan, J.T. Aufdenkampe, A. Kan, J.J., et al., (in review). Stemflow influences soil microbial community structure near tree trunks.

Rousseau, J. V. D., & Reid, C. P. P. (1991). Effects of phosphorus fertilization and mycorrhizal development on phosphorus-nutrition and carbon balance of Loblolly-Pine. *New Phytologist, 117*, 319–326.

Same, B. I., Robson, A. D., Abbott, L. K., et al. (1983). Phosphorus, soluble carbohydrates and endomycorrhizal infection. *Soil Biology and Biochemistry, 15*, 593–597.

Sande, E., & Young, D. R. (1992). Effect of sodium chloride on growth and nitrogenase activity in seedlings of *Myrica cerifera* L. *New Phytologist, 120*, 345–350.

Santi, C., Bogusz, D., Franche, C., et al. (2013). Biological nitrogen fixation in non-legume plants. *Annals of Botany, 111*, 743–767.

Schulze, K., Borken, W., Matzner, E., et al. (2011). Dynamics of dissolved organic ^{14}C in throughfall and soil solution of a Norway spruce forest. *Biogeochemistry, 106*, 461–473.

Schwärzel, K., Ebermann, S., Schalling, N., et al. (2012). Evidence of double-funneling effect of beech trees by visualization of flow pathways using dye tracers. *Journal of Hydrology, 470-471*, 184–192.

Shamseldin, A., & Werner, D. (2005). High salt and high pH tolerance of new isolated *Rhizobium etli* strains from Egyptian soils. *Current Microbiology, 50*, 11–16.

Smith, S. E., & Read, D. J. (2008). *Mycorrhizal symbiosis* (3rd ed., pp. 1–787). New York: Elsevier.

Smith, S. E., Jakobsen, I., Grønlund, M., Smith, F. A., et al. (2011). Roles of arbuscular mycorrhizas in plant phosphorus nutrition: interactions between pathways of phosphorus uptake in arbuscular mycorrhizal roots have important implications for understanding and manipulating plant phosphorus acquisition. *Plant Physiology, 156*, 1050–1057.

Sprent, J. I., & Parsons, R. (2000). Nitrogen fixation in legume and non-legume trees. *Field Crops Research, 65*, 183–196.

Staelens, J., De Schrijver, A., Verheyen, K., et al. (2007). Seasonal variation in throughfall and stemflow chemistry beneath a European beech (*Fagus sylvatica*) tree in relation to canopy phenology. *Canadian Journal of Forest Research, 37*, 1359–1372.

Surange, S., Wollum, A. G., II, Kumar, N., Nautiyal, C. S., et al. (1997). Characterization of rhizobium from root nodules of leguminous trees growing in alkaline soils. *Canadian Journal of Microbiology, 43*, 891–894.

Taylor, A. F. S., & Alexander, I. (2005). Ectomychorizal symbiosis: life in the real world. *Mycologist, 19*, 102–112.

Tonin, C., Vandenkoornhuyse, P., Joner, E. J., Straczek, J., & Leyval, C. (2001). Assessment of arbuscular mycorrhizal fungi diversity in the rhizosphere of *Viola calaminaria* and effect of these fungi on heavy metal uptake by clover. *Mycorrhiza, 10*, 161–168.

Valente, F., David, J. S., Gash, J. H. C., et al. (1997). Modelling interception loss for two sparse eucalypt and pine forests in central Portugal using reformulated Rutter and Gash analytical models. *Journal of Hydrology, 190*, 141–162.

Van Stan, J. T., Levia, D. F., Inamdar, S. P., Lepori-Bui, M., Mitchell, M. J., et al. (2012). The effects of phenoseason and storm characteristics on throughfall solute washoff and leaching dynamics from a temperate deciduous forest canopy. *Science of the Total Environment, 430*, 48–58.

Van Stan, J. T., Van Stan, J. H., Levia, D. F., et al. (2014). Meteorological influences on stemflow generation across diameter size classes of two morphologically distinct deciduous species. *International Journal of Biometeorology, 58*, 2059–2069.

Van Stan, J. T., Stubbins, A., Bittar, T., Reichard, J. S., Wright, K. A., Jenkins, R. B., et al. (2015). *Tillandsia usneoides* (L.) L. (Spanish moss) water storage and leachate characteristics from two maritime oak forest settings. *Ecohydrology, 8*, 988–1004.

Van Stan, J. T. (2012). *Controls and dynamics of canopy-derived dissolved organic matter from co-dominant broadleaved deciduous canopies to the soil of a temperate catchment in the Northeastern United States* (Dissertation). University of Delaware, 177 p.

Vierheilig, H., Bago, B., Albrecht, C., Poulin, M. P., & Piché, Y. (1998). Flavonoids and arbuscular mycorrhizal fungi. In J. A. Manthey, & B. S. Buslig (Eds.), *Flavonoids in the living system* (pp. 9–33). New York: Plenum Press.

Vierheilig, H., Lerat, S., & Piche, Y. (2003). Systemic inhibition of arbuscular mycorrhiza development by root exudates of cucumber plants colonized by *Glomus mosseae*. *Mycorrhiza, 13*, 167–170.

Wang, B., & Qiu, Y. L. (2006). Phylogenetic distribution and evolution of mycorrhizas in land plants. *Mycorrhiza, 16*, 299–363.

Wang, G. M., Stribley, D. P., Tinker, P. B., Walker, C., et al. (1993). Effects of pH on arbuscular mycorrhiza. I. field observations on the long-term liming experiments at Rothamsted and Woburn. *New Phytologist, 124*, 465–472.

Wardle, D. A., Nilsson, M. C., Gallet, C., Zackrisson, O., et al. (1998). An ecosystem-level perspective of allelopathy. *Biology Reviews, 73*, 305–319.

Wilkinson, S. C., & Anderson, J. M. (2001). Spatial patterns of soil microbial communities in a Norway spruce (*Picea abies*) plantation. *Microbial Ecology, 42*, 248–255.

Zahran, H. H. (1999). Rhizobium-legume symbiosis and nitrogen fixation under severe conditions and in an arid climate. *Microbiology and Molecular Biology Reviews, 63*, 968–989.

Zimmermann, A., Wilcke, W., & Elsenbeer, H. (2007). Spatial and temporal patterns of throughfall quantity and quality in a tropical montane forest in Ecuador. *Journal of Hydrology, 343*, 80–96.

Zimmermann, A., Zimmermann, B., Elsenbeer, H., et al. (2009). Rainfall redistribution in a tropical forest: spatial and temporal patterns. *Water Resources Research, 45*, W11413.

CHAPTER TEN

Soluble and Volatile Metabolites of Plant Growth-Promoting Rhizobacteria (PGPRs): Role and Practical Applications in Inhibiting Pathogens and Activating Induced Systemic Resistance (ISR)

Gul Shad Ali[1], David Norman, Ashraf S. El-Sayed
Mid-Florida Research and Education Center and Department of Plant Pathology, University of Florida/Institute of Food and Agricultural Sciences, Apopka, FL, USA
[1]Corresponding author: E-mail: gsali@ufl.edu

Contents

1. Overview of Current Disease Management Practices 242
2. Plant Growth-Promoting Rhizobacteria 244
   2.1 Overview 244
   2.2 Mode of Action of PGPRs 245
3. Beneficial Soluble Metabolites of PGPRs 247
   3.1 Soluble PGPR Metabolites Display Antibiosis against Phytopathogens 248
   3.2 Soluble PGPR Metabolites Induce ISR 251
   3.3 Plant Growth Promotion 252
      3.3.1 Nutrient Solubility and Availability to Plants *252*
      3.3.2 Plant Growth Regulators *253*
4. Volatile Organic Compounds 254
   4.1 Screening and Profiling Methods of PGPR VOCs 255
   4.2 VOC-Producing PGPRs 259
   4.3 Chemical Diversity of PGPR VOCs 259
   4.4 Phytopathogens Targeted by PGPR VOCs 263
   4.5 VOCs Display Antibiosis 265
      4.5.1 Alcohols *265*
      4.5.2 Ketones and Aldehydes *266*
      4.5.3 Alkanes and Alkenes *266*
      4.5.4 Sulphur Compounds *267*
   4.6 PGPR VOCs Are Involved in ISR 268
   4.7 VOCs Effect in Plant Disease Escape 270

Advances in Botanical Research, Volume 75
ISSN 0065-2296
http://dx.doi.org/10.1016/bs.abr.2015.07.004

5. Practical Uses in Agriculture 272
6. Outlook — Opportunities and Challenges 273
Acknowledgements 275
References 275

Abstract

Plant growth-promoting rhizobacteria (PGPRs) release soluble and volatile compounds, which are implicated in antibiosis, cell signalling, induction of resistance and tolerance in plants against biotic and abiotic stresses. Due to public concerns about the potential environmental hazards of chemical use in agriculture, exploration of PGPRs for alternative antimicrobials is picking up momentum. Biochemical and metabolomics analyses of diverse PGPRs have identified several potential compounds for development and deployment in agriculture. Chemical profiling and functional analyses have provided fundamental insights into the nature of soluble and volatile chemicals produced by PGPRs. These studies are providing a solid launching point for further studies into unravelling the chemical basis of activities of PGPRs. In this chapter, some of these findings will be discussed, which will hopefully stimulate much needed additional investigations focused on exploring the full spectrum of the antimicrobial potential of PGPRs. Future challenges in the field, including biosafety concerns and potential ecological consequences of PGPR-derived antimicrobials will also be discussed.

1. OVERVIEW OF CURRENT DISEASE MANAGEMENT PRACTICES

It is well known that world population will increase — estimated to reach 9 billion by 2050 — and that resources will be stretched to their limits to feed and meet other needs of such a large number of people. What is more frequently missing in these discussions is not just how to feed them, but how to feed them nutritious, healthy and safe food. More and more people, especially the younger generation, are concerned about food safety and nutrition. Given the current pace of technological advancements, we will most likely be able to meet the food needs of 9 billion people, but at what cost? Modern-day agriculture relies heavily on chemical inputs such as fertilizers, pesticides, herbicides and growth regulators. Despite using chemicals, diseases are still estimated to account for at least 15% of total harvestable crop losses, and in the absence of chemical control, losses would be even higher (Dangl, Horvath, & Staskawicz, 2013). Thus reliance on chemicals for productive agriculture, at least in the foreseeable future, is inevitable. As scary as it sounds, relying on chemical inputs is not always harmful, and not all synthetic chemicals are bad for the environment, and human and animal health. To be safe and acceptable to society, chemical use would require integrated efforts that rely not only on good research and development to ensure chemical safety, but would also require studying social, psychological and political aspects of the whole agriculture system. These studies should be backed by outreach and extension activities to educate the general public and policy-makers about the consequences of using or not using chemicals in agriculture. In our opinion, the starting point for taking an integrated approach would be to tap more vigorously into natural resources of chemicals – and what could be better resource than beneficial microorganisms that have been part of the natural ecosystems for millions of years? Just like organic commodities, products directly derived from naturally occurring beneficial microorganisms would be more readily acceptable to the general public and most likely safer.

Growth of disease-resistant crop cultivars and the application of antimicrobial chemicals are currently the most widely used methods for the management of plant diseases in modern agriculture (Strange & Scott, 2005). As attractive as they are, these two methods have their own limitations. Breeding for disease resistance is one of the cornerstones of most crop improvement programmes, and it has successfully led to the control of many important diseases, which otherwise could have led to epidemics and famines (Strange & Scott, 2005; Vanderplank, 2012). However, breeding is a prophylactic approach requiring prior knowledge of the genetic structure and virulence status of the pathogen populations and availability of useful resistance (*R*) genes preferably in a sexually compatible plant variety or wild relatives. Since pathogen populations are usually shaped by selection pressure exerted by *R* genes in host plants over several generations, understanding the virulence dynamics and evolutionary mechanisms that modulate pathogen populations requires extensive and continuous efforts employing tools of population genetics. These studies by themselves demand intensive efforts and monitoring of virulence in pathogen populations at multiple geographical locations. Similarly, on the host side, availability of *R* genes in suitable relatives of a crop is also essential. Identification and genetic characterization often involve intensive screening and genetic approaches. Although molecular markers have raised hopes of accelerating breeding *R* genes into crop varieties, it is still a time-consuming process taking several generations of genetic crosses and multiyear and multilocation trials before a variety is released (Strange & Scott, 2005; van der Vossen et al., 2005). *R* genes are often introgressed from wild relatives into cultivated crop varieties, leading to the introgression of undesirable traits, which could be hard to get rid of, especially if they are linked with *R* genes (Vanderplank, 2012). Sometimes, breeding a desirable *R* gene from a sexually incompatible wild relative often involve complex bridge crosses further complicating and prolonging the breeding process. On the top of all that, durability of *R* genes is dependent on highly dynamic coevolutionary genetic factors in both the host and the pathogen and, therefore, most *R* genes are highly unstable and last only a few generations before they are overcome by new strains of the pathogen. This requires search for additional *R* genes. Breeding for disease resistance is often very successful in annual crops such as most vegetables and cereal crops. In perennial crops, such as fruits trees, which require several decades to breed and establish, emergence of new pathogen could be devastating. Under these circumstances, saving established orchards from new diseases requires immediate interventions primarily through chemicals. Similarly, use of chemical as convenient and on-demand solver, is not perceived favourably by consumers. Reality is that most successful pathogens have adapted to infecting plants for millions of years and they will continue to evolve and infect our crops. To be productive will require the use of chemicals. Chemicals produced by microbes are perhaps the best option available to us. Live PGPRs for agriculture has been used but due to their dependence on environmental, soil and other biological factors, they are not always successful. One of the best options of utilizing PGPRs is perhaps to isolate their antimicrobial compounds and use them ex vivo for controlling plant pathogens and augmenting plant defences. In this chapter, we will discuss the chemical nature and potential application of these PGPR products.

2. PLANT GROWTH-PROMOTING RHIZOBACTERIA

2.1 Overview

The rhizosphere, the nutrient-enriched zone in the immediate vicinity of the soil—root interface, teems with diverse groups of bacterial species, collectively known as rhizobacteria. Although hidden from plain sight, the rhizosphere is an ecological system of epic proportion, swarmed by thousands of biological and physical entities interlocked in a complex web of interdependent interactions (Bais, Weir, Perry, Gilroy, & Vivanco, 2006; Mendes, Garbeva, & Raaijmakers, 2013). From a bird's-eye view prospective, the rhizosphere consists of plant roots, microbes and soil. However, when looked at more deeply from a biochemical perspective, the rhizosphere is enriched with numerous chemical moieties dominated by root exudates that are enriched with free amino acids, phenolic compounds and energy-rich sugars that attract rhizobacteria and other organisms. It is estimated that plants secrete about 20% of its metabolites into the root zone (Walker, Bais, Grotewold, & Vivanco, 2003), and it has become clear that this process is highly regulated and is affected not only by the physical and chemical characteristics of the rhizosphere but also by the types of living microbial species that reside in this zone (Bonito et al., 2014). Similarly, plant types also affect soil microbiomes, selectively recruiting and excluding rhizospheric microbial communities (Badri, Chaparro, Zhang, Shen, & Vivanco, 2013). In nature, most organisms have evolved to spend the least amount of energy and other resources to fulfil their needs. From this perspective, it is obvious that plants are not giving off their exudates for free. They help attract beneficial microbes, which help with obtaining nutrition and warding off biotic and abiotic stresses. Communication in the rhizosphere is of paramount importance, and both plants and microbes have evolved multiple mechanisms of producing and delivering chemicals to each other (Lakshmanan, Selvaraj, & Bais, 2014; Rout & Southworth, 2013). In nature, all organisms including host plants and microbes, both pathogenic and beneficial, release chemicals, which, depending on the purpose of the interaction, serve as foe or friendly signals with the ultimate goal to better adapt, survive and successfully reproduce. The primary method of delivering these chemicals is through diffusion through aqueous medium as soluble compounds. However, numerous recent reports suggest that volatiles through aerial routes are also utilized (Audrain, Farag, Ryu, & Ghigo, 2015; Schulz & Dickschat, 2007). Since traditionally most investigations have focused on soluble products, the role of volatile compounds was largely overlooked. However, if trends of the past one decade of employing analytical tools complemented with phenotypic and physiological analyses continue, we expect to see more and more insights into the role of volatiles in plant—microbe interactions.

2.2 Mode of Action of PGPRs

Soon after the observation that some soils display disease-suppressive ability, studies focussed on investigating the underlying causes firmly correlated PGPRs with this phenomenon (Kinkel, Bakker, & Schlatter, 2011; Lugtenberg & Kamilova, 2009; Mendes et al., 2011; Schroth & Hancock, 1982). Hundreds of investigations over the last three to four decades have established that PGPRs could suppress diseases using one or more of the following mechanisms (Kloepper, Leong, Teintze, & Schroth, 1981;

Lugtenberg & Kamilova, 2009): (1) direct antibiosis, (2) induction of induced systemic resistance (ISR), (3) competition with pathogens and (4) disease escape by promoting plant growth through mineral solubilization, and phytostimulation through plant hormones, such as auxins, cytokinins, gibberellins, ethylene and abscisic acid (Figure 1) (Maksimov, Abizgil'dina, & Pusenkova, 2011). For most PGPRs, disease suppression appears to be achieved by a combination of all these mechanisms. The relative contribution of each of these mechanisms depends on the PGPR and the host plant. Some PGPRs are excellent producers of antibiotics, whereas others are well suited for induction of ISR, nutrient mobilization and production of growth regulators (Lugtenberg & Kamilova, 2009). Although, a majority of studies have focused on the physiological mode of action of PGPRs, many other studies have focused on identifying and characterizing their bioactive compounds (see Sections 3 and 4 below). In this chapter, we will restrict our discussion to reports that have focussed on dissecting the chemical mode of action of PGPRs. Results from these studies point to a multicomponent mode of action for PGPRs. Driven by strong interest in identifying and developing ecofriendly compounds for pest management, a majority of the latest studies have adapted metabolomics and bioactivity-driven chemical profiling approaches. Collectively these studies have revealed tremendous functional and structural diversity of compounds, which diffuse

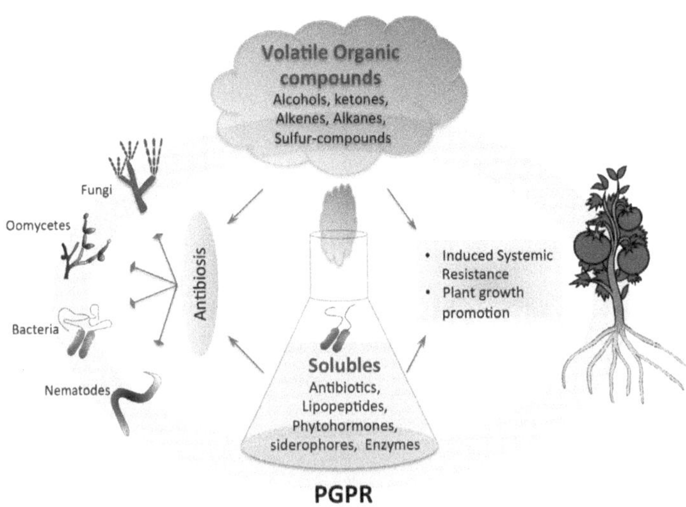

Figure 1 Overview of soluble and volatile organic compounds produced by plant growth-promoting rhizobacteria (PGPRs). (See colour plate)

either in aqueous soluble form or as volatiles through the rhizosphere and perhaps throughout the aerial plant canopy.

3. BENEFICIAL SOLUBLE METABOLITES OF PGPRs

For many decades, research activities in the PGPR field were focused on the identification and physiological characterization of the causal microbes with a hope of potential application in agriculture for biological control of plant pathogens and improvement of plant growth. These investigations led to the identification of a wide range of microbial species, primarily fungi and bacteria. Attention was then focussed on the biochemical principle(s) using antimicrobial assays primarily in in vitro dual-culture methods. One of the main observations in these studies was the presence of distinct growth inhibition zones, which clearly showed that PGPRs can affect their target microbes without coming in physical contact, indicating that soluble metabolites might be responsible for the bioactivity of PGPRs. Indeed, subsequent fractionation, biochemical and metabolic profiling revealed that PGPRs secrete different antimicrobial compounds including cell wall-degrading enzymes such as chitinases and a chemically diverse blend of secondary metabolites into the surrounding medium (Berendsen, Pieterse, & Bakker, 2012; De Vleesschauwer & Hofte, 2009; Lugtenberg & Kamilova, 2009; Raaijmakers & Mazzola, 2012). This was further verified with using cell-free culture filtrates of PGPRs grown in synthetic media against the target pathogens in numerous studies. Identification and profiling of PGPR-soluble metabolites was accomplished employing several variations of mass spectroscopy (MS) such as matrix-assisted laser desorption ionization time-of-flight tandem mass spectrometry (MALDI-TOF-MS/MS) and electrospray ionization mass spectroscopy (ESI-MS). Major peaks in mass spectra were identified and their antimicrobial activities characterized using commercially available individual chemicals. In the following, we have focussed on the soluble metabolites of PGPRs with an emphasis on their role in antibiosis and ISR, and their potential for development and deployment in agriculture. Collectively hundreds of different soluble compounds have been identified in the culture filtrates of PGPRs (Lemfack, Nickel, Dunkel, Preissner, & Piechulla, 2014; Raaijmakers & Mazzola, 2012). It is worth mentioning that these compounds were mostly discovered in a few dozen PGPRs that can be cultured in vitro. It is estimated that only 1–5% of soil microbiome can be cultured whereas the rest are not culturable (Berendsen et al., 2012; Mendes et al., 2011). Since the full genetic and biochemical repertoire of these unculturable microbes is not known, it is possible that they may produce many orders of magnitude more diverse compounds. Using metagenomics approaches, more than 33,000 bacterial and archaeal microbes were detected in disease-suppressive soils (Mendes et al., 2011). Comparative analyses showed that they were more abundant in suppressive soils than disease-conducive soils. Functional analyses showed that disease suppression was associated with *Burkholderiaceae*, *Xanthomonadales*, *Actinobacteria* and *Firmicutes*, which harbour genera and species with activity against plant pathogenic fungi and bacteria (Mendes et al., 2011). Given the diversity of soil microbiome, the currently known compounds represent just the tip of an iceberg, and a lot more with novel structure and function are yet to be discovered. Nevertheless, known soluble compounds just like their producing PGPRs can suppress plant pathogens and diseases through direct antibiosis, ISR and plant growth promotion as discussed below.

3.1 Soluble PGPR Metabolites Display Antibiosis against Phytopathogens

PGPR-derived compounds that inhibit growth of plant pathogens can be roughly categorized into the following mains groups: (1) enzymes and proteins, (2) secondary metabolite antibiotics and (3) lipopeptides (CLPs). Enzymes and proteins include glucanases, chitinases, bacteriocins, lantibiotics, bacilysin and several others produced by PGPRs (Raaijmakers & Mazzola, 2012). In earlier studies aimed at identifying the bioactive mechanism of PGPRs, small molecular weight antibiotics were some of the first compounds identified in MS analyses of the soluble cell-free culture of PGPRs. Prominent examples include classical antibiotics such as 2,4-diacetylphloroglucinol (2,4-DAPG), pyrrolnitrin, pyoluteorin, phenazines, 2,5-dialkylresorcinol, quinolones, gluconic acid and rhamnolipids (Raaijmakers & Mazzola, 2012). These compounds inhibit microbial growth in vitro and some of them have also been demonstrated to function in plants. In addition, siderophores, which are iron-binding proteins, are implicated in suppressing plant pathogens primarily by depriving them of available iron in the rhizosphere (Kloepper, Leong, Teintze, & Schroth, 1980). The first two groups of PGPR compounds have been reviewed several times, and readers are referred to several recently published reviews (De Vleesschauwer & Hofte, 2009; Raaijmakers & Mazzola, 2012). CLPs are one of the most prominent and emerging PGPR compounds with strong potential for deployment in the management of agriculture pests, and they will be discussed here in detail.

CLPs are produced by a large and varied class of PGPRs, primarily in the genera *Pseudomonas*, *Bacillus* and *Streptomyces* (Raaijmakers, De Bruijn, Nybroe, & Ongena, 2010). In addition to rhizospheric PGPRs, CLP may also be produced from phyllosphere. For example, *Bacillus liquefaciens* isolated from wheat spikes infected with *Fusarium graminearum* also produces CLPs with strong inhibitory activities against the same fungus, which causes *Fusarium* head blight of wheat (Gong et al., 2015). CLPs, as their names indicate, contain a lipid moiety linked to an amino acid peptide. The types and lengths of both lipids and peptides vary considerably, providing tremendous structural and functional diversity to CLPs. Hundreds of different CLPs have been discovered in PGPRs, and newer ones with slight variations of amino acids position in the peptide or in the lipid type are being discovered in new strains of PGPRs regularly (Pathak, Bose, & Keharia, 2014; Raaijmakers et al., 2010; Zachow et al., 2015). Given the diversity of PGPRs and the number of unresolved peaks on MS chromatograms of many existing reports, perhaps even more with novel features are not yet discovered and characterized. Due to their subtle structural diversity, classification of CLPs is not fixed and is regularly updated to accommodate CLPs with new amino acid and fatty acid configurations (Zachow et al., 2015).

CLPs produced by *Bacillus* spp. are grouped into iturins, fengycin and surfactins (Ongena & Jacques, 2008; Stein, 2005). All three classes have a cyclic peptide ring that contains unusual amino acids D-, O-methyl, N-methyl and β-amino acids (Figure 1). The cyclic peptide ring is linked to a fatty acid side chain of variable lengths: C14—C17 in iturins, C12—C16 in surfactins and C14—C18 in fengycins. CLPs of *Pseudomonas* spp. are perhaps the most varied, and they are classified into at least six families — viscosin, syringomycin amphisin, putisolvin, tolaasin and syringopeptin — each with their own characteristic cyclic peptides and variable length fatty acid tails (Raaijmakers et al., 2010). Similarly, CLPs produced by *Streptomyces* spp. also vary considerably in the configuration of peptides and fatty acid tails. In some *Streptomyces* CLPs, certain amino acids are covalently modified by oxidation or methylation providing an additional layer of diversity in their structure.

Most CLPs are synthesized by large multidomain nonribosomal peptide synthetases (NRPS) or by a combination of polyketide synthases and NRPS (Koglin & Walsh, 2009; Roongsawang, Washio, & Morikawa, 2010). Since some PGPR strains produce a complex mixture of CLPs, it would be interesting to determine how only a few NRPS accomplish so much diversity or what determines specificity of CLPs produced by a single strain. Sequence analyses have revealed multiple copies of NRPS genes in some PGPRs, which could partially account for producing structurally diverse CLPs (Muller et al., 2013).

CLPs primarily target cell membranes. They are amphipathic, which allows them to arrange across cell membranes creating pores or channels leading to membrane permeabilization, ion leakage and free flow of other cellular metabolites, which are otherwise tightly regulated (Patel, Tscheka, Edwards, Karlsson, & Heerklotz, 2011). Since membranes regulate many diverse cellular processes including ion homoeostasis, protein export and import, and signal recognition and transduction, CLP interference with their integrity disrupts one or more of these processes leading to either stress responses or cell death. CLP treatment in target organisms produces a variety of growth abnormalities in fungal hyphae including blebbing, excessive branching and abnormal hyphal swellings. CLPs affect not only mycelial growth but are also known to inhibit reproductive stages such as spore production, release and germination (van de Mortel, Tran, Govers, & Raaijmakers, 2009; Tran, Ficke, Asiimwe, Hofte, & Raaijmakers, 2007).

Studies with CLP have primarily been conducted in vitro on synthetic media. They are primarily secreted into the culture medium but have also been isolated in functional form from cell pellets (Yuan et al., 2014). Literature documenting antibiosis against a large number of fungal, bacterial and oomycetes is vast and all of them cannot be cited here. In summary, target organisms include *Rhizoctonia solani, Phoma lingam, Alternaria brassicae, Sclerotinia sclerotiorum, Botrytis cinerea, Ophiostoma ulmi, Aspergillus* spp., *Fusarium* spp., *Penicillium* spp., *Magnaporthe grisea* (Ongena & Jacques, 2008), *Phytophthora infestans, Phytophthora capsici* and *Pythium ultimum* (van de Mortel et al., 2009; Zachow et al., 2015), and Gram-negative and Gram-positive bacteria (Li et al., 2013; Raaijmakers & Mazzola, 2012). Activity against oomycetes primarily involves zoospore lysis (de Bruijn et al., 2007; De Souza, De Boer, De Waard, Van Beek, & Raaijmakers, 2003), zoospore encystment (van de Mortel et al., 2009) and zoospore immobilization. A recently discovered new CLP, poaeamide, displayed strong anti-oomycete activity. In addition to in vitro antimicrobial activity, evidence for the in vivo growth inhibition by most CLPs is also reported. Most of these studies have relied on ectopic application and on genetic approaches using CLP-defective or -overexpresser PGPR strains (Arrebola, Jacobs, & Korsten, 2010; Bais, Fall, & Vivanco, 2004; Leclere et al., 2005; Ongena & Jacques, 2008; Zachow et al., 2015). In addition to bacteria, fungi and oomycetes, CLPs also display activity against enveloped viruses (Vollenbroich, Ozel, Vater, Kamp, & Pauli, 1997), most likely by compromising the integrity of lipid envelope of viruses leading to unstable viral particles. Although CLPs have been shown to display activity against enveloped viruses of animal, no activity against plant viruses has been demonstrated yet.

3.2 Soluble PGPR Metabolites Induce ISR

Since the first discovery that colonization of plant roots by rhizobacteria can lead to systemic priming of defence responses in distal noninoculated tissues termed as ISR (Alstrom, 1991; Gang, Kloepper, & Tuzun, 1991; reviewed in Pieterse et al., 2014), many studies using genetic and biochemical analyses have implicated a variety of PGPR signals including proteins, lipids, CLPs, carbohydrates and secondary metabolites in activating ISR (De Vleesschauwer & Hofte, 2009). Role of cell wall carbohydrates, proteins such as flagellin, EF-Tu, has been reviewed very well and they will not be discussed here (De Vleesschauwer & Hofte, 2009; Pieterse et al., 2014). In the following, we will focus mostly on CLPs for their role in ISR.

Based on genetic and biochemical analyses, surfactins and fengycins, which are well studied for their antibiosis activity, are also suggested to induce ISR in many different plants including melons, beans, tobacco and tomato (Cawoy et al., 2014; Garcia-Gutierrez et al., 2013; Henry, Deleu, Jourdan, Thonart, & Ongena, 2011; KyungSeok et al., 2013; Ongena et al., 2007; Rahman, Uddin, & Wenner, 2014). The ISR-inducing ability was positively correlated with fengycin production capacity of several *Bacillus* strains (Cawoy et al., 2014). Ectopic application of surfactin showed enhanced production of oxylipin genes in host plants, and *Bacillus subtilis* mutants defective in surfactin production were compromised in ISR (Cawoy et al., 2014). In addition to surfactin, iturin may also induce ISR as shown by induction of chitinase and β-1,3-glucanase gene expression in strawberry leaves (Yamamoto, Shiraishi, & Suzuki, 2015). However, induction was very modest (at the most 1.5 times that of control) and at very high concentrations of iturin (500 μg/ml), suggesting that iturins might not be strong inducers of ISR. More convincing evidence for the ISR activity of the CLP fengycin and surfactin came from enhanced gene expression and enzymatic activities of several defence-related proteins such as *glucanase, chitinase, peroxidase* and *lipoxygenase* in *Citrus sinensis* cv. valencia fruit (Waewthongrak, Leelasuphakul, & McCollum, 2014). Similarly, CLPs of other groups such as massetolide A also induced ISR in different plant species against fungi and oomycetes (Tran et al., 2007). In a study assaying differential effects of CLPs on defence pathways under healthy and infection conditions, an iturin activated *LOX (lipoxygenase), AOS (allene oxide synthase)* and *PR1* genes in healthy tissues, whereas in response to *Blumeria graminis* f. sp. *tritici*, only a surfactin induced these genes (Khong et al., 2013). These results show that in addition to displaying antibiosis, these compounds could also provide protection against pathogens indirectly by priming the natural defences of plants. Whether or not other groups of CLPs also induce ISR is not well characterized yet, but given the recent resurgence in interest in CLPs, it would not be too long before we see reports addressing these questions.

CLP synthesis seems to be regulated by signals such as polysaccharides of plants and as yet unknown signals from pathogens (Cawoy et al., 2015). A substantial fraction of genomes of some *Bacillus* spp. is dedicated to the synthesis of antibiotics including CLPs (Chen, Koumoutsi, et al., 2007; Ruckert, Blom, Chen, Reva, & Borriss, 2011), suggesting that CLPs play a major role in the biological control of PGPRs. After showing that CLPs are involved in ISR, the next obvious question is to unravel the signalling components that are involved in ISR. Since live PGPRs act in jasmonate (JA)- and ethylene (Et)-dependent but salicylic acid (SA)-independent pathways, it is plausible to hypothesize that ISR-inducing CLPs and other metabolites likely also use these pathways. However, in a recent study, a purified surfactin activated SA pathway, whereas an iturin activated both SA- and JA-pathways (Farace et al., 2015), suggesting that CLPs might utilize multiple modes of signalling in activating ISR.

3.3 Plant Growth Promotion

3.3.1 Nutrient Solubility and Availability to Plants

Although not directly involved in suppressing pathogens, promotion of plant growth by PGPRs could help plants tolerate or escape adverse effects of pathogenic microbes. Acquisition of nutrients from soil by plants is very complex and dependent on several environmental and biological factors (Lakshmanan et al., 2014). First, nutrients in the soil must be available in a form that can be readily absorbed by roots, and second, plants must possess specific nutrient transporters for the uptake of each element. PGPRs promote root growth and crop productivity, most likely by converting unavailable nutrients to a form that can be readily absorbed by plants. Evidence exists, where PGPRs have been shown to facilitate solubilization and availability of Fe, Cu, N_2, P and S (Haas & Defago, 2005). Genetic and biochemical analyses have linked enzymatic activities in the metabolism of these nutrients to the plant growth-promoting activities of PGPRs (Pieterse et al., 2014).

3.3.2 Plant Growth Regulators

Four major phytohormones viz., auxins (Duca, Lorv, Patten, Rose, & Glick, 2014), cytokinins (Choi, Choi, Lee, Ryu, & Hwang, 2011), gibberellins (Gutierrez-Manero et al., 2001) and abscisic acid (Cohen, Bottini, & Piccoli, 2008) are produced by several PGPRs. In most cases, biosynthesis genes of these hormones are also encoded in the genomes of sequenced PGPRs. Although, PGPRs do not directly produce ethylene, several PGPRs encode gene for 1-aminocyclopropane-1-carboxylate (ACC) deaminase, which cleaves the plant ethylene precursor, ACC, thus leading to reduced ethylene production and enhanced plant growth (Glick, 2014). Regulation of these hormones in plants by PGPRs and their subsequent effect on plant metabolism is also suggested as one explanation for plant growth promotion (Cernadas & Benedetti, 2009; Glick, 2014; Lebuhn, Heulin, & Hartmann, 1997). These hormones, in addition to affecting plant development, are also well known to affect plant defence responses in a complex manner with auxin playing a negative role, whereas cytokinins playing mostly a positive role in plant defence (Choi et al., 2011), but also note that pathogen-derived cytokinins delay senescence and perhaps suppress plant defences (Walters & McRoberts, 2006). Therefore PGPR-produced cytokinin could be assumed to contribute to priming plant defences through activating ISR.

Based on genetic and molecular data, auxin signalling has been shown to affect plant–microbe interaction in both beneficial interactions and pathogenic interactions (Ali, Sabri, Ljung, & Hasnain, 2009; Duca et al., 2014; Fu et al., 2011; Patten & Glick, 2002). Application of PGPRs promotes plant growth partially because they produce auxins. In general, pathogens either produce auxin themselves or induce auxin signalling in their host plants for disease development perhaps by promoting stomatal opening and loosening cell walls to facilitate entry of bacteria in host tissues (Chen, Agnew, et al., 2007; Ding et al., 2008; Fu et al., 2011). How auxin production by PGPRs contributes to plant defence is at odds with these observations, which show that auxin is involved in susceptibility and that in response to flg22 or SA application, plants actively repress auxin signalling (Navarro et al., 2006; Wang, Pajerowska-Mukhtar, Culler, & Dong, 2007). Although possible explanations could include auxin concentration-dependent differential regulation of defence responses, further research is needed to fully understand how these seemingly antagonizing pathways are regulated in pathogenic and beneficial plant–microbe interactions. In addition to direct production, evidence for regulation of endogenous plant hormone signalling that affect plant growth and defence also exists (reviewed in Pieterse et al., 2014). In summary, phytohormones either produced or induced in plants by PGPRs might contribute to strengthening host tolerance to plant pathogens.

4. VOLATILE ORGANIC COMPOUNDS

Volatile organic compounds (VOCs) are inherently present in the atmosphere and they affect tropospheric ozone production and secondary organic aerosol formation, which affect climate, human health and other biological processes on a global scale (Andreae & Crutzen, 1997; Chameides, Lindsay, Richardson, & Kiang, 1988). Approximately 90% of total atmospheric VOCs have a biogenic origin (Guenther et al., 1995), and they are produced by species in all domains of life (Audrain et al., 2015; Penuelas et al., 2014). Biogenic VOCs primarily serve as foe or friendly signals, among many other functions, for inter- and intraspecies communications across kingdoms (Audrain et al., 2015; Davis, Crippen, Hofstetter, & Tomberlin, 2013; Effmert, Kalderas, Warnke, & Piechulla, 2012; Hughes & Sperandio, 2008; Kai et al., 2009; Lemfack et al., 2014). Although their existence has been known for hundreds of years, it is only during the last few decades that their role in shaping ecology at the subcellular, organismal and population levels is being extensively investigated (Audrain et al., 2015; Farag, Zhang, & Ryu, 2013; Schulz & Dickschat, 2007). A steady flow of new findings based on metabolite profiling and functional analyses is now a common theme in many reports that are yielding fundamental new insights into the chemical diversity and function of VOCs. A recently released comprehensive database based on information in published reports shows that microbial species altogether, including PGPRs emit, approximately 1000 VOCs (Lemfack et al., 2014). VOCs in this database are categorized into different major chemical classes, which allow users to browse and search a specific VOC based on its class, chemical name, species of origin, activity and chemical formula. Many reports now clearly show that VOCs are the cornerstones of both beneficial and antagonistic plant–microbe interactions (Audrain et al., 2015; Effmert et al., 2012; Junker & Tholl, 2013). In recent years, we have seen a renaissance in interest in microbial VOCs, not only because of their role in airborne communications but also because of their potential usefulness in industry and agriculture (Effmert et al., 2012;

Junker & Tholl, 2013). Given the prospect of the development and easy availability of analytical tools with greater compound separation capabilities combined with powerful bioinformatics infrastructure, and genetic, genomics, proteomics, metabolomics and large-scale functional phenomics tools, we are expecting exciting new findings in the field of VOCs in the not so distant future. These investigations will not only shed light on the function of VOCs but will also hopefully provide new leads in the development of agrochemicals that are sustainable, economical and safe for the environment and health. Several recent reviews have thoroughly discussed and catalogued VOCs produced by plants, animals and a wide range of microbes (Audrain et al., 2015; Davis et al., 2013; Effmert et al., 2012; Hughes & Sperandio, 2008; Kai et al., 2009; Lemfack et al., 2014). Their discussion is beyond the scope of this chapter and readers are referred to these reviews for specific information. In this chapter, we focus on those PGPR-derived VOCs that have been implicated in the biological control of plant pathogens either through direct antibiosis and competition with pathogenic microbes or through indirect ISR and priming of defence responses.

4.1 Screening and Profiling Methods of PGPR VOCs

VOCs are gas-phase low molecular weight (<300 Da) hydrocarbons that have low boiling points and low vapour pressure (0.01 kPA) under normal conditions. They are mostly lipophilic and, therefore, are soluble in organic solvents but not in aqueous solutions. They could be emitted in a gaseous phase or secreted into liquids, which, due to their low vapour pressure, then readily change to vapour phase and quickly dissipate into the surrounding space. Their easy conversion to gaseous forms poses challenges in their collection and concentration for subsequent analytical and functional studies. Perhaps due to this challenge many researchers have shied away from investigating VOCs, and research in VOC profiling and functional analyses has lagged far behind than other projects focussed on soluble microbial products. These challenges, however, have not deterred several groups from taking on this challenge, and since the first extensive VOC profiling of a PGPR by Ryu et al. (2003), investigations in numerous other laboratories have led to several fundamental insights and have laid out the basic framework for further investigations and practical applications (Choi, Song, Yi, & Ryu, 2014; Cortes-Barco, Goodwin, & Hsiang, 2010; Cortes-Barco, Hsiang, & Goodwin, 2010; Dandurishvili et al., 2011; Fernando, Ramarathnam, Krishnamoorthy, & Savchuk, 2005; Groenhagen et al., 2013; Hernandez-Leon et al., 2015; Huang et al., 2012; Hunziker et al., 2015; Kishimoto,

Matsui, Ozawa, & Takabayashi, 2007; Lee et al., 2012; Meldau et al., 2013; Park, Lee, Kloepper, & Ryu, 2013; Popova et al., 2014; Rudrappa et al., 2010; Ryu et al., 2004; Velazquez-Becerra et al., 2013; Yuan, Raza, Shen, & Huang, 2012; Zhao, Yang, Li, Mu, & Liu, 2011).

Most of these investigations have followed more or less the following workflow: (1) identification and culturing of VOC-producing rhizobacteria on artificial media in an enclosed container, (2) collection and concentration of volatiles, (3) analyses using gas chromatography—mass spectroscopy (GC—MS) and identification of VOCs using databases of known compounds such as the NIST database (www.nist.gov). Some studies have also reported analyses using nuclear magnetic resonance (NMR) spectroscopy for the identification of unknown VOCs. (4) Positive identification is ideally followed by bioactivity-driven functional analyses using commercially available individual metabolites for their effect on phytopathogens and/or induction of ISR and growth promotion in plants. At each of these steps, investigators are faced with several challenges in objectively designing experiments that yield insightful findings, which would accurately or at least as closely as possible, reflect natural situations in which PGPRs reside. A necessary prerequisite for following this workflow is availability of PGPRs with VOCs of the desired and known bioactivity such as antifungal and ISR activity. Information about VOC-emitting PGPRs is often gleaned from publications, which have analyzed effect of volatiles on plant growth promoting and/or antibiosis of a specific PGPR. Several recent reports have also screened large numbers of PGPR strains freshly isolated from the rhizosphere or phyllosphere for their ability to produce antimicrobial VOCs using one or more phytopathogens as target organisms (Hernandez-Leon et al., 2015; Hunziker et al., 2015; Popova et al., 2014). Plant growth promotion or antimicrobial bioactivities have been conducted primarily using a two-chamber Petri-plate design in which the test rhizobacteria, and the target pathogen or plant species are grown in physically separate compartments (split-plate method) or two inverted Petri-plates. In the split-plate setup, each compartment is filled with media appropriate for the rhizobacterium and the target organism under investigation (Figure 2) (Hunziker et al., 2015; Ryu et al., 2004, 2003). This design allows for free exchange of gases without letting rhizobacteria and the target pathogen or plant physically contact each other. Prospective VOC-producing PGPRs are identified by their ability to inhibit growth of the target pathogen or by promotion of plant growth. Candidate PGPRs are then further tested in additional rounds of screening. The next step in the workflow is the collection and concentration of VOCs in sufficiently large

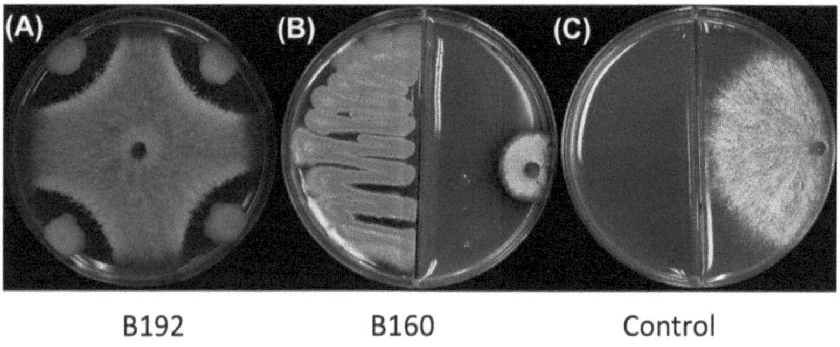

Figure 2 Antimicrobial activity select rhizobacteria against *Rhizoctonia solani* in vitro. (A) Soluble metabolites produced by B192 (*Bacillus cereus*) inhibit mycelialgrowth. (B and C) Split-plate setup showing that volatile metabolities produced by B160 (*Bacillus* sp.) inhibit *R. solani*. Control shows that *R. solani* in the absence of B160 grows normally.

quantity for GC−MS analyses. This is perhaps one of the most challenging steps in the VOCs profiling procedures and is prone to introduction of artefacts in the analysis pipeline such as unintentional sampling of irrelevant chemicals either from the media, atmosphere or other biological activities in the surrounding (Farag et al., 2013). This could introduce noise in the data and would inevitably affect the quality and interpretation of results. Similarly, inadequate sampling not covering the entire growth phases, ranging from the lag period through logarithmic to stationary phases, could also lead to missing rare VOCs that might be emitted at a specific growth stage. Currently, the best control to avoid or minimize variation includes parallel profiling of the headspace of control media without the test rhizobacteria, otherwise treated, incubated and processed exactly in the same manner as with rhizobacteria. Similarly, composition of culture media, particularly C-source such as sugars and proteins and other constituents also affect VOCs (Farag et al., 2013).

VOC profiles are primarily examined by first extracting headspace volatiles followed by GC−MS. Headspace VOCs are mainly collected using two methods: closed-loop-stripping analysis (CLSA) and solid-phase microextraction (SPME). Both these methods have been used for rhizobacterial VOC profiling with satisfactory results. Both these methods have pros and cons, and choice between the two methods is mostly dictated by the availability and access to the infrastructure. In CLSA, volatiles are first collected by adsorption onto a filter, which is then rinsed with a solvent such CH_2Cl_2, and then injected into a GC−MS instrument (Meruva, Penn, & Farthing,

2004). CLSA requires adsorption/desorption of samples into a solvent prolonging and complicating sample preparation. Despite these limitations, however, it has been utilized for profiling rhizobacterial VOCs in several studies yielding useful information (e.g. Groenhagen et al., 2013; Hunziker et al., 2015). In contrast, SPME is a relatively simple, quicker and more sensitive method (parts per trillion detection limit) (Vas & Vekey, 2004), and is becoming a method of choice for VOCs profiling in rhizobacteria (e.g. Dandurishvili et al., 2011; Farag, Ryu, Sumner, & Pare, 2006; Hernandez-Leon et al., 2015). Volatiles are collected without solvent extraction by adsorption onto fused silica-based fibre, which is coated with a solid or liquid stationary phase (Goupry, Rochut, Robins, & Gentil, 2000; Vas & Vekey, 2004). The VOC-loaded fibre is inserted into the input port of GC—MS instruments, where high temperature releases VOCs from the fibres thus extremely simplifying sample preparation. One drawback of the SPME method is that the types of stationary phase fibre coatings selectively include/exclude certain compounds limiting the range of compounds that can be detected in one run. However, several different kinds of fibre with different polarity and absorption properties could be utilized to capture a wider spectrum of VOCs from a sample.

Since VOC emission by PGPRs is dependent on several physical (e.g. media nutrients, pH, temperature) and biological factors (e.g. plants and other microbes), future studies aimed at elucidating VOCs emission under the natural rhizospheric and phyllospheric conditions would suffer from variation in metabolic profiling data. Such variation could be minimized by using more advanced analytical tools such as the proton transfer reaction—time-of-flight—mass spectrometer (PTR-TOF-MS), which has been used for studying the highly dynamic flux of VOCs between soil and atmosphere in several studies (Graus, Müller, & Hansel, 2010; Holzinger, Kasper-Giebl, Staudinger, Schauer, & Röckmann, 2010; Jordan et al., 2009). Given the pros and cons, it is advisable to use different complementary methods to capture as much diversity as is possibly emitted by rhizobacteria. This will inevitably increase the cost per sample, but the decreasing costs of services offered by core analytical labs in academics and commercial settings, and increased collaborations among laboratories with complementary capabilities will hopefully lead to useful comprehensive analyses.

Differences between rhizobacterial strains or other conditions are analyzed using principal component analysis, hierarchical cluster analysis and/or partial least square discriminate analysis (Audrain et al., 2015). With increased interest in VOCs, we expect more complex investigations in the future involving multiple physical, environmental and biological factors leading to the generation of metadata. These complex metadata analyses will require more robust multivariate analyses and will require collaboration with statisticians and computer scientists.

In many VOC profiling studies, several peaks with substantial intensity on MS chromatograms are unresolved indicating that they may represent novel compounds (Hernandez-Leon et al., 2015). These compounds are in most cases ignored for subsequent functional analyses, because they cannot be purchased in pure form. Identification of these compounds through 1D and 2D NMR spectroscopy is warranted for obtaining information about their chemical structure and further investigations aimed at functional analyses.

4.2 VOC-Producing PGPRs

VOCs are produced by many different Gram-positive and Gram-negative rhizobacterial species (Farag et al., 2013) (Table 1). Logically most of these studies used PGPRs with relatively well-characterized antimicrobial activity. A majority of these studies had originally focused on the antimicrobial activity of the whole volatile blends produced by these bacteria. However, many other studies have profiled VOCs using GC–MS (Table 1). Major PGPRs for which VOCs have been profiled include *B. subtilis*, *Bacillus amyloliquefaciens*, *Bacillus cereus*, *Paenibacillus* spp., *Pseudomonas* spp., *Burkholderia ambifaria*, *Streptomyces* spp., *Anthrobactor* and *Serratia* spp. (Table 1). In addition to PGPRs, certain strains of plant pathogens such as *Agrobacterium tumefaciens*, *Pectobacterium carotovorum*, *Pseudomonas phaseolicola*, *Pseudomonas syringae*, *Xanthomonas campestris* pv. *vesicatoria* have been reported to emit VOCs with antifungal activity against plant fungi (Kai et al., 2009). Many VOCs produced by PGPRs are also produced by plants when wounded by herbivores (Matsui, 2006). Some of these VOCs such as 2,3-butanediol, hexenal and 1-octene-3-ol are also produced by plant growth-promoting fungi suggesting that they may have an ancient evolutionary origin and general cytotoxicity (Kishimoto et al., 2007; Zhao et al., 2011).

4.3 Chemical Diversity of PGPR VOCs

Using the split-plate or the double-plate method, many studies have reported antimicrobial and plant growth-promoting activities associated with PGPRs (Table 1; Figure 2). Collectively, these studies have revealed that most PGPRs produce a blend of chemically diverse compounds in different proportions. In general, VOCs of PGPRs consist of simple

Table 1 Volatile Organic Compounds Produced by Plant Growth-Promoting Rhizobacteria (PGPRs) with known Roles in Defence Against Pathogens

Compound	Function	Producing PGPRs	Target Pathogens	References
Alcohols				
2,3-Butanediol	PGP and ISR	*Bacillus subtilis* GB03, *Bacillus amyloliquefaciens*, *Pseudomonas chlororaphis* O6	*Erwinia carotovora* pv. *carotovora*, *Colletotrichum orbiculare*, *Microdochium nivale*, *Rhizoctonia solani*, *Sclerotinia homoeocarpa*	Ryu et al. (2004), Cortes-Barco, Goodwin, et al. (2010) and Cortes-Barco, Hsiang, et al. (2010)
1-Hexanol	Antibiosis	*Pseudomonas* sp.	*Sclerotinia sclerotiorum*	Fernando et al. (2005)
1-Octen-3-ol	ISR	*Paenibacillus polymyxa* BMP-11	*Phytophthora capsici*, *Alternaria brassicicola*, *Botrytis cinerea*, *Colletotrichum capsici*, *Fusarium oxysporum*	Zhao et al. (2011) and Kishimoto et al. (2007)
3-Pentanol	ISR in pepper and cucumber	*B. amyloliquefaciens* strain IN937a	*Xanthomonas axonopodis* pv. *vesicatoria*	Choi et al. (2014)
2-ethyl 1-hexanol	Antibiosis	*Pseudomonas* sp.	*S. sclerotiorum*	Fernando et al. (2005) and Yuan et al. (2012)
Cyclohexanol	Antibiosis	*Pseudomonas* sp.	*S. sclerotiorum*	Fernando et al. (2005)
Citronellol	Antibiosis	*P. polymyxa* strain BMP-11	*P. capsici*, *Tribolium castaneum*, *B. cinerea*	Zhao et al. (2011)
Ketones				
Acetoin	ISR in *Arabidopsis*	*B. subtilis* strain FB17	*Pseudomonas syringae* pv. *tomato*	Rudrappa et al. (2010)

2-Nonanone	Antibiosis	*Burkholderia ambifaria; P. chlororaphis* strain 449; *Serratia proteamaculans* strain 94; *B. amyloliquefaciens* NJN-6	*Alternaria alternata; B. cinerea; R. solani; F. oxysporum* f. sp. *cubense; Agrobacterium tumefaciens*	Groenhagen et al. (2013), Popova et al. (2014) and Yuan et al. (2012)
2-Heptanone	Antibiosis	*P. chlororaphis* strain 449; *S. proteamaculans* strain 94	*A. tumefaciens*	Popova et al. (2014)
2-Undecanone	Antibiosis	*P. chlororaphis* strain 449; *S. proteamaculans* strain 94	*R. solani*	Groenhagen et al. (2013) and Popova et al. (2014)
4-Octanone; acetophenone; 1-phenyl-1,2-propanedione	Antibiosis	*B. ambifaria*	*R. solani*	Groenhagen et al. (2013)
Aldehydes				
n-Decanal; nonanal	Antibiosis	*Pseudomonas* sp.	*S. sclerotiorum*	Fernando et al. (2005) and Yuan et al. (2012)
Sulphur Compounds				
Dimethyl disulphide	Antibiosis; PGP and ISR in tobacco, *Medicago truncatula*, tomato	*Bacillus cereus; Pseudomonas* sp.; *Serratia plymuthica* IC1270	*B. cinerea; R. solani; A. tumefaciens; Agrobacterium vitis*	(Huang et al. (2012), Meldau et al. (2013), Hernandez-Leon et al. (2015), Groenhagen et al. (2013), Dandurishvili et al. (2011) and Popova et al. (2014)
Dimethyl trisulphide	Antibiosis	*Pseudomonas* sp.; *B. ambifaria*	*S. sclerotiorum; R. solani*	Fernando et al. (2005) and Groenhagen et al. (2013)

(*Continued*)

Table 1 Volatile Organic Compounds Produced by Plant Growth-Promoting Rhizobacteria (PGPRs) with known Roles in Defence Against Pathogens—cont'd

Compound	Function	Producing PGPRs	Target Pathogens	References
Benzothiazole	Antibiosis	*Pseudomonas* sp.; *P. polymyxa* strain BMP-11	*P. capsici; S. sclerotiorum; T. castaneum; B. cinerea*	Fernando et al. (2005), Yuan et al. (2012) and Zhao et al. (2011)
Alkanes				
Dimethylhexadecylamine	Antibiosis	*Arthrobacter agilis*; *Pseudomonas fluorescens*	*B. cinerea*; *Phytophthora cinnamomi*	Velazquez-Becerra et al. (2013) and Hernandez-Leon et al. (2015)
Tridecane	PGP and ISR in *Arabidopsis*	*P. polymyxa E681*	*Pseudomonas syringae*	Lee et al. (2012)
Alkenes				
1-Undecene	Antibiosis	*Pseudomonas* sp.	*Phytophthora infestans*; *Agrobacterium* sp.	Dandurishvili et al. (2011) and Hunziker et al. (2015)
Ethers				
1-Hexadecane	ISR in *Arabidopsis*	*P. polymyxa*	*Pectobacterium carotovorum* subsp carotovora; *P. syringae* pv maculicola	Park et al. (2013)

PGP, plant growth promotion; ISR, induced systemic resistance.

hydrocarbons such as alcohols, ketones, aldehydes, sulphur-containing compounds, alkanes, alkenes, ethers and organic acids (Table 1). In addition, several other derivative compounds, which fall into multiple chemical categories, have also been detected in the headspace of PGPRs. This diversity, however, is suggested to be gross underestimation due to several reasons. First, metabolism of microbes is strongly regulated by growth conditions such as media type, temperature, growth duration and aeration (Farag et al., 2013). In most of the studies conducted so far, analyses are usually limited to a narrow range of highly artificial growth conditions including synthetic media, growth temperature and lighting, and therefore the spectrum of VOC obtained under these conditions might be different from natural conditions. Second, sampling and detection methods that are currently employed are not sufficient to capture and resolve the full spectrum of VOCs. These technical limitations could be resolved by using multiple complementary analytical tools. Finally, VOC profiling has so far been restricted to PGPRs that can be cultured on synthetic media. Since the diversity of microbes in soil is several orders of magnitude more than what has been analyzed so far, many novel compounds still await to be discovered. To tap into these PGPRs, conditions as close to natural as possible would be required. In addition to organic volatiles, inorganic profiling is usually restricted to a few known compounds such HCN, [NH^{4+}] and H_2S (Farag et al., 2013). These compounds have been detected in several PGPRs and their role is associated with general growth inhibition or plant growth promotion (Audrain et al., 2015; Hunziker et al., 2015).

4.4 Phytopathogens Targeted by PGPR VOCs

Due to their gas phase, VOCs can easily diffuse through soil and air and therefore they are expected to impact both below- and above-ground ecological systems. Consistently with this notion, VOCs are known to inhibit both soil-borne and foliar pathogens. Their activity has been reported against true fungi including *Alternaria, Botrytis, Colletotrichum, Fusarium, Sclerotinia, Verticillium, Rhizoctonia,* and oomycetes, such as *Phytophthora* and *Pythium* spp. and bacteria (Table 1). Similarly, some volatiles especially the sulphur-containing dimethyl disulphide (DMDS) inhibit several life stages of various phytopathogenic nematodes such as *Meloidogyne incognita.* In addition to fungal, bacterial, oomycetes and nematodes (Coosemans, 2005), some VOCs also display herbicidal and insecticidal activities (Dugravot et al., 2003; Freeman, Rideout, & Wimer, 2009; McAvoy & Freeman, 2013). On the anatomical level, VOCs have been reported to inhibit mycelial growth and spore germination in many fungal pathogens. In other cases other epidemiological factors, such as spore production, are also reported to be affected by VOCs (Popova et al., 2014). Currently available data do not allow researchers to draw general conclusions about the specificity of VOCs produced. These kinds of studies will be important because in certain situations high specificity will be desired such as in controlling only pathogenic fungi but not the host plants and their associated beneficial mycorrhizae. In other cases, it would be desirable to have a general biocide volatile or blend of VOCs with broad-spectrum activity against soil-borne plant pathogens and weeds for preplanting treatment in a manner very similar to methyl bromide.

The question of how specific VOCs are will also need to be addressed. Although not directly addressing this question, many studies that have analyzed the effect of VOCs of multiple PGPRs on multiple pathogens suggest that target pathogens display variability in their response to VOCs (Hunziker et al., 2015). An overview of data from these studies indicates that some fungal species are inhibited strongly but not necessarily by VOCs from different PGPRs (Hunziker et al., 2015). On the species levels, both on the PGPR and target pathogen side, results vary considerably. In some cases, VOCs from the same PGPR species inhibited one fungus but not another, whereas in other cases, VOCs from different PGPR species inhibited the same fungal species more strongly (Dandurishvili et al., 2011; Groenhagen et al., 2013; Hernandez-Leon et al., 2015; Huang et al., 2012; Meldau et al., 2013; Popova et al., 2014; Zhao et al., 2011). In a recent study with more than 32 PGPR strains belonging to diverse rhizobacterial genera, *P. infestans* was shown to respond very strongly to VOCs from certain *Pseudomonas* spp. than were fungi including *R. solani* and *Fusarium oxysporum*. Similarly, *Pseudomonas* spp. also varied greatly with some showing no activity against *P. infestans*. VOC-mediated *P. infestans* inhibition was attributed to both HCN production and other compounds mostly ketones such as 1-undecene, which were in abundance in the headspace of strains with anti-oomycete activity but not in strains with no anti-oomycete activity. While blends of VOCs from different PGPRs might display some specificity against their target pathogens (Hunziker et al., 2015), it is not clear if any of the known VOCs show any specificity. Many of the prominently detected VOCs, such as such as DMDS, 1-octen-3-ol, benzothiazole, 1-undecene, several ketones and citronellol display broad-spectrum activities against species in different kingdoms such as prokaryotes, fungi, oomycetes, algae, insects, nematodes and plants (Dandurishvili et al., 2011; Dugravot et al., 2003; Groenhagen et al., 2013; Hernandez-Leon et al.,

2015; Huang et al., 2012; Meldau et al., 2013; Popova et al., 2014; Zhao et al., 2011). For determining specificity, it is important to assay many different VOCs individually and in combination ideally in the same experimental setup using a battery of target species from different fungal, bacterial, oomycetes and nematodes. It would also be useful to include some host plants, algae and insects in these studies to investigate the complete spectrum of activity of these VOCs.

4.5 VOCs Display Antibiosis

For the following discussion, antibiosis is defined as the inhibition, reversible or irreversible, of a living organism by specific or nonspecific metabolites produced by a PGPR. A vast majority of VOCs produced by PGPRs have been studied for their antimicrobial activity against bacteria, fungi, nematodes, insects, and in some cases against plants and algae. Antibiosis caused by total VOCs has been reported for several PGPRs and a variety of phytopathogens. Similarly, activity of many individual VOCs has also been reported. Results from several major studies employing a large number of individual VOCs are summarized in Table 1. The types of VOCs that display antibiosis roughly fall into those categories that were detected in the headspace of PGPRs. In general, they include alcohols, ketones, aldehydes, alkanes, alkenes and sulphur-containing compounds. Some compounds exhibit both antibiosis and ISR activity alluding to their multimode mechanism of action (see below).

4.5.1 Alcohols

Several types of alcohols were detected in the headspace of PGPRs (Table 1). Collectively alcohols displayed broad-spectrum activity against bacteria and fungi. 2,3-butanediol was one of the most abundant of all alcohols in many diverse PGPRs and is also extensively analyzed in vitro assays. In yeast (Gonzalez et al., 2000) and several bacterial species including *B. subtilis* (Hohn-Bentz & Radler, 1978; Zhang et al., 2014), 2,3-butanediol is synthesized from acetoin by the acetoin reductase/2,3-butanediol dehydrogenase (Xiao & Xu, 2007). This reaction is reversible with alkaline conditions favouring oxidation (2,3-butanediol to acetoin) and acidic or neutral conditions favouring reductase (acetoin to 2,3-butanediol) activity (Zhang et al., 2014), suggesting that soil pH might regulate relative proportions of 2,3-butanediol and acetoin. Since acetoin also displays antibiosis, this could expand bioactivity zones to a wider pH range, which differs quite considerably in agricultural soils.

4.5.2 Ketones and Aldehydes

Several ketones are regularly detected in the headspace of many diverse PGPRs when cultured in vitro (Table 1). They are extracted with both SPME and CLSA methods (Groenhagen et al., 2013; Popova et al., 2014). Similar to alcohols, they are also active against diverse fungal and bacterial plant pathogens. Among the ketones, acetoin is also unique in that it not only displays antibiosis but is also known to induce ISR (Rudrappa et al., 2010). Whether other ketones that display antibiosis also have ISR activity has not been reported yet. Acetoin can be synthesized from α-acetolactate by α-acetolactate decarboxylase or from diacetyl by diacetyl-acetoin reductase (Xiao & Xu, 2007). Another ketone, 2,3-butanedione is also produced from α-acetolactate, suggesting that the synthesis of diverse ketones might be regulated in concert to expand their spectrum of activity. In addition to the well-characterized biosynthesis genes of acetoin in *Bacillus* spp., genes that code for enzymes in the biosynthetic pathways of several other ketones are also present in the sequenced genomes of several PGPRs (Ali, Norman, & Brennan, unpublished results). These genes could be cloned and utilized for manipulation of their respective ketones biosynthesis for their optimal production and enhancement of bioactivity of PGPRs.

4.5.3 Alkanes and Alkenes

Several long-chain hydrocarbons are emitted by a diverse set of PGPRs (Hernandez-Leon et al., 2015; Velazquez-Becerra et al., 2013). Dimethyl-hexadecylamine (DMHDA) and tridecane are probably the most prominent that were shown to display antimicrobial activity against fungi and bacteria. Several alkenes were also observed in different proportions by different *Pseudomonas* spp. Antimicrobial analyses showed that these compounds are general inhibitors, as they inhibited growth of several fungi, bacteria, a cyanobacterium, nematode and *Drosophila* at a comparable micromolar concentration (Popova et al., 2014). Some of these compounds showed stronger activity against one type of organism than another pointing to a certain level of specificity (Popova et al., 2014). For example, 1-undecene seems to display activity against oomycetes, nematodes and insects as it strongly inhibited *P. infestans* (Hunziker et al., 2015), *Caenorhabditis elegans* and *Drosophila*, but not *R. solani* and *Agrobacterium* (Popova et al., 2014). In a large-scale screen of rhizosphere and phyllosphere PGPRs, 1-undecene stood out as one of the prominent VOCs with anti-*Phytophthora* activity. It inhibited several different developmental stages of *P. infestans* including mycelial growth, sporangial production and zoospore release in a dose-dependent manner

(Hunziker et al., 2015). The molecular mechanism of how 1-undecene exerts its broad-spectrum activity against different life stages of *P. infestans* is currently not known and addressing this issue would be very useful for developing this compound for application in agriculture.

4.5.4 Sulphur Compounds

Several sulphur-containing compounds, primarily DMDS, dimethyl trisulphide and dimethyl sulphide, and to a lesser extent benzothiazole are frequently detected in GC-profiles of VOCs of a large number of different PGPRs (Table 1). Collectively they display broad-spectrum antibiosis against fungal and bacterial pathogens. Similar to most other VOCs, the exact mode of action of sulphur-containing VOCs is not known, however, in insects, DMDS was shown to interfere with the function of mitochondria and potassium channels (Dugravot et al., 2003). Similarly, the biosynthesis pathways of these sulphur compounds in PGPRs are also not known but based on results from *lactobacillus* spp. and truffles they may involve the methionine catabolism pathways (Liu et al., 2013). They could be synthesized by spontaneous chemical oxidation of methanethiol, which in turn could be formed from L-methionine by L-methioninase (El-Sayed, 2010) or by demethiolation of 4-methylthio-2-oxobutanoic acid (Liu et al., 2013). The production of methanethiol can be enhanced by using immobilized fungal L-methioninase (El-Sayed & Shindia, 2011), thus providing a more efficient synthesis platform.

Since not many studies have attempted to ascertain if VOCs are biostatic or biocide, in their mode of action, a generalization about the antibiosis nature of VOCs cannot be made. Bacterial suppressor activity of two *Pseudomonas, Pseudomonas chlororaphis* 205 and *Pseudomonas fluorescens* B-4117, and *Serratia plymuthica* strain IC1270 was reported to be bacteriostatic (Popova et al., 2014). In the same study, VOCs from all tested *Pseudomonas* and *Serretia* strains acted as biocide against a cyanobacterium, but VOCs only from *S. plymuthica* strain IC1270 was lethal to a nematode resulting in the death of larvae. Activity of VOCs from both *Pseudomonas* sp. and *Serratia* sp. against three fungal species *R. solani, Helminthosporium sativum* and *S. sclerotiorum* was fungistatic as upon transfer to no-VOCs conditions these fungi resumed growth. However, in another study activities of total VOCs produced by four different *Pseudomonas* spp., *P. chlororaphis, P. fluorescens, Pseudomonas corrugata* and *Pseudomonas aurantiaca*, against *S. sclerotiorum* were fungicidal because transfer of mycelial plugs or sclerotia from the VOC exposed plate to fresh plate did not resume growth (Fernando et al.,

2005). In this study, benzothiazole, cyclohexanol, n-decanal, dimethyl trisulphide, 2-ethyl 1-hexanol and nonanal produced by *Pseudomonas* sp. strongly inhibited *S. sclerotiorum*, but fungicidal or fungistatic status of individual compounds was not tested. It would be very interesting to conduct a comprehensive study of the most effective volatiles identified in these studies against a battery of different fungal, bacterial and nematode species to determine their biostatic or biocide nature. This kind of study will also address the issue of specificity of individual VOC, which has not been thoroughly investigated or has provided somewhat variable results.

In their natural environment, many different beneficial organisms interact with each other. It is plausible to imagine that VOCs might play a role in these interactions, and therefore VOCs from beneficial PGPRs might be detrimental to other PGPRs or other beneficial organisms such as mycorrhizae and free-living fungi. Since several VOCs display general inhibition against multiple organisms, including beneficial ones such as *Trichoderma* spp. (Hernandez-Leon et al., 2015; Velazquez-Becerra et al., 2013), future studies should also look into their effects on beneficial organisms before they are developed and deployed in disease-management programmes.

4.6 PGPR VOCs Are Involved in ISR

Several plant-produced volatile compounds such as methyl jasmonate and ethylene and the genetic pathways associated with these airborne hormones are implicated in ISR (Kloepper, Ryu, & Zhang, 2004; van Loon, Bakker, & Pieterse, 1998; Pieterse et al., 2014). Since PGPRs induce ISR, several studies explored the possibility of bacterial VOCs as possible inducer of ISR. Data from several such investigations suggest that in addition to several known diffusible PGPR metabolites, VOCs are also involved in inducing ISR (Farag et al., 2006; Han et al., 2006; Lee et al., 2012; Ryu et al., 2004). Roles for some of the known individual VOCs in ISR are summarized in Table 1. Alcohols, ketones, alkanes and sulphur-containing compounds have been shown to induce ISR in various plants including *Arabidopsis*, tobacco, tomato, pepper and cucumber against bacterial, fungal and oomycete pathogens (Table 1).

Although, by now substantial evidence for the positive role of VOCs in ISR exists, genetic and molecular mechanism of this phenomenon is not clear. We do not know if VOCs are recognized in the roots or aerial plant parts and by what type(s) of protein receptors if at all. In addition, we also do not know the signalling component that could be unequivocally associated with VOC-induced ISR. Nevertheless, based on several studies that have employed various mutants in the salicylic acid (SA), jasmonic acid (JA) and ethylene (Et) pathways, it appears that some VOCs might be acting through one or more of these pathways. It is also possible that other VOCs might be acting completely independent of these pathways. Below we summarize some of these studies that have shed some light on individual VOC-induced ISR.

In general, similar to live PGPRs, a majority of VOCs appear to exert their action through the JA/Et pathway independent of SA, but also see Lee et al. (2012), where along with the JA and ET signalling markers *ChiB* and *VSP2*, the SA marker *PR1* was also induced by VOCs of *Paenibacillus polymyxa*. Based on expression analyses of GUS driven by promoters of *PDF1.2* (a marker for the JA/Et pathway), *JIN1* (JA pathway) and *PR1* (SA pathway), Ryu et al. (2004) showed that VOCs produced by *B. subtilis* GB03 and *B. amyloliquefaciens* IN937a induced *PDF1.2* but not *JIN1* and *PR1*, suggesting that only the ethylene signalling pathway is involved in VOC-induced ISR. Using a battery of specific *Arabidopsis* defence mutants, these observations were verified by showing that GB03 VOC-induced defence against *Erwinia carotovora* subsp. carotovora was lost only in *ein2* (ethylene-insensitive) mutant, but not in *coi1* (coronatine-insensitive 1), NahG (degrades SA), *cpr1* (constitutively produces salicylic acid) and *npr1* (nonexpresser of *PR* genes). Additional evidence for the involvement of ethylene pathway in VOC-induced ISR came from a proteomic study, which revealed that several proteins in the ethylene biosynthesis and response pathways were upregulated in *Arabidopsis* exposed to bacterial VOCs (Kwon et al., 2010).

In GC–MS and NMR-MS, chemical and functional profiling studies of more than 32 VOCs, 2,3-butanediol stood out as the major inducer of ISR (Farag et al., 2006; Han et al., 2006). Similar studies in tobacco with VOCs of *P. chlororaphis* O6 also indicate that 2,3-butanediol is the major inducer of ISR (Han et al., 2006). Used either alone or in combination with other volatiles, 2,3-butanediol significantly reduced symptoms caused by fungal and bacterial pathogens in tobacco (Table 1). Reduced symptoms were correlated with enhanced expression of basic *PR* genes in the JA/Et pathway (Cortes-Barco, Goodwin, et al. 2010; Cortes-Barco, Hsiang, et al., 2010). Conclusive evidence for whether these pathways are involved in other VOC-mediated ISR would require detailed analyses using traditional set of mutants in these pathways such as *jar1, coi1* (JA), *etr* (ethylene) and *eds1, ics1, npr1* (SA). In addition to the 2,3-butanediol, other alcohols such as 1-octen-3-ol and 3-pentanol have also been implicated in ISR in *Arabidopsis* and pepper (Choi et al., 2014; Kishimoto et al., 2007).

Acetoin is another VOC, which was shown to induce ISR in *Arabidopsis* in an ethylene-dependent manner against *P. syringae* pv. DC3000. Acetoin is a ketone, which in GC–MS profiling analyses has been detected in the headspace of several PGPRs (Farag et al., 2006; Goupry et al., 2000; Lee et al., 2012; Rudrappa et al., 2010; Ryu et al., 2004). In fact, it is produced by diverse soil bacteria in the *Proteobacteria* and *Firmicutes* groups and is suggested to regulate NAD/NADH ratios and used as a storage compound (Xiao & Xu, 2007).

In addition to low molecular weight molecules described above, long-chain VOCs (C13 and C16) emitted by *P. polymyxa* E681 have also been shown to induce ISR in *Arabidopsis* against a hemibiotrophic (*P. syringae* pv. *maculicola*) and necrotrophic pathogen (*P. carotovorum* subsp. *carotovorum*) in *Arabidopsis* (Lee et al., 2012; Park et al., 2013). Taken together, these data suggest that some VOCs are likely specific, whereas others inhibit both biotrophic and necrotrophic pathogens. Interestingly, these long-chain compounds displayed stronger ISR activities compared to the short-chain compounds such as 2,3-butanediol and acetoin (Park et al., 2013).

DMDS, although mostly associated with antibiosis activity, has also been shown to induce ISR (Huang et al., 2012). The molecular mechanism of how DMDS induce ISR is not known, and it would be interesting to determine if its ISR activity is related to its antibiosis activity, which is proposed to involve dysfunction of mitochondrial activity and activation of K(ATP) channel in insects (Dugravot et al., 2003).

From the above discussion, it is clear that although substantial progress has been made in cataloguing VOCs that induce ISR, much more research is needed to unravel molecular mechanisms and pathways of how these VOCs function in inducing ISR. To discover any novel pathways, differential gene expression analyses of host plants treated with different VOCs using high-throughput approaches such as RNA-seq and proteomics are needed and warranted.

4.7 VOCs Effect in Plant Disease Escape

Growth promotion, although not directly related to enhanced defence against pathogens, could help plants tolerate or escape plant pathogens. Several studies using sealed containers have shown that different PGPRs produce volatile blends that promote plant growth (reviewed in Bailly & Weisskopf, 2012; Farag et al., 2006; Gutierrez-Luna et al., 2010; Kai et al., 2010, 2009; Lugtenberg & Kamilova, 2009; Meldau, Long, & Baldwin, 2012; Vespermann, Kai, & Piechulla, 2007; Velázquez-Becerra et al., 2011; Zou, Li, & Yu, 2010). Although studies with individual VOCs effect on plant growth are scarce, 2,3-butanediol and DMDS, which have been shown to prime different plants such as *Arabidopsis*, *Nicotiana benthamiana*, *Medicago truncatula* and tomato for defence against bacterial, fungal and oomycete pathogens, also promoted plant growth (Meldau et al., 2013; Ryu et al., 2003). Using a split-plate system where plants and PGPRs are grown in physically separate compartments in enclosed Petri dishes sharing the same airspace, Ryu et al. (2003) showed that a blend of VOCs produced by *B. subtilis* GB03 and *B. amyloliquefaciens* IN937a enhanced *Arabidopsis* growth. GC−MS analyses of the headspace of these two PGPRs showed that 3-hydroxy-2-butanone (acetoin) and its downstream product 2,3-butanediol were present in these two strains but not in other bacterial strains that did not display VOC-induced plant growth promotion. In *B. subtilis*, 2,3-butanediol is synthesized from acetoin through the *als* operon, which consists of *alsS* (acetolactate synthase), which converts pyruvate to acetolactate, and *alsD* (acetolactate decarboxylase), which converts acetolactate to acetonin (Xiao & Xu, 2007). Insertional inactivation of the acetolactate synthase (*als*) operon led to reduced 2,3-butanediol production and VOC-induced plant growth promotion providing genetic evidence that this compound was the primary plant growth stimulator (Cruz Ramos et al., 2000; Ryu et al., 2003). Dose-dependent enhancement of *Arabidopsis* growth with exogenous application of commercial 2,3-butanediol further corroborated this notion (Ryu et al., 2003).

Recently, *Bacillus* sp. B55 was shown to significantly enhance plant growth in tobacco and *Arabidopsis* plants (Meldau et al., 2012, 2013). In this study, DMDS was detected as the most likely major contributor to plant growth promotion. Furthermore, using genetic and biochemical analyses, it was shown that DMDS provided sulphur nutrition to plants especially under low-sulphur conditions. These results, however, are different from those reported elsewhere (Kai et al., 2010), where growth reduction was observed in *Arabidopsis* when exposed to DMDS. Differences between these two studies could be due to difference in growth media and other experimental conditions. These apparently contradictory results underline caution in generalization of results obtained under laboratory conditions, which are obviously very different from the natural conditions. VOCs from *P. polymyxa* E681, *B. subtilis* GB03 significantly enhanced fresh weight and total leaf area in *Arabidopsis* plants (Lee et al., 2012). This growth promotion was not seen in an *ein2.5* (Et-insensitive) mutant plants suggesting that ethylene signalling is involved in this response. Tridecane, a long-chain alkane, was shown to induce ISR but it decreased plant growth suggesting that other VOCs might be causing enhanced growth in *Arabidopsis*. In summary, more research is needed to clearly establish roles for individual VOCs in enhancing plant growth and tolerance to pathogens.

5. PRACTICAL USES IN AGRICULTURE

Until recently growers relied heavily on methyl bromide for preplanting treatment of soil, which helped in suppressing many soil-borne pathogens and weeds. However, now that methyl bromide because of its adverse climate impact is banned in the USA, growers are in desperate need of alternative chemicals that are as effective and are safe for the environment and climate. Investigations into the biochemical, antimicrobial and herbicidal activities of VOCs isolated from several different PGPRs have yielded encouraging results. DMDS either alone or in combination with another compound has attracted considerable attention as an alternative to methyl bromide. In has emerged as one of the best options (Auger & Charles, 2003), as it display general growth suppression activity against fungi (Kai et al., 2009), bacteria (Freeman et al., 2009), weeds (Freeman et al., 2009; McAvoy & Freeman, 2013) and nematodes (Coosemans, 2005). Several trial with vegetables have been conducted with MDMS under field and greenhouses conditions in tomato (McAvoy & Freeman, 2013), melon (Sasanelli, Dongiovanni, Santori, & Myrta, 2014), strawberry (López-Aranda, 2012), pepper, cucumber (Zanón, Gutiérrez, & Myrta, 2014) and ornamentals (Kokalis-Burelle, Rosskopf, & Hartman, 2010; Rosskopf, Kokalis-Burelle, McSorley, & Skvarch, 2010). Initial results from these trials are very encouraging and indicate that DMDS provides control of the target organisms comparable to MeBr whenever it was included in the trials. In addition to preplant, DMDS has also been tested post-planting for the control of several important yield-limiting nematodes in open-field grape vineyards with encouraging results (Cabrera, Wang, Gerik, & Gan, 2014). DMDS has been developed by Cerexagri (King of Prussia, PA), a subsidiary of ATOFINA Chemicals Inc. (Philadelphia, PA), and is marketed under the trade name Paladin® (Arkema, Inc.) for control of *Verticillium*, *Fusarium*, *Pythium*, *Sclerotinia*, *Rhizoctonia*, nutsedge (purple and yellow), chickweed, lambsquarters, purslane, grasses and several nematodes (IR4-PALADIN, 2015).

To the best of our knowledge, no other VOCs have been registered or are under development for the control of plant diseases. Among the many compounds, DMDS, DMHDA, acetoin, butanediol and tridecane stand out very promising for the control of a wide variety of plant pathogens (Table 1). Based on research findings about the efficacy of these VOCs, they will hopefully be developed and registered for use in agriculture. More importantly, similar to the live PGPRs that most likely exert their bioactivity through a combination of soluble and volatiles, future studies should test VOCs together with solubles. In this regard, in laboratory and green house studies, combination of DMDS and dazomet synergistically reduced nematodes, two fungal species and two weed species (Mao et al., 2014). Subsequent growth of cucumber on treated soil leads to enhanced cucumber production compared to untreated control. This study underlines the importance of testing multiple VOCs in combination, which may yield better result with lower dose rates.

Most PGPR formulations consist of living organisms, which are highly dependent on environmental factors such as soil pH, temperature and light (Bloemberg & Lugtenberg, 2001; Yu & Lee, 2013). For example the biological control activities of *P. fluorescence* against *Fusarium* were reduced when pH was changed from 7.5 to 6.5 (Duuff, Erkelens, Bakker, & Schippers, 1995). Dependence on these environmental factors often leads to unsuccessful establishment of PGPRs in soils and unsatisfactory results (Fravel, 2005; Weller & Thomashow, 1994). One way to resolve this issue could be to develop and utilize bioactive compounds of PGPRs, instead of live formulation. Potency and efficacy will be important factors in deciding on selecting appropriate PGPR metabolites for development and deployment. Instead of looking to enhance the potency of naturally found biopesticides through chemical modification, it would be wise to look for potent mixtures of natural products. In our opinion, using mixtures could produce synergistic results. In addition, mixtures of different chemistries would delay the onset of resistance development and would prolong the usable life of chemicals.

6. OUTLOOK – OPPORTUNITIES AND CHALLENGES

1. Opportunities for deployment of secreted chemicals from PGPRs are numerous. In addition to developing the secreted metabolites as biopesticides as discussed above, the biosynthesis genes of PGPR metabolites

can be introduced into crops using genetic engineering. Genetic modification, however, will be more appropriate for those products that require one or very few genes as opposed to those that require multigene clusters which are often involved in the biosynthesis of PGPR-derived soluble metabolites.

2. How specific are PGPR metabolites with respect to a specific pathogen? Since PGPRs come in contact, chemically speaking, in their environment with many other organisms including other beneficial ones belonging either to the same broader class such as bacteria or phylogenetically distant organisms such as mycorrhizae, it is important to conduct ecotoxicity studies to ascertain antimicrobial activities against beneficial organisms.
3. What are the risks, if any, of resistance development in pathogens against PGPR metabolites? With higher specificity inherently comes the risk of resistance development because minor modifications in the molecular targets are all that is needed to render an antibiotic ineffective. These investigations would require using genetic, molecular and physiological approaches to monitor and document risk levels in the target pathogens. The underlying genetic mutation can be cloned using careful genetic crosses and several of the latest high-throughput approaches such as cloning-by-sequencing the nucleotide changes that are associated with resistance.
4. What are the molecular targets of PGPR metabolites? Despite substantial progress in metabolite profiling and bioactivity of PGPR metabolites, our knowledge about their molecular targets is limited. Using various approaches such as comparative associative genomics along with targeted mutagenesis and biochemical approaches will be needed to address this issue.
5. Perhaps more importantly it is critical to thoroughly evaluate cytotoxicity of PGPR metabolites to mammalian and other animal cells. 1-octene-ol, one PGPR bioactive metabolite, is a prime example. It is implicated in impairing human health and inducing inflammation response in *Drosophila* in several studies (Inamdar & Bennett, 2014; Inamdar et al., 2013). Similar studies with other potential soluble and VOCs are needed to avoid repetition of mistakes made with the now defunct chemicals such as methyl bromide and DDT.
6. Utilizing extracted PGPR metabolites is attractive, but when applied ex vivo it might imbalance the natural balance exerted by these chemicals on live organisms. For example, DMDS is known to be degraded by certain *B. cereus* strains (Liang, An, Li, & Zhang, 2014), and perhaps this keeps

DMDS-producing strains from entirely taking a whole ecosystem. Therefore, due consideration should be given to ecological consequences of using PGPR metabolites in pure form.

7. And lastly and more importantly, as with many other chemicals, there would be questions of uncertainty about the health impact of PGPR metabolites, which will require detailed research and public relations endeavours to quell any controversy that might arise from sporadic observation of their detrimental effects. This will also require striking a balance between uncertainty and beneficial impacts of PGPR metabolites, and taking public and policy-makers into confidence.

ACKNOWLEDGEMENTS

Research in Dr Ali lab is supported by the University of Florida. Work in Dr Norman lab is additionally funded by American Floral Endowment.

REFERENCES

Ali, B., Sabri, A. N., Ljung, K., & Hasnain, S. (2009). Auxin production by plant associated bacteria: impact on endogenous IAA content and growth of *Triticum aestivum* L. *Letters in Applied Microbiology, 48*(5), 542–547.

Ali, G. S., Norman, D., & Brennan, M. Comparison of cyclic lipopeptide genes in plant growth promoting rhizobacteria, unpublished results.

Alstrom, S. (1991). Induction of disease resistance in common bean susceptible to halo blight bacterial pathogen after seed bacterization with rhizosphere pseudomonads. *Journal of General and Applied Microbiology, 37*(6), 495–501.

Andreae, M. O., & Crutzen, P. J. (1997). Atmospheric aerosols: biogeochemical sources and role in atmospheric chemistry. *Science, 276*(5315), 1052–1058.

Arrebola, E., Jacobs, R., & Korsten, L. (2010). Iturin A is the principal inhibitor in the biocontrol activity of *Bacillus amyloliquefaciens* PPCB004 against postharvest fungal pathogens. *Journal of Applied Microbiology, 108*(2), 386–395.

Audrain, B., Farag, M. A., Ryu, C. M., & Ghigo, J. M. (2015). Role of bacterial volatile compounds in bacterial biology. *FEMS Microbiology Reviews, 39*(2), 222–233.

Auger, J., & Charles, P. (2003). Biogenic emission, biological origin, and mode of action of DMDS, a natural, ubiquitous fumigant. In *Annual International Research Conference on Methyl Bromide Alternatives and Emissions Reductions* (pp. 131–138). MBAO.

Badri, D. V., Chaparro, J. M., Zhang, R., Shen, Q., & Vivanco, J. M. (2013). Application of natural blends of phytochemicals derived from the root exudates of *Arabidopsis* to the soil reveal that phenolic-related compounds predominantly modulate the soil microbiome. *Journal of Biological Chemistry, 288*(7), 4502–4512.

Bailly, A., & Weisskopf, L. (2012). The modulating effect of bacterial volatiles on plant growth: current knowledge and future challenges. *Plant Signaling & Behavior, 7*(1), 79–85.

Bais, H. P., Fall, R., & Vivanco, J. M. (2004). Biocontrol of *Bacillus subtilis* against infection of Arabidopsis roots by *Pseudomonas syringae* is facilitated by biofilm formation and surfactin production. *Plant Physiology, 134*(1), 307–319.

Bais, H. P., Weir, T. L., Perry, L. G., Gilroy, S., & Vivanco, J. M. (2006). The role of root exudates in rhizosphere interactions with plants and other organisms. *Annual Review of Plant Biology, 57*, 233–266.

Berendsen, R. L., Pieterse, C. M., & Bakker, P. A. (2012). The rhizosphere microbiome and plant health. *Trends in Plant Science, 17*(8), 478–486.

Bloemberg, G. V., & Lugtenberg, B. J. (2001). Molecular basis of plant growth promotion and biocontrol by rhizobacteria. *Current Opinion in Plant Biology, 4*(4), 343–350.

Bonito, G., Reynolds, H., Robeson, M. S., 2nd, Nelson, J., Hodkinson, B. P., Tuskan, G., et al. (2014). Plant host and soil origin influence fungal and bacterial assemblages in the roots of woody plants. *Molecular Ecology, 23*(13), 3356–3370.

de Bruijn, I., de Kock, M. J., Yang, M., de Waard, P., van Beek, T. A., & Raaijmakers, J. M. (2007). Genome-based discovery, structure prediction and functional analysis of cyclic lipopeptide antibiotics in *Pseudomonas* species. *Molecular Microbiology, 63*(2), 417–428.

Cabrera, J. A., Wang, D., Gerik, J. S., & Gan, J. (2014). Spot drip application of dimethyl disulfide as a post-plant treatment for the control of plant parasitic nematodes and soilborne pathogens in grape production. *Pest Management Science, 70*(7), 1151–1157.

Cawoy, H., Debois, D., Franzil, L., De Pauw, E., Thonart, P., & Ongena, M. (2015). Lipopeptides as main ingredients for inhibition of fungal phytopathogens by *Bacillus subtilis/amyloliquefaciens*. *Microbial Biotechnology, 8*(2), 281–295.

Cawoy, H., Mariutto, M., Henry, G., Fisher, C., Vasilyeva, N., Thonart, P., et al. (2014). Plant defense stimulation by natural isolates of Bacillus depends on efficient surfactin production. *Molecular Plant Microbe Interactions, 27*(2), 87–100.

Cernadas, R. A., & Benedetti, C. E. (2009). Role of auxin and gibberellin in citrus canker development and in the transcriptional control of cell-wall remodeling genes modulated by *Xanthomonas axonopodis* pv. citri. *Plant Science, 177*(3), 190–195.

Chameides, W. L., Lindsay, R. W., Richardson, J., & Kiang, C. S. (1988). The role of biogenic hydrocarbons in urban photochemical smog – Atlanta as a case-study. *Science, 241*(4872), 1473–1475.

Chen, X. H., Koumoutsi, A., Scholz, R., Eisenreich, A., Schneider, K., Heinemeyer, I., et al. (2007). Comparative analysis of the complete genome sequence of the plant growth-promoting bacterium *Bacillus amyloliquefaciens* FZB42. *Nature Biotechnology, 25*(9), 1007–1014.

Chen, Z., Agnew, J. L., Cohen, J. D., He, P., Shan, L., Sheen, J., et al. (2007). *Pseudomonas syringae* type III effector AvrRpt2 alters *Arabidopsis thaliana* auxin physiology. *Proceedings of the National Academy of Sciences of the United States of America, 104*(50), 20131–20136.

Choi, H. K., Song, G. C., Yi, H. S., & Ryu, C. M. (2014). Field evaluation of the bacterial volatile derivative 3-pentanol in priming for induced resistance in pepper. *Journal of Chemical Ecology, 40*(8), 882–892.

Choi, J., Choi, D., Lee, S., Ryu, C. M., & Hwang, I. (2011). Cytokinins and plant immunity: old foes or new friends? *Trends in Plant Science, 16*(7), 388–394.

Cohen, A. C., Bottini, R., & Piccoli, P. N. (2008). *Azospirillum brasilense* Sp 245 produces ABA in chemically-defined culture medium and increases ABA content in arabidopsis plants. *Plant Growth Regulation, 54*(2), 97–103.

Coosemans, J. (2005). Dimethyl disulphide (DMDS): a potential novel nematicide and soil disinfectant. *Acta Horticulturae, 698*, 57–64.

Cortes-Barco, A. M., Goodwin, P. H., & Hsiang, T. (2010). Comparison of induced resistance activated by benzothiadiazole, (2R,3R)-butanediol and an isoparaffin mixture against anthracnose of *Nicotiana benthamiana*. *Plant Pathology, 59*(4), 643–653.

Cortes-Barco, A. M., Hsiang, T., & Goodwin, P. H. (2010). Induced systemic resistance against three foliar diseases of *Agrostis stolonifera* by (2R,3R)-butanediol or an isoparaffin mixture. *Annals of Applied Biology, 157*(2), 179–189.

Cruz Ramos, H., Hoffmann, T., Marino, M., Nedjari, H., Presecan-Siedel, E., Dreesen, O., et al. (2000). Fermentative metabolism of *Bacillus subtilis*: physiology and regulation of gene expression. *Journal of Bacteriology, 182*(11), 3072–3080.

Dandurishvili, N., Toklikishvili, N., Ovadis, M., Eliashvili, P., Giorgobiani, N., Keshelava, R., et al. (2011). Broad-range antagonistic rhizobacteria *Pseudomonas fluorescens* and *Serratia plymuthica* suppress *Agrobacterium* crown gall tumours on tomato plants. *Journal of Applied Microbiology, 110*(1), 341–352.

Dangl, J. L., Horvath, D. M., & Staskawicz, B. J. (2013). Pivoting the plant immune system from dissection to deployment. *Science, 341*(6147), 746–751.

Davis, T. S., Crippen, T. L., Hofstetter, R. W., & Tomberlin, J. K. (2013). Microbial volatile emissions as insect semiochemicals. *Journal of Chemical Ecology, 39*(7), 840–859.

De Souza, J. T., De Boer, M., De Waard, P., Van Beek, T. A., & Raaijmakers, J. M. (2003). Biochemical, genetic, and zoosporicidal properties of cyclic lipopeptide surfactants produced by *Pseudomonas fluorescens*. *Applied and Environmental Microbiology, 69*(12), 7161–7172.

De Vleesschauwer, D., & Hofte, M. (2009). Rhizobacteria-induced systemic resistance. In *Plant Innate Immunity* (Vol. 51). Academic Press.

Ding, X., Cao, Y., Huang, L., Zhao, J., Xu, C., Li, X., et al. (2008). Activation of the indole-3-acetic acid-amido synthetase GH3-8 suppresses expansin expression and promotes salicylate- and jasmonate-independent basal immunity in rice. *Plant Cell, 20*(1), 228–240.

Duca, D., Lorv, J., Patten, C. L., Rose, D., & Glick, B. R. (2014). Indole-3-acetic acid in plant-microbe interactions. *Antonie van Leeuwenhoek, 106*(1), 85–125.

Dugravot, S., Grolleau, F., Macherel, D., Rochetaing, A., Hue, B., Stankiewicz, M., et al. (2003). Dimethyl disulfide exerts insecticidal neurotoxicity through mitochondrial dysfunction and activation of insect K(ATP) channels. *Journal of Neurophysiology, 90*(1), 259–270.

Duuff, B. J., Erkelens, A., Bakker, P. A. H. M., & Schippers, B. (1995). Influence of pH on suppression of fusarium wilt of carnation by *Pseudomonas fluorescens* WCS417r. *Journal of Phytopathology, 143*(4), 217–222.

Effmert, U., Kalderas, J., Warnke, R., & Piechulla, B. (2012). Volatile mediated interactions between bacteria and fungi in the soil. *Journal of Chemical Ecology, 38*(6), 665–703.

El-Sayed, A. S. (2010). Microbial L-methioninase: production, molecular characterization, and therapeutic applications. *Applied Microbiology and Biotechnology, 86*(2), 445–467.

El-Sayed, A. S., & Shindia, A. A. (2011). Characterization and immobilization of purified *Aspergillus flavipes* l-methioninase: continuous production of methanethiol. *Journal of Applied Microbiology, 111*(1), 54–69.

Farace, G., Fernandez, O., Jacquens, L., Coutte, F., Krier, F., Jacques, P., et al. (2015). Cyclic lipopeptides from *Bacillus subtilis* activate distinct patterns of defence responses in grapevine. *Molecular Plant Pathology, 16*(2), 177–187.

Farag, M. A., Ryu, C. M., Sumner, L. W., & Pare, P. W. (2006). GC-MS SPME profiling of rhizobacterial volatiles reveals prospective inducers of growth promotion and induced systemic resistance in plants. *Phytochemistry, 67*(20), 2262–2268.

Farag, M. A., Zhang, H., & Ryu, C. M. (2013). Dynamic chemical communication between plants and bacteria through airborne signals: induced resistance by bacterial volatiles. *Journal of Chemical Ecology, 39*(7), 1007–1018.

Fernando, W. D., Ramarathnam, R., Krishnamoorthy, A. S., & Savchuk, S. C. (2005). Identification and use of potential bacterial organic antifungal volatiles in biocontrol. *Soil Biology and Biochemistry, 37*(5), 955–964.

Fravel, D. R. (2005). Commercialization and implementation of biocontrol. *Annual Review of Phytopathology, 43*, 337–359.

Freeman, J., Rideout, S., & Wimer, A. (2009). Dimethyl disulfide use for bacterial wilt management and weed control in Virginia tomatoes. *Hort Science, 44*, 571.

Fu, J., Liu, H., Li, Y., Yu, H., Li, X., Xiao, J., et al. (2011). Manipulating broad-spectrum disease resistance by suppressing pathogen-induced auxin accumulation in rice. *Plant Physiology, 155*(1), 589–602.

Gang, W., Kloepper, J. W., & Tuzun, S. (1991). Induction of systemic resistance of cucumber to colletotrichum orbiculare by select strains of plant growth-promoting rhizobacteria. *Phytopathology, 81*(12), 1508–1512.

Garcia-Gutierrez, L., Zeriouh, H., Romero, D., Cubero, J., de Vicente, A., & Perez-Garcia, A. (2013). The antagonistic strain *Bacillus subtilis* UMAF6639 also confers protection to melon plants against cucurbit powdery mildew by activation of jasmonate-and salicylic acid-dependent defence responses. *Microbial Biotechnology, 6*(3), 264–274.

Glick, B. R. (2014). Bacteria with ACC deaminase can promote plant growth and help to feed the world. *Microbiological Research, 169*(1), 30–39.

Gong, A. D., Li, H. P., Yuan, Q. S., Song, X. S., Yao, W., He, W. J., et al. (2015). Antagonistic mechanism of iturin a and plipastatin a from *Bacillus amyloliquefaciens* S76–3 from wheat spikes against *Fusarium graminearum*. *PLoS One, 10*(2), e0116871.

Gonzalez, E., Fernandez, M. R., Larroy, C., Sola, L., Pericas, M. A., Pares, X., et al. (2000). Characterization of a (2R,3R)-2,3-butanediol dehydrogenase as the *Saccharomyces cerevisiae* YAL060W gene product. Disruption and induction of the gene. *Journal of Biological Chemistry, 275*(46), 35876–35885.

Goupry, S., Rochut, N., Robins, R. J., & Gentil, E. (2000). Evaluation of solid-phase microextraction for the isotopic analysis of volatile compounds produced during fermentation by lactic acid bacteria. *Journal of Agricultural and Food Chemistry, 48*(6), 2222–2227.

Graus, M., Müller, M., & Hansel, A. (2010). High resolution PTR-TOF: quantification and formula confirmation of VOC in real time. *Journal of the American Society for Mass Spectrometry, 21*(6), 1037–1044.

Groenhagen, U., Baumgartner, R., Bailly, A., Gardiner, A., Eberl, L., Schulz, S., et al. (2013). Production of bioactive volatiles by different *Burkholderia ambifaria* strains. *Journal of Chemical Ecology, 39*(7), 892–906.

Guenther, A., Hewitt, C. N., Erickson, D., Fall, R., Geron, C., Graedel, T., et al. (1995). A global-model of natural volatile organic-compound emissions. *Journal of Geophysical Research Atmospheres, 100*(D5), 8873–8892.

Gutierrez-Luna, F. M., Lopez-Bucio, J., Altamirano-Hernandez, J., Valencia-Cantero, E., de la Cruz, H. R., & Macias-Rodriguez, L. (2010). Plant growth-promoting rhizobacteria modulate root-system architecture in *Arabidopsis thaliana* through volatile organic compound emission. *Symbiosis, 51*(1), 75–83.

Gutierrez-Manero, F. J., Ramos-Solano, B., Probanza, A., Mehouachi, J., Tadeo, F. R., & Talon, M. (2001). The plant-growth-promoting rhizobacteria *Bacillus pumilus* and *Bacillus licheniformis* produce high amounts of physiologically active gibberellins. *Physiologia Plantarum, 111*(2), 206–211.

Haas, D., & Defago, G. (2005). Biological control of soil-borne pathogens by fluorescent pseudomonads. *Nature Reviews Microbiology, 3*(4), 307–319.

Han, S. H., Lee, S. J., Moon, J. H., Park, K. H., Yang, K. Y., Cho, B. H., et al. (2006). GacS-dependent production of 2R, 3R-butanediol by *Pseudomonas chlororaphis* O6 is a major determinant for eliciting systemic resistance against *Erwinia carotovora* but not against *Pseudomonas syringae* pv. *tabaci* in tobacco. *Molecular Plant Microbe Interactions, 19*(8), 924–930.

Henry, G., Deleu, M., Jourdan, E., Thonart, P., & Ongena, M. (2011). The bacterial lipopeptide surfactin targets the lipid fraction of the plant plasma membrane to trigger immune-related defence responses. *Cellular Microbiology, 13*(11), 1824–1837.

Hernandez-Leon, R., Rojas-Solis, D., Contreras-Perez, M., Orozco-Mosqueda, M. D., Macias-Rodriguez, L. I., Reyes-de la Cruz, H., et al. (2015). Characterization of the antifungal and plant growth-promoting effects of diffusible and volatile organic compounds produced by *Pseudomonas fluorescens* strains. *Biological Control, 81*, 83–92.

Hohn-Bentz, H., & Radler, F. (1978). Bacterial 2,3-butanediol dehydrogenases. *Archives of Microbiology, 116*(2), 197–203.

Holzinger, R., Kasper-Giebl, A., Staudinger, M., Schauer, G., & Röckmann, T. (2010). Analysis of the chemical composition of organic aerosol at the Mt. Sonnblick observatory using a novel high mass resolution thermal-desorption proton-transfer-reaction mass-spectrometer (hr-TD-PTR-MS). *Atmospheric Chemistry and Physics, 10*(20), 10111–10128.

Huang, C.-J., Tsay, J.-F., Chang, S.-Y., Yang, H.-P., Wu, W.-S., & Chen, C.-Y. (2012). Dimethyl disulfide is an induced systemic resistance elicitor produced by *Bacillus cereus* C1L. *Pest Management Science, 68*(9), 1306–1310.

Hughes, D. T., & Sperandio, V. (2008). Inter-kingdom signalling: communication between bacteria and their hosts. *Nature Reviews Microbiology, 6*(2), 111–120.

Hunziker, L., Bonisch, D., Groenhagen, U., Bailly, A., Schulz, S., & Weisskopf, L. (2015). Pseudomonas strains naturally associated with potato plants produce volatiles with high potential for inhibition of *Phytophthora infestans*. *Applied and Environmental Microbiology, 81*(3), 821–830.

Inamdar, A. A., & Bennett, J. W. (2014). A common fungal volatile organic compound induces a nitric oxide mediated inflammatory response in *Drosophila melanogaster*. *Scientific Reports, 4*, 3833.

Inamdar, A. A., Hossain, M. M., Bernstein, A. I., Miller, G. W., Richardson, J. R., & Bennett, J. W. (2013). Fungal-derived semiochemical 1-octen-3-ol disrupts dopamine packaging and causes neurodegeneration. *Proceedings of the National Academy of Sciences of the United States of America, 110*(48), 19561–19566.

IR4-PALADIN. (2015). http://ir4.rutgers.edu/FoodUse/FUWorkshop/IndustryTalks/2010 FUWcompany presentations/UPIPaladinfumigantIR-4updateFUW2010.pdf Accessed 08.03.2015.

Jordan, A., Haidacher, S., Hanel, G., Hartungen, E., Märk, L., Seehauser, H., et al. (2009). A high resolution and high sensitivity proton-transfer-reaction time-of-flight mass spectrometer (PTR-TOF-MS). *International Journal of Mass Spectrometry, 286*(2), 122–128.

Junker, R. R., & Tholl, D. (2013). Volatile organic compound mediated interactions at the plant-microbe interface. *Journal of Chemical Ecology, 39*(7), 810–825.

Kai, M., Crespo, E., Cristescu, S. M., Harren, F. J., Francke, W., & Piechulla, B. (2010). Serratia odorifera: analysis of volatile emission and biological impact of volatile compounds on *Arabidopsis thaliana*. *Applied Microbiology and Biotechnology, 88*(4), 965–976.

Kai, M., Haustein, M., Molina, F., Petri, A., Scholz, B., & Piechulla, B. (2009). Bacterial volatiles and their action potential. *Applied Microbiology and Biotechnology, 81*(6), 1001–1012.

Khong, N. G., Randoux, B., Deravel, J., Tisserant, B., Tayeh, C., Coutte, F., et al. (2013). Induction of resistance in wheat by bacterial cyclic lipopeptides. *Communications in Agricultural and Applied Biological Sciences, 78*(3), 479–487.

Kinkel, L. L., Bakker, M. G., & Schlatter, D. C. (2011). A coevolutionary framework for managing disease-suppressive soils. *Annual Review of Phytopathology, 49*, 47–67.

Kishimoto, K., Matsui, K., Ozawa, R., & Takabayashi, J. (2007). Volatile 1-octen-3-ol induces a defensive response in *Arabidopsis thaliana*. *Journal of General Plant Pathology, 73*(1), 35–37.

Kloepper, J. W., Leong, J., Teintze, M., & Schroth, M. N. (1980). Enhanced plant-growth by siderophores produced by plant growth-promoting rhizobacteria. *Nature, 286*(5776), 885–886.

Kloepper, J. W., Leong, J., Teintze, M., & Schroth, M. N. (1981). Enhanced plant-growth by siderophores produced by plant growth-promoting rhizobacteria. *Phytopathology, 71*(2), 231.

Kloepper, J. W., Ryu, C. M., & Zhang, S. A. (2004). Induced systemic resistance and promotion of plant growth by *Bacillus* spp. *Phytopathology, 94*(11), 1259–1266.

Koglin, A., & Walsh, C. T. (2009). Structural insights into nonribosomal peptide enzymatic assembly lines. *Natural Product Reports, 26*(8), 987–1000.

Kokalis-Burelle, N., Rosskopf, E. N., & Hartman, R. D. (2010). Evaluation of soil treatments for control of *Meloidogyne arenaria* in Caladium tubers (*Caladium* × *hortulanum*) and nematode susceptibility of selected cultivars. *Nematropica, 40*(2), 177–189.
Kwon, Y. S., Ryu, C. M., Lee, S., Park, H. B., Han, K. S., Lee, J. H., et al. (2010). Proteome analysis of *Arabidopsis* seedlings exposed to bacterial volatiles. *Planta, 232*(6), 1355–1370.
KyungSeok, P., YungEun, C., Balaraju, K., SeWeon, L., JinWoo, P., Reddy, M., et al. (2013). Induction of systemic resistance against *Phytophthora capsici* and *Pectobacterium carotovorum* SCC1 by treatment of iturin derivatives in chili-pepper. In , *Proceedings of 3rd Asian Conference on Plant Growth-Promoting Rhizobacteria (PGPR) and Other Microbials, Manila, Philippines, 21–24 April, 2013Recent advances in biofertilizers and biofungicides (PGPR) for sustainable agriculture* (pp. 342–349). Asian PGPR Society for Sustainable Agriculture.
Lakshmanan, V., Selvaraj, G., & Bais, H. P. (2014). Functional soil microbiome: belowground solutions to an aboveground problem. *Plant Physiology, 166*(2), 689–700.
Lebuhn, M., Heulin, T., & Hartmann, A. (1997). Production of auxin and other indolic and phenolic compounds by *Paenibacillus polymyxa* strains isolated from different proximity to plant roots. *FEMS Microbiology Ecology, 22*(4), 325–334.
Leclere, V., Bechet, M., Adam, A., Guez, J. S., Wathelet, B., Ongena, M., et al. (2005). Mycosubtilin overproduction by *Bacillus subtilis* BBG100 enhances the organism's antagonistic and biocontrol activities. *Applied and Environmental Microbiology, 71*(8), 4577–4584.
Lee, B., Farag, M. A., Park, H. B., Kloepper, J. W., Lee, S. H., & Ryu, C. M. (2012). Induced resistance by a long-chain bacterial volatile: elicitation of plant systemic defense by a C13 volatile produced by *Paenibacillus polymyxa*. *PLoS One,* 7(11), e48744.
Lemfack, M. C., Nickel, J., Dunkel, M., Preissner, R., & Piechulla, B. (2014). mVOC: a database of microbial volatiles. *Nucleic Acids Research, 42*(Database issue), D744–D748.
Li, W., Rokni-Zadeh, H., De Vleeschouwer, M., Ghequire, M. G., Sinnaeve, D., Xie, G. L., et al. (2013). The antimicrobial compound xantholysin defines a new group of *Pseudomonas* cyclic lipopeptides. *PLoS One, 8*(5), e62946.
Liang, Z., An, T., Li, G., & Zhang, Z. (2014). Aerobic biodegradation of odorous dimethyl disulfide in aqueous medium by isolated *Bacillus cereus* GIGAN2 and identification of transformation intermediates. *Bioresource Technology, 175C*, 563–568.
Liu, R. S., Zhou, H., Li, H. M., Yuan, Z. P., Chen, T., & Tang, Y. J. (2013). Metabolism of L-methionine linked to the biosynthesis of volatile organic sulfur-containing compounds during the submerged fermentation of *Tuber melanosporum*. *Applied Microbiology and Biotechnology, 97*(23), 9981–9992.
van Loon, L. C., Bakker, P. A., & Pieterse, C. M. (1998). Systemic resistance induced by rhizosphere bacteria. *Annual Review of Phytopathology, 36*, 453–483.
López-Aranda, J. (2012). Methyl bromide alternatives (chemical and non-chemical) for strawberry cultivation in conventional areas of production. In *VII International Strawberry Symposium 1049* (pp. 77–88).
Lugtenberg, B., & Kamilova, F. (2009). Plant-growth-promoting rhizobacteria. *Annual Review of Microbiology, 63*, 541–556.
Maksimov, I. V., Abizgil'dina, R. R., & Pusenkova, L. I. (2011). Plant growth promoting rhizobacteria as alternative to chemical crop protectors from pathogens (review). *Applied Biochemistry and Microbiology, 47*(4), 333–345.
Mao, L., Yan, D., Wang, Q., Li, Y., Ouyang, C., Liu, P., et al. (2014). Evaluation of the combination of dimethyl disulfide and dazomet as an efficient methyl bromide alternative for cucumber production in China. *Journal of Agricultural and Food Chemistry, 62*(21), 4864–4869.
Matsui, K. (2006). Green leaf volatiles: hydroperoxide lyase pathway of oxylipin metabolism. *Current Opinion in Plant Biology, 9*(3), 274–280.

McAvoy, T., & Freeman, J. H. (2013). Yellow nutsedge (*Cyperus esculentus*) control with reduced rates of dimethyl disulfide in combination with totally impermeable film. *Weed Technology, 27*(3), 515–519.

Meldau, D. G., Long, H. H., & Baldwin, I. T. (2012). A native plant growth promoting bacterium, *Bacillus* sp. B55, rescues growth performance of an ethylene-insensitive plant genotype in nature. *Frontiers in Plant Science, 3*, 112.

Meldau, D. G., Meldau, S., Hoang, L. H., Underberg, S., Wunsche, H., & Baldwin, I. T. (2013). Dimethyl disulfide produced by the naturally associated bacterium *bacillus* sp B55 promotes *Nicotiana attenuata* growth by enhancing sulfur nutrition. *Plant Cell, 25*(7), 2731–2747.

Mendes, R., Garbeva, P., & Raaijmakers, J. M. (2013). The rhizosphere microbiome: significance of plant beneficial, plant pathogenic, and human pathogenic microorganisms. *FEMS Microbiology Reviews, 37*(5), 634–663.

Mendes, R., Kruijt, M., de Bruijn, I., Dekkers, E., van der Voort, M., Schneider, J. H., et al. (2011). Deciphering the rhizosphere microbiome for disease-suppressive bacteria. *Science, 332*(6033), 1097–1100.

Meruva, N. K., Penn, J. M., & Farthing, D. E. (2004). Rapid identification of microbial VOCs from tobacco molds using closed-loop stripping and gas chromatography/time-of-flight mass spectrometry. *Journal of Industrial Microbiology & Biotechnology, 31*(10), 482–488.

van de Mortel, J. E., Tran, H., Govers, F., & Raaijmakers, J. M. (2009). Cellular responses of the late blight pathogen *Phytophthora infestans* to cyclic lipopeptide surfactants and their dependence on G proteins. *Applied and Environmental Microbiology, 75*(15), 4950–4957.

Muller, H., Zachow, C., Alavi, M., Tilcher, R., Krempl, P. M., Thallinger, G. G., et al. (2013). Complete genome sequence of the sugar beet endophyte *Pseudomonas poae* RE*1-1-14, a disease-suppressive bacterium. *Genome Announcements, 1*(2), e0002013.

Navarro, L., Dunoyer, P., Jay, F., Arnold, B., Dharmasiri, N., Estelle, M., et al. (2006). A plant miRNA contributes to antibacterial resistance by repressing auxin signaling. *Science, 312*(5772), 436–439.

Ongena, M., & Jacques, P. (2008). *Bacillus lipopeptides*: versatile weapons for plant disease biocontrol. *Trends in Microbiology, 16*(3), 115–125.

Ongena, M., Jourdan, E., Adam, A., Paquot, M., Brans, A., Joris, B., et al. (2007). Surfactin and fengycin lipopeptides of *Bacillus subtilis* as elicitors of induced systemic resistance in plants. *Environmental Microbiology, 9*(4), 1084–1090.

Park, H. B., Lee, B., Kloepper, J. W., & Ryu, C. M. (2013). One shot-two pathogens blocked: exposure of *Arabidopsis* to hexadecane, a long chain volatile organic compound, confers induced resistance against both *Pectobacterium carotovorum* and *Pseudomonas syringae*. *Plant Signaling & Behavior, 8*(7), e24619.

Patel, H., Tscheka, C., Edwards, K., Karlsson, G., & Heerklotz, H. (2011). All-or-none membrane permeabilization by fengycin-type lipopeptides from *Bacillus subtilis* QST713. *Biochimica et Biophysica Acta-Biomembranes, 1808*(8), 2000–2008.

Pathak, K. V., Bose, A., & Keharia, H. (2014). Characterization of novel lipopeptides produced by *Bacillus tequilensis* P15 using liquid chromatography coupled electron spray ionization tandem mass spectrometry (LC-ESI-MS/MS). *International Journal of Peptide Research and Therapeutics, 20*(2), 133–143.

Patten, C. L., & Glick, B. R. (2002). Role of *Pseudomonas putida* indoleacetic acid in development of the host plant root system. *Applied and Environmental Microbiology, 68*(8), 3795–3801.

Penuelas, J., Asensio, D., Tholl, D., Wenke, K., Rosenkranz, M., Piechulla, B., et al. (2014). Biogenic volatile emissions from the soil. *Plant Cell and Environment, 37*(8), 1866–1891.

Pieterse, C. M., Zamioudis, C., Berendsen, R. L., Weller, D. M., Van Wees, S. C., & Bakker, P. A. (2014). Induced systemic resistance by beneficial microbes. *Annual Review of Phytopathology, 52*, 347–375.

Popova, A. A., Koksharova, O. A., Lipasova, V. A., Zaitseva, J. V., Katkova-Zhukotskaya, O. A., Eremina, S., et al. (2014). Inhibitory and toxic effects of volatiles emitted by strains of *Pseudomonas* and *Serratia* on growth and survival of selected microorganisms, *Caenorhabditis elegans*, and *Drosophila melanogaster*. *BioMed Research International, 2014*, 125704.

Raaijmakers, J. M., De Bruijn, I., Nybroe, O., & Ongena, M. (2010). Natural functions of lipopeptides from *Bacillus* and *Pseudomonas*: more than surfactants and antibiotics. *FEMS Microbiology Reviews, 34*(6), 1037–1062.

Raaijmakers, J. M., & Mazzola, M. (2012). Diversity and natural functions of antibiotics produced by beneficial and plant pathogenic bacteria. *Annual Review of Phytopathology, 50*, 403–424.

Rahman, A., Uddin, W., & Wenner, N. G. (2014). Induced systemic resistance responses in perennial ryegrass against *Magnaporthe oryzae* elicited by semi-purified surfactin lipopeptides and live cells of *Bacillus amyloliquefaciens*. *Molecular Plant Pathology*. http://dx.doi.org/10.1111/mpp.12209.

Roongsawang, N., Washio, K., & Morikawa, M. (2010). Diversity of nonribosomal peptide synthetases involved in the biosynthesis of lipopeptide biosurfactants. *International Journal of Molecular Sciences, 12*(1), 141–172.

Rosskopf, E. N., Kokalis-Burelle, N., McSorley, R., & Skvarch, E. (2010). *Optimizing alternative fumigant applications for ornamental production in Florida*.

Rout, M. E., & Southworth, D. (2013). The root microbiome influences scales from molecules to ecosystems: the unseen majority. *American Journal of Botany, 100*(9), 1689–1691.

Ruckert, C., Blom, J., Chen, X., Reva, O., & Borriss, R. (2011). Genome sequence of *B. amyloliquefaciens* type strain DSM7(T) reveals differences to plant-associated *B. amyloliquefaciens* FZB42. *Journal of Biotechnology, 155*(1), 78–85.

Rudrappa, T., Biedrzycki, M. L., Kunjeti, S. G., Donofrio, N. M., Czymmek, K. J., Pare, P. W., et al. (2010). The rhizobacterial elicitor acetoin induces systemic resistance in *Arabidopsis thaliana*. *Communicative & Integrative Biology, 3*(2), 130–138.

Ryu, C. M., Farag, M. A., Hu, C. H., Reddy, M. S., Kloepper, J. W., & Pare, P. W. (2004). Bacterial volatiles induce systemic resistance in *Arabidopsis*. *Plant Physiology, 134*(3), 1017–1026.

Ryu, C. M., Farag, M. A., Hu, C. H., Reddy, M. S., Wei, H. X., Pare, P. W., et al. (2003). Bacterial volatiles promote growth in *Arabidopsis*. *Proceedings of the National Academy of Sciences of the United States of America, 100*(8), 4927–4932.

Sasanelli, N., Dongiovanni, C., Santori, A., & Myrta, A. (2014). Control of the root-knot nematode *Meloidogyne incognita* by dimethyl disulfide (DMDS) applied in drip irrigation on melon and tomato in Apulia and Basilicata (Italy). *Acta Horticulturae, 1044*, 401–404.

Schroth, M. N., & Hancock, J. G. (1982). Disease-suppressive soil and root-colonizing bacteria. *Science, 216*(4553), 1376–1381.

Schulz, S., & Dickschat, J. S. (2007). Bacterial volatiles: the smell of small organisms. *Natural Product Reports, 24*(4), 814–842.

Stein, T. (2005). *Bacillus subtilis* antibiotics: structures, syntheses and specific functions. *Molecular Microbiology, 56*(4), 845–857.

Strange, R. N., & Scott, P. R. (2005). Plant disease: a threat to global food security. *Annual Review of Phytopathology, 43*, 83–116.

Tran, H., Ficke, A., Asiimwe, T., Hofte, M., & Raaijmakers, J. M. (2007). Role of the cyclic lipopeptide massetolide A in biological control of *Phytophthora infestans* and in colonization of tomato plants by *Pseudomonas fluorescens*. *New Phytologist, 175*(4), 731–742.

Vanderplank, J. E. (2012). *Disease resistance in plants*. Elsevier.

Vas, G., & Vekey, K. (2004). Solid-phase microextraction: a powerful sample preparation tool prior to mass spectrometric analysis. *Journal of Mass Spectrometry, 39*(3), 233–254.

Velázquez-Becerra, C., Macías-Rodríguez, L. I., López-Bucio, J., Altamirano-Hernández, J., Flores-Cortez, I., & Valencia-Cantero, E. (2011). A volatile organic compound analysis from *Arthrobacter agilis* identifies dimethylhexadecylamine, an amino-containing lipid modulating bacterial growth and *Medicago sativa* morphogenesis in vitro. *Plant and Soil, 339*(1–2), 329–340.

Velazquez-Becerra, C., Macias-Rodriguez, L. I., Lopez-Bucio, J., Flores-Cortez, I., Santoyo, G., Hernandez-Soberano, C., et al. (2013). The rhizobacterium *Arthrobacter agilis* produces dimethylhexadecylamine, a compound that inhibits growth of phytopathogenic fungi in vitro. *Protoplasma, 250*(6), 1251–1262.

Vespermann, A., Kai, M., & Piechulla, B. (2007). Rhizobacterial volatiles affect the growth of fungi and *Arabidopsis thaliana*. *Applied and Environmental Microbiology, 73*(17), 5639–5641.

Vollenbroich, D., Ozel, M., Vater, J., Kamp, R. M., & Pauli, G. (1997). Mechanism of inactivation of enveloped viruses by the biosurfactant surfactin from *Bacillus subtilis*. *Biologicals, 25*(3), 289–297.

van der Vossen, E. A., Gros, J., Sikkema, A., Muskens, M., Wouters, D., Wolters, P., et al. (2005). The Rpi-blb2 gene from *Solanum bulbocastanum* is an Mi-1 gene homolog conferring broad-spectrum late blight resistance in potato. *Plant Journal, 44*(2), 208–222.

Waewthongrak, W., Leelasuphakul, W., & McCollum, G. (2014). Cyclic LIPopeptides from *Bacillus subtilis* ABS-S14 elicit defense-related gene expression in citrus fruit. *PLoS One, 9*(10), e109386.

Walker, T. S., Bais, H. P., Grotewold, E., & Vivanco, J. M. (2003). Root exudation and rhizosphere biology. *Plant Physiology, 132*(1), 44–51.

Walters, D. R., & McRoberts, N. (2006). Plants and biotrophs: a pivotal role for cytokinins? *Trends in Plant Science, 11*(12), 581–586.

Wang, D., Pajerowska-Mukhtar, K., Culler, A. H., & Dong, X. N. (2007). Salicylic acid inhibits pathogen growth in plants through repression of the auxin signaling pathway. *Current Biology, 17*(20), 1784–1790.

Weller, D. M., & Thomashow, L. S. (1994). Current challenges in introducing beneficial microorganisms into the rhizosphere. In F. O'Gara, D. N. Dowling, & B. Boesten (Eds.), *Molecular ecology of rhizosphere microorganisms: Biotechnology and the release of GMOs* (pp. 1–18).

Xiao, Z., & Xu, P. (2007). Acetoin metabolism in bacteria. *Critical Reviews in Microbiology, 33*(2), 127–140.

Yamamoto, S., Shiraishi, S., & Suzuki, S. (2015). Are cyclic lipopeptides produced by *Bacillus amyloliquefaciens* S13-3 responsible for the plant defence response in strawberry against *Colletotrichum gloeosporioides? Letters in Applied Microbiology, 60*(4), 379–386.

Yuan, J., Raza, W., Shen, Q., & Huang, Q. (2012). Antifungal activity of *Bacillus amyloliquefaciens* NJN-6 volatile compounds against *Fusarium oxysporum* f. sp. *cubense*. *Applied and Environmental Microbiology, 78*(16), 5942–5944.

Yuan, J., Zhang, F., Wu, Y., Zhang, J., Raza, W., Shen, Q., et al. (2014). Recovery of several cell pellet-associated antibiotics produced by Bacillus amyloliquefaciens NJN-6. *Letters in Applied Microbiology, 59*(2), 169–176.

Yu, S. M., & Lee, Y. H. (2013). Effect of light quality on *Bacillus amyloliquefaciens* JBC36 and its biocontrol efficacy. *Biological Control, 64*(3), 203–210.

Zachow, C., Ghazaleh, J., de Bruijn, I., Song, C., Ianni, F., Pataj, Z., et al. (2015). The novel lipopeptide poaeamide of the endophyte *Pseudomonas poae* RE*1-1-14 is involved in pathogen suppression and root colonization. *Molecular Plant Microbe Interactions*. http://dx.doi.org/10.1094/MPMI-12-14-0406-R.

Zanón, M., Gutiérrez, L., & Myrta, A. (2014). Spanish experiences with dimethyl disulfide (DMDS) on the control of root-knot nematodes, *Meloidogyne* spp., in fruiting vegetables in protected crops. In *VIII International Symposium on Chemical and Non-Chemical Soil and Substrate Disinfestation 1044* (pp. 421–425).

Zhang, X., Bao, T., Rao, Z., Yang, T., Xu, Z., Yang, S., et al. (2014). Two-stage pH control strategy based on the pH preference of acetoin reductase regulates acetoin and 2,3-butanediol distribution in *Bacillus subtilis*. *PLoS One, 9*(3), e91187.

Zhao, L. J., Yang, X. N., Li, X. Y., Mu, W., & Liu, F. (2011). Antifungal, insecticidal and herbicidal properties of volatile components from *Paenibacillus polymyxa* strain BMP-11. *Agricultural Sciences in China, 10*(5), 728–736.

Zou, C., Li, Z., & Yu, D. (2010). *Bacillus megaterium* strain XTBG34 promotes plant growth by producing 2-pentylfuran. *The Journal of Microbiology, 48*(4), 460–466.

CHAPTER ELEVEN

Flying Whispers of Inter-kingdom Conversation: A Complementary Perspective of Plant and Bacterial Volatile Signals

Gopinath Selvaraj
Department of Plant and Soil Sciences, University of Delaware, Newark, DE, USA
E-mail: gopiselva@gmail.com

Contents

1. Introduction 286
2. Volatile Organic Compounds from Plants 287
3. Influence of Plant VOCs on Rhizosphere and Phyllosphere Bacteria 290
4. VOCs from Bacteria and Their Effects on Plants 292
5. Bacterial Volatiles in QS and QQ 299
6. Commonness and Distinctions among Plant and Bacterial VOCs 300
7. Future Perspectives 303
Acknowledgement 304
References 304

Abstract

Volatile organic compounds (VOCs) are released and utilized by various organisms starting from bacteria to higher organisms. The VOCs produced from plants are involved in self-repairing against biological and nonbiological insults as well as in communicating with neighbouring plants and other organisms for reasons such as priming and pollination. The VOCs produced by plant growth-promoting rhizobacteria have immense capability of modulating plant growth and development, eliciting induced systemic resistance against biotic and abiotic stressors, and conferring stress tolerance. These bacterial volatiles also have direct anti-phytopathogenic (bacterial and fungal) and anti-phytoparasitic nematode activities, which in turn benefit plants indirectly. Additionally, bacterial volatiles are found to be involved in Quorum Sensing (QS), a coordinated multicellular behaviour. Sometimes, the QS system interferes with the production of bacterial VOCs. The metabolic investment of the bacteria in providing such benefits to plants is in expectation of obtaining mutual advantage for space and other resources. Recent studies demonstrate that bacterial interaction can contribute to the characteristic volatile composition of plant essential oils, which were previously thought to be solely of plant origin. Bacterial interaction can also manipulate emission

Advances in Botanical Research, Volume 75
ISSN 0065-2296
http://dx.doi.org/10.1016/bs.abr.2015.08.001

of plant stress volatiles. Increasing evidence shows the multiple potentials of volatile signalling in plant–bacteria interactions. Recent research efforts aim to investigate a 'Volatile Conversation' in plant–bacterium interactions, in which dynamic and responsive exchanges of volatiles occur. The present chapter aims to address the important roles of plant and bacterial volatiles during interaction, and also discusses the future directions to better utilize and understand these valuable flying signals.

1. INTRODUCTION

Throughout evolution, biological entities tend to be highly diversified and thus enrich the earth with distinct organisms of various characteristics. At the same time, they communicate and interact for the benefit of their own selves and others in achieving better survival of different communities in the same territory. In this cooperative survival, every organism interacts in multiple ways, including emission of volatile organic compounds (VOCs) for communication. VOCs have unique characteristics of being less susceptible to various forms of inactivation and having a higher sphere of influence, or diffusion radius, when compared to non-VOCs (Stotzky & Schenck, 1976). VOCs are delineated as compounds of low molecular weight (<300 g/mol), higher vapour pressures under normal conditions, and include alcohols, aldehydes, ketones, terpenes, fatty acid derivatives and other hydrocarbons (Dudareva, Klempien, Muhlemann, & Kaplan, 2013; Dudareva, Pichersky, & Gershenzon, 2004; Ortíz-Castro, Contreras-Cornejo, Macías-Rodríguez, & López-Bucio, 2009).

Many of these VOCs can be perceived by the human nose, while others cannot be sensed. From prokaryotic to eukaryotic organisms, the VOCs play crucial roles in intra-organismal and inter-organismal communications, and help to maintain the balance of the ecosystem (Peñuelas et al., 2014). Plant VOCs are the most diverse and are released into the atmosphere in bulk volumes (some of them are released as waste, and others mediate ecological functions). To date, >1700 VOCs have been identified from 90 different plants (Baldwin, Halitschke, Paschold, von Dahl, & Preston, 2006; Knudsen, Eriksson, Gershenzon, & Stahl, 2006). Beyond defining the chemical landscape of numerous ecosystems, plant VOCs are involved in many activities such as attracting pollinators and seed dispersers, above- and below-ground defence against herbivores, protection against biotic and abiotic stresses, and other plant–plant signalling (Dudareva et al., 2013; Maffei, Gertschb, & Appendinoc, 2011). More importantly, it was realized from studies of transcriptome and signal cascade analysis that plants might be eavesdropping on messages embedded in the emitted VOCs from herbivore-attacked neighbouring plants to prime direct and indirect defences (Baldwin et al., 2006).

Among many inter-kingdom interactions, bacterial volatiles, especially of plant-associated bacteria, have immense potential to regulate plant growth and metabolism in many ways. More than 1000 volatile compounds were reported to be released by different bacteria, and they influence other kingdoms such as fungi and animals (Kai et al., 2009; Lemfack, Nickel, Dunkel, Preissner, & Piechulla, 2014; Schulz & Dickschat, 2007). In plants, bacterial VOCs have shown to promote plant growth, to promote resistance to biotic and abiotic stresses, and also to suppress the growth of pathogens (Cho et al., 2008; Rudrappa, Splaine, Biedrzycki, & Bais, 2008; Ryu et al., 2004, 2003; Zhang, Kim et al., 2008). The production of VOCs is found to be strain-specific from tested plant growth-promoting rhizobacteria (PGPRs) (Farag, Ryu, Sumner, & Paré, 2006) and opens up a possibility to use these specific profiles of VOCs for classification of the beneficial bacteria (Farag, Zhang, & Ryu, 2013). Likewise, the profile of VOCs from soil could potentially be exploited for detecting microbial composition (Insam & Seewald, 2010).

Thus, from many scientific attempts, it became obvious that both plants and bacteria have the ability to produce VOCs for many purposes including influencing and interacting with other organisms. A comprehensive list of plant and bacterial volatiles may also be obtained from the publically available 'Superscent' and 'mVOC' databases (Dunkel et al., 2009; Lemfack et al., 2014). However, the commonalities and distinctions of plant and bacterial VOCs, especially during interactions, their potentials and mutual benefits were not explored towards finding ways to better exploit them for agricultural and other beneficial applications. Hence, the focus of this chapter is on the comparative analysis of these inter-kingdom VOCs, their signalling mechanisms and future directions. This chapter starts with separate sections on plant and bacterial VOCs, and then on the comparative aspects in relevance to plant–bacteria interactions along with their functions. Finally, the chapter ends with future perspectives.

2. VOLATILE ORGANIC COMPOUNDS FROM PLANTS

Plants and their flowers can fill the surrounding atmosphere with specific smells. These smells are due to a specific blend of VOCs released from the floral or vegetative parts of plants. Further, the essential oils derived from plants also represent volatile components that give unique flavours. The term 'plant volatilome' collectively describes the essential oils and VOCs emitted by a plant (Bicchi & Maffei, 2012; Maffei, Mithofer, & Boland, 2007), as being noninvasive biomarkers that can be utilized for phenotyping of live plants during different stages or conditions such as ripening, senescence, storage, interactions and stress (biotic and abiotic) (Niederbacher, Winkler, & Schnitzler, 2015). Plant volatiles are released into the atmosphere in large volumes, especially from forests and dense vegetation. As a result, they occupy a major portion of biologically synthesized VOCs in the atmosphere (Goldstein & Galbally, 2007). The emission of VOCs by plants is a major contributor to the earth's atmosphere, accounting for 1 Pg C per year, in which half the volume is occupied by isoprene (a monoterpene) alone (Guenther et al., 2012). Some of the reactive species from the breakdown of isoprene is responsible for the production of ozone, while monoterpenes along with sesquiterpenes lead to the formation of cloud condensation nuclei, thus becoming a crucial factor in the earth's atmosphere and its composition (Loreto, Dicke, Schnitzler, & Turlings, 2014). The total VOCs released from vegetation is greater than the VOCs from petroleum production and use, and plant VOCs are likely the reason behind the blue haze found above many forest regions. Every part of a plant has the ability to synthesize volatile compounds including roots (Baetz & Martinoia, 2014; Maffei et al., 2011) and the profile of VOC production changes according to each developmental stage (Niederbacher et al., 2015; Niinemets e al., 2010). The floral VOCs mainly aid in attracting pollinators, while at the same time, they have also shown other functions such as antimicrobial activities. The plant roots are also capable of releasing a considerable amount of VOCs during normal and stress conditions, and establish a multi-trophic network hub (Bais, Weir, Perry, Gilroy, & Vivanco, 2006; Chen et al., 2004; Peñuelas et al., 2014; Rasmann et al., 2005; Robert et al., 2012; Steeghs et al., 2004).

Most of the plant volatiles can be categorized into any of the following classes: terpenoids, fatty acid derivatives including oxylipin products, green leaf volatiles (GLVs), benzenoids and phenylpropanoids, methyl salicylate (MeSA), methyl jasmonate (MeJA), indole, C_5-branched compounds and various nitrogen- and sulphur-containing compounds (Dudareva et al., 2004; Holopainen & Blande, 2012; Maffei et al., 2011). GLVs are six-carbon aldehydes, alcohols and esters. Ethylene is another important volatile plant hormone involved in various functions, especially, in fruit ripening and senescence. MeSA and MeJA are derived from the diffusible plant hormones, namely, salicylic acid and jasmonic acid. However, these derivatives of plant hormones are highly involved in defence mechanisms rather than growth regulation like their diffusible counterparts (Holopainen & Blande, 2012; Delano-Frier, Pearce, Huffaker, & Stratmann, 2013). Likewise, indole was very recently found to be an emerging and rapid herbivore-induced volatile priming signal (Erb et al., 2015). GLVs, terpenoids, MeSA and MeJA are the major plant defence-related volatiles that function during physical damage, as well as herbivore and pathogen attacks (Holopainen & Blande, 2012). Most of the volatile signals are released with the intent to carry messages across long distances.

Recent scientific and technical advancements on plant volatiles unveiled multifaceted potentials and applications. Plant volatiles, their biosynthesis, functions and signalling mechanisms were extensively reviewed (Baldwin et al., 2006; Dudareva et al., 2013, 2004; Loreto et al., 2014; Holopainen & Blande, 2012; Maffei, 2010; Maffei et al., 2011). A special issue was recently (i.e. in 2010) dedicated in the journal 'Trends in Plant Science' for plant volatiles (Volume-15; Issue — 3, Title: Induced biogenic volatile organic compounds from plants). During stress conditions, some plant VOCs act as infochemicals (excluding the VOCs that entered the open atmosphere) and transmitted specific messages to adjacent organisms for priming, inviting pollinators and defending against subsequent attacks. Although many studies have been undertaken in volatile-based plant—plant interactions and other tri-trophic interactions, the ecological relevance of laboratory experiments in real field conditions are required for translating laboratory work to future agricultural applications. Baldwin et al. (2006) proposed that comparing wild type with mutants and transgenic lines deficient in one or more VOC syntheses in response to biotic or abiotic stress under field conditions would clarify much. There are few studies in this direction carried out to use VOC deficient and constitutively (exogenous and endogenous) expressed gene mutant lines of plants for investigating herbivore attack, priming activities and attractiveness to predators of herbivores (Kos et al., 2013; Christensen et al., 2013). These mutants (of *Arabidopsis thaliana*, *Solanum lycopersicum* and *Zea mays*) are mainly in LOX (lipoxygenase) pathway products (GLVs and jasmonic acid) and terpenoids that are considered to be involved in herbivore-/pathogen-induced defence and priming. It was also clearly shown that herbivore attack in a plant host induces the release of certain VOCs, such as GLVs, which attract carnivorous predators against the herbivores (Bais et al., 2006). Those induced volatiles due to herbivore damage were shown to possess antimicrobial activities, which is likely to discourage microbial invasion and colonization in the wounded sites. Plants compromised in these volatiles are found to be highly vulnerable to herbivore and microbial attacks (Delano-Frier et al., 2013).

There are few studies that aim to elucidate specific receptors involved in perception and metabolism of VOCs. To date, most research has focused on elucidating the release of VOCs from plants. Shulaev, Silverman, and Raskin (1997) demonstrated that exposure of tobacco plants (infected with tobacco mosaic virus) to the VOC MeSA activates resistance in neighbouring uninfected plants, in which the perceived MeSA was converted back to its diffusible non-methylated form, salicylic acid that induces resistance. Similarly, Sugimoto et al. (2014) demonstrated that a herbivore-(common cutworm) infested plant host released a GLV, (*Z*)-3-hexenol, which was received and transformed directly into (*Z*)-3-hexenylvicianoside, which becomes responsible for the host resistance against the herbivore. A recent review summarizes the catabolism of plant volatiles in the context of plant health and interactions (Oikawa & Lerdau, 2013). Apart from the native production of VOCs from plants, a recent study demonstrated that some of the key components of the essential oil can be supplied by a bacterial partner (Del Giudice et al., 2008). In this study, phylogenetically diverse root-associated bacteria are responsible for the presence of certain in vivo volatile constituents of commercially valuable essential oil from Vetiver (*Vetiveria zizanioides* (L.) Nash). In support of this view, these bacteria were also able to metabolize or biotransform oil sesquiterpenes (of raw Vetiver oil) and release many essential oil constituents (especially of a volatile nature) found in many commercial vetiver oils. This is an entirely different scenario, where a direct metabolic contribution of VOCs is delivered to plants, bringing into question the evolutionary relationships in these tightly linked interactions. Hence, the area of plant volatile perception and subsequent metabolism deserves further research with high-throughput techniques. Another emerging field of study related to plant VOCs is their role in self and non-self-recognition, which is very crucial in an ecological context (Holopainen & Blande, 2012).

3. INFLUENCE OF PLANT VOCs ON RHIZOSPHERE AND PHYLLOSPHERE BACTERIA

To date, most studies have focused mainly on the negative effects of plant volatiles on bacterial growth and development, as well as the utility of VOCs in defence against bacterial infections, rather than focussing on the positive influence that some VOCs have on microbes. Plant volatiles and essential oils exhibit significant antibacterial activity, which are not discussed in this chapter. Instead other positive aspects of plant volatiles on bacteria and their symbiotic relationship are considered and discussed in detail, mostly in the rhizosphere and phyllosphere regions. The rhizosphere is a rich habitat supporting multi-kingdom interactions, influenced mainly by root exudates. The sphere of influence by volatile components (long distance diffusion) is larger than the one formed by nonvolatile components. Many of the VOCs from root exudates exert allelopathic effects like carvacrol and D-limonene, which inhibit the growth and seed germination of neighbouring plants (Maffei et al., 2011). In contradiction, root exudate volatiles can also mediate, support and regulate certain beneficial interactions. For instance, MeJA was found to aid and regulate the nodulation of legume roots by *Sinorhizobium meliloti* directly by inducing nod genes in the bacteria and indirectly by inducing (iso)flavonoid biosynthesis genes that in turn promote nod factor synthesis (Hause & Schaarschmidt, 2009). MeJA is also involved in the autoregulation of nodulation to maintain the symbiotic balance that was evidently demonstrated by the suppression of nodulation, when MeJA was applied in shoots of *Lotus japonicas* (Hause & Schaarschmidt, 2009; Nakagawa & Kawaguchi, 2006). Similarly, MeJA plays a key role by direct enhancement of fungal colonization and also indirectly by inducing enzymes involved in carbohydrate partitioning during the establishment of arbuscular mycorrhizae (Hause & Schaarschmidt, 2009).

Advancing from the previous knowledge of nonvolatile compounds involved in tri-trophic interactions, recent studies freshly unveiled important volatile and semi-volatile compounds of root exudates in interactions such as 1,8-cineole and rhizathalene A (Baetz & Martinoia, 2014). Kobayashi (2015) has recently demonstrated that MeSA, being a pathogen-induced root volatile, could provide a rhizobacterial friend (*Bacillus subtilis*) with 'the armor' (induced defence by means of capsule formation) and 'the sword' (induced production of antibiotics) to protect itself and the host plant against phytopathogens, thus forming a tightly interwoven relation. Two more important findings in this study include (i) the regulatory nature of MeSA on the colony morphology; (ii) MeSA was not induced in plants by *B. subtilis*, implying the exclusive origin of MeSA from the host in a pathogenic attack. In a recent study, successful enhancement of growth and tolerance against severe draught was shown by applying rhizosphere bacteria to wheat seedlings, in which the expression of plant stress volatiles was monitored

(Timmusk et al., 2014). From this study, three volatiles, that is, benzaldehyde, β-pinene and geranyl acetone were found to be differentially emitted. These volatiles were emitted at modest levels (equivalent to healthy control plants) in drought stressed plants that had been inoculated with bacteria, whereas these volatiles were released at higher concentrations in plants under drought stress alone. Thus, the VOCs strongly exhibit the potential to be used as a better, noninvasive biomarker to validate the effect of such bacterial applications. Another recent study demonstrated rhizobacteria-induced root production of a defence plant volatile (*E*)-β-caryophyllene, which is plausibly involved in reducing attack by a herbivore (Santos et al., 2014). Conversely, plant root exudate can tune the production of VOCs from a beneficial bacterium for the benefit of a plant host itself. In a recent proteomic analysis, maize root exudate was found to induce the highest fold of an enzyme, acetolactate synthase of a PGPR that is involved in the synthesis of acetoin (refer to Table 1 for acetoin's functions), by which a mutual support can be served to plant through growth promotion and induced systemic resistance (ISR) induction (Kierul et al., 2015). Sometimes, plant leaf volatiles such as *R*,*S*-linalool, 2-phenylethanol and nonanal were found to exhibit a dual role of enhancing nonpathogenic epiphytic bacteria (that can metabolize these volatiles possibly to obtain energy) and suppressing the fungal phytopathogens simultaneously (Abanda-Nkpwatt, Krimm, Coiner, Schreiber, & Schwab, 2006). These three volatiles were also produced by bacteria, among them 2-phenylethanol and nonanal used by the bacteria to suppress fungal growth (Table 2 and mVOC). Likewise, leaves of aromatic Mediterranean perennial plants had higher epiphytic bacterial populations than did nonaromatic plants, which deviates with the earlier view of antimicrobial activity of essential oils (Yadav, Papatheodorou, Karamanoli, Constantinidou, & Vokou, 2008). These observations convey how plants and their volatiles can impact the associated microbes and their release of VOCs for their own benefits. At the same time, the associated rhizobacteria have also shown to regulate the production of plant volatiles that helps both the host and bacterial partner. Future research would benefit from more detailed examination of this topic.

4. VOCs FROM BACTERIA AND THEIR EFFECTS ON PLANTS

After the first report on the beneficial growth effects of VOCs released from PGPR on plant growth (Ryu et al., 2003), there have been many

Table 1 Major Bacterial Volatiles (Mostly of Relevant Direct Growth Promotion or Successful Trials in Open Field Conditions), Their Function in Plant Health and Their Presence or Absence in the Plant Volatilome

Name of the VOC	Function of VOCs in Plant Growth and Development	Production in Plants[a]	References
2,3-Butanediol	• Growth promotion • Elicitation of ISR against drought, salt stress and pathogenic (bacteria and fungi) attack • Potent regulator of auxin homeostasis • Induction of plant iron acquisition by direct and indirect strategies such as direct acidification and indirect enhancement of root proton release • Enhancement of photosynthetic capacity • Elicitation of ISR in open field against fungal phytopathogens	No	Cho et al. (2008) Cortes-Barco, Goodwin, and Hsiang (2010) Cortes-Barco, Hsiang, and Goodwin (2010) Han et al. (2006) Ryu et al. (2003) Ryu et al. (2004) Schulz and Dickschat (2007) Zhang et al. (2007) Zhang, Xie, et al. (2008) Zhang et al. (2009)
DMDS	• Growth promotion • Growth inhibition • Suppression of crown gall tumours by direct suppression of *Agrobacterium* spp. • Elicitation of ISR against fungal pathogen • Enhancement of sulphur nutrition • Potential role in iron acquisition and nodule formation • QS quenching • Phytoparasitic nematicidal activity	Yes	Blom et al. (2011) Dandurishvili et al. (2010) Huang et al. (2012) Meldau et al. (2013) Kai et al. (2010) Groenhagen et al. (2013) Orozco-Mosqueda, Macías-Rodríguez, et al. (2013) Chernin et al. (2011) Popova et al. (2014)

(*Continued*)

Table 1 Major Bacterial Volatiles (Mostly of Relevant Direct Growth Promotion or Successful Trials in Open Field Conditions), Their Function in Plant Health and Their Presence or Absence in the Plant Volatilome—cont'd

Name of the VOC	Function of VOCs in Plant Growth and Development	Production in Plants[a]	References
Indole	• Growth promotion • Root development • Effectors of plant defence response	Yes	Bailly et al. (2014) Bhattacharyya, Garladinne, and Lee (2015) Blom et al. (2011) Yu and Lee (2013)
Dimethyl hexadecylamine	• Growth promotion and morphogenesis • Enhancement of iron acquisition by direct impact on plant iron metabolism • Potential rhizosphere acidifying agent • Potential role in nodule formation	No	Orozco-Mosqueda, Macías-Rodríguez, et al. (2013) Orozco-Mosqueda, Velázquez-Becerra, et al. (2013) Velázquez-Becerra et al. (2011)
1-Hexenol	• Growth promotion • Growth inhibition • Effectors of plant defence response	Yes	Blom et al. (2011)
Pentadecane	• Growth promotion • Effectors of plant defence response	Yes	Blom et al. (2011) Fernando, Ramarathnama, Krishnamoorthy, and Savchuk (2005)
Ammonia (inorganic volatile)	• Growth inhibition • Potential alkylating agent in the rhizosphere region	Yes (reported only in two plant species)	Kai et al. (2010) Weise et al. (2013) Wenke et al. (2012) Kai and Piechulla (2010)

Acetophenone	• Potent phytoparasitic nematicidal activity • Potential growth promoter by modulating root architecture • Anti-phytopathogenic fungal activity	Yes	Gu, Mo, Zhou, Zou, and Zhang (2007) Gutiérrez-Luna et al. (2010) Groenhagen et al. (2013)
o-Amino acetophenone	• Growth promotion	No	Groenhagen et al. (2013)
3-Hexanone		Yes	
2,5-Dimethylpyrazine		Yes (reported only in two plant species)	
2-Butanone	• Elicitation of ISR and induction of plant stress volatiles in open field against fungal phytopathogens and insect damage	Yes	Song and Ryu (2013) Farag et al. (2006)
Tridecanal	• Potential growth promoter by modulating root architecture	Yes (rarely occur)	Gutiérrez-Luna et al. (2010)
Tetradecanal		No	
6,10,14-Trimethyl-2-pentadecanone		Yes (single report)	
2-Pentylfuran	• Growth promotion	Yes (reported only in two plant species)	Zou, Li, and Yu (2010)
3-Phenylpropionic acid	• Growth inhibition	No	Chung et al. (2010)
Hexadecane	• Induction of ISR	No	Park et al. (2013)
Tridecane	• Induction of ISR	Yes	Lee et al. (2012)
Unidentified volatiles	• Enhancement of starch accumulation	NA	Ezquer et al. (2010)

[a]'Superscent' database was browsed for production from plants.

Table 2 Common Plant Floral Scents and Insect Volatiles (Schiestl, 2010) Produced by Bacteria Having Direct and Indirect Benefits to Plant Growth and Development

Name of the VOC	Number of Plant Families	Number of Insect Families	Functions Relevant to Plant Growth and Development	Corresponding and Example References for Bacterial Origin and Its Function Relevant to Plant Health
Benzaldehyde	57	13	Phytoparasitic nematicidal activity Anti-phytopathogenic fungal activity Soil fungistasis Potent player in the increase of primary root length and lateral root number	Gu et al. (2007) Kloepper et al. (1999) Schulz and Dickschat (2007) Tracey and Britz (1989) Zou, Mo, Gu, Zhou, and Zhang (2007) Gutiérrez-Luna et al. (2010)
Phenylacetaldehyde	29	8	Soil fungistasis Anti-phytopathogenic fungal activity	Blom et al. (2011) Gu et al. (2007) Schulz and Dickschat (2007) Zou et al. (2007)
2-Phenylethanol	49	15	Growth suppression Potent anti-phytopathogenic fungal activity	Blom et al. (2011) Gu et al. (2007) Kai, Effmert, Berg, and Piechulla (2007) Weise et al. (2014) Wenke et al. (2012)
3-Hydroxy-2-butanone (acetoin)	16	0	Growth promotion ISR against phytopathogen Potent regulator of auxin homeostasis	Blom et al. (2011) Rudrappa, Biedrzycki, Kunjeti, Donofrio, Czymmek, et al. (2010) Ryu et al. (2003) Schulz and Dickschat (2007) Zhang et al. (2007)
Nonanal	32	16	Anti-phytopathogenic fungi	Fernando et al. (2005)
Decanal	24	4	Phytoparasitic nematicidal activity	Gu et al. (2007)

attempts made to investigate other potentials of the bacterial VOCs on plant growth and metabolism. This first study reports the growth promotion ability of two bacterial VOCs, that is, 2,3-butanediol and acetoin (emitted by PGPRs *Bacillus subtilis* GB03 and *B. amyloliquefaciens* IN937a), on *Arabidopsis thaliana* exclusively by aerial contact. In further research, it was confirmed that bacterial volatiles have the ability to directly inhibit phytopathogens or indirectly trigger plant defence against them through aerial contact alone (for reviews see Kai et al., 2009; Farag et al., 2013; e.g. Groenhagen et al., 2013; Ryu et al., 2004). The detailed collection of bacterial VOCs (mostly airborne only) impacting plant growth is enlisted in Table 1. Major plant beneficial bacterial volatiles include 2,3-butanediol, acetoin, dimethyl disulphide (DMDS), indole, dimethyl hexadecylamine, 2-pentylfuran, 2-butanone, acetophenone, aminoacetophenone and 3-hexanone. Out of these, 2,3-butanediol, acetoin and DMDS are the most reported and studied compounds (Table 1). Some of the PGPRs found to suppress plant growth possibly do so through the production of volatile cyanide, 3-phenylpropionic acid and other growth suppressing compounds (Rudrappa et al., 2008; Chung, Park, Park, Ahn, & Chung, 2010; Blom et al., 2011; Table 1).

The plant growth promotion by bacterial volatiles is achieved in terms of plant biomass, leaf area, photosynthetic ability, root growth, number of lateral roots and improved nutrition (Bailly et al., 2014; Bhattacharyya, Yu, & Lee, 2014; Groenhagen et al., 2013; Meldau et al., 2013; Ryu et al., 2003). Many studies involving bacterial volatiles for plant growth (in normal and stress conditions) attempted to find possible pathways involved. Now, the scientists started realizing that these VOCs could interfere and regulate most of the conventionally known plant hormone pathways to manipulate plants for growth and defence purposes. They were also found to interfere with the biosynthesis of many important plant VOCs and other secondary metabolites, by which manipulating the way plants interact with a third trophic partner. From the 15 years of research on bacterial VOCs and their effect on plant health, diverse groups of rhizobacteria were reported. Of which, major bacteria include *Bacillus* spp., *Pseudomonas* spp., *Burkholderia* spp., *Serratia* spp., *Paenibacillus* sp., *Arthrobacter* sp., *Collimonas* sp., *Xanthomonas* sp., *Azospirillum* sp. and *Sinorhizobium* sp. More recently, many plant hosts have been tested for growth response in the presence of bacterial volatiles, in contrast to previous research, which was largely limited to the model plant, *Arabidopsis thaliana.* Through this diversified approach, we have gained an understanding that volatile effects seem to be species-specific, structure-specific, and ambiguousness messages can be embedded in a single volatile signal (Farag et al., 2006; Han et al., 2006; Table 1). For instance, DMDS is involved in both plant growth promotion and growth inhibition in different plant species (Table 1). This ambiguity will possibly be clarified into a single functional sense by who releases the signal (bacterial species), where it was released (habitat), who receives the signal (plant species) and finally why it was released (reason behind VOC emission). To date, the differential function of constitutively emitted VOCs and induced VOCs was not well understood, and needs a different analytical experimental strategy. Another common factor is the concentration of the selected VOC. In general, higher concentrations of any selected VOC can have a deleterious effect on plant growth (e.g. Blom et al., 2011).

Bacterial VOCs can modify the surrounding pH by either acidification or alkalization (Weise, Kai, & Piechulla, 2013; Zhang et al., 2009; Orozco-Mosqueda, Macías-Rodríguez, Santoyo, Farías-Rodríguez, & Valencia-Cantero, 2013; Orozco-Mosqueda, Velázquez-Becerra, Macías-Rodríguez, Santoyo, Flores-Cortez, et al., 2013). Such changes in the pH of the rhizosphere will surely influence this region, which will enhance iron acquisition by the host or sometimes inhibit the growth of the plant or phytopathogen. Rhizobacterial volatiles involved in iron uptake are 2,3-butanediol, DMDS and dimethyl hexadecylamine (Orozco-Mosqueda, Macías-Rodríguez, et al., 2013; Orozco-Mosqueda, Velázquez-Becerra, et al., 2013). Sometimes, they manipulate the host plant root to acidify its surroundings through physiological modification, which is an indirect way of aiding iron uptake. They also significantly induce the plant's metabolic activities involved in iron uptake (Zhang et al., 2009). Likewise, the growth inhibitory effect of bacterial inorganic volatile ammonia might be due to alkalization of the rhizosphere. Since the pH of the soil and rhizosphere is a key factor in determining microbiome composition (Lakshmanan, Selvaraj, & Bais, 2014), bacterial volatiles could play a crucial role in the dynamics of microbial populations by transforming the niche. Rhizobacterial volatiles have been shown to directly contribute to sulphur nutrition by providing reduced sulphur in the form of DMDS through exclusive aerial contact, which can be potentially useful to plants under sulphur deficiency (Meldau et al., 2013). Apart from these common growth effects, which are frequently reported, there are some noticeable features of bacterial VOCs recently realized that include suppression of crown gall disease, root architectural manipulation, potential role in nodule formation, and quorum sensing (QS) and quenching (Chernin et al., 2011; Dandurishvili et al., 2010; Gutiérrez-Luna et al., 2010; Orozco-Mosqueda, Macías-Rodríguez, et al., 2013).

Interestingly, a glimpse of a 'volatile conversation' between plant and bacteria becomes apparent in recent studies, in which the volatile profile was transformed during interaction, differing distinctly from the individual profiles of the plant or bacteria alone (Orozco-Mosqueda, Macías-Rodríguez, et al., 2013; Santoro, Zygadlo, Giordano, & Banchio, 2011). Santoro et al. (2011) demonstrated that the exposure of rhizobacterial VOCs could enhance the growth and essential oils of the aromatic plant, peppermint, implying how a volatile-based signalling of bacteria is being responded with enhanced plant volatile emission, thus having commercial importance. Similarly, the emission of VOCs is also sometimes regulated by a QS system, which would be of ecological relevance, as discussed in the following section.

5. BACTERIAL VOLATILES IN QS AND QQ

PGPRs exhibit plant growth-related activities by means of forming biofilms, which is highly regulated by a multicellular coordination system known as 'quorum sensing' (Hartmann, Rothballer, Hense, & Schröder, 2014; Lakshmanan & Bais, 2014; Steindler et al., 2009). In particular, biofilm formation in roots is the key success either to mutual or pathogenic relationships. QS networks support beneficial bacteria in plant growth promotion and/or protection, whereas QQ by beneficial bacteria is mainly used to suppress the QS of pathogens (Steindler et al., 2009; Chernin et al., 2011). Few studies showed that QS does not affect VOC emission (Groenhagen et al., 2013; Blom et al., 2011). In contradiction, the important and common bacterial volatile signal, 2,3-butanediol was reported to elicit a QS-dependent ISR in tobacco (Ryu et al., 2013). However, another important volatile DMDS is found to suppress the expression of a QS signal molecule, exhibiting a clear QQ activity (Chernin et al., 2011). Another rhizobacterial VOC dimethyl hexadecylamine produced by growth modulating rhizobacteria is structurally similar to QS signal molecules, which induces growth and iron uptake in *Medicago truncatula*, and could possibly prepare the host for nodulation (Velázquez-Becerra et al., 2011; Orozco-Mosqueda, Macías-Rodríguez, et al., 2013). The above contradictory observations might imply a species-specific regulation either dependent or independent of a QS system.

A population-dependent autoregulation was also demonstrated with bacterial dimethyl hexadecylamine, similar to usual QS signals (AHLs (acyl-homoserine lactones)), (Velázquez-Becerra et al., 2011). Muller et al.

(2009) showed that QS signals, AHLs, are involved in the regulation of antagonistic VOC emission in beneficial bacteria against fungal phytopathogens. Other bacterial VOCs such as indole, 2-aminoacetophenone and trimethylamine are shown to act like QS signals, controlling many intercellular communications (Bandyopadhaya et al., 2012; Létoffé, Audrain, Bernier, Delepierre, & Ghigo, 2014). Indole is also a herbivore-induced priming signal involved in both systemic tissues and neighbouring plants (Erb et al., 2015). Likewise, other bacterial volatiles, namely, 2-nonanone, 2-heptanone and 2-undecanone exhibit significant QQ activity (Plyuta, Popova, Koksharova, Kuznetsov, & Khmela, 2014). All these findings together indicate the potential of bacterial VOCs as being QS signals, QS regulators and QQ molecules in rhizospheric bacteria—bacteria and plant—bacteria interactions. This field potentially has a lot of application-oriented outcomes in both agriculture and medicine upon further exploration.

6. COMMONNESS AND DISTINCTIONS AMONG PLANT AND BACTERIAL VOCs

Both plants and bacteria have the ability to produce volatile compounds. There are several VOCs, which are common to plant and bacteria, and several others are not. For instance, acetoin, one of the active bacterial volatile compounds having plant growth-promoting ability, was already reported in many of the terrestrial and also mangrove species. Similarly, many bacteria are capable of synthesizing volatile compounds, which are normally synthesized by plants (Schulz & Dickschat, 2007). However, a derivative of acetoin, known as 2,3-butanediol, is only produced by bacteria, not by plants, but both the compounds positively affect the growth of plants. To understand commonness and contrasts between bacterial and plant VOCs, databases such as 'Superscent' and 'mVOC' were utilized and analysed (both VOCs and references). Further, one of the review articles from the journal 'Ecology Letters' was considered, which has a detailed compilation of volatiles widespread in floral scents and insects. The key findings of this paper depict that both plants and insects converge in the production of volatiles for pollination and defence with different classes of VOCs. This paper provides an evolutionary aspect on the most common volatiles found between floral scents and insects to see how the evolution of floral scents is being influenced by insect communications (Schiestl, 2010). Apart from them, some of the reviews and research articles were used to obtain the current position of the VOCs from plant and bacteria. The differences in the

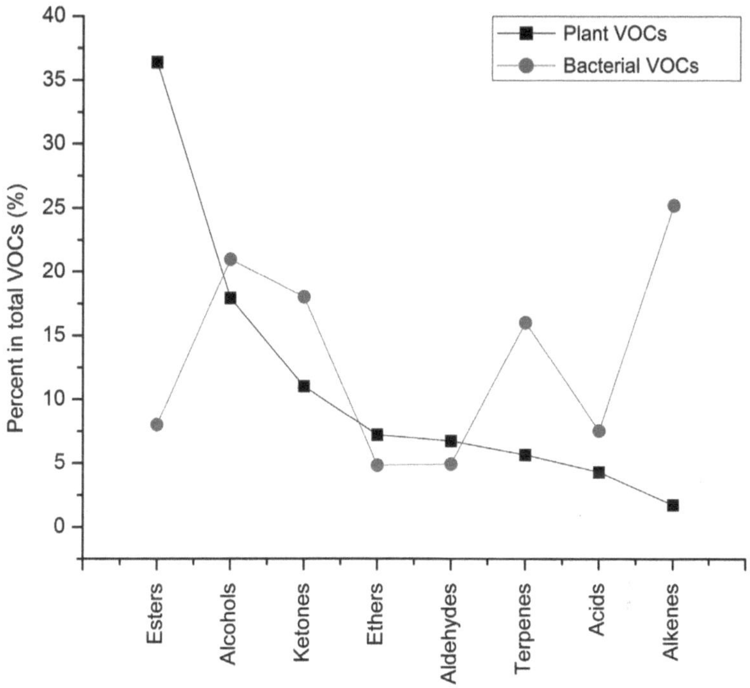

Figure 1 Difference between plant and bacterial VOCs in terms of chemical classification *(data obtained from Superscent and mVOC)*. Note: Only major chemical classes were considered for calculation, having at least ≥5% in either plant or bacterial volatiles. (See colour plate)

major chemical classes are depicted in Figure 1. Esters account for the maximum percentage of total plant VOCs, whereas alkenes occupy the maximum in the total bacterial VOCs. Further, the terpene percentage is higher in total bacterial VOCs, when compared to the terpene percentage in total plant VOCs. Alcohols, ethers, ketones, aldehydes and acids are almost equal. This result indicates the diversity of VOCs gained during evolution. As far as the diversity of chemical classes is concerned, bacterial VOCs are more diverse than fungal and plant VOCs. But the number of reported plant VOCs (≈2000) is double the total number of bacterial and fungal VOCs (≈1000) reported together (Superscent and mVOC databases).

Table 1 gives the list of bacterial volatiles, their functions relevant to plants and their occurrence in the plant volatilome. Table 2 compares only bacterial volatiles having a direct or indirect impact on plant growth, which are also widespread in floral scents and insects. Table 1 was discussed in the previous section on bacterial volatiles. However, the commonness of bacterial and plant volatiles alone was taken into account. From the 19 VOCs listed, 14 (≈74%) were found to be released by plants. Out of five VOCs of exclusive rhizobacterial origin, 2,3-butanediol and dimethyl hexadecylamine are very important in mediating volatile-based plant—microbe interactions and act as multifunctional infochemicals (Table 1). o-Aminoacetophenone (2'-aminoacetophenone) is another important and potential volatile candidate, since the compound exhibits plant growth promotion (Létoffé et al., 2014). As mentioned above, 2-aminoacetophenone was reported elsewhere, in the medical field, as a novel QS signal molecule that could successfully modulate a host immune response to facilitate pathogen persistence, while enabling host tolerance to infection (Bandyopadhaya et al., 2012). This molecule is responsible for the sweet 'grape-like' odour in cultures of *P. aeruginosa* and acts as a noninvasive biomarker for detecting this species. The remaining two VOCs include 3-phenylpropionic acid and hexadecane, which showed plant growth inhibition and induction of ISR, respectively (Chung et al., 2010; Park, Lee, Kloepper, & Ryu, 2013). It seems that the rhizobacteria evolved to use certain unique volatile compounds, which cannot be produced by plants, and mutually plants tend to perceive the information from these compounds to initiate a specific signalling cascade. However, this assumption is not complete concerning such volatile mediated interactions. The remaining 14 compounds are common to plants and bacteria. Among these, DMDS, indole, 1-hexenol, pentadecane and acetophenone have multiple functions in plant health (Table 1). Thus, they can use common VOCs during the interaction, which might have common properties, but would be used to influence each other in different spatiotemporal circumstances. For example, DMDS was found to influence fungal pathogens and was released from flowers as a defence. At the same time, rhizhobacteria can produce DMDS to protect host roots against fungal phytopathogens and phytoparasitic nematodes. Based on the above comparisons, a Venn diagram represents the exclusive plant and bacterial VOCs, and volatiles shared by both of them during their interactions (Figure 2).

Schiestl (2010) compiled and compared 71 VOCs that are most commonly found in floral scents and insects. Out of 71 compounds, 42 (59%) were reported to be released by bacteria. However, out of these 42, only six (0.8%) compounds were listed, because only these compounds exhibited functions related to plant health either by direct or indirect means (Table 2). Acetoin (3-hydroxy-2-butanone) is the most widely reported common VOC in many plant species and bacteria, but not in insects.

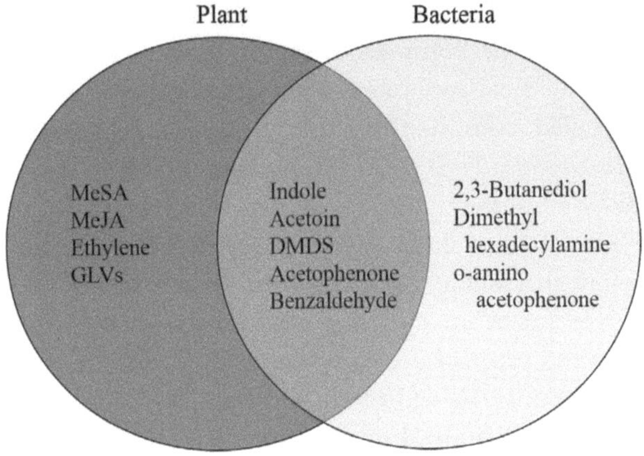

Figure 2 Venn diagram representing volatiles released during plant–bacteria interaction by both partners. These VOCs represent exclusive plant origin (plant volatilome) inside the pink circle, exclusive bacterial origin (bacterial volatilome) inside the sandal circle and VOCs common to both plant and bacteria in the overlapping region. (See colour plate)

Acetoin, a precursor of 2,3-butanediol, has similar activities in plant health by direct or indirect impacts (Table 2). Next to this, benzaldehyde is the most widespread VOC reported to be released from all three kingdoms. Although they have multiple functions, there are direct and strong plant growth effects. It may be possible that many new VOCs, from the 42 bacterial compounds, can be made that in turn will reveal potent plant growth-promoting and ISR inducing activities. Thus, a comparative and bi-directional aspects of volatile-based plant–bacteria interactions should be considered in every future attempt to draw a complementary understanding.

7. FUTURE PERSPECTIVES

Since many new volatile mutants of model plants are available now, an understanding on the central role of important plant volatiles on the microbial community development and establishment can be brought forward by coupling plant genetics and metagenomics of microbiome analysis. In the same way, these mutants can be used to investigate the individual plant–PGPR interactions. Another crucial scientific challenge, which lies ahead is to find the exact receptors in membranes that perceive these volatile signals to initiate signal transduction process. To achieve this objective, computer-assisted molecular docking can be used with subsequent validation in in vitro and in planta analyses. Likewise, the role of beneficial bacterial (epiphytic and endophytic) volatile compounds during plant viral infection remains unexplored. Finally, field application of volatile chemicals is a big challenge and needs a detailed scientific focus to succeed. This might be possible by establishing new methods (i) to use these compounds in open fields with a chemical or physical technique for aiding slow release; (ii) to preexpose the seeds of interest for a short period to either liquid or gaseous form in a sealed chamber that might need less volume of a selected VOC; (iii) to apply a diffusible precursor chemical along with selected PGPR, which readily converts the precursor to a bioactive VOC. However, most of the studies expose seedlings instead of seeds; hence, it would be beneficial to test the effect of VOCs from the seed stage as well. It is also foreseen that genetically engineered plants incorporated with genes of bacterial VOCs will be analysed. Further, the bacterial motility, chemotaxis and substrate utilization ability in response to plant VOCs should be examined to deepen our understanding.

ACKNOWLEDGEMENT

G.S acknowledges the United States–India Educational Foundation for a Fulbright-Nehru postdoctoral fellowship (grant no. 1847/FNPDR/2013).

REFERENCES

Abanda-Nkpwatt, D., Krimm, U., Coiner, H. A., Schreiber, L., & Schwab, W. (2006). Plant volatiles can minimize the growth suppression of epiphytic bacteria by the phytopathogenic fungus *Botrytis cinerea* in co-culture experiments. *Environmental and Experimental Botany, 56*, 108–119.

Baetz, U., & Martinoia, E. (2014). Root exudates: the hidden part of plant defense. *Trends in Plant Science, 19*, 90–98.

Bailly, A., Groenhagen, U., Schulz, S., Geisler, M., Eberl, L., & Weisskopf, L. (2014). The inter-kingdom volatile signal indole promotes root development by interfering with auxin signaling. *The Plant Journal*. http://dx.doi.org/10.1111/tpj.12666.

Bais, H. P., Weir, T. L., Perry, L. G., Gilroy, S., & Vivanco, J. M. (2006). The role of root exudates in rhizosphere interactions with plants and other organisms. *Annual Reviews of Plant Biology, 57*, 233–266.

Baldwin, I. T., Halitschke, R., Paschold, A., von Dahl, C. C., & Preston, C. A. (2006). Volatile signaling in plant–plant interactions: "Talking Trees" in the genomics era. *Science, 311*, 812.

Bandyopadhaya, A., Kesarwani, M., Que, Y.-A., He, J., Padfield, K., Tompkins, R., et al. (2012). The quorum sensing volatile molecule 2-amino acetophenone modulates host immune responses in a manner that promotes life with unwanted guests. *PLoS Pathogen, 8*, e1003024.

Bhattacharyya, D., Garladinne, M., & Lee, Y. H. (2015). Volatile indole produced by rhizobacterium *Proteus vulgaris* JBLS202 stimulates growth of *Arabidopsis thaliana* through auxin, cytokinin, and brassinosteroid pathways. *Journal of Plant Growth Regulation*. http://dx.doi.org/10.1007/s00344-014-9453-x.

Bhattacharyya, D., Yu, S.-M., & Lee, Y. H. (2014). Volatile compounds from *Alcaligenes faecalis* JBCS1294 confer salt tolerance in *Arabidopsis thaliana* through the auxin and gibberellin pathways and differential modulation of gene expression in root and shoot tissues. *Plant Growth Regulation*. http://dx.doi.org/10.1007/s10725-014-9953-5.

Bicchi, C., & Maffei, M. (2012). The plant volatilome: methods of analysis. In J. Normanly (Ed.), *Methods in molecular biology: Vol. 918. High-throughput phenotyping in plants: Methods and protocols* (pp. 289–310).

Blom, D., Fabbri, C., Connor, E. C., Schiestl, F. P., Klauser, D. R., Boller, T., et al. (2011). Production of plant growth modulating volatiles is widespread among rhizosphere bacteria and strongly depends on culture conditions. *Environmental Microbiology, 13*, 3047–3058.

Chen, F., Ro, D.-K., Petri, J., Gershenzon, J., Bohlmann, J., Pichersky, E., et al. (2004). Characterization of a root-specific *Arabidopsis* terpene synthase responsible for the formation of the volatile monoterpene 1,8-Cineole. *Plant Physiology, 135*, 1956–1966.

Chernin, L., Toklikishvili, N., Ovadis, M., Kim, S., Ben-Ari, J., Khmel, I., et al. (2011). Quorum-sensing quenching by rhizobacterial volatiles. *Environmental Microbiology Reports, 3*, 698–704.

Cho, S. M., Kang, B. R., Han, S. H., Anderson, A. J., Park, J.-Y., Lee, Y.-H., et al. (2008). 2R,3R-butanediol, a bacterial volatile produced by *Pseudomonas chlororaphis* O6, is involved in induction of systemic tolerance to drought in *Arabidopsis thaliana*. *Molecular Plant–Microbe Interactions, 21*, 1067–1075. http://dx.doi.org/10.1094/MPMI-21-8-1067.

Christensen, S. A., Nemchenko, A., Borrego, E., Murray, I., Sobhy, I. S., Bosak, L., et al. (2013). The maize lipoxygenase, ZmLOX10, mediates green leaf volatile, jasmonate and herbivore-induced plant volatile production for defense against insect attack. *The Plant Journal, 74*, 59–73.

Chung, E. J., Park, J. H., Park, T. S., Ahn, J.-W., & Chung, Y. R. (2010). Production of a phytotoxic compound, 3-phenylpropionic acid by a bacterial endophyte, *Arthrobacter humicola* YC6002 isolated from the root of *Zoysia japonica*. *The Plant Pathology Journal, 26*, 245–252.

Cortes-Barco, A. M., Goodwin, P., & Hsiang, T. (2010). Comparison of induced resistance activated by benzothiadiazole, (2R,3R)-butanediol and an isoparaffin mixture against anthracnose of *Nicotiana benthamiana*. *Plant Pathology, 59*, 643–653.

Cortes-Barco, A. M., Hsiang, T., & Goodwin, P. H. (2010). Induced systemic resistance against three foliar diseases of *Agrostis stolonifera* by (2*R*,3*R*)-butanediol or an isoparaffin mixture. *Annals of Applied Biology, 157*, 179–189.

Dandurishvili, N., Toklikishvili1, N., Ovadis, M., Eliashvili, P., Giorgobiani, N., Keshelava, R., et al. (2010). Broad-range antagonistic rhizobacteria *Pseudomonas fluorescens* and *Serratia plymuthica* suppress *Agrobacterium* crown gall tumours on tomato plants. *Journal of Applied Microbiology, 110*, 341–352.

Del Giudice, L., Massardo, D. R., Pontieri, P., Bertea, C. M., Mombello, D., Carata, E., et al. (2008). The microbial community of vetiver root and its involvement into essential oil biogenesis. *Environmental Microbiology, 10*, 2824–2841.

Delano-Frier, J. P., Pearce, G., Huffaker, A., & Stratmann, J. W. (2013). Systemic wound signaling in plants. In F. Baluska (Ed.), *Long-distance systemic signaling and communication in plants, signaling and communication in plants* (pp. 323–362).

Dudareva, N., Klempien, A., Muhlemann, J. K., & Kaplan, I. (2013). Biosynthesis, function and metabolic engineering of plant volatile organic compounds. *New Phytologist, 198*, 16–32.

Dudareva, N., Pichersky, E., & Gershenzon, J. (2004). Biochemistry of plant volatiles. *Plant Physiology, 135*, 1893–1902.

Dunkel, M., Schmidt, U., Struck, S., Berger, L., Gruening, B., Hossbach, J., et al. (2009). Super scent – a database of flavors and scents. *Nucleic Acids Research, 37*, D291–D294.

Erb, M., Veyrat, N., Robert, C. A. M., Xu, H., Frey, M., Ton, J., et al. (2015). Indole is an essential herbivore-induced volatile priming signal in maize. *Nature Communications, 6*, 6273. http://dx.doi.org/10.1038/ncomms7273.

Ezquer, I., Li, J., Ovecka, M., Baroja-Fernandez, E., Jose Munoz, F., Montero, M., et al. (2010). Microbial volatile emissions promote accumulation of exceptionally high levels of starch in leaves in mono- and dicotyledonous plants. *Plant and Cell Physiology, 51*, 1674–1693.

Farag, M. A., Ryu, C.-M., Sumner, L. W., & Paré, P. W. (2006). GC–MS SPME profiling of rhizobacterial volatiles reveals prospective inducers of growth promotion and induced systemic resistance in plants. *Phytochemistry, 67*, 2262–2268.

Farag, M. A., Zhang, H., & Ryu, C.-M. (2013). Dynamic chemical communication between plants and bacteria through airborne signals: Induced resistance by bacterial volatiles. *Journal of Chemical Ecology, 39*, 1007–1018.

Fernando, W. G. D., Ramarathnama, R., Krishnamoorthy, A. S., & Savchuk, S. C. (2005). Identification and use of potential bacterial organic antifungal volatiles in biocontrol. *Soil Biology & Biochemistry, 37*, 955–964.

Goldstein, A. H., & Galbally, I. E. (2007). Known and unexplored organic constituents in the earth's atmosphere. *Environmental Science & Technology, 41*, 1514–1521.

Groenhagen, U., Baumgartner, R., Bailly, A., Gardiner, A., Eberl, L., Schulz, S., et al. (2013). Production of bioactive volatiles by different *Burkholderia ambifaria* strains. *Journal of Chemical Ecology, 39*, 892–906. http://dx.doi.org/10.1007/s10886-013-0315-y.

Gu, Y.-Q., Mo, M.-H., Zhou, J.-P., Zou, C.-S., & Zhang, K.-Q. (2007). Evaluation and identification of potential organic nematicidal volatiles from soil bacteria. *Soil Biology & Biochemistry, 39*, 2567–2575.

Guenther, A. B., Jiang, X., Heald, C. L., Sakulyanontvittaya, T., Duhl, T., Emmons, L. K., et al. (2012). The model of emissions of gases and aerosols from nature version 2.1 (MEGAN2.1): an extended and updated framework for modeling biogenic emissions. *Geoscientific Model Development, 5*, 1471–1492.

Gutiérrez-Luna, F. M., López-Bucio, J., Altamirano-Hernández, J., Valencia-Cantero, E., de la Cruz, H. R., & Macías-Rodríguez, L. (2010). Plant growth-promoting rhizobacteria modulate root-system architecture in *Arabidopsis thaliana* through volatile organic compound emission. *Symbiosis, 51*, 75–83.

Han, S. H., Lee, S. J., Moon, J. H., Park, K. H., Yang, K. Y., & Cho, B. H. (2006). GacS-Dependent production of 2R, 3R-butanediol by *Pseudomonas chlororaphis* O6 is a major determinant for eliciting systemic resistance against *Erwinia carotovora* but not against *Pseudomonas syringae* pv. *tabaci* in tobacco. *Molecular Plant-Microbe Interactions, 8*, 924–930.

Hartmann, A., Rothballer, M., Hense, B. A., & Schröder, P. (2014). Bacterial quorum sensing compounds are important modulators of microbe–plant interactions. *Frontiers in Plant Science, 5*, 131.

Hause, B., & Schaarschmidt, S. (2009). The role of jasmonates in mutualistic symbioses between plants and soil-born microorganisms. *Phytochemistry, 70*, 1589–1599.

Holopainen, J., & Blande, J. (2012). Molecular plant volatile communication. In C. Lopez-Larrea (Ed.), *Sensing in nature* (pp. 17–31). US: Springer.

Huang, C.-J., Tsay, J.-F., Chang, S.-Y., Yang, H.-P., Wub, W.-S., & Chen, C.-Y. (2012). Dimethyl disulfide is an induced systemic resistance elicitor produced by *Bacillus cereus* C1L. *Pest Management Science, 68*, 1306–1310.

Insam, H., & Seewald, M. S. (2010). Volatile organic compounds (VOCs) in soils. *Biology and Fertility of Soils, 46*, 199–213.

Kai, M., Crespo, E., Cristescu, S., Harren, F., Francke, W., & Piechulla, B. (2010). *Serratia odorifera*: analysis of volatile emission and biological impact of volatile compounds on *Arabidopsis thaliana*. *Applied Microbiology Biotechnology, 88*, 965–976.

Kai, M., Effmert, U. E., Berg, G., & Piechulla, B. (2007). Volatiles of bacterial antagonists inhibit mycelial growth of the plant pathogen *Rhizoctonia solani*. *Archives in Microbiology, 187*, 351–360.

Kai, M., Haustein, M., Molina, F., Petri, A., Scholz, B., & Piechulla, B. (2009). Bacterial volatiles and their action potential. *Applied Microbiology and Biotechnology, 81*, 1001–1012.

Kai, M., & Piechulla, B. (2010). Impact of volatiles of the rhizobacteria *Serratia odorifera* on the moss Physcomitrella patens. *Plant Signaling & Behavior, 5*, 444–446.

Kierul, K., Voigt, B., Albrecht, D., Chen, X. H., Carvalhais, L. C., & Borriss, R. (2015). Influence of root exudates on the extracellular proteome of the plant growth-promoting bacterium *Bacillus amyloliquefaciens* FZB42. *Microbiology, 161*, 131–147.

Kloepper, J. W., Rodriguez-Kabana, R., Zehnder, G. W., Murphy, J. F., Sikora, E., & Fernandez, C. (1999). Plant root–bacterial interactions in biological control of soilborne diseases and potential extension to systemic and foliar diseases. *Australasian Plant Pathology, 28*, 21–26.

Knudsen, J. T., Eriksson, R., Gershenzon, J., & Stahl, B. (2006). Diversity and distribution of floral scent. *Botanical Review, 72*, 1–120.

Kobayashi, K. (2015). Plant methyl salicylate induces defense responses in the rhizobacterium *Bacillus subtilis*. *Environmental Microbiology, 17*, 1365–1376.

Kos, M., Houshyani, B., Overeem, A.-J., Bouwmeester, H. J., Weldegergis, B. T., van Loon, J. J. A., et al. (2013). Genetic engineering of plant volatile terpenoids: effects on a herbivore, a predator and a parasitoid. *Pest Management Science, 69*, 302–311.

Lakshmanan, V., & Bais, H. P. (2014). Factors other than root secreted malic acid that contributes toward *Bacillus subtilis* FB17 colonization on *Arabidopsis* roots. *Plant Signaling & Behavior, 8*, e27277.

Lakshmanan, V., Selvaraj, G., & Bais, H. P. (2014). Functional soil microbiome: belowground solutions to an aboveground problem. *Plant Physiology, 166*, 689–700.

Lee, B., Farag, M. A., Park, H. B., Kloepper, J. W., Lee, S. H., & Ryu, C.-M. (2012). Induced resistance by a long-chain bacterial volatile: elicitation of plant systemic defense by a C13 volatile produced by *Paenibacillus* polymyxa. *PLoS One, 7*, e48744.

Lemfack, M. C., Nickel, J., Dunkel, M., Preissner, R., & Piechulla, B. (2014). mVOC: a database of microbial volatiles. *Nucleic Acids Research, 42*, D744–D748.

Létoffé, S., Audrain, B., Bernier, S. P., Delepierre, M., & Ghigo, J.-M. (2014). Aerial exposure to the bacterial volatile compound trimethylamine modifies antibiotic resistance of physically separated bacteria by raising culture medium pH. *mBio, 5*, e00944.

Loreto, F., Dicke, M., Schnitzler, J.-P., & Turlings, T. C. J. (2014). Plant volatiles and the environment. *Plant, Cell & Environment, 37*, 1905–1908.

Maffei, M. E. (2010). Sites of synthesis, biochemistry and functional role of plant volatile. *South African Journal of Botany, 76*, 612–631.

Maffei, M. E., Gertschb, J., & Appendinoc, G. (2011). Plant volatiles: production, function and pharmacology. *Natural Product Reports, 28*, 1359–1380.

Maffei, M. E., Mithofer, A., & Boland, W. (2007). Insects feeding on plants: rapid signals and responses preceding the induction of phytochemical release. *Phytochemistry, 68*, 2946–2959.

Meldau, D. G., Meldau, S., Hoang, L. H., Underberg, S., Wünsche, H., & Baldwin, I. T. (2013). Dimethyl disulfide produced by the naturally associated bacterium *Bacillus* sp. B55 promotes *Nicotiana attenuate* growth by enhancing sulfur nutrition. *The Plant Cell, 25*, 2731–2747.

Muller, H., Westendorf, C., Leitner, E., Chernin, L., Riedel, K., Schmidt, S., et al. (2009). Quorum-sensing effects in the antagonistic rhizosphere bacterium *Serratia plymuthica* HRO-C48. *FEMS Microbiology Ecology, 67*, 468–478.

Nakagawa, T., & Kawaguchi, M. (2006). Shoot-applied MeJA suppresses root nodulation in *Lotus japonicas*. *Plant and Cell Physiology, 47*, 176–180.

Niederbacher, B., Winkler, J. B., & Schnitzler, J. P. (2015). Volatile organic compounds as non-invasive markers for plant phenotyping. *Journal of Experimental Botany*. http://dx.doi.org/10.1093/jxb/erv219.

Niinemets, U., Arneth, A., Kuhn, U., Monson, R. K., Penuelas, J., & Staudt, M. (2010). The emission factor of volatile isoprenoids: stress, acclimation, and developmental responses. *Biogeosciences, 7*, 2203–2223.

Oikawa, P. Y., & Lerdau, M. T. (2013). Catabolism of volatile organic compounds influences plant survival. *Trends in Plant Science, 18*, 695–703.

Orozco-Mosqueda, M. C., Macías-Rodríguez, L. I., Santoyo, G., Farías-Rodríguez, R., & Valencia-Cantero, E. (2013). *Medicago truncatula* increases its iron-uptake mechanisms in response to volatile organic compounds produced by *Sinorhizobium meliloti*. *Folia Microbiologica, 58*, 579–585.

Orozco-Mosqueda, M. C., Velázquez-Becerra, C., Macías-Rodríguez, L. I., Santoyo, G., Flores-Cortez, I., Alfaro-Cuevas, R., et al. (2013). *Arthrobacter agilis* UMCV2 induces iron acquisition in *Medicago truncatula* (strategy I plant) *in vitro* via dimethylhexadecylamine emission. *Plant and Soil, 362*, 51–66.

Ortíz-Castro, R., Contreras-Cornejo, H. A., Macías-Rodríguez, L., & López-Bucio, J. (2009). The role of microbial signals in plant growth and development. *Plant Signaling & Behavior, 4*, 701–712.

Park, H. B., Lee, B., Kloepper, J. W., & Ryu, C.-M. (2013). One shot-two pathogens blocked: exposure of *Arabidopsis* to hexadecane, a long chain volatile organic compound, confers induced resistance against both *Pectobacterium carotovorum* and *Pseudomonas syringae*. *Plant Signalling and Behavior, 8*, e24619.

Peñuelas, J., Asensio, D., Tholl, D., Wenke, K., Rosenkranz, M., Piechulla, B., et al. (2014). Biogenic volatile emissions from the soil. *Plant, Cell and Environment, 37*, 1866–1891.

Plyuta, V. A., Popova, A. A., Koksharova, O. A., Kuznetsov, A. E., & Khmela, I. A. (2014). The ability of natural ketones to interact with bacterial quorum sensing systems. *Molecular Genetics, Microbiology and Virology, 29*, 167–171.

Popova, A. A., Koksharova, O. A., Lipasova, V. A., Zaitseva, J. V., Katkova-Zhukotskaya, O. A., Eremina, S. I., et al. (2014). Inhibitory and toxic effects of volatiles emitted by strains of *Pseudomonas* and *Serratia* on growth and survival of selected microorganisms, *Caenorhabditis elegans*, and *Drosophila melanogaster*. *BioMed Research International, 2014*. Article ID 125704.

Rasmann, S., Köllner, T. G., Degenhardt, J., Hiltpold, I., Töpfer, S., Kuhlmann, U., et al. (2005). Recruitment of entomopathogenic nematodes by insect-damaged maize roots. *Nature, 434*, 732–737.

Robert, C. A. M., Erb, M., Hibbard, B. E., French, B. W., Zwahlen, C., & Turlings, T. C. J. (2012). A specialist root herbivore reduces plant resistance and uses an induced plant volatile to aggregate in a density-dependent manner. *Functional Ecology, 26*, 1429–1440.

Rudrappa, T., Biedrzycki, M. L., Kunjeti, S. G., Donofrio, N. M., Czymmek, K. J., Pare, P. W., et al. (2010). The rhizobacterial elicitor acetoin induces systemic resistance in *Arabidopsis thaliana*. *Communicative & Integrative Biology, 3*, 130–138.

Rudrappa, T., Splaine, R. E., Biedrzycki, M. L., & Bais, H. P. (2008). Cyanogenic Pseudomonads influence multitrophic interactions in the rhizosphere. *PLoS One, 3*, e2073.

Ryu, C.-M., Choi, H. K., Lee, C.-H., Murphy, J. F., Lee, J.-K., & Kloepper, J. W. (2013). Modulation of quorum sensing in acyl-homoserine lactone-producing or-degrading tobacco plants leads to alteration of induced systemic resistance elicited by the rhizobacterium *Serratia marcescens* 90–166. *Plant Pathology Journal, 29*, 182–192.

Ryu, C.-M., Farag, M. A., Hu, C.-H., Reddy, M. S., Kloepper, J. W., & Pare, P. W. (2004). Bacterial volatiles induce systemic resistance in *Arabidopsis*. *Plant Physiology, 134*, 1017–1026.

Ryu, C.-M., Farag, M. A., Hu, C.-H., Reddy, M. S., Wei, H.-X., Paré, P. W., et al. (2003). Bacterial volatiles promote growth in *Arabidopsis*. *Proceedings of the National Academy of Sciences of the United States of America, 100*, 4927–4932.

Santoro, M. V., Zygadlo, J., Giordano, W., & Banchio, E. (2011). Volatile organic compounds from rhizobacteria increase biosynthesis of essential oils and growth parameters in peppermint (*Mentha piperita*). *Plant Physiology and Biochemistry, 49*, 1177e1182.

Santos, F., Penaflor, M. F. G. V., Pare, P. W., Sanches, P. A., Kamiya, A. C., & Tonelli, M. (2014). A novel interaction between plant-beneficial rhizobacteria and roots: colonization induces corn resistance against the root herbivore *Diabrotica speciosa*. *PLoS One, 9*, e113280.

Schiestl, F. P. (2010). Evolution of floral scent and insect chemical communication. *Ecology Letters, 13*, 643–656.

Schulz, S., & Dickschat, J. S. (2007). Bacterial volatiles: the smell of small organisms. *Natural Product Reports, 24*, 814–842.

Shulaev, V., Silverman, P., & Raskin, I. (1997). Airborne signalling by methyl salicylate in plant pathogen resistance. *Nature, 385*, 718–721.

Song, G. C., & Ryu, C.-M. (2013). Two volatile organic compounds trigger plant self-defense against a bacterial pathogen and a sucking insect in cucumber under open field conditions. *International Journal of Molecular Science, 14*, 9803–9819.

Steeghs, M., Bais, H. P., de Gouw, J., Goldan, P., Kuster, W., Northway, M., et al. (2004). Proton-transfer-reaction mass spectrometry (PTR-MS) as a new tool for real time analysis of root-secreted volatile organic compounds (VOCs) in *Arabidopsis thaliana*. *Plant Physiology, 135*, 47–58.

Steindler, L., Bertani, I., De Sordi, L., Schwager, S., Eberl, L., & Venturi, V. (2009). LasI/R and RhlI/R Quorum sensing in a strain of *Pseudomonas aeruginosa* beneficial to plants. *Applied and Environmental Microbiology, 75*, 5131–5140.

Stotzky, G., & Schenck, S. (1976). Volatile organic compounds and microorganisms. *CRC Critical Reviews in Microbiology, 4*, 333–382.

Sugimoto, K., Matsui, K., Iijima, Y., Akakabe, Y., Muramoto, S., Ozawab, R., et al. (2014). Intake and transformation to a glycoside of (*Z*)-3-hexenol from infested neighbors reveals a mode of plant odor reception and defense. *PNAS, 19*, 7144–7149.

Timmusk, S., Abd El-Daim, I. A., Copolovici, L., Tanilas, T., Kannaste, A., Behers, L., et al. (2014). Drought-tolerance of wheat improved by rhizosphere bacteria from harsh environments: enhanced biomass production and reduced emissions of stress volatiles. *PLoS One, 9*, e96086.

Tracey, R. P., & Britz, T. J. (1989). Freon 11 extraction of volatile metabolites formed by certain lactic acid bacteria. *Applied Environmental Microbiology, 55*, 1617–1623.

Velázquez-Becerra, C., Macías-Rodríguez, L. I., López-Bucio, J., Altamirano-Hernández, J., Flores-Cortez, I., & Valencia-Cantero, E. (2011). A volatile organic compound analysis from *Arthrobacter agilis* identifies dimethylhexadecylamine, an amino-containing lipid modulating bacterial growth and *Medicago sativa* morphogenesis in vitro. *Plant and Soil, 339*, 329–340.

Weise, T., Kai, M., & Piechulla, B. (2013). Bacterial ammonia causes significant plant growth inhibition. *PLoS One, 8*, e63538.

Weise, T., Thurmer, A., Brady, S., Kai, M., Daniel, R., Gottschalk, G., et al. (2014). VOC emission of various *Serratia* species and isolates and genome analysis of *Serratia plymuthica* 4Rx13. *FEMS Microbiology Letters, 352*, 45–53.

Wenke, K., Wanke, D., Kilian, J., Berendzen, K., Harter, K., & Piechulla, B. (2012). Volatiles of two growth-inhibiting rhizobacteria commonly engage AtWRKY18 function. *The Plant Journal, 70*, 445–459.

Yadav, R. K. P., Papatheodorou, E. M., Karamanoli, K., Constantinidou, H.-I. A., & Vokou, D. (2008). Abundance and diversity of the phyllosphere bacterial communities of Mediterranean perennial plants that differ in leaf chemistry. *Chemoecology, 18*, 217–226.

Yu, S.-M., & Lee, Y. H. (2013). Plant growth promoting rhizobacterium *Proteus vulgaris* JBLS202 stimulates the seedling growth of Chinese cabbage through indole emission. *Plant and Soil, 370*, 485–495.

Zhang, H., Kim, M.-S., Krishnamachari, V., Payton, P., Sun, Y., Grimson, M., et al. (2007). Rhizobacterial volatile emissions regulate auxin homeostasis and cell expansion in Arabidopsis. *Planta, 226*, 839–851.

Zhang, H., Kim, M.-S., Sun, Y., Dowd, S. E., Shi, H., & Paré, P. W. (2008). Soil bacteria confer plant salt tolerance by tissue-specific regulation of the sodium transporter HKT1. *Molecular Plant–Microbe Interactions, 21*, 737–744. http://dx.doi.org/10.1094/MPMI-21-6-0737.

Zhang, H., Sun, Y., Xie, X., Kim, M.-S., Dowd, S. E., & Pare, P. W. (2009). A soil bacterium regulates plant acquisition of iron via deficiency-inducible mechanisms. *The Plant Journal, 58*, 568–577.

Zhang, H., Xie, X., Kim, M.-S., Kornyeyev, D. A., Holaday, S., & Pare, P. W. (2008). Soil bacteria augment *Arabidopsis* photosynthesis by decreasing glucose sensing and abscisic acid levels in planta. *The Plant Journal, 56*, 264–273.

Zou, C.-S., Li, Z., & Yu, D. (2010). *Bacillus megaterium* strain XTBG34 promotes plant growth by producing 2-pentylfuran. *Journal of Microbiology, 48*, 460–466.

Zou, C.-S., Mo, M.-H., Gu, Y.-Q., Zhou, J.-P., & Zhang, K.-Q. (2007). Possible contributions of volatile-producing bacteria to soil fungistasis. *Soil Biology & Biochemistry, 39*, 2371–2379.

SUBJECT INDEX

'*Note:* Page numbers followed by "f" indicate figures, "t" indicate tables.'

A

Abscisic acid (ABA), 69–70
Aceria tulipae, 197
Acetoin, 266, 300–301
Allelopathic compounds, 224–225
Allopolyploidy, 159–160, 179
1-Aminocyclopropane-1-carboxylate (ACC) deaminase, 12–13, 253
Antibiosis
  alcohols, 265
  alkanes and alkenes, 266–267
  defined, 265
  ketones and aldehydes, 266
  sulphur compounds, 267–268
Arabidopsis
  A. halleri, 60–62
  A. lyrata, 60–62
  A. thaliana, 2
Arbuscular mycorrhizal (AM), 225
Asclepias syriaca, 198
ATP-binding cassette (ABC), 165–166
Autopolyploidy, 159–160
Autoregulation of nodulation (AON), 156–157
Auxin, 69–70
Azospirillum
  A. brasilense, 5–6
  A. lipoferum, 5–6

B

Bacillus subtilis strain N11, 5–6
Biofertilizers
  iron absorption, 89–90
  nitrogen fixation, 89–90
  phosphorus solubilization, 90–91
  phytostimulators, 92
    cytokinin, 94
    ethylene, 95
    gibberellin, 94
    IAA, 92–94
Biofilm formation, 7–8
Biological nitrogen fixation (BNF), 89–90
Biopesticides, 96–97
Bioremediators, 97–99
2,3-Butanediol, 300–301

C

Caenorhabditis elegans, 194
Canopy precipitation partitioning
  potential throughfall and stemflow impacts, 230–231
    Acacia, 224–225
    allelopathic compounds, 224–225
    B deficiency, 222–223
    Ca spiking, 223–224
    endo- and ectomycorrhizal fungi host root signalling, 225–230
    FeMoCo, 222–223
    Frankia, 219–220
    K, 223–224
    mechanisms, 220–221, 220f
    N-fixing bacteria, 219–220
    nodulation, 220–221
    optimal pH, 221–222
    soil salinity, 221–222
  symbiotic root microbial communities
    edaphic soil conditions, 217–219, 218f
    elements, 217–219
    microbial clades, 219
    stemflow, 216–217
    throughfall, 216–217
Carbon sequestration, 84–86
Cardamine hirsuta, 60–62
Centers for Disease Control and Prevention (CDC), 116
Chemotaxis, 3–5
Chitinases, 247–248
Closed-loop-stripping analysis (CLSA), 257–258
Cytokinins, 69–70, 94

D

2,4-Diacetylphloroglucinol (DAPG), 12–13
Dimethyl disulphide (DMDS), 292–297

E

Ectomycorrhizal (ECM), 225
Entomopathogenic nematodes, 195–196
Ethylene, 95
Exopolysaccharides (EPS), 7–8, 31
Extracellular plant growth promoting rhizobacteria (ePGPR), 87–88

F

FeMoCo enzyme, 222–223
Flagellin, 25–26
Flavonoids, 163–164
  ABC, 165–166
  chemotaxis, 166–167
  MATE transporters, 165–166
  Nod factor synthesis, 164
  signalling spaces, 164–165
Frankia, 219–220

G

Genomic shock, 159–160
Gibberellin, 69–70, 94
Green leaf volatiles (GLVs), 288–289

H

HAP2, 143
Herbivore-induced plant volatiles (HIPVs), 196–203
Heterorhabditis bacteriophora, 198
Human pathogen
  attachment and attraction mechanisms
    E. coli and *Salmonella*, 129
    S. typhimurium, 127–129
    transmission electron microscopy image, 129, 130f
    types of interactions, 127–129
  circadian rhythm, 123–125
  control and prevention, 125–126
    AHLs, 126–127
    PGPR, 126, 127f
    stomata closure, 126, 128f
  merging plant science and food science, 116, 117t
  phytobacteria and rhizobacteria
    aperture opening size, stomata, 122–123, 123f
    Bacillus subtilis spp., 122
    cryo-scanning electron microscopy images, 122–123, 122f
    E. coli O157:H7, 120–122
    environmental factors, 121–122
    internalization, 120
    Salmonella, 120–121
  on plants
    Arabidopsis, 118–119
    bacterial relationship, 116–118
    human norovirus, 118
    hypersensitive reaction, 119–120
    JA and SA signalling, 119–120
    leafy greens, 116–118
    MAPK, 118
    SAR pathway, 119–120
  plant stomata, 123–125
Hyperparasites, 208
Hypersensitive response (HR), 24, 119–120

I

Indirect plant defence, 196–200
Indole-3-acetic acid (IAA), 92–94
Induced systemic resistance (ISR), 22–23
  soluble metabolites
    CLPs, 252
    fengycin, 251–252
    surfactin, 251–252
  three-step process, 23
  VOCs
    acetoin, 270
    DMDS, 270
    GC–MS and NMR-MS, 269–270
    JA/Et pathway, 269
Intracellular plant growth promoting rhizobacteria (iPGPR), 87
Iron absorption, 89–90

K

Klebsiella oxytoca C1036, 38

L

Lipopeptides (LPs), 34–35, 44–45
Lipopolysaccharides (LPS), 27
L-malate effect, 8–9

M

MAMPs-triggered immunity (MTI), 139
Mass spectroscopy (MS), 247–248
MATE transporters, 165–166
Methyl jasmonate (MeJA), 288–289
Methyl salicylate (MeSA), 288–289
Microbe associated molecular patterns (MAMPs), 24
  molecular patterns and cognate receptors
    Ax21, 27–28
    flagellin, 25–26
    flg22 and BAK1, 25–26
    FLS2, 25–26
    LPS, 27
    O-antigen, 27
    peptidoglycan, 26–27
    XA21, 27–28
  physiological responses and signalling, 28–29
Microbiota, 58–59
Mitogen-activated protein kinase (MAPK), 118
MYB72, 45–46

N

N-acyl-homoserine lactones (AHLs), 35, 126–127
N-alkylated benzylamine derivative (NABD), 35–36
National Institutes of Health (NIH), 118–119
Nematodes
  agricultural improvement
    A. thaliana, 204–205
    EβC, 205–206
    lepidopteran parasitism, 204
    TPS23, 204–205
  C. elegans, 194
  cuticular sense organs, 194
  entomopathogenic nematodes, 195–196
  functional types, 193
  indirect defences and HIPVs, 196
    A. syriaca, 198
    A. tulipae, 197
    B. juncea, 203
    entomopathogenic nematodes, 199
    gas chromatography, 201
    H. bacteriophora, 198
    Pellioditis, 199–200
    PTR-MS technology, 202
    SPME, 201
    T. tetraophthalmus, 198
    VOCs, 197
  multitrophic interactions, 206–209
  paradox surrounding belowground indirect defences, 206–209
  phasmids, 194
  roots, 192–193
  soil matrix, 192–193
Nitrogen-fixing clade (NFC), 170
Nod factor receptors (NFRs), 155
Nod factors (NFs), 137–138
Nodule Inception (NIN), 143
Nonribosomal peptide synthetases (NRPS), 249–250

O

O-antigen, 27

P

Pellioditis, 199–200
Peptidoglycan (PGN), 26–27
PGPR. *See* Plant growth-promoting rhizobacteria (PGPR)
Phasmids, 194
Phosphatidic acid (PA), 29
Phosphorus solubilization, 90–91
Phytohormones, 229–230
Phytostimulators, 92
  cytokinin, 94
  ethylene, 95
  gibberellin, 94
  IAA, 92–94
Plant growth promoting rhizobacteria (PGPR)
  A. thaliana, 2
  attraction
    A. brasilense, 5–6
    A. lipoferum, 5–6

Plant growth promoting rhizobacteria (PGPR) (*Continued*)
bacterial motility, 3–5
B. subtilis N11, 5–6
chemotaxis, 3–5
P. fluorescens, 3–5
characteristics, 83, 244–245
colonization, 88–89
biofilm formation, 7–8
B. subtilis, 9
exopolysaccharides, 7–8
L-malate effect, 8–9
methyl-salicylate, 9–10
quorum sensing, 8
tomato roots exudates, 8–9
commercialization, 101t
Azospirillum inoculants, 99–102
B. subtilis, 99
inoculation, 102–103
competence, 88–89
disease management practices, 242–244
general effects
A. brasilense, 10–11
flavonoids, 11
FZB42, 11
Paenibacillus polymyxa E681, 12
SbpA, 10–11
jasmonic acid signalling pathway, 2
mode of actions, 83, 84f, 85t, 245–247, 246f
nutrition
A. brasilense, 6
P. fluorescens, 7
P. putida, 6–7
phytohormones, 83–84
plant compounds, 12–14
plant immunity elicitation
glucanase, 22–23
ISR, 22–23
N-acetyl glucosaminidase, 22–23
rhizosphere
carbon sequestration, 84–86
ePGPR, 87–88
hydrogen cyanide/ethylene, 87
iPGPR, 87
root exudate, 84–86
root colonization, steps, 3, 4f
soluble metabolites
antibiosis, phytopathogens, 248–251
chitinases, 247–248
ISR, 251–252
mass spectroscopy, 247–248
plant growth promotion, 252–254
for sustainable agriculture, 89
biofertilizers, 89–95
biopesticides, 96–97
bioremediators, 97–99
VOCs, 254–255
antibiosis, 265–268
chemical diversity, 259–263
ISR, 268–270
phytopathogens, 263–265
plant disease escape, 270–272
screening and profiling methods, 255–259
Plant immunity elicitation, 39t–40t
alkyl chain-containing metabolites, 34–36
antibiotic function, 36–37
exopolysaccharides, 31
iron-regulated metabolites, 31–34
Klebsiella oxytoca C1036, 38
MAMPs
cognate receptors, 25–28
molecular patterns, 25–28
physiological responses, 28–29
signalling, 28–29
microbe-induced plant immunity, 24–25
PGPR/strains, 41t–42t
CHA0, 43–44
defence-related events, 46–47
downstream early signalling, 46–47
glucanase, 22–23
ISR, 22–23
N-acetyl glucosaminidase, 22–23
plant cell surface, 44–46
TCV, 38–43
WCS358, 38–43
WCS374, 38–43
surface-exposed MAMPs
LPS, 30–31
PGN, 30
Pseudomonas fluorescens WSC358, 29–30
P. syringae pv. *tomato* DC3000, 29–30
volatiles, 37–38

Plant volatilome, 287–288
Polyploidy, 150, 176–178
  flavonoids, 163–167
  Nod factor receptors, 167
  nodulation and rhizobial signalling
    Chamaecrista, 174–175
    legume phylogeny, 170, 171f
    M. truncatula genome, 173
    NFC, 170
    papilionoid lineage, 172
    predisposition, 172
  plant growth and cytokinins, 168–170
  resistance genes, 167–168
  species-level signalling spaces and allopolyploidy outcomes, 162–163, 163f
Potential throughfall/stemflow impacts
  dye tracer, 230–231
  endo- and ectomycorrhizal fungi host root signalling
    allelopathic compounds, 229–230
    arbuscular mycorrhizal, 225
    Cd and Zn, 228–229
    ectomycorrhizal, 225
    phytohormones, 229–230
    soil P, 227–228
    soil pH, 226–227
  N-fixing nodule-forming bacterial signalling pathways
    Acacia, 224–225
    allelopathic compounds, 224–225
    B deficiency, 222–223
    Ca spiking, 223–224
    FeMoCo, 222–223
    Frankia, 219–220
    K, 223–224
    mechanisms, 220–221, 220f
    N-fixing bacteria, 219–220
    nodulation, 220–221
    optimal pH, 221–222
    soil salinity, 221–222
Probiotics, 125–127
Proton-transfer-reaction mass spectrometry (PTR-MS), 202
Proton transfer reaction-time-of-flight-mass spectrometer (PTR-TOF-MS), 258
Pseudobactins, 31–32
Pseudomonas
  P. fluorescens, 3–5, 7
  P. putida, 6–7
Pyoverdines, 31–32

Q

Quorum sensing (QS), 8

R

Receptor-like kinases (RLKs), 24
Receptor-like proteins (RLPs), 24
R genes, 243–244
Rhamnolipids (RLs), 34–35
Rhizobial nodulation symbiosis
  genotype-by-genotype interactions, 178–179
  polyploidy, 150
    flavonoids, 163–167
    Nod factor receptors, 167
    nodulation and rhizobial signalling, 170–175
    plant growth and cytokinins, 168–170
    resistance genes, 167–168
    species-level signalling spaces and allopolyploidy outcomes, 162–163, 163f
  signalling space, 151, 152f
  symbiotic interactions, 152–153
    AON, 156–157
    L. japonicus, 155
    LysM, 155
    NFRs, 155
    NodD, 154
    Nod factors, 154–155
    papilionoid legume systems, 154
    Rfg1 and *Rj2* alleles, 156
    root hair infection, 153–154, 153f
    TIR-NBS-LRR, 156
  WGD
    allopolyploidy, 159–160
    autopolyploidy, 159–160
    neofunctionalization, 159
    NSP1, 158
    pollination interactions, 161
    RAM1, 158
    subfunctionalization, 159

Rhizosphere
carbon sequestration, 84–86
ePGPR, 87–88
hydrogen cyanide/ethylene, 87
iPGPR, 87
root exudate, 84–86
Ribosomal RNA (rRNA), 59
Root microbiome assemblage
below belowground rhizospheric microbiome, 70–72, 71f
host developmental stage
Arabidopsis, 63–64
bradyrhizobia and cyanobacteria, 64–65
phenolics/amino acids, 64–65
P. syringae, 63–64
host genome
A. thaliana, 60–62
OTU, 62
16S rRNA gene sequencing, 60–62
host health and fitness, 67–69
microbiota, 58–59
in plant signalling, 69–70
rRNA, 59
Root-nodule (RN) symbiosis
flavonoids, 137–138
innate immunity
MAMPs, 139
MTI, 139
NGR234, 140
succinoglycans, 139–140
T3SS, 140
isoflavonoids, 137–138
miRNAs, 143–144
missing regulators
limitations, 142
transcriptional reprogramming, 141
VAPIRYN, 142
Nod factors, 137–138

S

Salicylic acid (SA), 32–34
Salmonella typhimurium, 118
Signalling space, 151, 152f
Sinorhizobium meliloti, 290–291
Soil-dwelling organisms, 191–192
Solid-phase microextraction (SPME), 201, 257–258
Soluble metabolites
antibiosis, phytopathogens
categorization, 248–249
CLPs, 248–249
nonribosomal peptide synthetases, 249–250
chitinases, 247–248
ISR
CLPs, 252
fengycin, 251–252
surfactin, 251–252
mass spectroscopy, 247–248
plant growth promotion
availability to plants, 252–253
nutrient solubility, 252–253
plant growth regulators, 253–254
Steinernema glaseri, 195–196
Succinoglycans, 139–140
Sustainable agriculture, PGPR, 89
biofertilizers, 89–95
biopesticides, 96–97
bioremediators, 97–99
Symbiotic root microbial communities
edaphic soil conditions, 217–219, 218f
elements, 217–219
microbial clades, 219
stemflow, 216–217
throughfall, 216–217
Systemic acquired resistance pathway (SAR) pathway, 119–120

T

Tetraopes tetraophthalmus, 198
Throughfall, 216–217
Toll-interleukin receptor/nucleotide-binding site/leucine-rich repeat (TIR-NBS-LRR), 156
Type III secretion system (T3SS), 140

V

Volatile organic compounds (VOCs), 254–255
acetoin, 302–303
antibiosis
alcohols, 265

alkanes and alkenes, 266–267
defined, 265
ketones and aldehydes, 266
sulphur compounds, 267–268
A. thaliana, 297–298
bacterial volatiles, 299–300
benzaldehyde, 302–303
2,3-butanediol, 300–301
chemical classification, 300–301, 301f
chemical diversity, 259–263
defence against pathogens, 259, 260t–262t
DMDS, 292–297
ISR
acetoin, 270
DMDS, 270
GC–MS and NMR-MS, 269–270
JA/Et pathway, 269
P. aeruginosa, 301–302
phytopathogens
fungi and oomycetes, 263–264
1-undecene, 264–265
plant disease escape, 270–272
from plants
ethylene, 288–289
plant volatilome, 287–288
roots, 287–288
smells, 287–288
Vetiver oil, 290
quorum sensing, 298
rhizosphere and phyllosphere bacteria
bacterial volatiles, 293t–295t
1,8-cineole and rhizathalene A, 291–292
MeJA, 290–291
MeSA, 291–292
2-phenylethanol and nonanal, 291–292
S. meliloti, 290–291
screening and profiling methods, 255–256
antimicrobial activity, 256–257, 257f
CLSA, 257–258
nuclear magnetic resonance, 256–257
PTR-TOF-MS, 258
split-plate method, 256–257
SPME, 257–258
Venn diagram, 303f

W

Whole genome duplication (WGD)
allopolyploidy, 159–160
autopolyploidy, 159–160
neofunctionalization, 159
NSP1, 158
pollination interactions, 161
RAM1, 158
subfunctionalization, 159

AUTHOR INDEX

'*Note:* Page numbers followed by "f" indicate figures, "t" indicate tables.'

A

Aaron, J., 226
Aarts, H.J., 121–122
Abanda Nkpwatt, D., 291–292
Abbo, S., 62
Abbott, L.K., 227–228
Abbott, R., 176–177
Abd El Daim, I.A., 291–292
Abdala, G., 32–33
Abdeddayem, R., 221–222
Abdelmajid, K., 222–223
Abedon, B.G., 63–64
Abell, C., 32–33
Abizgil'dina, R.R., 245–247
Abou-Shanab, R.I., 97–99
Abuzinadah, R.A., 226–227
Acebo-Guerrero, Y., 98t
Aceto, H.W., 121–122
Achouak, W., 2, 12–13
Adam, A., 34–36, 39t–42t, 44, 96, 250–252
Adesemoye, A., 89–90
Adesemoye, A.O., 58–59
Adio, A.M., 63–64
Agerer, R., 227–228
Agnew, J.L., 253–254
Agrawal, A.A., 191–192
Aguilar, J.M.M., 166–167
Aguirre de Carcer, D., 9
Aharoni, A., 204–205
Ahemad, M., 83–84, 89–90
Ahmad, S., 5, 11, 67
Ahmadinejad, N., 59, 62–63, 70–72
Ahmed, I., 32–33
Ahmer, B.M., 120–121
Ahn, J.W., 292–297, 293t–295t, 301–302
Ait Barka, E., 22–23, 47, 87–88
Akakabe, Y., 290
Akashi, T., 164
Akiyama, K., 226
Alabouvette, C., 60–62
Alavi, M., 249–250
Albert, I., 25–26
Alborn, H.T., 197, 201–202, 204, 206–208
Albrecht, C., 226, 229–230
Albrecht, D., 11, 291–292
Albrektsen, A.S., 156–157
Al-Daoud, F., 63–64
Aldrich, J.T., 138–139, 142
Alemano, S., 32–33
Alexander, I., 225
Alexandre, G., 3–6, 96–97, 166–167
Alfaro-Cuevas, R., 293t–295t, 298
Ali, B., 253–254, 266
Ali, G.S., 266
Ali, J.G., 192–194, 197, 200–202, 204, 206–208
Ali, N.I., 36, 43–44
Ali, S., 32–33
Allan, D.L., 90–91
Allison, J.D., 207–208
Almeida, A., 159–160
Al-Soud, W.A., 64–65
Alston, M., 60–62
Alstrom, S., 66, 251
Altamirano-Hernandez, J., 270–271, 293t–296t, 298–300
Althabegoiti, M.J., 166–167
Alunni, B., 219–220
Alvarado-Affantranger, X., 143–144
Alvarez, D., 32–33
Alves, B.J.R., 87–88
Amara, U., 32–33
Ambert, A., 127–129
Ambrosini, A., 2–3, 66
Amiri, G.Z., 216–217, 230–231
An, C., 63–64
An, T., 274–275
Andersen, A.W., 195

Anderson, A.J., 37, 41t–42t, 287, 293t–295t
Anderson, J., 156–157, 216–217
Anderson, R.C., 226
Andrade, Z.D., 226–227
André Levesque, C., 59
André, F., 221–222, 226–227
Andreae, M.O., 254–255
Andreote, F.D., 59–60
Andreotti, S., 3, 7–8
Andrio, E., 29
Ané, J.M., 137–138, 142, 157–158
Angelli, A., 216–217
Angle, J.S., 97–99
Angulo, F.J., 116
Annapurna, K., 95
Antolín-Llovera, M., 25–26
Antonyuk, L.P., 100t
Antoun, H., 90–91
Ao, Y., 26–27
Aoki, K., 168–169
Aoki, T., 164
Appelbaum, E.R., 164
Appendinoc, G., 286–291
Aratchige, N.S., 197
Araújo, W.L., 58–59
Ares, A., 223–224
Ares-Orpel, F., 140
Arkhipova, T.N., 94
Armitage, J., 166–167
Arneth, A., 287–288
Arnold, B., 253–254
Arrebola, E., 250–251
Arrighi, J.F., 137–138, 155
Ascher, J., 216–217
Asensio, D., 254–255, 286–288
Ash field, T., 167–168
Ashby, A.M., 166–167
Ashford, D.A., 89–90
Asiimwe, T., 34–35, 39t–42t, 250–252
Aslam, S.N., 26–27, 139–140
Assenza, F., 59, 62–63, 70–72
Atieno, M., 99–102
Attarod, P., 216–217, 230–231
Atzorn, R., 119–120
Aubert, D., 217–219, 228–229
Audenaert, K., 33, 36–37, 39t–42t, 43–44
Audrain, B., 244–245, 254–255, 258–263, 299–302
Auger, J., 272
Aulakh, M.S., 66–67
Ausubel, F.M., 63–64, 118
Ausubel, F.N., 63–64
Avila, A., 217–219, 228–229
Avni, A., 27–28
Ayabe, S., 164
Ayax, C., 157–158
Ayers, T., 116
Ayyampalayam, S., 158
Azam, S., 173
Azarova, T., 6–7, 88–89
Azcon, R., 100t
Azcon-Aguilar, C., 228–229
Azevedo, J.L., 58–60
Aziz, A., 33, 41t–42t, 43–44

B

Backhaus, H., 2, 60–62
Backman, P.A., 99–102
Backnäs, S., 220–221, 223–224, 227–228
Badamo, J.M., 118
Badri, D.V., 2, 5–6, 10–11, 63–67, 69–72, 84–86, 244–245
Baetz, U., 287–288, 291–292
Bago, B., 226, 229–230
Bague, G., 208–209
Bahat-Samet, E., 9–10
Bahnweg, G., 35, 39t–40t, 96
Bailey, D.J., 138–139, 141–142
Bailey, M.J., 64–65
Bailey-Serres, J., 143
Baillieul, F., 35
Bailly, A., 255–267, 260t–262t, 270–271, 292–299, 293t–295t
Baird, L., 127–129
Bais, H.P., 3–6, 8–10, 58–60, 67–72, 84–86, 88, 118, 122–125, 122f–123f, 244–245, 287–290, 292–299
Bajic, V.B., 97–99, 100t
Bakker, M.G., 58–59, 64–67, 69, 84–86
Bakker, P.A., 22–23, 29–33, 38–43, 39t–42t, 46–47, 247–248, 251–254, 268

Bakker, P.A.H.M., 5–6, 10–11, 23, 33–34, 36, 39t–42t, 44–46, 58–59, 69–70, 83, 91–92, 96, 160–161, 245–247, 273
Balaraju, K., 34–35, 39t–40t, 251–252
Baldani, J.I., 7–8
Baldwin, I.T., 67–70, 192–193, 208, 255–256, 260t–262t, 264–265, 270–272, 286–287, 289–290, 293t–295t, 297–298
Balesdent, J., 59–62
Balestrini, R., 3, 7–8
Ballester, A., 83–84, 87–88
Balsanelli, E., 7–8
Bamberg, J., 160–161
Banba, M., 59
Banchio, E., 299
Bandyopadhaya, A., 299–302
Bao, T., 265
Bar, H., 166
Barahona, E., 9
Baraille, H., 160–161
Barak, J.D., 120–121, 129
Barbazuk, W.B., 143
Barber, A.L., 224–225
Barbour, W.M., 66, 166–167
Bardgett, R.D., 191–192, 219–220
Barea, J.M., 2–3, 66, 90–91, 100t, 221–222, 228–229
Bargmann, C.I., 194
Barka, E.A., 22–23
Barkei, J., 164
Barker, M.S., 158
Barnum, S.R., 89–90
Baroja Fernandez, E., 293t–295t
Baroni, M., 198–200, 205–206, 208
Barre, A., 137–138, 155
Barriuso Maicas, J., 30–31, 39t–40t
Barriuso, J., 22–23
Barron, D., 165–166
Bartels, J., 30
Barthelemy, J.P., 34–35, 39t–40t, 44–47
Bartley, J., 5–6
Bartolome-Esteban, H., 226
Bartsev, A.V., 140
Bartz, J.A., 127–129
Basavaraju, P., 98t
Bashan, Y., 5, 83–84, 87–88, 99–102
Basheer, C., 97–99, 100t
Basu, S., 163–164, 166
Batterman, S.A., 223–224
Batut, J., 219–220, 223–224
Bauchan, G., 129
Baudoin, E., 63–65
Bauer, W.D., 8, 166–167
Bauerle, T.L., 192–193, 197–198
Baumgarte, S., 59–60
Baumgartner, R., 255–258, 260t–262t, 264–266, 292–299, 293t–295t
Bazzicalupo, P., 194
Beatty, P.H., 179–180
Beauchamp, C.J., 90–91
Beaudry, T., 34–35
Beauregard, P.B., 6–9
Bécard, G., 226
Bec-Ferte, M.P., 164
Bechet, M., 250–251
Beck, M., 25
Beckendorf, S.K., 205
Becker, A., 11
Becker, J.D., 138–139
Bedini, E., 27
Bednarek, P., 63–64
Behers, L., 291–292
Beis, K., 129
Bélanger, R., 31–33
Belausov, E., 127–129
Belimov, A.A., 94
Bellenger, J.P., 222–223
Ben Amor, B., 137–138, 155
Ben Ari, J., 293t–295t, 298–299
Ben, C., 143
Bendixen, C., 59
Benedetti, C.E., 253
Beneduzi, A., 2–3, 66
Benhamou, N., 31–33, 91–92
Bennett, J.W., 274
Bennett, M.D., 159–160
Benson, C.E., 121–122
Berendsen, R.L., 23, 44–46, 58–59, 84–86, 160–161, 247–248, 251–254, 268
Berendzen, K., 293t–296t

Berg, G., 2–3, 59–64, 66–67, 84–86, 88, 96–97, 102–103, 296t
Berg, M.P., 191–194
Berge, O., 59–62
Bergelson, J., 64–65, 84–86
Berger, L., 287
Berger, S., 124–125
Bergström, G., 203
Bernier, S.P., 299–302
Bernstein, A.I., 274
Berrabah, F., 139
Bersoult, A., 137–138, 155
Bertani, I., 299
Bertea, C.M., 290
Bertrand, B., 160–161
Bertsche, U., 26–27, 30, 39t–40t
Beuchat, L.R., 116–118, 120
Beuter, C., 25–26
Beyeler, M., 33
Bhasin, A., 195
Bhat, S., 216–217
Bhatnagar, S., 58–59, 63, 70–72
Bhattacharyya, D., 293t–295t
Bhattacharyya, P.N., 89–91, 96–97
Bian, Z., 127–129
Bianciotto, V., 3, 7–8
Bicchi, C., 287–288
Biedrzycki, M.L., 37, 39t–42t, 58–59, 255–256, 260t–262t, 266, 270, 287, 292–297, 296t
Bigeard, F., 139
Bigirimana, J., 39t–42t, 43–44
Bihn, E.A., 121–122
Bilat, J., 201–202, 207–208
Bilgin, D.D., 139, 143–144
Bilgrami, A.L., 193–196
Billeter, R., 168–169
Bilyeu, K.D., 156–157
Bingham, E., 175–176
Binladen, J., 59
Birchler, J.A., 159–160
Bird, D.M., 156–157
Biro, B., 100t
Bischoff, S., 217–219, 224–225
Bisseling, T., 138–139, 143, 158, 169, 219–220, 223–224
Biswas, B., 156–157
Biswas, R., 26–27
Bittar, T., 217–219
Black, E.P., 129
Blaise, M., 155–156
Blanc, G., 159
Blanco, F.A., 143
Blande, J., 288–290
Blank, A.F., 60–64
Blaser, M.J., 160–161
Bloem, J., 193–194
Bloemberg, G.V., 2–7, 82–83, 85t, 88, 96, 273
Blom, D., 292–299, 293t–296t
Blom, J., 252
Boatwright, J., 172
Bobbink, R., 226–227
Boddey, R.M., 87–88
Bode, E., 195
Boff, M.I.C., 192–193, 195–197
Bogusz, D., 219–220, 223–224
Bohlmann, J., 287–288
Böhm, H., 25–26
Boiero, M.L., 94
Boland, W., 287–288
Bolaños, L., 222–223
Bolaños-Vásquez, M.C., 164
Boll, T., 139–140
Bollback, J.P., 59
Bollen, W.B., 216–217
Boller, T., 24–29, 39t–40t, 140, 292–299, 293t–296t
Bombarely, A., 176
Bonfante, P., 3, 7–8
Bongers, T., 194
Bonilla, I., 222–223
Bonisch, D., 255–267, 260t–262t
Bonito, G., 244–245
Bonkowski, M., 12–13, 191–192
Bookland, R., 164
Boon, J., 30–31
Boopathi, E., 91–92
Boor, K.J., 118
Bordiec, S., 47
Borken, W., 217–219
Borlotti, A., 25–26
Borrego, E., 289–290
Borriss, R., 11, 252, 291–292

Boruvka, L., 221–222, 228–229
Bosak, L., 289–290
Bose, A., 249
Bottini, R., 92, 94, 253
Boualem, A., 143
Boudesocque, L., 35, 39t–40t
Bouffaud, M.-L., 2–3, 60–62, 83–84, 84f, 87–88, 92–94
Bouillant, M.L., 96–97
Boukli, N.M., 140
Bourque, S., 28
Boursault, A.E., 208–209
Bouten, W., 217–219
Boutet Buchala, E.A., 96
Boutet, E., 36, 39t–42t, 43–44
Boutin, R., 222–223
Bouton, C.E., 191–193
Bouw, E., 121–122
Bouwman, L.A., 193–194
Bouwmeester, H.J., 204–205
Bouwmeester, K., 29–30, 34–35, 45–46
Bowen, B.A., 164
Bowen, G.D., 228–229
Bowers, J.E., 159, 167–168
Boyer, R.R., 120–121
Boyle, M., 228–229
Brad field, G.E., 222–223
Braden, C.R., 116
Bradley Day, R., 219–220
Brady, S., 296t
Brandao, M.M., 67
Brandl, M.T., 120–122, 125–129
Brans, A., 34–35, 39t–42t, 96, 251–252
Brauer, D.K., 223–224
Brazil, G.M., 97–99
Break field, N., 58–59, 69–70
Breakspear, A., 141
Brechenmacher, L., 137–139, 142–144
Bredemeier, R., 157–158
Brencic, A., 224–225
Brennan, M., 266
Bres, C., 157–158
Brigitte, B., 97–99
Brill, L.M., 155–156
Brill, W.J., 175–176
Brimecombe, M.J., 87
Britz, T.J., 296t
Brodribb, T.J., 123–124
Broeckling, C.D., 5, 64–65, 84–86
Broghammer, A., 155–156
Brombacher, E., 127–129
Brooks, D.M., 119–120
Broughton, W.J., 140, 151, 156
Brown, A.H., 160–161
Brown, D., 193–194
Brown, M.E., 83
Broz, A.K., 64–65, 84–86
Bruinsma, M., 208–209
Brun, A., 226–227
Bruneau, A., 172
Brunelle, A., 13
Brunner, F., 26–27, 30, 39t–40t
Bryant, M.L., 216–217
Brysting, A.K., 160–161, 177–178
Bubenickova, L., 221–222, 228–229
Bucciarelli, B., 138–139
Buchala, A., 33, 36, 39t–42t, 43–44
Buchan, A., 3–5
Buchner, A., 59–60, 63–64, 66
Budzikiewicz, H., 31–32, 35–36, 39t–42t, 44
Buee, M., 59–60, 63–64, 226
Bueno, C., 66–67
Buggs, R.J., 159–161
Bulgarelli, D., 2, 58–59, 62–63, 70–72
Bull, C.T., 3
Bulloch, E.M.M., 32–33
Burd, G.I., 97–99, 100t
Burdman, S., 87–88
Burke, P.V., 178–179
Burner, D.M., 223–224
Buschena, C.A., 228–229
Bustos-Sanmamed, P., 143
Butterfield, J.E., 120–121
Büttner, D., 140
Buyer, J.S., 98t
Buysens, S., 88–89
Buzas, D.M., 156–157

C

Caballero-Mellado, S., 85t, 92, 99–102
Cabellos, G.G., 97–99
Cabrera, J.A., 272
Cacco, G., 168–169

Caetano-Anollés, G., 166–167
Cai, M.Y., 90–91
Cairney, J.W.G., 226–229
Calderon-Villalobos, L.I.A., 92–94
Callanan, M., 97–99
Callaway, R.M., 84–86
Cameron, R.K., 63–64
Camire, C., 226–227
Campbell, C.D., 84–86
Campbell, J.F., 195–196, 205, 222–223
Campos-Herrera, R., 201–202, 204, 207–209
Camut, S., 141
Caniard, A., 26
Cannon, S.B., 158, 170–175, 171f
Cantor, A., 226
Cao, Q., 155, 169, 172
Cao, S., 8–9
Cao, Y., 140, 253–254
Capieau, K., 33, 39t–42t
Caplan, J.L., 58–59, 67–69, 122, 127f
Carata, E., 290
Cardenas, L., 137–138
Cardoso, D., 172
Cardoso, E.J.B.N., 66
Carella, P., 63–64
Carlyle-Moses, D.E., 216–219, 221–222
Carreri, A., 118
Carrillo-Castañeda, G., 91–92
Carroll, B.J., 155–156
Carroll, S., 152–153
Carvalhais, L.C., 2, 11, 69–70, 291–292
Carviel, J.L., 63–64
Casneuf, T., 159
Cassán, F.D., 92, 94
Castaneda, R., 8–9
Castellanos, T., 59–60
Castro, H., 59, 62–63
Castro, P.M.L., 221–222
Castro-Sowinski, S., 9–10
Catana, V., 63–64
Catrice, O., 219–220
Catroux, G., 99
Cattelan, A.J., 90–91
Cawoy, H., 13, 34–35, 251–252
Ceccherini, M.T., 216–217
Cenci, A., 160–161
Cernadas, R.A., 253
Certini, G., 216–217
Cesco, S., 154
Cevallos-Cevallos, J.M., 127–129
Chabaux, F., 222–223
Chabot, R., 90–91
Chacón, N., 223–224
Chai, Y., 3, 6–9, 67–69
Chalhoub, B., 157–158
Chalot, M., 226–227
Chameides, W.L., 254–255
Chan, P., 156–157
Chanderbali, A.S., 158
Chandler, D., 85t
Chandrasekar, G., 99–102
Chaney, R.L., 97–99
Chang, S.C., 216–219
Chang, S.Y., 37–38, 39t–40t, 255–256, 260t–262t, 264–265, 270, 293t–295t
Channer, S., 64–65
Chaparro, J.M., 2, 5–6, 10–11, 59–60, 63–67, 69–72, 84–86, 244–245
Chapman, B.A., 159, 167–168
Chappelka, A.H., 228–229
Charles, P., 272
Charles, T.C., 95
Charpentier, E., 118
Charpentier, M., 28, 143, 157–158
Charvat, I., 228–229
Chaston, J.M., 195
Chavez, R., 95
Che, F.S., 26
Chedly, A., 222–223
Chen, C.S., 216–217
Chen, C.-Y., 33, 37–38, 39t–40t, 255–256, 260t–262t, 264–265, 270, 293t–295t
Chen, F., 287–288
Chen, M., 177–178
Chen, N.W., 167–168
Chen, S., 29
Chen, T., 267
Chen, W., 100t
Chen, X.H., 11, 29, 252, 291–292
Chen, X.L., 66
Chen, Y., 3, 8–9, 67–69, 99–102

Chen, Z.J., 159–160, 229–230, 253–254
Cheng, J., 138–139, 142
Cheng, Z., 12–13
Cherdyntseva, T.A., 92–94
Chernin, L., 293t–295t, 298–300
Cheron, A., 219–220
Chew, K., 82–83
Chilvers, G.A., 225
Chin A-Woeng, T.F., 3, 82–83, 85t, 88, 96
Chinchilla, D., 25–26, 139–140
Chitwood, D.J., 193–194
Chiurazzi, M., 58–59
Cho, B.H., 37, 41t–42t, 268–270, 293t–295t, 297–298
Cho, I., 160–161
Cho, K.S., 100t
Cho, M.J., 168–169
Cho, S.M., 287, 293t–295t
Cho, Y.E., 34–35, 39t–40t
Choi, D., 253
Choi, H.K., 37–38, 39t–40t, 255–256, 260t–262t, 269–270, 299
Choi, J., 140, 253
Choi, O., 98t
Choudhary, D.K., 46
Choudhury, A.T.M.A., 99–102
Chovanec, P., 164
Chowdhary, P.K., 8
Christensen, S.A., 289–290
Chrzanowski, Ł., 35
Chung, E.J., 292–297, 293t–295t, 301–302
Chung, S., 98t
Chung, Y.R., 292–297, 293t–295t, 301–302
Churchland, C., 225
Cividini, D., 222–223
Clardy, J., 8–9
Clark, R.B., 226–227
Clauss, T.R., 138–139, 142
Clay, N.K., 63–64
Clément, C., 22–23, 35–36, 39t–40t, 87–88
Clemente, M.R., 143–144
Cloots, L., 10–11
Coaker, G.L., 120–121
Coate, J.E., 166, 176–177
Coenders-Gerrits, A.M.J., 216–217
Cognat, V., 28, 143
Cohen, A.C., 253–254
Cohn, J., 219–220
Coiner, H.A., 291–292
Colcombet, J., 139
Coleman, D.C., 192–194
Colon-Carmona, A., 2, 64–65
Colpaert, J.V., 226–227
Combes, M.C., 160–161
Combes-Meynet, E., 84–86
Combier, J.P., 143
Compant, S., 22–23, 87–88
Cong, P.T., 99–102
Connor, E.C., 292–299, 293t–296t
Constantinidou, H.I.A., 291–292
Contreras Cornejo, H.A., 286
Contreras-Perez, M., 255–259, 260t–262t, 264–268
Cook, D., 137–138
Cooley, M.B., 119
Coosemans, J., 263–264, 272
Copolovici, L., 291–292
Corbell, N.A., 88–89
Corberand, T., 64–65
Cornelis, P., 33, 36–37, 39t–42t, 43–44
Cornwell, W.K., 172, 174–175
Correa, O.S., 5
Cortes-Barco, A.M., 255–256, 260t–262t, 269–270, 293t–295t
Coruzzi, G.M., 142, 168–169
Costa, R., 66
Couderchet, M., 33, 41t–42t, 43–44
Courteaux, B., 35, 39t–40t
Courty, P.E., 225
Coutte, F., 251–252
Coventry, D.R., 221–222
Cowles, K.N., 120–121
Cox, C.E., 120–122, 125–126
Coxson, D.S., 217–219
Crannell, W.K., 223–224
Crawford, N.M., 168–169
Crespi, M., 143, 156–157, 270–272, 293t–295t
Crippen, T.L., 254–255

Cristescu, S.M., 202–203, 270–272, 293t–295t
Crist-Estes, D.K., 166–167
Crombie, B., 3–5
Croonenborghs, A., 92–94
Crossley, D.A., Jr., 192–193
Crowley, D.E., 83–84, 87–89, 91–92, 100t
Crutzen, P.J., 254–255
Cruz Ramos, H., 270–271
Cubero, J., 34–35, 251–252
Cuenca, G., 226–227
Culler, A.H., 253–254
Cullimore, J., 137–138, 154–155
Czaczyk, K., 35
Czaja, L.F., 141
Czapp, M., 30
Czarnota, M., 5–6
Czymmek, K.J., 5–6, 8–9, 37, 39t–42t, 58–59, 67–69, 88, 122, 255–256, 260t–262t, 266, 270, 296t

D

D' Haeze, W., 154–155
D' Hont, A., 159–160
da Costa, P.B., 66
da Silva, T.F., 60–64
Daayf, F., 31–32, 91–92
Dadon, T., 96–97
Dahmen, J.L., 138–139, 142
Damayanti, T.A., 83–84
Dandurishvili, N., 255–258, 260t–262t, 264–265
Dandurishvili, N., 293t–295t, 298
Dangl, J.L., 2, 24, 29, 59, 62–63, 70–72, 119–120, 242–243
Danhorn, T., 8
Daniel Hare, J., 207–208
Daniel, R., 296t
Daniel, V.L., 97–99
Danielson, R.M., 226–227
Danjo, Y., 11
Danna, C.H., 63–64
Danner, H., 202–203
Dardick, C., 27–28
Darvishsefat, A.A., 216–217, 230–231
Das, F., 10–11
Dash, S., 170–172, 171f, 174–175
Davey, L., 120
David, J.S., 216–217
Davidov, Y., 63–64
Davidson, G., 85t
Davies, K.A., 98t
Davis, T.S., 254–255
Dawson, J.O., 221–222
Day, M., 64–65
Day, R.B., 155–156
de Almeda Halfeld-Viera, B., 98t
de Silva, R.R., 98t
de Baura, V.A., 7–8
de Billy, F., 141, 143
De Bodt, S., 158–159
De Boer, W., 59–60, 63–64, 250–251
de Brogniez, A., 44–45
de Bruijn, I., 34–35, 58–59, 245–251
de Carvalho-Nieble, F., 141
De Castro, C., 26–27
De Deyn, G.B., 191–194
de Gouw, J., 202, 287–288
De Hoff, P.L., 155–156
De Keyser, A., 143–144
de Kock, M.J., 250–251
De Kogel, W.J., 91–92
de la Cruz, H.R., 270–271
De La Fuente, L., 7
de La Vega, O.M., 67–69
De Leij, F.A.A.M., 87
de Lima, H.C., 172
de Lorenzo, V., 97–99
De Luis, A., 28, 143
De Meyer, G., 33, 39t–42t, 43–44
De Mita, S., 155, 158, 169, 172
De Moraes, C.M., 208
De Mot, R., 10–11
de Nys, R., 8
De Pauw, E., 252
De Queiroz, K., 172, 178–179
de Queiroz, L.P., 172
de Ruiter, P.C., 191–194
De Rycke, R., 143–144
De Schrijver, A., 217–219
De Sordi, L., 299
De Souza, J.T., 250–251
de Souza, R., 66

de Vasconcellos, R.L.F., 66
de Vicente, A., 34–35, 251–252
De Vleeschouwer, M., 250–251
De Vleesschauwer, D., 22–23, 31–32, 36–44, 39t–42t, 247–249, 251
de Vos, R.C.H., 29–30, 34–35, 45–46
de Waard, P., 250–251
De Weert, S., 3–5
de Weger, L.A., 7
De Werra, P., 12–13
de Zamaroczym, M., 13–14
Deaker, R., 99
Deakin, W.J., 156
Deakiv, W.J., 140
Dean, C.H., 116–118
DeAngelis, K.M., 64–65
Deans, J.D., 221–222
de-Bashan, L.E., 87
Debellé, F., 137–138, 143, 170, 173
Debois, D., 44–45, 252
Deepak, S.A., 98t
Défago, G., 31–33, 41t–42t, 43–44, 91–92, 252–253
DeFlaun, M.F., 88
Degenhardt, J., 67–70, 191–192, 197–199, 201, 204–206, 208, 287–288
Degenkolb, T., 35
Dekkers, E., 29–30, 34–35, 45–46, 58–59, 245–248
Dekkers, L.C., 88
Del Giudice, L., 290
Del Nero, M., 217–219, 228–229
Del Val, C., 228–229
dela Cruz, H.R., 293t–296t, 298
Delano-Frier, J.P., 288–290
De-la-Pena, C., 67
Delaux, P.M., 142
Delepierre, M., 299–302
Deleu, M., 44–45, 251–252
DellaChiesa, V., 33
Delvasto, P., 83–84, 87–88
DeMarco, S.F., 199
Den Herder, G., 143–144
den Ouden, E.M., 30–33, 38–43, 39t–42t
den Ouden, F.M., 41t–42t
Dénarié, J., 137–138
Deng, K., 129
Deng, Y., 143
Denison, R.F., 177–178
Dennis, P.G., 2, 11, 63, 69–70
Denno, R.F., 191–192
Denoux, C., 63–64
Depuydt, S., 168–169
Deravel, J., 251–252
Desaki, Y., 27, 155
Desbrosses, G., 2–3, 83–84, 84f, 87–88, 92–94
Deslandes, L., 141
Devers, E.A., 143–144
Deyholos, M.K., 170–172, 171f, 174–175
Dezzeo, N., 223–224
Dhaoui, S., 221–222
Dharmasiri, N., 253–254
Dhondt, S., 47
Diaz-Perez, J.C., 120
Dick, R., 11
Dicke, M., 67–70, 191–193, 197, 204–205, 287–290
Dickschat, J.S., 244–245, 254–255, 287, 293t–296t, 300–301
Diez-Gonzalez, F., 120–122, 127–129
Dillman, A.R., 199, 208
Dillon, A.B., 195
Dilworth, M.J., 221–222
Dinakar, S., 31, 39t–40t
Ding, G.C., 59–64
Ding, X., 25–26, 253–254
Dirkx, F.P.M., 30–33, 38–43, 39t–42t
Divito, G.A., 223–224
Dixon, D.G., 97–99, 100t
Dixon, R.A., 165–166, 179–180
Dixon, R.K., 228–229
Djavaheri, M., 22–23, 31–33, 38–43, 39t–42t
Djordjevic, M.A., 156–157
Dobbelaere, S., 84–87, 92–94
Dodd, A.N., 124–125
Dodd, I.C., 94, 221–222
Doerge, R., 177–178
Dohrmann, A.B., 59–60
Domanski, G., 2
Dombrowski, N., 60–63
Domenech, J., 30–31, 39t–40t

Dommes, J., 31–32, 34–36, 39t–40t, 44–47
Dong, X., 23–24, 119–120, 253–254
Dongiovanni, C., 272
Donofrio, N.M., 37, 39t–42t, 58–59, 255–256, 260t–262t, 266, 270, 296t
Doornbos, R.F., 58–59, 64–65, 69–70
Doraiswamy, S., 99–102
Dorey, S., 35–36
Doty, S.L., 97–99
Dou, Z., 121–122
Douds, D.D., 67–69
Dougherty, E.C., 199–200
Dow, J.M., 27
Dowd, J.M., 172
Dowd, S.E., 287, 293t–295t, 298
Dowling, D.N., 97–99
Downie, J.A., 143, 151, 153–154, 156–158
Doyle, J.J., 120, 152–153, 157–161, 166, 170–172, 171f, 176–177
Doyle, M.P., 120–122
Drabek, O., 221–222, 228–229
Drancourt, M., 58–59
Dreesen, O., 270–271
Drevon, J.J., 223–224
Dridi, B., 58–59
Driscoll, C.T., 231
Drnevich, J., 139, 143–144
Dröge, J., 59, 62–63
Drogue, B., 84–86
du Jardin, P., 35–36
Duan, J., 95
Duarte-Gardea, M., 91–92
Dubery, I.A., 30–31, 46–47
Duby, F., 34–35, 39t–40t, 44–47
Duc, G., 157–158
Duca, D., 253–254
Dudareva, N., 197, 286–290
Dudeja, S.S., 221–222, 224–225
Duffy, B.K., 22–23, 91–92
Dugravot, S., 263–265, 267, 270
Duhl, T., 287–288
Duijff, B.J., 39t–40t, 91–92
Dun field, K.E., 66–67
Dunand, C., 29
Duncan, L.W., 201–202, 204, 207–209
Duncker, B.P., 12–13
Dung, T.D., 99–102
Dunkel, M., 247–248, 254–255, 287
Dunleavy, J., 99
Dunn, D.C., 208–209
Dunoyer, P., 253–254
Dunwell, J.M., 170, 173–174
Duran, J.D., 25–26
Durner, J., 30–31, 46–47
Durrant, W.E., 119–120
Dutta, S., 3, 12
Duuff, B.J., 273

E

Eberl, L., 8, 255–258, 260t–262t, 264–266, 292–299, 293t–295t
Ebermann, S., 216–217, 230–231
Edger, P.P., 142
Edqvist, J., 226–227
Edwards, A.C., 84–86, 143, 158
Edwards, J., 58–59, 63, 70–72
Edwards, K., 250
Edwards, P.J., 168–169
Effmert, U.E., 254–255, 296t
Egan, A.N., 158, 167–168, 176
Eggermont, K., 3–5
Eggert, K., 204
Ehlers, R.U., 205
Ehrhard, D.W., 137–138
Eilers, K.G., 64–65
Eisalou, H.K., 221–222, 228–229
Eisenreich, A., 252
El Alawi, Y., 97–99, 100t
El-Borai, F.E., 208–209
Elfstrand, M., 226–227
El-Hamdaoui, A., 222–223
Elhottovā, D., 58–59
Eliashvili, P., 255–258, 260t–262t, 293t–295t, 298
Elmerich, C., 219–220, 222–223
El-Sayed, A.S., 267
Elsenbeer, H., 216–217, 230–231
El-Yahyaoui, F., 143
Emadzade, K., 150
Emmons, L.K., 287–288
Endre, G., 139, 157–158

Engelen, K., 10–11
Ennis, D.E., 195
Enyedi, A.J., 119–120
Erb, M., 67–70, 191–192, 196, 198–199, 201–202, 207–208, 287–289, 299–300
Erbs, E., 139
Erbs, G., 25–27, 29–30, 139–140
Erdem, A.L., 129
Eremina, S., 255–257, 260t–262t, 263–268, 293t–295t
Erickson, D., 254–255
Erickson, M.C., 120, 124–125
Eriksson, K.B., 66
Eriksson, R., 286–287
Erkelens, A., 273
Espinosa-Urgel, M., 7–9
Estelle, M., 92–94, 253–254
Estill, J.C., 159, 167–168
Etzler, M.E., 155–156
Evanno, G., 219–220
Evans, J., 221–222
Evguenieva-Hackenberg, E., 35
Ewing, N.N., 155–156
Ezenwa, V.O., 160–161
Ezquer, I., 293t–295t

F

Fabbri, C., 292–299, 293t–296t
Falkengren-Grerup, U., 217–219
Falkinham, J.O., 120–121
Fall, R., 3, 8, 250–251, 254–255
Fan, B., 11
Fan, L., 25–26
Fan, T., 5–6
Farace, G., 252
Farag, M.A., 22–23, 33, 37–38, 39t–40t, 46, 69–70, 87–88, 96, 244–245, 254–263, 260t–262t, 268–272, 287, 292–298, 293t–296t
Farías Rodríguez, R., 293t–295t, 298–299
Farmer, A.D., 139, 143–144, 158, 174–175
Farrand, S.K., 30–31, 39t–40t
Farthing, D.E., 257–258
Fathizadeh, O., 216–217, 230–231
Fatica, M., 120–121
Fauconnier, M.L., 35–36
Fawcett, J.A., 160–161
Fedorova, E., 219–220
Fedoseyenko, D., 11
Feijo, J.A., 137–138
Fekete, A., 12
Felix, G., 24–29
Felle, H.H., 137–138, 155
Felten, J., 229–230
Feltus, F.A., 159, 167–168
Fendrik, I., 5–6, 13–14
Feng, H., 6, 8, 67–69
Ferguson, B.J., 154–155, 175–176
Fernandez, C., 126, 296t
Fernandez, M.R., 265
Fernandez, O., 35–36, 44–45, 252
Fernando, W.D., 255–256, 260t–262t, 267–268
Fernando, W.G.D., 99, 293t–296t
Ferrari, G., 168–169
Ferris, H., 193–194
Ficke, A., 34–35, 39t–42t, 250–252
Fierer, N., 64–65
Fiesselmann, A., 11
Fink, R.C., 120–122, 127–129
Finlay, B.B., 129
Firestone, M.K., 66–67
Fischer, M., 5
Fischer, S.E., 9–10
Fisher, C., 34–35, 251–252
Fisher, R.F., 154
Flagel, L.E., 159–160
Flagiello, A., 26–27
Flemming, H.-C., 7–8
Flis, S.E., 221–222
Flores Cortez, I., 67, 217–219, 293t–295t, 298–300
Fochesato, S., 12–13
Fonty, E., 172
Forchetti, G., 32–33
Ford, E.D., 221–222
Formey, D., 143
Fortes, E.D., 121–122
Fournier-Level, A., 165–166
Foxworthy, A., 6–7
Fraga, R., 83–84, 87–88
Frago, E., 69–70

Franc, A., 225
Franche, C., 219–220, 222–224
Francis, I., 22–23
Francke, W., 270–272, 293t–295t
Frank, J.F., 129
Franz, E., 121–122
Franzil, L., 44–45, 252
Fraser, E., 226–227
Fravel, D.R., 273
Fredeen, A.L., 222–223
Freeling, M., 158–160
Freeman, J.H., 263–264, 272
Freer, E., 129
Freestone, P., 228–229
Freitas, H., 97–99, 100t
French, B.W., 287–288
French, R.J., 60–62
Frey, M., 11, 288–289, 299–300
Friesen, J., 231
Frioni, L., 225
Froidure, S., 141
Frost, E.E., 216–219, 221–224
Frugier, F., 143, 156–157
Fterich, A., 221–222
Fu, F., 138–139
Fu, J., 253–254
Fu, S., 193–194
Fu, Y., 143
Fu, Z.Q., 23–24
Fuentes, S.I., 143–144
Fuentes-Ramirez, L.E., 85t, 92, 99–102
Fuhrmann, J.J., 90–91
Fujihara, S., 59
Fukai, E., 155
Fukui, R., 66–67
Fulchieri, M., 94
Fuqua, C., 8
Furlan, V., 226–227
Furnholm, T.R., 223–224
Furukawa, T., 26

G

Gaeta, R.T., 177–178
Gafny, R., 5
Gagne, S., 229–230
Gagnon, H., 154
Gaikwad, K., 95
Gaillard, M.D.P., 201–202, 207–208
Gajardo, G., 83–84, 87
Galbally, I.E., 287–288
Galiart, R.J.F.H., 206
Gallagher, A., 166–167
Gallagher, J., 159–160
Gallegos, M.-T., 11
Gallet, C., 217–219
Gamalero, E., 64–65
Gan, J., 272
Gang, W., 251
Ganio, L., 222–223
Gantner, S., 35, 39t–40t, 96
Gao, M., 6–8, 156
Gao, X., 29
Garbaye, J., 225
Garbeva, P., 2–3, 58–59
García, A.F., 5
Garcia-Brugger, A., 28, 35, 39t–40t
Garcia-Gutierrez, L., 34–35, 251–252
Gardener, B.B.M., 58–59
Gardiner, A., 255–258, 260t–262t, 264–266, 292–299, 293t–295t
Gardner, C.B., 222–223
Garladinne, M., 293t–295t
Garrido-Oter, R., 59, 62–63
Garsmeur, O., 159–160
Gash, J.H.C., 216–217
Gaskins, M.H., 83
Gaude, N., 143–144
Gaugler, R., 193–196, 199–200, 205
Gauthier-Lafaye, F., 217–219, 228–229
Gavrilenko, T., 160–161
Gehring, J., 58–59, 62–63, 69–72
Geisler, M., 293t–295t, 297–298
Gentil, E., 257–258, 270
Georgis, R., 199–200
Georgopoulos, C., 127–129
Geraats, B.P.J., 69–70
Gerardo, N.M., 160–161
Gerba, C.P., 88
Gerber, I.B., 30–31, 46–47
Gerik, J.S., 272
Germaine, K.J., 97–99
Germer, S., 223–224
Germida, J.J., 66–67
Geron, C., 254–255

Gerrits, A.M.J., 230–231
Gershenzon, J., 197–199, 204–205, 286–290
Gerstel, U., 30
Gertschb, J., 286–291
Geurts, R., 157–158, 170, 173, 219–220
Ghazaleh, J., 249–251
Ghequire, M.G., 250–251
Ghigo, J.M., 244–245, 254–255, 258–263, 299–302
Ghim, S.Y., 12, 38, 39t–40t, 69–70
Ghosh, S., 95
Gianinazzi, S., 90–91
Gianinazzi-Pearson, V., 39t–40t
Gibson, A.H., 221–222
Gierl, A., 11
Giger, A., 31–32
Gilbert, M.T., 59
Gildon, A., 228–229
Gilroy, S., 3–5, 59–60, 84–86, 244–245, 287–290
Gimsing, A.L., 200–201
Giordano, W., 299
Giorgobiani, N., 255–258, 260t–262t, 293t–295t, 298
Giraud, E., 219–220, 223–224
Giri, R., 221–222, 224–225
Giron, D., 69–70
Giron, J.A., 129
Gisch, N., 27
Gish, J., 156–157
Givskov, M., 8
Glazebrook, J., 63–64, 69–70
Gleason, C.A., 143, 157–158
Glenn, A.R., 221–222
Glevarec, G., 69–70
Glick, B.R., 2–3, 12–13, 87–88, 91–95, 97–99, 100t, 253–254
Gobbato, E., 158
Gogala, N., 228–229
Gojon, A., 168–169
Gokbulak, F., 221–222, 228–229
Goldan, P., 287–288
Golding, B., 63–64
Goldstein, A.H., 287–288
Goljevscek, K., 228–229
Gols, R., 69–70
Gomes, N.C.M., 66
Gomez, C., 165–166
Gomez, E., 91–92
Gomez-Bolea, A., 217–219, 228–229
Gomez-Gomez, L., 25–26, 39t–40t
Gong, A.D., 249
Gong, F., 6–7
Gonzalez, E., 265
Gonzalez, H., 222–224
Gonzalez, J.E., 8
Gonzalez, J.F., 12–13
Gonzalez, J.M., 222–224
González, T., 83–84, 87–88
Gonzalez-Guerrero, M., 222–223
Gonzalez-Rizzo, S., 156–157
Good, A.G., 179–180
Goodwin, P.H., 255–256, 260t–262t, 269–270, 293t–295t
Gool, A.P.V.A.N., 7–8
Gopher, A., 62
Gordon-Weeks, R., 5, 11, 67
Gorfer, M., 91–92
Gottel, N., 59, 62–63
Gottschalk, G., 88, 296t
Götz, F., 26–27
Goudeau, D.M., 121–122
Gough, C., 143, 154–155
Gould, S.J., 152–153
Goupry, S., 257–258, 270
Gourion, B., 139
Goussard, N., 90–91
Govers, F., 250–251
Govindasamy, V., 95
Grabherr, H.M., 26–27, 30, 39t–40t, 155
Graedel, T., 254–255
Graham, J.H., 222–224
Graham, P.H., 137–138
Gram, L., 8
Granhall, U., 92, 94
Grant, M., 69–70
Grant, W.P., 85t
Grantcharova, N., 7–8
Graus, M., 258
Gray, E.J., 87
Grayston, S., 84–86, 225
Greaves, J., 85t
Green, A.F., 160–161

Greenberg, B.M., 97–99, 100t
Greengart, A., 199–200
Greenspoon, P.B., 158
Greer, S., 6
Gresshoff, P.M., 156–157
Grewal, P.S., 195–196, 205
Grichko, V.P., 95
Griesbach, R., 163–164
Griffin, C.T., 195
Griffiths, B., 191–192
Grimson, M., 293t–296t
Grimsrud, P.A., 138–139, 141–142
Grishkan, I., 59
Gritsenko, M.A., 138–139, 142
Groenhagen, U., 255–267, 260t–262t, 292–299, 293t–295t
Grohn, Y.T., 121–122
Grolleau, F., 263–265, 267, 270
Grønlund, M., 155, 227–228
Gross, P., 191–193, 243–244
Grotewold, E., 244–245
Grover, C., 159–160
Groznik, N., 228–229
Gruening, B., 287
Grusell, M.-C., 217–219
Gu, G., 127–129
Gu, Y.-H., 7
Gu, Y.Q., 293t–296t
Guckert, A., 64–65
Guefrachi, I., 221–222
Guenther, A., 254–255
Guenther, A.B., 287–288
Guevara-García, A., 67–69
Guez, J.S., 250–251
Guillermin, M.L., 208
Guillod, L., 29–30, 34–35, 45–46
Gunter, L., 59, 62–63
Guo, J.-H., 98t
Guo, W., 168–169
Guo, Ya-H., 98t
Gurska, J., 97–99, 100t
Gust, A.A., 26–27, 155
Guswa, A.J., 216–217, 230–231
Gutiérrez Luna, F.M., 293t–296t, 298
Gutierrez Manero, F.J., 22–23, 30–31, 39t–40t
Gutiérrez, L., 272
Gutiérrez, R.A., 168–169
Gutierrez-Luna, F.M., 270–271
Gutierrez-Manero, F.J., 253
Gutjahr, C., 137–138

H

Haapalainen, M., 118
Haas, D., 33, 252–253
Had field, J., 158
Hadar, Y., 60–62, 66–67
Hafeez Adam, H., 97–99, 100t
Haga, M., 156
Hagen, M.J., 88–89
Haggag, W.M., 3
Hahm, M.S., 38, 39t–40t
Haichar, F.E.Z., 2, 12–13, 59–62
Haidacher, S., 258
Hain, R., 99–102
Haines, D.C., 8
Haines, K., 63–64
Hale, A., 226
Halitschke, R., 191–192, 286–287, 289–290
Halkmanb, A.K., 116–118
Hallem, E.A., 199, 208
Hallmann, J., 30–31, 39t–40t, 58–59
Hamburg, S.P., 224–225
Hammar, M., 127–129
Han, K.S., 269
Han, Q., 34–35, 39t–40t
Han, S.H., 37–38, 39t–42t, 268–270, 287, 293t–295t, 297–298
Han, S.W., 27–28
Hanano, Y., 11
Hancock, J.G., 66–67, 245–247
Hane, E.N., 224–225
Hanel, G., 258
Hann, D.R., 140
Hansel, A., 258
Hara, A.H., 199–200
Harada, K., 156
Harbert, R.S., 160–161
Hardtke, C.S., 168–169
Hare, J.D., 196, 206–208
Harish, S., 98t
Harkess, A., 170–172, 171f, 174–175
Harmand, J.M., 223–224

Haro, A., 97–99
Harper, J.E., 168–169
Harren, F.J., 270–272, 293t–295t
Harris, P.J., 226
Harrison, M.J., 229–230
Harrison, R., 195–196
Hartel, P.G., 90–91
Harter, K., 293t–296t
Hartley, J., 228–229
Hartman, R.D., 272
Hartmann, A., 12, 63–64, 99–102, 253, 299
Hartungen, E., 258
Hartwig, U.A., 137–138
Harvey, E.N., 199–200
Harwood, C.S., 5
Hase, C., 31–32, 41t–42t, 43–44
Haselwandter, K., 228–229
Hasnain, S., 253–254, 266
Hass, H., 172
Hassan, A.N., 129
Hassan, S., 229–230
Hassink, J., 193–194
Hata, S., 139
Hattermann, D.R., 66, 166–167
Hause, B., 69–70, 290–291
Haustein, M., 254–255, 259, 287, 292–297
Hawkins, B.A., 191–192
Hayashi, H., 226
Hayashi, M., 100t, 152–153, 156–158
Hayashi, N., 25–26
Hayashi, S., 154–155, 175–176
Hayat, R., 32–33
He, J., 299–302
He, P., 25, 28, 253–254
He, S.Y., 119–120, 123–124, 140
He, W.J., 249
He, X.Z., 165–166
Heald, C.L., 287–288
Heard, W., 25
Heath, K.D., 178–179
Heath, R.R., 203
Heavens, D., 60–62
Heckmann, A.B., 143
Hedin, L.O., 223–224
Heeb, S., 33
Heerklotz, H., 250
Hegarty, M., 176–177
Heijne, B., 226–227
Heil, G.W., 226–227
Heil, M., 39t–40t, 198, 207–208
Heimovaara, T.J., 217–219
Heinemeyer, I., 252
Heinrich, D., 5
Heinsbrock, M., 41t–42t
Hejzlar, J., 222–224
Held, M., 198–199, 204–205
Helder, J., 192–194, 200–201
Heller, G., 226–227
Hellriegel, H., 82–83, 99
Helm, R.F., 166–167
Helmisaari, H.S., 222–223
Hendrickx, N., 3–5
Hendson, M., 66–67
Henkels, M.D., 88–89, 91–92
Henry, G., 34–35, 39t–40t, 44–47, 251–252
Hense, B.A., 299
Henson, B.J., 89–90
Hepper, C.M., 226
Herbert, S.J., 66
Herendeen, P.S., 172
Hernandez-Lauzardo, A.N., 98t
Hernandez-Leon, R., 255–259, 260t–262t, 264–268
Hernandez-Reyes, C., 35, 39t–40t, 126–127
Hernandez-Rodriguez, A., 98t
Hernandez-Soberano, C., 255–256, 260t–262t, 266–268
Herrera-Estrella, L., 67–69
Herrmann, L., 99–102
Herschkovitz, Y., 63–64
Hervé, C., 137–138
Hess, D., 5
Hetherington, A.M., 123–124
Hetong, Y., 99
Heu, S., 120–121, 127–129
Heuer, H., 59–64, 66–67
Heulin, T., 2, 253
Heungens, K., 88–89
Hewitt, C.N., 254–255

Heydrich-Perez, M., 98t
Hibbard, B.E., 201–202, 207–208, 287–288
Hien, N.T., 99–102
Hien, T.M., 99–102
Hijmans, R.J., 160–161
Hildreth, S., 166–167
Hilliard, M.A., 194
Hiltpold, I., 196–202, 204–208, 287–288
Hirai, H., 26
Hirani, T.A., 156–157
Hirata, K., 100t
Hirayama, T., 168–169
Hirkala, D.L., 98t
Hirneisen, K.A., 118
Hirooka, K., 11
Hirose, N., 168–169
Hirota, A., 139
Hirsch, A.M., 155–156
Hirsch, S., 143
Hirt, H., 118, 139
Hiscock, S., 176–177
Hixson, K.K., 138–139, 142
Hoang, L.H., 255–256, 260t–262t, 264–265, 270–272, 293t–295t, 297–298
Hodkinson, B.P., 244–245
Hoekstra, R.M., 116
Hofferek, V., 143–144
Hoffmann, T., 270–271
Hofstetter, R.W., 254–255
Höfte, M., 22–23, 29–44, 39t–42t, 88–89, 247–252
Hogan, J.P., 97–99
Hogekamp, C., 141
Hohn-Bentz, H., 265
Holaday, S., 293t–295t
Holdaway-Clarke, T.L., 137–138
Holden, N., 118
Holguin, G., 5
Hollmer, S., 205
Holopainen, J., 288–290
Holovachov, O., 22–23, 194
Holsters, M., 154–155
Holt, D.B., 28, 143
Holterman, M., 194
Holzinger, R., 258
Hong, A.V., 199
Hooiveld, G., 138–139
Hopp, L., 216–217
Horan, D.P., 225
Horvath, D.M., 242–243
Hossain, M.M., 274
Hossbach, J., 287
Hou, Z., 129
Houlden, A., 64–65
Houlton, B.Z., 231
Houshyani, B., 289–290
Hoy, M.A., 205
Hsiang, T., 255–256, 260t–262t, 269–270, 293t–295t
Hsueh, Y.H., 67–69, 122, 127f
Hu, C.H., 22–23, 33, 37, 39t–40t, 87–88, 96, 122, 255–257, 260t–262t, 268–271, 287, 292–298, 293t–296t
Hu, J., 127–129
Hua, S., 168–169
Huang, C.-J., 37–38, 39t–40t, 255–256, 260t–262t, 264–265, 270, 293t–295t
Huang, L., 253–254
Huang, Q., 5, 255–256, 260t–262t
Huang, X.D., 97–99, 100t
Huang, X.F., 59–60, 70–72
Hubbell, D.H., 83
Hubert, J., 35, 39t–40t
Hue, B., 263–265, 267, 270
Huffaker, A., 288–290
Hugenholtz, P., 7–8
Hughes, D.T., 254–255
Huhman, D., 165–166
Huhndorf, S., 59
Hull-Sanders, H.M., 161, 163–164
Humphrey, P.T., 69–70
Hunter, M.D., 191–192, 196
Hunziker, L., 255–267, 260t–262t
Hurek, T., 5–6
Husband, B.C., 160–161
Huser, A., 12–13
Hutchins, W.C., 120–121
Hwang, I., 253
Hynes, M.F., 166–167
Hynes, R.K., 98t

I

Ianni, F., 249–251
Iavicoli, A., 36, 39t–42t, 43–44, 96
Ibekwe, A.M., 120–121
Ibrahim, R.K., 154
Ichinose, Y., 27–28
Ignatov, V.V., 100t
Igual, J.M., 83–84, 87–88
Ihring, A., 35, 39t–40t, 96
Iijima, Y., 290
Ikeda, M., 156
Ikeda, S., 59
Illig, T., 27
Ilut, D., 158, 174–175
Imfeld, G., 59–60
Inagaki, H., 26
Inagaki, Y., 27–28
Inamdar, A.A., 274
Inamdar, S.P., 216–219, 221–224
Inceoglu, O., 64–66
Indrasumunar, A., 154–156, 167, 175–176
Ingham, E.R., 193–194
Ingham, R.E., 193–194
Ingle, R.A., 124–125
Iñiguez, L.P., 138–139, 143–144
Iniguez-Luy, F., 177–178
Innan, H., 159
Inokura, Y., 221–222, 226–227
Inouye, D.W., 160–161
Insam, H., 287
Insunza, V., 66
Isaac, M.E., 223–224
Isaacs, M., 63–64
Ishibashi, N., 195–196
Ishiga, Y., 27–28
Ishikawa, K., 25–26
Ishizuka, J., 156
Islam, M., 120–122
Ito, A., 230–231
Iturbe-Ormaetxe, I., 156–157
Izaguirre, M.J., 32–33

J

Jabbouri, S., 151
Jackson, S.A., 138–139
Jaco, V., 97–99
Jacobs, J.M., 216–217
Jacobs, R., 250–251
Jacobsen, C.S., 87–88
Jacoud, C., 96–97
Jacquens, L., 252
Jacques, P., 31–32, 34–35, 91–92, 249–252
Jadin, V., 34–35
Jakobsen, I., 227–228
James, E.K., 58–59, 70–72, 150
Jander, G., 63–64
Jang, T.S., 150
Jasper, D.A., 227–228
Jauneau, A., 226
Jay, F., 253–254
Jeffries, P., 90–91
Jenkins, R.B., 217–219
Jersáková, J., 161–162
Jha, D.K., 89–91, 96–97
Jha, A.K., 118
Jiang, Q., 156–157
Jiang, S., 168–169
Jiang, X., 120–122, 287–288
Jiang, Z.-Q., 98t
Jiao, Y., 158
Jikumaru, Y., 27
Jin, J., 64–66
Jin, Y., 118
Jin, Z., 2, 59, 62–63, 70–72
JinWoo, P., 251–252
Jofré, E., 9–10
John, J., 224–225
Johnson, A., 12–13
Johnson, C., 58–59, 63, 70–72
Johnson, M.S., 220–221
Johnson, R.H., 161, 163–164
Johnson, S.N., 192–193
Johnston, A., 166–167
Johri, B.N., 46, 91–92, 99–102
Jonard, M., 221–222, 226–227
Joner, E.J., 228–229
Jones, A., 6–7
Jones, C.D., 59, 62
Jones, E.E., 64–65
Jones, J., 192–194
Jones, J.B., 120–121
Jones, J.D.G., 24–26, 29, 69–70, 119–120, 167–168

Jones, K.M., 139–140
Jongema, Y., 208–209
Joosten, M.H., 24
Jordan, A., 258
Joris, B., 34–35, 39t–42t, 96, 251–252
Jorquera, M.A., 83–84, 87
Jose Munoz, F., 293t–295t
Joseph, C.M., 137–138, 154–155
Joshi, R.S., 164
Jourda, C., 159–160
Jourdan, E., 34–36, 39t–42t, 44–47, 96, 251–252
Jousset, A., 12–13
Jovanovic, M., 143
Joy, J., 120–121
Juárez Muños, J., 91–92
Junge, H., 99–102
Junker, R.R., 254–255
Jurkevitch, E., 59–60, 63–64, 87–88
Jurkiewicz, A., 155

K

Kai, M., 254–255, 259, 270–272, 287, 292–298, 293t–296t
Kaiser, S., 59–60, 63–64, 66
Kaku, H., 27
Kalderas, J., 254–255
Kalisz, S., 226
Kaló, P., 143, 157–158
Kalsi, G., 155–156
Kam, J., 156–157
Kamilova, F., 2–3, 88–89, 96, 245–248, 270–271
Kamiya, A.C., 291–292
Kamiya, Y., 27
Kamo, K., 163–164
Kamp, R.M., 250–251
Kanamori, H., 156
Kaneko, T., 140, 174–175
Kang, B.R., 287, 293t–295t
Kania, A., 67–69
Kannaste, A., 291–292
Kape, R., 164, 166–167
Kaplan, F., 201–202, 204, 207
Kaplan, I., 69–70, 191–192, 196, 207–208, 286–287, 289–290
Kappers, I.F., 204–205
Kapulnik, Y., 5
Karamanoli, K., 291–292
Karaturovic, D.M., 83–84
Karban, R., 192–193
Karim, B.H., 222–223
Karlsson, G., 250
Karpati, E.V.A., 13–14
Kasper-Giebl, A., 258
Katagiri, F., 119–120
Katkova Zhukotskaya, O.A., 255–257, 260t–262t, 263–268, 293t–295t
Kato, J., 5
Kattge, J., 172, 174–175
Kauffinann, S., 35, 39t–40t, 47
Kaushik, S., 167–168
Kavino, M., 98t
Kavitha, K., 99
Kawaguchi, M., 156–157, 168–169, 290–291
Kaya, H., 193–196
Kaya, H.K., 195–196, 199–200
Kearns, D.B., 67–69, 122, 127f
Kech, C., 35–36, 39t–42t
Kecskés, K.L., 99–102
Keel, C., 12–13
Keen, N., 151
Keharia, H., 249
Keijers, V., 5, 10–11
Keim, R.F., 216–219, 221–222, 230–231
Kelley, W.D., 228–229
Kemmerling, B., 25–27
Kempster, V.N., 98t
Kene fick, L., 97–99
Kennedy, I.R., 99
Kenny, B., 129
Kenton, P., 119–120
Kerbarh, O., 32–33
Kerber, N.L., 5
Kereszt, A., 155–158, 167
Kernaghan, G., 226–227
Kerp, H., 172
Kesarwani, M., 299–302
Keshavan, N.D., 8
Keshelava, R., 255–258, 260t–262t, 264–265
Keshelava, R., 293t–295t, 298

Kessler, A., 67–69, 191–192, 198, 207–208
Kevei, Z., 157–158
Khalid, R., 32–33
Khan, G.H., 36, 43–44
Khan, M.S., 83–84, 89–90, 100t
Khan, W., 154
Khmel, I., 293t–295t, 298–299
Khmela, I.A., 299–300
Khong, N.G., 251–252
Kiang, C.S., 254–255
Kiba, T., 168–169
Kidd, B.N., 69–70
Kieber, J.J., 156–157
Kiers, E.T., 172, 174–175, 177–178
Kierul, K., 11, 291–292
Kilian, J., 293t–296t
Kilian, M., 99–102
Kim, B., 208
Kim, D.H., 173
Kim, I.S., 22–23, 38, 39t–40t
Kim, J., 89–90, 98t, 143, 158
Kim, K.Y., 37–38, 39t–42t
Kim, M.S., 287, 293t–296t, 298
Kim, S., 293t–295t, 298–299
Kim, S.D., 98t
Kim, S.H., 98t
Kim, Y.C., 22–23, 38, 39t–42t
Kinkel, L.L., 245–247
Kirkegaard, J.A., 60–62
Kishimoto, K., 255–256, 259, 260t–262t, 269–270
Kishore, G.K., 82–83, 91–92
Kiss, G.B., 157–158
Kiss, P., 13–14
Kissinger, J.C., 159, 167–168
Kistner, C., 174–175
Kitto, S.L., 67–69, 122, 127f
Klauser, D.R., 292–299, 293t–296t
Klaus-Heisen, D., 141
Klempien, A., 286–287, 289–290
Klimova, S.Y., 92–94
Klironomos, J.N., 191–192
Kloepper, J., 82–83, 89–90
Kloepper, J.W., 2–3, 22–23, 31–34, 37–38, 39t–42t, 58–59, 69–70, 83, 90–91, 96–97, 98t, 99–102, 122, 126, 245–249, 251, 255–256, 260t–262t, 268–272, 287, 292–297, 293t–296t, 299, 301–302
Kløve, B., 220–221, 223–224, 227–228
Knappe, C., 35, 39t–40t
Knappe, C., 96
Knee, E., 6–7
Kniel, K.E., 118, 122–125, 122f–123f
Knight, R., 64–65
Knirel, Y.A., 27
Knowlton, S., 221–222
Knuckles, M.E., 116–118
Knudsen, J.T., 286–287
Kobayashi, D., 27
Kobayashi, H., 140
Kobayashi, K., 9–10, 13, 291–292
Kobe, B., 156–157
Koczan, J., 123–124
Kodesova, R., 221–222, 228–229
Kogel, K.H., 126–127, 128f
Koglin, A., 249–250
Kojima, M., 156–157, 168–169
Kokalis-Burelle, N., 272
Koksharova, O.A., 255–257, 260t–262t, 263–268, 293t–295t, 299–300
Kolb, D., 26–27, 30, 39t–40t
Köllner, T.G., 197–199, 201, 204–206, 208, 287–288
Kolter, R., 3, 6–9, 67–69
Komatsu, A., 5
Kondo, E., 195–196
Kondorosi, A., 137–138
Kondorosi, E., 137–138
Kondrashov, F., 159
Kong, H., 98t
Koo, S.Y., 100t
Kopacek, J., 222–224
Koren, O., 2, 59, 62–63, 70–72
Kornyeyev, D.A., 293t–295t
Korsten, L., 250–251
Kos, M., 289–290
Kosslak, R.M., 164
Kost, C., 207–208
Kothe, E., 221–222, 224–225
Kouchi, H., 139, 156
Koumoutsi, A., 252

Kraepiel, A.M.L., 222–223
Krajinski, F., 143–144
Kravchenko, L.V., 6–7, 88–89
Krebs, B., 99–102
Krell, T., 11
Krempl, P.M., 249–250
Kreuzwieser, J., 66–67
Krier, F., 252
Krimm, U., 291–292
Krishnamachari, V., 293t–296t
Krishnamoorthy, A.S., 255–256, 260t–262t, 267–268, 293t–296t
Krishnan, H.B., 156, 164
Krištůfek, V., 58–59
Kroes, A., 69–70
Kropacova, M., 164
Kroppenstedt, R.M., 63–64, 66–67
Krouk, G., 168–169
Kroupitski, Y., 121–122, 127–129
Kruijt, M., 58–59, 245–248
Krusell, L., 155–156
Kubátová, B., 161–162
Kubesova, M., 160–161, 177–178
Kuc, J., 119–120
Kudo, T., 168–169
Kudoyarova, G.R., 94
Kuffner, M., 91–92
Küfner, I., 3–5, 25–26
Kuhlmann, U., 197–201, 204–206, 208, 287–288
Kuhn, U., 287–288
Kuiper, I., 6–7
Kuiper, M., 159
Kulikova, O., 157–158
Kumagai, H., 139
Kumamoto, K., 11
Kumar, A.S., 58–59, 122
Kumar, J., 223–224
Kumar, N., 98t, 221–222
Kumar, S., 127–129
Kunikane, S., 11
Kunin, W.E., 160–161
Kunjeti, S.G., 37, 39t–42t, 58–59, 255–256, 260t–262t, 266, 270, 296t
Kunkel, B.N., 119–120
Kunkel, J.G., 137–138
Kunze, G., 26
Kuramae, E.E., 69–70
Kuromori, T., 168–169
Kurzt, B., 205–206
Kus, J.V., 63–64
Kush, J.S., 228–229
Kuster, W., 287–288
Kuznetsov, A.E., 299–300
Kuzyakov, Y., 2
Kwon, Y.S., 269
KyungSeok, P., 251–252

L

Lopez Bucio, J., 293t–295t, 299–300
Labbé, J., 59, 62–63
Lacroix, H., 47
Laffont, C., 143
Lafont, F., 5–6
Lagares, A., 9–10
Laine-Kaulio, H., 220–221, 223–224, 227–228
Lajunen, H.M., 26–27
Lakshman, D.K., 98t
Lakshmanan, V., 8–9, 58–60, 67–72, 84–86, 122, 127f, 244–245, 252–253, 298–299
Lalande, R., 90–91
Laloux, M., 35–36
Laluk, K., 25–26
Lambrecht, M., 3–5, 10–12
Lamm, P., 141
LaMoreaux, J., 200–201
Lamotte, O., 28
Lamparter, T., 124–125
Landgraf, R., 69–70
Landherr, L., 158
Landini, P., 127–129
Lane, A., 5–6
Langley, R.J., 139, 143–144
Laniya, T.S., 156–157
Lankau, R.A., 67
Lanoue, A., 12–13
Lapeyrie, F.F., 225
Laporte, P., 143
Larose, G., 229–230
Larroy, C., 265
Lashermes, P., 160–161

Laskey, J.G., 156–157
Lauber, C.L., 64–65
Lavania, S., 163–164, 166
Lavania, U.C., 163–164, 166
Lavin, M., 172
Ławniczak, Ł., 35
Le Mellec, A., 217–219
Le Roux, J.J., 160–161, 177–178
Lea, U.S., 168–169
Lebeck, L.M., 199–200
Lebeis, S.L., 2, 58–59, 62–63, 69–72
Lebuhn, M., 253
Leclere, V., 250–251
Lecourieux, D., 28
Lee, B.L., 37–38, 39t–40t, 67–70, 255–256, 260t–262t, 268–272, 293t–295t, 301–302
Lee, C.H., 299
Lee, D.H., 120–121, 127–129
Lee, D.L., 194
Lee, H.S., 177–178
Lee, J.H., 208, 269
Lee, J.J., 177–178, 199–200
Lee, J.K., 299
Lee, M.C., 37, 41t–42t
Lee, M.K., 167–168
Lee, S.H., 22–23, 34–35, 37–38, 39t–40t, 67–70, 173, 253, 255–256, 260t–262t, 268–272, 293t–295t
Lee, S.J., 37, 41t–42t, 268–270, 293t–295t, 297–298
Lee, S.W., 27–28, 34–35, 38, 39t–42t
Lee, Y.H., 12, 273, 287, 293t–295t, 297–298
Leelasuphakul, W., 251–252
Leeman, M., 30–33, 38–43, 39t–42t
Lefebvre, B., 137–138
Legates, D.R., 217–219
Legué, V., 229–230
Lehmann, J., 220–221
Lei, Z., 67
Leija, A., 143–144
Leitner, E., 299–300
Lejeune, P., 127–129
LelandaisBrière, C., 143
Lemanceau, P., 39t–40t, 58–59
Lemarchand, D., 222–223
Lemfack, M.C., 247–248, 254–255, 287
Lenk, C., 67–70, 198–199, 204–205
Lenz, H.D., 26–27
Leon, E., 177–178
Leon, M., 5
Leonard, R.T., 222–224, 227–228
Leong, J., 2–3, 83, 245–249
Leonova-Erko, E., 6–7
Leon-Reyes, A., 24–25, 69–70
Lepori-Bui, M., 217–219, 223–224
Leps, W.T., 175–176
Lerat, S., 226
Lerdau, M.T., 290
Lerner, A., 63–64
Lesna, I., 197
Lester, S., 164, 175–176
Lesueur, D., 99–102
Létoffé, S., 299–302
Letort, S., 141
Leung, G.C., 98t
Leveau, J.H., 120–121
Levia, D.F., 58–59, 122, 216–219, 221–224
Levin, D.A., 163–164, 166, 179
Lévy, J., 157–158
Levy, M., 163–164, 179
Lewin, A., 155
Lewis, E., 195–196
Lewis, G.P., 172
Lewis, W.J., 197, 208
Leyval, C., 228–229
Li, C., 170, 173–174, 222–223
Li, D., 167
Li, G., 274–275
Li, H.M., 143, 267
Li, H.P., 249
Li, J., 8, 120–121, 293t–295t
Li, J.D., 87–88
Li, J.F., 26–27
Li, Q.G., 95, 170, 173–174
Li, S., 5–6, 8
Li, S.-M., 98t
Li, W., 25–26, 250–251
Li, X.Y., 216–217, 253–256, 259, 260t–262t, 264–265
Li, Y.T., 69–70, 168–169, 216–217

Li, Z.N., 26–27, 31–32, 38–43, 39t–42t, 253–254, 270–271, 273, 293t–295t
Liang, Y., 140
Liang, Z., 274–275
Liao, S., 140
Libault, M., 137–139, 143–144
Lichtenzveig, J., 156–157
Licy, O., 97–99
Lieberei, R., 67–69
Lillo, A., 169
Lillo, C., 168–169
Lim, J.A., 120–121, 127–129
Limpens, E., 138–139, 169, 223–224
Lin, H., 216–217
Lin, M., 154–156, 175–176
Lin, W., 29
Lin, X., 168–169
Lin, Y., 154–155, 175–176
Lindsay, R.W., 254–255
Lindsay, W.L., 221–222
Lindström, K., 219–220, 222–223
Ling, N., 5
Liparoti, V., 26–27, 255–257, 260t–262t, 263–268
Lipasova, V.A., 293t–295t
Lipka, V., 25–26
Lisa, L., 164
Litt, G., 222–223
Little, R.D., 95
Liu, B., 26–27
Liu, C., 141
Liu, F., 255–256, 259, 260t–262t, 264–265
Liu, H., 3, 8, 67–69, 253–254
Liu, J.D., 64–66
Liu, L., 99–102
Liu, P., 273
Liu, R.S., 267
Liu, W., 155, 172
Liu, X.B., 26–27, 30, 39t–40t, 64–66, 97–99, 159–160
Liu, Y.H., 5–6, 8, 34–35, 67–69, 167–168
Liu, Z., 25–26
Ljung, K., 253–254, 266
Lloret, J., 100t
Loake, G.J., 166–167
Lodge, D.J., 225
Lohar, D.P., 139–140, 156–157
Long, H.H., 270–272
Long, S.R., 60–62, 89–90, 137–138, 154, 164, 175–176
Long, T.A., 222–223
Loon, L., 64–65
Loon, L.C., 33, 38–43, 41t–42t
Loper, J.E., 33–34, 88–89
Lopez Bucio, J., 286, 293t–296t, 298
López-Aranda, J., 272
Lopez-Bucio, J., 67–69, 255–256, 260t–262t, 266–268, 270–271
López-García, S.L., 166–167
Lopez-Velasco, G., 120–121
Loreto, F., 287–290
Lorv, J., 253–254
Losick, R., 3, 6–9, 67–69
Lottmann, J., 63–64, 66–67
Love, J., 124–125
Lowry, E., 164, 175–176
Loyola-Vargas, V.M., 67
Lu, D., 29
Lu, J., 196
Lu, K.C., 216–217
Lu, S., 99–102
Lucangeli, C., 94
Lucas, M., 129
Lucci, G., 168–169
Luciano, A.K., 166
Luckerhoff, L.L.P., 204–205
Lucy, M., 91–92
Lucy, S., 60–64
Lugtenberg, B.J.J., 2–3, 6–7, 22–23, 82–83, 85t, 88–89, 96, 245–248, 270–271, 273
Lumyong, S., 60–62
Lundberg, D.S., 2, 58–59, 62–63, 69–72
Lundquist, J., 231
Luo, C., 34–35
Lurie, E., 58–59, 63, 70–72
Luxen, A., 35–36, 39t–42t
Lydon, J., 98t
Lynch, J.M., 87
Lyons, W.B., 222–223

M

Ma, J., 5, 120–121
Ma, L., 120
Ma, Y., 97–99, 100t
Mabood, F., 154
Macagnan, D., 98t
Macarisin, D., 129
Macherel, D., 263–265, 267, 270
Macho, A.P., 24–26, 44
Macías Rodríguez, L.I., 255–259, 260t–262t, 264–268, 270–271, 286, 293t–296t, 298–300
Macinnis-Ng, C.M.O., 217–219
Mackova, M., 100t
Madamanchi, N.R., 119–120
Madlung, A., 150, 177–178
Madsen, E.B., 155
Madsen, L.H., 155–157
Maeda, J., 138–139, 141–142
Maere, S., 158–159
Maffei, M.E., 286–291
Mage, S.M., 216–217, 221–222
Mahaffee, W.F., 58–59
Mahdhi, M., 221–222
Maillet, F., 137–138, 141, 158
Makarov, O.E., 100t
Makarova, N., 6–7, 88–89
Maki, S.L., 158, 174–175
Maksimov, I.V., 245–247
Malfatti, S., 59, 62–63, 70–72
Mälkönen, E., 222–223
Mallon, C.A., 58–59
Malnoë, P., 140
Mandrell, R.E., 119
Manefield, M., 8
Mangoni, L., 27
Manjula, K., 22–23
Manter, D.K., 5–6, 10–11, 64–67, 69–72, 84–86
Manukian, A., 203
Mao, L., 273
Maolanon, N., 155–156
Marchal, K., 10–11
Marino, D., 29
Marino, M., 270–271
Marinova, K., 165–166
Mariutto, M., 34–36, 251–252
Märk, L., 258
Markland, S.M., 122f–123f, 122–125
Markmann, K., 28, 143
Marol, C., 59–62
Marques, J.M., 60–64
Marschner, H., 226–227
Marschner, P., 67–69
Marsh, J., 143
Marsh, J.F., 158
Martin, F., 59–60, 63–64, 229–230
Martin, S.L., 160–161
Martinenko, E.V., 94
Martinez, E.A., 141
Martínez-Gil, M., 7–8
Martínez-Granero, F., 9
Martinez-Meza, E., 216–217, 230–231
Martinez-Vaz, B.M., 120–122, 127–129
Martinez-Viveros, O., 83–84, 87
Martinoia, E., 287–288, 291–292
Masciarelli, O.A., 32–33, 94
Massardo, D.R., 290
Masson-Boivin, C., 219–220, 223–224
Mathesius, U., 63–64, 152–153, 229–230
Matschonat, G., 217–219
Matsui, K., 255–256, 259, 260t–262t, 269–270, 290
Matsuoka, H., 11
Matsuzaki, K.I., 226
Matthiadis, A., 222–223
Matz, T., 200–201
Matzner, E., 216–219
Maurhofer, M., 12–13, 31–33, 41t–42t, 43–44
Mavrodi, D.V., 7, 36, 39t–42t
Maximilien, R., 8
Maxwell, C.A., 137–138, 154–155
May, G.D., 137–138, 158, 174–175
Mayak, S., 95
Mayer, A.M., 96–97
Mayer, C., 26–27
Mayer, P., 200–201
Mayerhofer, M.S., 226–227
Maynard, J.A., 100t
Mayrose, I., 158
Mazzola, M., 7, 247–251
Mazzotta, S., 25–26
Mbengue, M., 25, 141

McAdam, S.A., 123–124
McAlvin, C.B., 140
McAvoy, T., 263–264, 272
McClintock, B., 159–160, 176–177
McCollum, G., 251–252
McConkey, B.J., 12–13
McDonald, M., 58–59, 69–70
McDowell, W.H., 231
McGuire, T.R., 205
McIntyre, D.D., 217–219
McKain, M.R., 170–172, 171f, 174–175
McNeill, A.M., 2–3, 66
McPheron, B.A., 191–193
McRoberts, N., 253
McSorley, R., 272
Mears, J.A., 164
Medina, M., 160–161
Medrinna, A., 143–144
Meesenburg, H., 217–219
Meharg, A.A., 226–229
Mehouachi, J., 253
Mei, R., 99–102
Mei, X., 5, 8, 67
Meier, S., 228–229
Meincke, R., 59–62
Meldau, D.G., 255–256, 260t–262t, 264–265, 270–272, 293t–295t, 297–298
Meldau, S., 69–70, 255–256, 260t–262t, 264–265, 270–272, 293t–295t, 297–298
Melentiev, A.I., 94
Mellersh, D., 63–64
Melotto, M., 123–124
Men, A.E., 155–157
Mendes, R., 2–3, 58–59, 70–72, 244–248
Mendonca-Hagler, L., 66
Meneses, C.H., 7–8, 226–227
Meng, H., 168–169
Menge, J.A., 222–224, 227–228
Mënne-Loccoz, Y., 84–86
Mercado-Blanco, J., 22–23, 33, 38–43, 41t–42t
Mercure, M., 172
Merg, K.F., 161
Mergaert, P., 219–220
Meruva, N.K., 257–258
Metraux, J.P., 31–33, 36, 39t–42t, 43–44, 96
Meuwly, P., 31–32, 41t–42t, 43–44
Meyer, A., 30–31, 39t–40t, 46–47
Meyer, F.H., 225
Meyer, G.A., 161, 163–164
Meyer, J., 66
Meyer, J.M., 33, 38–43, 41t–42t
Meyers, B.C., 143, 167–168
Meziane, H., 29–32, 38–43, 39t–42t
Micallef, S.A., 2, 64–65
Michael, R., 97–99, 100t
Michalzik, B., 217–219, 224–225, 231
Miché, L., 96–97
Michiels, K.W., 7–8, 10–11
Mieczkowski, P., 59, 62
Miersch, O., 119–120
Miguel, M.J., 9–10
Mihacea, S., 157–158
Miiller, K.M., 95
Milillo, S.R., 118
Miller, G.W., 274
Miller, L.D., 166–167
Miller, R.W., 223–224
Miller, S.S., 138–139
Miller, W.G., 119
Millet, Y.A., 63–64
Millner, P., 120–122
Minasian, H., 225
Minchew, K.J., 166
Minz, D., 60–62, 66–67
Mirabella, R., 137–138, 155
Miranda, J.D., 226
Misra, N.K., 163–164, 166
Missiakas, D., 127–129
Mitchell, M.J., 216–219, 221–224
Mithofer, A., 287–288
Mitra, R.M., 143, 157–158
Miya, A., 27
Miyagi, M., 167
Miyahara, A., 156–157
Mo, M.H., 293t–296t
Moeder, W., 63–64
Moënne-Loccoz, Y., 2–3, 60–64, 83–84, 84f, 87–88, 92–94
Mohammad, A., 63–64
Mohney, B.K., 200–201

Molgaard, J., 100t
Molina, F., 254–255, 259, 287, 292–297
Molinari, J., 231
Moling, S., 138–139
Molitor, A., 126–127, 128f
Mombello, D., 290
Mongiardini, E.J., 166–167
Monson, R.K., 287–288
Montero, M., 293t–295t
Moon, J.H., 37, 41t–42t, 268–270, 293t–295t, 297–298
Moore, D.M., 216–219, 221–222, 227–228
Moore, F.P., 97–99
Moore, J.C., 193–194
Moore, R.D., 216–217
Mora, M.L., 83–84, 87
Morandi, D., 226
Moreau, S., 141, 143
Morel, F.M.M., 222–223
Morgan, D.R., 172
Morgan, J., 120–122
Mori, G.B., 9–10
Mori-Hosokawa, S., 156
Morikawa, M., 249–250
Moriwaki, H., 230–231
Morrissey, K.L., 139–140
Mou, Z., 63–64
Mougel, C., 64–65
Moutoglis, P., 229–230
Moyne, A.L., 32–33, 98t
Mu, W., 255–256, 259, 260t–262t, 264–265
Mubarik, N.R., 83–84
Mueller, M.J., 124–125
Muhlemann, J.K., 286–287, 289–290
Mukai, Y., 163–164, 166
Mulchandani, A., 100t
Mulder, L., 174–175
Mulders, I., 88
Mulders, I.H.M., 3–5
Mulle, D., 2–3, 60–62, 83–84, 84f, 87–88, 92–94
Muller, H., 96–97, 102–103, 217–219, 249–250, 299–300
Müller, M., 258
Mullin, B.C., 172
Mumm, R., 197
Münch, P.C., 59, 62–63
Munnik, T., 29
Muñoz, A., 143
Munoz, J.A., 83–84, 87–88
Münzbergová, Z., 161–162
Muoghalu, J.I., 223–224
Mur, L.A., 119–120
Muramoto, S., 290
Muratova, A.Y., 100t
Murooka, Y., 100t
Murphy, J.B., 155–156
Murphy, J.F., 99–102, 126, 296t, 299
Murray, I., 289–290
Murray, J.D., 151
Musarrat, J., 100t
Muscolo, A., 224–225
Musgrave, A., 29
Muskens, M., 243–244
Mutch, K.E., 89–90
Mutch, L.A., 89–90
Myrold, D.D., 223–224
Myrta, A., 272

N

Nadkarni, N.M., 216–217
Nadler, S., 195–196
Nagahashi, G., 67–69
Naismith, J.H., 129
Nakagawa, T., 290–291
Nakanishi, A., 221–222, 226–227
Nakano, Y., 156
Nakao, T., 221–222, 226–227
Nakashimada, Y., 5
Nakayama, T., 66
Nakkeeran, S., 99–102
Nam, Y.W., 156–157
Nanda, A.K., 29
Nandety, R.S., 167–168
Nanko, K., 230–231
Nannipieri, P., 88
Narasimhan, K., 97–99, 100t
Näthe, K., 217–219
Naureen Zakira, P., 97–99, 100t
Nautiyal, C.S., 221–222
Návar, J., 222–224
Navarro, L., 253–254

Navazo, A., 9
Nazaret, S., 63–64
Neal, A.L., 5, 11, 67
Nedjari, H., 270–271
Nelson, J., 244–245
Nelson, L.M., 98t
Nelson, M.N., 170–172, 171f, 174–175
Nelson, W.L., 82–83
Nemchenko, A., 289–290
Nemcova, M., 164
Netrusov, A.I., 92–94
Neumann, C., 35, 39t–40t, 126–127
Neumann, G., 67–69, 154
Neumann, M., 63–64
Neumann, R., 2, 60–62
Newell, P.C., 83–84
Newman, E.I., 225
Newman, M.A., 25–27, 29–30, 139–140
Nguyen, C.D., 167
Nguyen, K., 208–209
Nguyen, T.H.N., 138–139, 142
Nicander, B., 92, 94
Nickel, J., 247–248, 254–255, 287
Niebel, A., 141, 143
Niederbacher, B., 287–288
Niehaus, K., 26, 30–31, 39t–40t, 46–47
Nielsen, J.T., 25, 29–30, 139
Nielsen, R., 59
Nielsen, U.N., 192–193
Niemann, G.J., 30–31, 39t–40t
Nightingale, K.K., 121–122
Nigon, V., 199–200
Nihorimbere, V., 13
Niinemets, U., 287–288
Nikodem, A., 221–222, 228–229
Nilsson, M.C., 217–219
Nimnoi, P., 60–62
Niranjan Raj, S., 98t
Niu, Y., 29
Noel, J.T., 120–121
Noguti, Y., 168–169
Nomura, K., 123–124
Norden, U., 222–223
Norman, D., 266
Normark, S., 127–129
Northway, M., 287–288
Nova-Franco, B., 143–144
Novak, K., 164
Novak, O., 94
Nowak, J., 22–23, 87–88
Nuismer, S.L., 161
Nun, N.B., 96–97
Nürnberger, T., 24–26, 155
Nybroe, O., 34–35, 249
Nygren, C.M.R., 226–227

O

O'Callaghan, M., 64–65
O'Rourke, J.A., 138–139
Oakes, M., 156–157
Oakhumen, A., 223–224
Oberpichler, I., 124–125
Oelmüller, R., 28
Oestergaard, J., 205
Ofek, M., 60–62, 66–67
Offre, P., 64–65
Ofner, I., 62
Ogden, F.L., 222–223
Oger, P., 30–31, 39t–40t
Oikawa, P.Y., 290
Oka, H., 168–169
Okalebo, R., 99–102
Okazaki, S., 140
Okeke, I.N., 129
Okon, Y., 5, 9–10, 63–64, 84–88
Oku, S., 5
Okubo, T., 59
Olbryt, M., 155
Oldroyd, G.E.D., 137–138, 142–143, 151, 153–158, 170, 173, 175–176, 179–180, 220f
Oliveira, R.S., 221–222
Olsson, P.A., 226–227
Ona, O., 12
Onda, Y., 230–231
Ongena, M., 13, 31–32, 34–36, 39t–42t, 44–45, 91–92, 96, 249–252
Ono, H., 100t
Onodera, S.-I., 231
Opden Camp, R.H., 155, 169, 172
Orians, C.M., 179–180
Orosz, L., 13–14

Orozco-Mosqueda, M.C., 255–259, 260t–262t, 264–268, 293t–295t, 298–299
Ortíz-Castro, R., 286
Osborn, T.C., 177–178
Oter, R.G., 60–63
Otomo, I., 27
Ott, T., 137–138, 141
Otuka, T., 168–169
Ouedraogo, O., 96–97
Ouimet, R., 226–227
Ouyang, C., 273
Ouziad, F., 228–229
Ovadis, M., 255–258, 260t–262t, 293t–295t, 298–299
Ovecka, M., 293t–295t
Overeem, A.J., 289–290
Oves, M., 83–84, 89–90
Owen, H.A., 161, 163–164
Owens, T.G., 166
Ozawa, R., 255–256, 259, 260t–262t, 269–270
Ozawab, R., 290
Ozel, M., 250–251

P

Paaren, H.E., 164
Pad field, K., 299–302
Painter, J.A., 116
Pajerowska-Mukhtar, K., 253–254
Pan, Y., 59
Pandey, S.P., 119–120
Pandit, M.K., 160–161
Panitz, F., 59
Pánková, H., 161–162
Panuccio, M., 224–225
Papatheodorou, E.M., 291–292
Paquis, S., 47
Paquot, M., 34–35, 39t–42t, 96, 251–252
Par, P.W., 88
Parales, R.E., 5
Pardede, H., 83–84
Paré, P.W., 5–6, 8–9, 22–23, 33, 37, 39t–42t, 58–59, 67–69, 87–88, 96, 122, 208, 255–258, 260t–262t, 266, 268–271, 287, 291–298, 293t–295t
Paredes, S.H., 2, 58–59, 62–63, 69–72
Pares, X., 265
Park, C.J., 27–28
Park, C.S., 98t
Park, H.B., 37–38, 39t–40t, 69–70, 255–256, 260t–262t, 268–272, 293t–295t, 301–302
Park, J.H., 292–297, 293t–295t, 301–302
Park, J.W., 34–35, 39t–40t
Park, J.Y., 38, 39t–42t, 287, 293t–295t
Park, K.H., 31, 37, 39t–42t, 268–270, 293t–295t, 297–298
Park, M.R., 22–23, 38, 39t–42t
Park, S.H., 12, 34–35, 39t–40t
Park, S.W., 84–86
Park, T.S., 292–297, 293t–295t, 301–302
Parker, C.T., 121–122
Parniske, M., 28, 137–138, 143, 157–158, 164, 166–167
Parra, J.R.P., 58–59
Parsons, R., 219–220
Parupalli, S., 173
Parzy, J., 59–60, 63–64, 66
Paschke, M.W., 67–69
Paschold, A., 286–287, 289–290
Passaglia, L.M.P., 2–3
Pataj, Z., 249–251
Patel, H., 250
Patel, J., 127–129
Paterson, A.H., 159, 167–168
Pathak, K.V., 249
Patten, C.L., 12, 92–94, 253–254
Pattery, T., 33, 36–37, 39t–42t, 43–44
Paul, E.A., 228–229
Pauli, G., 250–251
Paulitz, T.C., 31–33, 60–62, 91–92
Paulsen, I., 12–13
Pauly, N., 29
Paungfoo-Lonhienne, C., 59
Pavlu, L., 221–222, 228–229
Pawliszyn, J., 201
Pawlowska, T.E., 228–229
Pawlowski, K., 170
Payne, R.J., 32–33
Payton, P., 293t–296t
Pearce, G., 288–290

Péchy-Tarr, M., 12–13
Peck, M.C., 154
Peiffer, J., 2
Peiffer, J.A., 59, 62–63, 70–72
Pel, M.J.C., 29–30
Peleg, Z., 62
Penaflor, M.F.G.V., 291–292
Penmetsa, R.V., 137–138, 156–158
Penn, J.M., 257–258
Penna, C., 94
Pennington, R., 172
Penrose, D.M., 87–88, 97–99, 100t
Penterman, J.N., 95
Peñuela, S., 139
Peñuelas, J., 254–255, 286–288
Peoples, M.B., 60–62
Peralta-Videa, J.R., 91–92
Pereira, P.A., 138–139
Pereyra de la Iglesia, M.T., 30–31, 39t–40t
Perez-Garcia, A., 34–35, 251–252
Pérez-Giménez, J., 166–167
Perez-Jaramillo, J.E., 58–59, 70–72
Pericas, M.A., 265
Permentier, H., 7
Perotto, S., 3, 7–8, 90–91
Perret, X., 151, 219–220, 223–224
Perrig, D., 94
Perry, D., 222–223
Perry, L.G., 3–5, 59–60, 84–86, 244–245, 287–290
Peters, A., 195, 205
Petri, A., 254–255, 259, 287–288, 292–297
Petutsching, E.K., 25–26
Pfeil, B.E., 167–168
Pfister, L., 216–217, 230–231
Phatak, S.C., 120–122
Phelan, V.H., 120
Philippot, L., 59, 96–97, 102–103
Philips, D.A., 137–138, 154–155
Piatti, T., 29
Piccinetti, C., 166–167
Piccoli, P.N., 92, 94, 253
Piceno, Y., 59–62
Piche, Y., 226, 229–230
Pichersky, E., 197, 286–290
Piechulla, B., 247–248, 254–255, 259, 270–272, 286–288, 292–298, 293t–296t
Pierret, M.C., 222–223
Pieterse, C.M.J., 23–25, 29–30, 36, 39t–42t, 44–46, 58–59, 67–70, 83–86, 96, 160–161, 247–248, 251–254, 268
Pietramellara, G., 216–217
Pineda, A., 29–30, 34–35, 45–46, 67–69
Pinhasi van-Oss, R., 62
Pinior, A., 226
Pink, B., 124–125
Pinto, R., 127–129
Pinton, R., 88, 154
Piovanelli, C., 216–217
Pires, J.C., 142, 159–160, 177–178
Pirhonen, M., 118
Pizzirani-Kleiner, A.A., 58–59
Plant, R., 193–194
Plaut, J.A., 224–225
Plet, J., 143
Plotnikova, J.M., 118
Plyuta, V.A., 299–300
Pocock, M.J., 160–161
Podicheti, R., 167–168
Podile, A.R., 3, 12, 22–23, 82–83, 91–92
Podishetty, N.K., 58–59, 63, 70–72
Poelman, E.H., 69–70, 206, 208–209
Poinar, G., Jr., 195
Poinar, G.O., 199–200
Poinssot, B., 28
Poirier, M.-A., 60–62
Polone, E., 155, 172
Polston, J.E., 98t
Ponette, Q., 221–222, 226–227
Pongsilp, N., 60–62
Pontieri, P., 290
Ponyi, T., 13–14
Poole, P.S., 58–59, 70–72, 150–151
Popova, A.A., 255–257, 260t–262t, 263–268, 299–300
Popp, C., 137–138
Poppe, J., 88–89
Poppy, G.M., 198
Porcel, R., 221–222
Portyanko, V., 143

Postel, S., 25–26
Potnis, N., 120–121
Poulin, M.P., 226, 229–230
Pourcel, L., 165–166
Pourcelot, L., 217–219, 228–229
Poveda, K., 192–193, 197–198
Powell, A.F., 176
Powell, D., 58–59, 122
Pozdnyakova, L.I., 100t
Preissner, R., 247–248, 254–255, 287
Prell, J., 150
Prescott, C.E., 222–223
Presecan-Siedel, E., 270–271
Press, C.M., 32–34
Preston, C.A., 286–287, 289–290
Price, N., 155
Price, P.W., 191–193
Prigent-Combaret, C., 2–3, 66, 84–86, 127–129
Prinsen, E., 12, 94
Pritchard, L., 118
Prithiviraj, B., 67–69, 118
Probanza, A., 253
Probst, A., 217–219, 228–229
Promé, D., 155
Promé, J.C., 137–138, 164
Proost, P., 5, 10–11
Prosser, J.I., 59–62
Provart, N.J., 63–64
Pucheu, N.L., 5
Pueppke, S., 155, 164
Pugin, A., 31–32, 38–43, 41t–42t, 46–47
Puhler, A., 46–47
Puschenreiter, M., 91–92
Pusenkova, L.I., 245–247

Q

Qi, H., 34–35, 39t–40t
Qi, H.-Y., 98t
Qiao, J., 34–35
Qiu, M., 6, 8, 67–69
Qiu, Y.L., 225
Qu, J., 26–27
Quadt-Hallmann, A., 58–59
Que, Y.A., 299–302
Quecine, M.C., 58–59
Quelas, J.I., 166–167
Quinn, W.J., 9–10
Quinto, C., 137–138
Quistgaard, E.M., 155

R

Raaijmakers, C.E., 191–194, 206
Raaijmakers, J.M., 2–3, 34–35, 39t–42t, 58–62, 70–72, 244–245, 247–252
Rabenoelina, F., 35, 39t–40t
Radler, F., 265
Radutoiu, S., 155–157
Raedts, J., 143
Raes, J., 159
Ragan, M.A., 63
Rahman, A., 251–252
Rahme, L.G., 118
Rahn, A., 129
Raina, S., 127–129
Rajkumar, M., 97–99, 100t
Rakwalska, M., 155
Ralph, P.E., 158
Ramakrishnan, G., 99–102
Ramakrishnan, K., 60–62
Ramarathnam, R., 255–256, 260t–262t, 267–268, 293t–296t
Ramos Solano, B., 30–31, 39t–40t
Ramos, H.C., 30
Ramos, J.L., 11
Ramos-Solano, B., 253
Rasmann, S., 200–201
Ramsey, J., 160–161, 168–169, 176–177
Ramsey, T.S., 160–161, 168–169, 176–177
Ran, L.X., 31–34, 38–43, 39t–42t
Randoux, B., 251–252
Ranf, S., 26–27
Rangel-Castro, J.I., 59–62
Rani, T.S., 12
Ranjard, L., 64–65
Ranjeva, R., 226
Rankin, S.C., 121–122
Rao, K.S., 91–92
Rao, Z., 265
Raoult, D., 58–59
Raskin, I., 119–120, 290
Rasmann, S., 191–194, 196–199, 201, 204–206, 208, 287–288

Rasouly, A., 124–125
Rastogi, G., 120–121
Ratet, P., 139
Ratnaparkhe, M.B., 167–168
Ratnayake, M., 227–228
Rauhut, T., 26–27
Raza, W., 5, 8, 67, 250–251, 255–256, 260t–262t
Read, D.J., 225–227, 230–231
Reardon, K.F., 59–60, 70–72
Rébeillé, F., 32–33
Recourt, K., 88–89
Reddy, M.S., 22–23, 32–33, 37, 39t–40t, 87–88, 96, 98t, 122, 251–252, 255–257, 260t–262t, 268–271, 287, 292–298, 293t–296t
Redondo-Nieto, M., 222–223
Reed, E., 91–92
Rees, D.C., 89–90
Reguera, M., 222–223
Regvar, M., 228–229
Reichard, J.S, 216–219, 221–222, 227–228
Reid, C.P.P., 227–228
Reid, D.E., 154–157, 175–176
Reimann, C., 33
Reinhard, A., 25–26
Reinhold, B., 5–6
Reis, V., 87–88
Reiter, B., 87–88
Reitz, M., 30–31, 39t–40t
Rejili, M., 221–222
Relíc, B., 155
Remans, R., 92–94
Remy, W., 172
Ren, A., 34–35, 39t–40t
Renault, J.H., 35, 39t–40t
Rennenberg, H., 66–67
Renukadevi, P., 99
Reva, O., 252
Revellin, C., 99
Reyes-de la Cruz, H., 255–259, 260t–262t, 264–268
Reynolds, H., 244–245
Reynoso, M.A., 143
Richards, A.J., 166–167
Richardson, A.E., 2–3, 64–66, 90–91
Richardson, D.M., 160–161, 177–178
Richardson, J.R., 254–255, 274
Richardson, S.M., 127–129
Rideout, S., 263–264, 272
Ried, M.K., 25–26
Riedel, K., 299–300
Riefler, M., 94
Riely, B.K., 157–158
Rieseberg, L.H., 158
Rigo, L.U., 7–8
Rimando, A., 5–6
Rivilla, R., 222–223
Ro, D.K., 287–288
Robatzek, S., 25–26
Robert, C.A.M., 59, 198–199, 201–202, 207–208, 287–289, 299–300
Roberts Fauzia, Y., 97–99, 100t
Roberts, K.J., 226
Roberts, N.J., 155–156
Robert-Seilaniantz, A., 69–70
Robeson, M.S., 244–245
Robin, C., 64–65
Robins, R.J., 257–258, 270
Robinson, J.B., 8
Robinson, N., 63
Robson, A.D., 227–228
Rocha, U.N., 59–60
Rochat, L., 12–13
Rochepeau, P., 166–167
Rochetaing, A., 263–265, 267, 270
Rochut, N., 257–258, 270
Röckmann, T., 258
Roden, L.C., 124–125
Rodrigo, A., 217–219, 228–229
Rodriguez Kabana, R., 296t
Rodríguez, H., 83–84, 87–88
Rodriguez-Kabana, R., 126
Rodriguez-Saona, C., 201–202, 204, 207
Rogers, C., 141
Rogers, E.E., 63–64
Röhrich, C.R., 35
Rojas, N., 222–223
Rojas-Solis, D., 255–259, 260t–262t, 264–268
Rokni-Zadeh, H., 250–251
Rombauts, S., 143–144
Romero, D., 34–35, 251–252

Römling, U., 127–129
Romps, D.M., 231
Ron, E.Z., 124–125
Ron, M., 27–28
Ronald, P.C., 27–28
Rong, Y., 167–168
Roongsawang, N., 249–250
Rose, C.M., 138–139, 141–142
Rose, D., 253–254
Rosen, R., 124–125
Rosenkranz, M., 254–255, 286–288
Rosier, C.L., 216–219, 221–222, 227–228
Ross, A.F., 119–120
Rossignol, M., 226
Rosskopf, E.N., 272
Rostás, M., 204
Rothballer, M., 12, 63–64, 299
Rott, M., 2, 59, 62–63, 70–72
Roughley, R.J., 99
Roumet, C., 5–6
Rousseau, J.V.D., 227–228
Rout, M.E., 244–245
Routaboul, J.M., 165–166
Rouws, L.F., 7–8
Roy, M., 161–162
Roy, S., 141
Ruckert, C., 252
Rudrappa, T., 5–6, 8–10, 37, 39t–42t, 58–59, 67–69, 88, 255–256, 260t–262t, 266, 270, 287, 292–297, 296t
Ruffel, S., 168–169
Rui, X., 129
Ruiz, O.A., 94
Ruiz-Lozano, J.M., 100t, 221–222
Rullman, J., 199–200
Rumbo, M., 30
Runion, G.B., 228–229
Rusterucci, C., 63–64
Ryan, D., 97–99
Rydell, P., 225
Rydlová, J., 161–162
Ryu, C.M., 2–3, 12, 22–23, 31, 33, 37–38, 39t–42t, 46, 67–70, 87–88, 96–97, 98t, 122, 244–245, 253–263, 260t–262t, 268–272, 287, 292–299, 293t–296t, 301–302
Ryu, R.J., 12

S

Saad, M.M., 140
Sabelis, M.W., 192–193, 195–197
Sabri, A.N., 253–254, 266
Sadowsky, M.J., 120–122, 127–129
Sadras, V.O., 223–224
Saeki, K., 140
Saeki, Y., 156
Saez, A., 222–223
Safronova, V.I., 94
Sahr, T., 32–33
Saini, R., 221–222, 224–225
Sakakibara, H., 156–157, 168–169
Sakulyanontvittaya, T., 287–288
Sakurai, N., 138–139
Salas, A., 160–161
Saldana, Z., 129
Saleh, S., 97–99
Saleh-Lakha, S., 95
Sallé, G., 58–59, 96–97
Salles, J.F., 64–66
Sallet, E., 143
Samac, D., 139
Samans, B., 35, 39t–40t, 126–127
Same, B.I., 227–228
Samin, E., 141
Samiyappan, R., 98t
Samudrala, D., 202–203
Sanches, P.A., 291–292
Sanchez Chacon, E., 129
Sanchez, F., 137–138
Sanchez, L., 35, 39t–40t
Sanchez-Contreras, M., 100t
Sandal, N., 156–157
Sande, E., 221–222
Santaella, C., 2
Santi, C., 219–220, 223–224
Santner, A., 92–94
Santori, A., 272
Santoro, M.V., 299
Santos, F., 291–292
Santos, T.S., 167–168
Santoyo, G., 255–256, 260t–262t, 266–268, 293t–295t, 298–299
Santruckova, H., 222–224
Sarada, S., 224–225
Saranga, Y., 62

Saravanakumar, D., 98t
Sardanelli, S., 191–192
Sarkar, R., 63–64
Sasaki, K., 59
Sasaki, T., 156–157, 168–169
Sasaki-Yamagata, H., 156
Sasanelli, N., 272
Sassaki, G., 7–8
Sato, F., 221–222, 226–227
Sato, S., 140, 156–157, 174–175
Sauer, J., 155–156
Savchuk, S.C., 255–256, 260t–262t, 267–268, 293t–296t
Savenije, H.H.G., 216–217, 230–231
Sayama, T., 156
Sbodio, A., 120–121
Schaarschmidt, S., 69–70, 290–291
Schaefer, M., 35–36, 39t–42t, 44
Schafer, M., 35–36, 39t–42t
Schaff, J.E., 156–157
Schäffer, M., 27
Schalling, N., 216–217, 230–231
Scharf, B.E., 166–167
Schauer, G., 258
Scheick, C.E., 221–222
Schemske, D.W., 168–169
Schenck, N.C., 226
Schenck, S., 286
Schenk, P.M., 2, 69–70
Schenk, S.T., 35, 39t–40t, 126–127, 128f
Scheu, S., 12–13
Schiestl, F.P., 292–299, 293t–296t, 300–303
Schikora, A., 118, 126–127, 128f
Schikora, M., 35, 39t–40t, 126–127
Schippers, B., 30–31, 39t–42t, 91–92, 273
Schlaepfer, D.R., 168–169
Schlaeppi, K., 2, 58–63, 70–72, 84–86
Schlatter, D.C., 245–247
Schleiff, E., 157–158
Schmid, M., 12
Schmidli-Sacherer, P., 33
Schmidt, S., 63, 299–300
Schmidt, U., 287
Schmiedeknecht, G., 99–102
Schmülling, T., 94
Schnable, J.C., 159–160
Schnee, C., 204–205
Schneeweiss, G.M., 150
Schneider, J.H., 58–59, 245–248
Schneider, K.R., 120–121, 129, 252
Schnitzler, J.P., 287–290
Schoch, C.L., 59
Scholz, B., 254–255, 259, 287, 292–297
Scholz, R., 252
Schonfeld, J., 66
Schoonhoven, L., 191–192
Schreiber, L., 291–292
Schröder, J.M., 30
Schroder, P., 299
Schrom, J.O.S., 216–219, 221–222, 227–228
Schroth, M.N., 2–3, 66–67, 82–83, 245–249
Schuhegger, R., 35, 39t–40t
Schullehner, K., 11
Schuller, S., 129
Schultze, M., 137–138
Schulz, S., 244–245, 254–267, 260t–262t, 292–303, 293t–296t
Schulze, K., 217–219
Schulze-Lefert, P., 58–63, 70–72, 84–86
Schuster, D.J., 98t
Schwab, W., 291–292
Schwager, S., 299
Schwarz, M.T., 165–166, 217–219, 224–225
Schwarze, P., 179
Schwärzel, K., 216–217, 230–231
Schwedt, A., 25–26
Schwendenmann, L., 217–219
Scott, E.S., 98t
Scott, P.R., 243–244
Searle, I.R., 155–157, 167
Seehauser, H., 258
Seeley, J.T., 231
Seewald, M.S., 287
Segarra, G., 45–46
Segraves, K., 161
Seifert, K.A., 59
Sela, S., 121–122, 127–129
Selvaraj, G., 58–60, 70–72, 84–86, 244–245, 252–253, 298
Semenov, A.V., 64–65

Sengonul, K., 221–222, 228–229
Senthilkumar, M., 95
Seo, Y.S., 27–28
Seralathan, V., 166
Serengil, Y., 221–222, 228–229
Serino, L., 33
Serrato, R.V., 7–8
Sessitsch, A., 87–88, 91–92
Setälä, H., 191–192
Seul, K., 12
SeWeon, L., 251–252
Seyer, A., 13
Seymour, M., 60–62
Shadle, G., 165–166
Shakya, M., 59, 62–63
Shamseldin, A., 221–222
Shan, L., 25, 28, 253–254
Shaposhnikov, A.I., 6–7, 88–89
Sharapova, N., 139–140
Sharma, A., 91–92, 99–102
Sharma, M., 118, 127–129
Sharma, V.K., 129
Shaukat, S.S., 36, 39t–40t, 43–44
Shaw, C.H., 137–138, 166–167
Shaw, S.L., 164
Sheen, J., 253–254
Sheflin, A.M., 70–72
Shen, O., 5–6, 8, 59–60, 64–65, 67, 70–72, 244–245, 250–251, 255–256, 260t–262t
Shenker, M., 5–6
Sherman-Broyles, S., 176–177
Sherrier, D.J., 143
Shetty, H.S., 98t
Shi, H., 287, 293t–295t
Shi, L., 34–35, 39t–40t
Shi, S., 64–65
Shiaris, M.P., 2, 64–65
Shibata, D., 138–139
Shibata, H., 221–222, 226–227
Shiklomanov, A.N., 221–222
Shimomura, K., 139
Shindia, A.A., 267
Shinozaki, K., 168–169
Shiraishi, S., 251–252
Shiraishi, T., 27–28
Shiro, S., 156
Shitan, N., 161–162
Shubin, N., 152–153
Shuhegger, R., 96
Shulaev, V., 290
Sidari, M., 224–225
Siddiqui, I.A., 36, 43–44
Siddiqui, M.A., 39t–40t
Siegert, C.M., 216–217, 221–222
Siemens, J., 217–219, 224–225
Sierralta, W.D., 127–129
Signer, E.R., 7–8
Sikkema, A., 243–244
Sikora, E., 98t, 126, 296t
Silipo, A., 26–27
Silva, H.S.A., 98t
Silva-Filho, M.C., 67
Silverman, P., 119–120, 290
Silvoy, J.J., 120
Simko, I., 121–122
Simoes-Araujo, J.L., 7–8
Simon, M.D., 63–64
Simon, S.A., 143
Simons, M., 7
Sims, T., 118
Sinclair, J., 5
Singer, S.R., 158, 174–175
Singh, B.R., 100t
Sinnaeve, D., 250–251
Sirard, J.C., 30
Sirois, J.C., 223–224
Sivananthan, A., 129
Skaugset, A.E., 216–217, 230–231
Skorpil, P., 140
Skrdleta, V., 164
Skvarch, E., 272
Slimestad, R., 168–169
Smalla, K., 2, 59–60, 63–64, 66–67, 84–86
Smedvig, P., 168–169
Smit, P., 143
Smith, D.L., 87, 154
Smith, F.A., 227–228
Smith, J.N., 120–121
Smith, S.E., 225, 227–228, 230–231
Smits, P.H., 192–193, 195–197
Sobhy, I.S., 289–290
Sobsey, M.D., 116–118
Söderström, B., 226–227

Sola, L., 265
Solano, B.R., 22–23
Soler, R., 69–70, 191–192
Soltis, D.E., 159–161, 172, 178–179
Soltis, P.S., 159–161, 172, 178–179
Somers, E., 3–7, 59–60, 85t
Somssich, I.E., 119–120
Song, C., 249–251
Song, G.C., 37–38, 39t–40t, 255–256, 260t–262t, 269–270, 293t–295t
Song, X.S., 249
Songnuan, W., 63–64
Sood, S.G., 5
Soriano, L.C., 137–138, 155
Soto-Arias, J.P., 120–121
Souleimanov, A., 154
Southworth, D., 244–245
Soyano, T., 152–153, 156–158, 168–169
Spaepen, S., 58–59, 70–72, 84–86, 92–94
Spence, C., 70–72
Sperandio, V., 254–255
Splaine, R.E., 287, 292–297
Spooner, D.M., 160–161
Spor, A., 2, 59, 62–63, 70–72
Spouge, J.L., 59
Sprent, J.I., 150, 170–172, 174–175, 219–220
Springer, N.M., 159–160
Srinivasan, M., 3–7, 59–60, 85t
Sriprang, R., 100t
Sririyanum, M., 27–28
Srivastava, S., 163–164, 166
Stacey, G., 66, 137–140, 142, 156–157, 166–167, 219–220
Stacey, N., 141
Staelens, J., 217–219
Stahl, B., 286–287
Stam, J.M., 69–70
Stankiewicz, M., 263–265, 267, 270
Stanley-Wall, N.R., 9–10
Staskawicz, B.J., 151, 242–243
Staudinger, M., 258
Staudt, M., 287–288
Steeghs, M., 287–288
Steidle, J.L.M., 30–31, 208
Steijl, H., 30–33, 38–43, 39t–42t
Stein, A., 216–217, 230–231
Stein, E., 35, 39t–40t, 126–127, 128f
Stein, T., 249
Steinberg, C., 60–62
Steinberger, Y., 59
Steindler, L., 299
Steiner, U., 99–102
Steinle, L., 140
Stelinski, L.L., 197, 201–202, 206–208
Stephenson, S., 160–161
Sternberg, P.W., 208
Stewart, A., 64–65
Stille, P., 217–219, 228–229
Stiller, J., 156–157
Stinchcombe, J.R., 178–179
Stock, S.P., 195–196
Stotzky, G., 286
Stracke, S., 174–175
Straczek, J., 228–229
Strange, R.N., 243–244
Stratmann, J.W., 288–290
Strauch, O., 205
Strawn, L.K., 121–122
Streng, A., 155, 158, 172
Stribley, D.P., 226–227
Strnad, M., 94
Struck, S., 287
Stubbins, A., 217–219
Su, J., 26–27
Subramanian, S., 143
Subramoni, S., 12–13
Suda, J., 160–162, 177–178
Sudová, R., 161–162
Suen, G., 195
Sugawara, M., 129
Sugimoto, K., 290
Sugiyama, A., 5–6, 10–11, 63–67, 69, 138–139, 168–169
Sullivan, J.T., 155–156
Sumayo, M., 38, 39t–40t
Sumera, M.M., 216–217
Sumner, L.W., 67, 165–166, 257–258, 268–271, 287, 293t–295t, 297–298
Sun, Y., 287, 293t–296t, 298
Sundelin, T., 25, 29–30, 139
Suneja-Madan, P., 221–222, 224–225
Sunkar, R., 143

Surange, S., 221–222
Suslow, T.V., 120–121
Sussman, M.R., 137–138, 142
Suzaki, T., 156–157, 168–169
Suzuki, H., 138–139
Suzuki, S., 251–252
Swarbreck, D., 60–62
Swarup, S., 97–99, 100t
Swensen, S.M., 172
Szadkowski, E., 159–160
Szczyglowski, K., 155

T

Tabacchi, R., 12–13
Tabana, B.M., 116–118
Tabin, C., 152–153
Tadeo, F.R., 118
Taira, S., 253
Tajima, T., 5
Takabayashi, J., 255–256, 259, 260t–262t, 269–270
Takagi, M., 100t
Takahashi, S., 66
Takai, R., 26
Takanashi, H., 138–139
Takanashi, K., 138–139
Takase, H., 63–67
Takebayashi, N., 158
Takeda, N., 59, 157–158
Takei, K., 168–169
Talmont, F., 155
Talon, M., 253
Tan, S., 5, 8, 67
Tanaka, K., 140
Tanaka, Y., 223–224
Tang, F., 156
Tang, Y.J., 165–166, 267
Taniguchi, M., 168–169
Tanilas, T., 291–292
Tanja, B., 97–99
Tao, G.C., 90–91
Tarrant, R.F., 216–217
Tatchell, G.M., 85t
Tauxe, R.V., 116
Tayeh, C., 251–252
Taylor, A.F.S., 225–227
Taylor, T.N., 172
te Beest, M., 160–161, 177–178
Tebbe, C.C., 59–60
Tech, J.J., 120–121
Teintze, M., 2–3, 83, 245–249
Teplitski, M., 6–8, 120–122, 125–126, 129
Terán, W., 11
Terrier, N., 165–166
Těšitelová, T., 161–162
Testerink, C., 29
Thaler, J.S., 69–70, 204, 208
Thallinger, G.G., 249–250
Thenmozhi, P., 31, 39t–40t
Theocaris, A., 22–23
Thibivilliers, S., 140
Thieme, L., 217–219, 224–225
Thilmony, R., 119–120
Tholl, D., 254–255, 286–288
Thomas, B.C., 158–159
Thomashow, L.S., 3, 7, 58–59, 82–86, 273
Thomma, B.P., 24
Thompson, B.N., 88–89
Thompson, J.N., 161, 191–193
Thonart, P., 13, 31–32, 34–36, 39t–42t, 44–47, 91–92, 251–252
Thurmer, A., 296t
Thys, A., 92–94
Tian, L., 177–178
Tian, S.J., 90–91
Tiemann, K.J., 91–92
Tien, T.M., 83–84
Tietjen, J.H., 199–200
Tiffin, P., 178–179
Tikhonovich, I.A., 6–7
Tiktak, A., 217–219
Tilcher, R., 96–97, 102–103, 249–250
Tillberg, E., 92, 94
Timms-Wilson, T.M., 64–65
Timmusk, S., 3, 7–8, 92, 94, 291–292
Tinker, P.B., 226–229
Tirichine, L., 156–157
Tirosh, T., 95
Tisa, L.S., 223–224
Tisdale, S.L., 82–83
Tisserant, B., 251–252
Tiwari, K.N., 87
Toda, H., 221–222, 226–227
Toepfer, S., 197–201, 204–206, 208

Tojo, S., 11
Toklikishvili, N., 255–258, 260t–262t, 264–265, 293t–295t, 298–299
Tollsten, L., 203
Tomalak, M., 205
Tomasi, N., 154
Tomberlin, J.K., 254–255
Tompkins, R., 299–302
Ton, J., 5, 11, 39t–40t, 288–289, 299–300
Ton, J., 67, 191–192
Tonelli, M., 291–292
Tonin, C., 228–229
Töpfer, S., 287–288
Torbert, H.A., 58–59, 89–90
Torregrosa, L., 165–166
Torres, M.A., 29
Tortorello, M.L., 129
Toth, I., 118
Toth, J.D., 121–122
Touraine, B., 2–3, 83–84, 84f, 87–88, 92–94
Town, D., 139
Toyoda, K., 27–28
Tracey, R.P., 296t
Tracy, W.F., 63–64
Tran, H., 34–35, 39t–42t, 250–252
Traw, M.B., 226
Trick, M., 141
Trillas, I., 45–46
Tringe, S.G., 2, 59, 62–63, 70–72
Trofymow, J.A., 193–194
Trompeter, J., 199–200
Trotel-Aziz, P., 33, 41t–42t, 43–44
Tsavkelova, E.A., 92–94
Tsay, J.F., 37–38, 39t–40t, 255–256, 260t–262t, 264–265, 270, 293t–295t
Tscharntke, T., 191–192
Tscheka, C., 250
Tsuda, K., 26–27
Tsui, S., 58–59
Tsuyumu, S., 27
Tucker, S.L., 195
Tumlinson, J.H., 197, 208
Turek, J., 222–224
Turkovskaya, O.V., 100t
Turlings, T.C.J., 67–70, 191–192, 196–200, 204–206, 208, 287–290
Turnau, K., 90–91, 228–229
Turner, T.R., 58–62, 70–72
Tuskan, G., 244–245
Tuzun, S., 32–33, 251
Tydings, H.A., 120–121
Tyson, G.W., 2, 69–70

U

Uchihashi, K., 25–26
Uchiumi, T., 219–220
Udall, J.A., 159–160
Uddin, W., 251–252
Udvardi, M.K., 170, 173
Ueda, N., 168–169
Ueda, Y., 63–67
Uhde-Stone, C., 90–91
Umehara, Y., 155–156
Underberg, S., 255–256, 260t–262t, 264–265, 270–272, 293t–295t, 297–298
Underwood, W., 123–124
Urfus, T., 161–162
Uribe, P., 156–157
Urquiaga, S., 87–88
Usmani, S., 100t
Uygur, B., 221–222, 228–229

V

Vacheron, J., 2–3, 83–84, 84f, 87–88, 92–94
Vadassery, J., 28
Valdés-Lopez, O., 143–144
Valencia Cantero, E., 270–271, 293t–296t, 298–300
Valente, F., 216–217
Validov, S., 88
Valverde, A., 83–84, 87–88
Van Stan, J.T., 221–222
van Aarle, I.M., 226–227
van Arendonk, J.J., 88–89
Van Bastelaere, E., 5, 10–11
van Beek, T.A., 250–251
van Bezooijen, J., 192–193, 195–197
van Bruggen, A.H., 120–121, 127–129
van Dam, D., 226–227

van Dam, N.M., 191–192, 198, 202–203, 206
van de Mortel, J.E., 29–30, 34–35, 45–46, 250–251
Van de Peer, Y., 69–70, 158, 160–161
Van de Velde, W., 143–144
van den Elsen, S., 194
Van Der Does, D., 24–25
Van Der Ent, S., 45–46
van der Heijden, M.G.A., 219–220
van der Heijdt, W.H.W., 31–32, 38–43, 41t–42t, 46–47
van der Lelie, D., 11, 67, 84–86
Van der Luit, A.H., 29
van der Putten, W.H., 58–59, 191–194, 200–201
Van der Sluis, I., 29–32, 38–43, 39t–42t
van der Sommen, A.T.C., 192–193, 195–197
van der Voort, M., 58–59, 245–248
van der Vossen, E.A., 243–244
van der Wurff, A., 194
Van Dommelen, A., 5, 10–11
Van Doorn, A., 29
van Elsas, J.D., 2, 58–60, 63–66
van Gestel, K., 118
van Herpen, T., 204–205
van Hoek, A.H., 121–122
Van Impe, J., 12
van Laere, A., 226–227
Van Loon, J.J.A., 67–70, 191–192, 206, 208, 289–290
Van Loon, L.C, 22–23, 29–32, 36, 38–43, 39t–42t, 46–47, 69–70, 83–84, 87–88, 96, 268
van Megen, H., 194
Van Montagu, M., 159
Van Overbeek, L., 2, 59–60, 63–64, 66
Van Peer, R., 39t–40t
van Pelt, J.A., 30–33, 36, 38–43, 39t–42t
Van Puyvelde, S., 10–11
Van Stan, J.H., 216–217
Van Stan, J.T., 216–219, 221–224, 227–228, 231
Van Straalen, N.M., 219–220
van Themaat, E.V.L., 58–59, 62–63, 70–72
van Tol, R.W.H.M., 192–193, 195–197
Van Wees, S.C.M., 23–25, 44–46, 69–70, 251–254, 268
Van Wyk, B., 172
Vance, C.P., 90–91, 137–138, 150
Vande Broek, A., 3–5, 12, 92–94
Vandelle, E., 28
VandenBosch, K.A., 139–140
Vandenkoornhuyse, P., 228–229
vander Hofstad, G.A., 88–89
Vanderleyden, J.O.S., 3–8, 10–12, 59–60, 85t, 92–94
Vanderplank, J.E., 243–244
vanElsas, J.D., 59–60
Vanneste, K., 158
Vannette, R., 192–193, 197–198
Vanpeer, R., 30–31, 41t–42t
Varala, K., 142
Varanini, Z., 88
Vargas, L.K., 66
Varnier, A.L., 35, 39t–40t
Varshney, R.K., 173
Vas, G., 257–258
Vasilyeva, N., 34–35, 251–252
Vater, J., 250–251
Vatsa, P., 35, 39t–40t
Veitia, R.A., 159
Vekey, K., 257–258
Velázquez Becerra, C., 255–256, 260t–262t, 266–268, 270–271, 293t–295t, 298–300
Velazquez-del Valle, M.G., 98t
Venderleyden, J., 84–87
Venkatesh, B., 27
Venkateshwaran, M., 137–139, 141–142
Venturi, V., 299
Ver Loren van Themaat, E., 60–63, 84–86
Verdenaud, M., 141
Vereecke, D., 22–23
Verhagen, B.W., 33, 41t–42t, 43–44
Verheyen, K., 217–219
Verhoef, H.A., 191–194
Verma, J.P., 87
Vermeiren, H., 3–5, 10–11
Vernié, T., 158
Verries, C., 165–166
Versluis, C., 33, 38–43, 41t–42t

Veselov, S.U., 94
Vesely, S., 85t, 91–92, 95
Vespermann, A., 270–271
Vessey, J.K., 2–3
Vet, L.E.M., 191–193
Veyrat, N., 288–289, 299–300
Vialet, S., 165–166
Vidal, M.S., 7–8
Vidal, O., 127–129
Vierheilig, H., 226, 229–230
Villacieros, M., 100t
Villenave, C., 191–192
Vinall, K., 7–8
Virdi, J.S., 99–102
Visca, P., 33
Visser, S., 226–227
Vivanco, J.M., 2–6, 8, 10–11, 59–60, 63–72, 84–86, 118, 244–245, 250–251, 287–290
Vivas, A., 100t
Vlamakis, H., 6–9
Vogel, H.J., 217–219
Vogg, G., 35, 39t–40t, 96
Voigt, B., 11, 291–292
Vokou, D., 291–292
Volfovicz, R., 225
Volkening, J.D., 137–139, 141–142
Volko, S., 25–26
Vollaro, D., 231
Vollenbroich, D., 250–251
Vollu, R.E., 60–64
von Dahl, C.C., 286–287, 289–290
von Wirén, N., 11
Voronov-Goldman, M., 60–62
Vosátka, M., 221–222
Vrba, E.S., 152–153
Vugman, M., 124–125

W

Wäckers, F., 196–197
Waewthongrak, W., 251–252
Wagner, E.G.H., 7–8
Wagner, R.A., 116–118
Wais, R., 137–138
Walker, C., 226–227
Walker, G., 139–140
Walker, T.S., 244–245
Wall, D.H., 191–192
Walsh, C.T., 249–250
Walshaw, J., 60–62
Walters, D.R., 253
Wan, J., 137–138, 140
Wang, B., 225
Wang, D., 5–6, 8, 253–254, 272
Wang, E., 158
Wang, G.H., 64–66
Wang, G.M., 226–227
Wang, J., 177–178
Wang, L., 120–121
Wang, Q., 118, 273
Wang, S., 84–86, 129
Wang, X., 34–35, 39t–40t
Wang, Y., 8, 205
Wani, P.A., 100t
Wanke, D., 293t–296t
Wanner, G., 157–158
Wardle, D.A., 191–192, 217–219
Warembourg, F., 5–6
Warneke, C., 202
Warnke, R., 254–255
Washburn, J.D., 159–160
Washio, K., 249–250
Wassmann, R., 66–67
Wasternack, C., 119–120
Wathelet, B., 250–251
Wathelet, J.P., 35–36
Watson, B.S., 67
Watson, L.E., 89–90
Watson, M.D., 166–167
Watt, M., 7–8
Webb, A.A., 124–125
Webb, B.A., 166–167
Webb, C.C., 120
Weder, B., 165–166
Weger, L.A., 88–89
Wei, H.X., 255–257, 270–271, 292–298, 293t–296t
Wei, J., 118
Weiler, M., 216–217, 230–231
Weiman, A., 59, 62–63
Weinert, N., 59–62
Weir, J., 175–176
Weir, T.L., 3–5, 11, 59–60, 67, 84–86, 244–245, 287–290

Weis, A.E., 191–193
Weisbeek, P.J., 88–89
Weise, T., 293t–296t, 298
Weiskopf, L., 67–69, 154, 255–267, 260t–262t, 270–271, 293t–295t, 297–298
Weiss-Schneeweiss, H., 150
Weldegergis, B.T., 208–209, 289–290
Weller, D.M., 3, 7, 23, 36, 39t–42t, 44–46, 58–59, 82–86, 251–254, 268, 273
Wells, J.M., 120–121
Wendehenne, D., 31–32, 38–43, 41t–42t, 46–47
Wendel, J.F., 159–160, 177–178
Weng, J., 8
Wenhua, T., 99
Wenke, K., 254–255, 286–288, 293t–296t
Wenner, N.G., 251–252
Wentworth, T.R., 225
Werck-Reichhart, D., 63–64
Werner, D., 164, 166–167, 221–222
Werner, G.D., 172, 174–175
Westendorf, C., 299–300
Weston, L.A., 5–6, 63–64
Westphal, L., 27
Whelan, C., 100t
Whistler, C.A., 88–89
White, J., 150
White, P.F., 60–62
White, R., 7–8
Whiteman, N.K., 69–70
Whitfield, C., 129
Whitford, W.G., 216–217, 230–231
Wickett, N.J., 158
Wie, H.X., 87–88, 96
Wiedmann, M., 118
Wieland, G., 2, 59–64, 66
Wieshammer, G., 91–92
Wijffelman, C.A., 7
Wilcke, W., 217–219, 224–225, 230–231
Wilcox, D.S., 200–201
Wilfarth, H., 82–83, 99
Wilkinson, S.C., 216–217
Willems, L., 44–45
Willmann, R., 26–27, 30, 39t–40t, 155
Wilson, D.C., 63–64
Wilson, M., 32–33, 99–102
Wimer, A., 263–264, 272
Winans, S.C., 224–225
Wingender, J., 7–8
Winkler, J.B., 287–288
Winkler, R.D., 216–217
Wintermans, P.C., 58–59
Wisniewski-Dyé, F., 84–86
Wojciechowski, M.F, 172
Wolfe, K.H., 159
Wollum, A.G.,II, 221–222
Wolters, P., 243–244
Wood, T.E., 158
Wood, T.K., 100t
Woods, C., 228–229
Woodward, F.I., 123–124
Worobo, R.W., 121–122
Wouters, D., 243–244
Wright, K.A., 217–219
Wu, C.H., 100t
Wu, F., 34–35, 39t–40t, 67
Wu, G.J., 31–32, 38–43, 39t–42t, 139
Wu, S., 25, 28
Wu, T., 143, 216–219, 221–222, 227–228
Wu, W.S., 37–38, 39t–40t
Wu, W.-S., 255–256, 260t–262t, 264–265, 270
Wu, Y., 250–251
Wu, Y.H., 167–168
Wu, Y.S., 67–69, 122, 127f
Wub, W.S., 293t–295t
Wünsche, H., 255–256, 260t–262t, 264–265, 270–272, 293t–295t, 297–298
Wurzburger, N., 223–224
Wyss, U., 226

X

Xavier, J.B., 160–161
Xiang, T., 25–26
Xiao, J., 253–254
Xiao, Z., 265–266, 270–271
Xicohtencatl-Cortes, J., 129
Xie, G.H., 90–91
Xie, G.L., 250–251
Xie, X., 293t–295t, 298

Xie, Z.P., 140
Xin, D.W., 140
Xu, C., 253–254
Xu, H., 288–289, 299–300
Xu, P., 265–266, 270–271
Xu, Y., 64–66, 222–223
Xu, Z., 265

Y

Yaacov, O., 63–64
Yadav, J., 87
Yadav, R.K.P., 291–292
Yakushiji, S., 27–28
Yalpani, N., 119–120
Yamada, K., 25–26
Yamaguchi, K., 25–26
Yamakawa, T., 156
Yamamoto, S., 251–252
Yamane, H., 27
Yamashita, K., 66
Yamazaki, A., 120–121
Yamazaki, Y., 66
Yan, D., 273
Yan, F., 3, 8, 67–69
Yang, C.H., 5, 8, 67, 100t
Yang, H.-P., 37–38, 39t–40t, 255–256, 260t–262t, 264–265, 270, 293t–295t
Yang, J.W., 2–3, 38, 39t–40t, 69–70
Yang, K.Y., 37, 41t–42t, 268–270, 293t–295t, 297–298
Yang, M., 250–251
Yang, S.Y., 38, 39t–40t, 156, 265
Yang, T., 265
Yang, X.N., 255–256, 259, 260t–262t, 264–265
Yang, Z.K., 59, 62–63
Yang, Z.P., 216–217
Yano, J.M., 199
Yao, C., 98t
Yao, W., 249
Yaryura, P.M., 5
Yates, M.G., 7–8
Yazaki, K., 63–67, 161–162
Yeates, G.W., 193
Yee, D.C., 100t
Yeoh, Y.K., 59
Yi, H.S., 37–38, 39t–40t, 69–70, 255–256, 260t–262t, 269–270
Yokota, K., 157–158
Yokoyama, T., 164
Yoo, M., 159–160, 177–178
Yoshida, K.-I., 11
Yoshida, S., 174–175
Yoshiga, T., 195–196
Yoshimura, S., 25–26
Yost, C.K., 166–167
You, Y., 121–122
Young, D.R., 221–222
Young, J.P.W., 89–90
Young, N.D., 170, 173
Yourstone, S., 59, 62–63, 70–72
Yousef-Coronado, F., 7–9
Yu, D., 270–271, 293t–295t
Yu, H., 253–254
Yu, J., 59
Yu, O., 143
Yu, S.M., 273, 293t–295t, 297–298
Yuan, J., 250–251, 255–256, 260t–262t
Yuan, Q.S., 249
Yuan, Z.P., 267
YungEun, C., 251–252

Z

Zachow, C., 96–97, 102–103, 249–251
Zackrisson, O., 217–219
Zahran, H.H., 83–84, 89–90, 220–224
Zaidi, A., 83–84, 89–90, 100t
Zaidi, S., 100t
Zaitseva, J.V., 255–257, 260t–262t, 263–268, 293t–295t
Zamioudis, C., 23–25, 44–46, 69–70, 251–254, 268
Zanetti, M.E., 143
Zanón, M., 272
Zarkani, A.A., 35
Zaton, K., 63–64
Zeder, M.A., 82
Zehnder, G.W., 98t, 126, 296t
Zeidler, D., 30–31, 46–47
Zeier, J., 124–125
Zeriouh, H., 34–35, 251–252
Zeto, S.K., 226–227
Zézé, A., 89–90

Zhang, F., 250–251
Zhang, G., 120
Zhang, H., 46, 254–257, 259–263, 287, 292–298, 293t–296t
Zhang, J.Q, 25–26, 139–140, 250–251
Zhang, K.Q., 293t–296t
Zhang, L., 170, 173–174
Zhang, M., 167–168
Zhang, N., 5–6, 8, 67–69
Zhang, Q., 64–66
Zhang, R., 5–6, 8, 34–35, 59–60, 64–65, 70–72, 244–245
Zhang, S.A., 22–23, 32–33, 96–97, 222–223, 268
Zhang, W., 129
Zhang, X., 265
Zhang, Y.J., 199
Zhang, Y.M., 170, 173–174
Zhang, Z., 274–275
Zhao, J., 165–166, 253–254
Zhao, L.J., 255–256, 259, 260t–262t, 264–265
Zhao, Z., 63–64
Zheng, S.J., 67–69
Zheng, X., 5
Zhou, H., 267
Zhou, J.P., 293t–296t
Zhou, X., 67
Zhou, Y., 121–122
Zhu, F., 208–209
Zhu, H., 156
Zhu, J.K., 143
Zhu, M., 138–139, 142
Zhulin, I., 6
Zimmermann, A., 216–217, 230–231
Zimmermann, B., 216–217
Zinovkina, N.Y., 94
Zipfel, C., 24–26, 44, 139
Zobel, R.W., 226–227
Zock, A., 59–60, 63–64, 66
Zoomer, H.R., 191–194
Zoon, F.C., 195–197
Zou, C., 270–271
Zou, C.S., 293t–296t
Zuccaro, A., 126–127, 128f
Zushi, T., 63–67
Zwahlen, C., 287–288
Zwart, K.B., 193–194
Zygadlo, J., 299

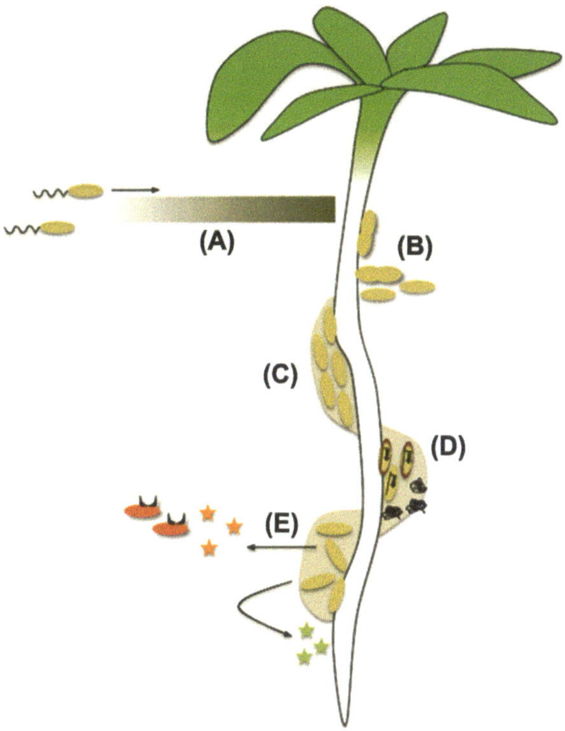

Pascale B. Beauregard, Figure 1 Schematic representation of the different steps for root colonization by plant growth-promoting rhizobacteria (PGPR) and how the plant can modulate them. (A) Root exudates secrete various molecules attracting the bacteria via chemotaxis. (B) Many of these molecules can also serve as nutrients to favour bacterial growth in the rhizosphere. (C) Once the PGPR are in contact with the root, they will establish a robust colonization via biofilm formation and matrix secretion, a process that can also be stimulated by the plant. (D) The plant also have general effects on the PGPR, both at the transcriptional and translational levels and (E) can stimulate the plant-beneficial properties of the PGPR such as antimicrobials secretion.

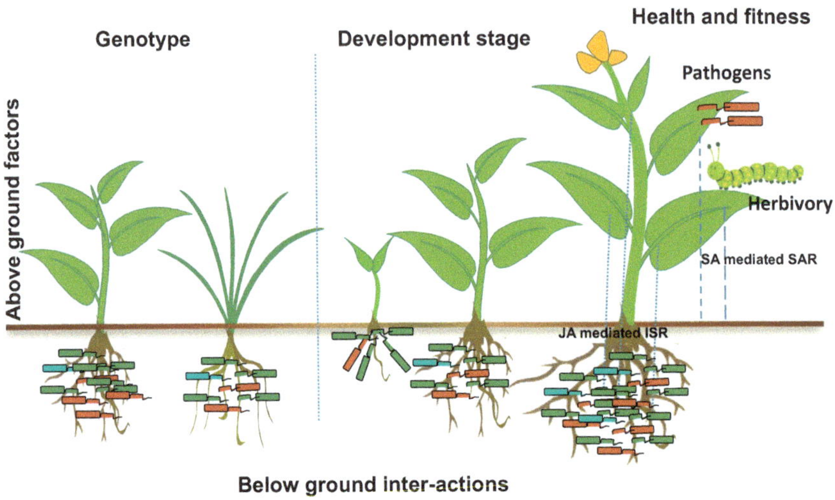

Venkatachalam Lakshmanan, Figure 1 The schematic depicts the impact of the host factors on assemblage of belowground rhizospheric microbiome in plants. It is shown that host genotype, developmental stage and plant health may alter the plant signalling and composition of root secretions which in turn facilitates rhizospheric composition of microbes.

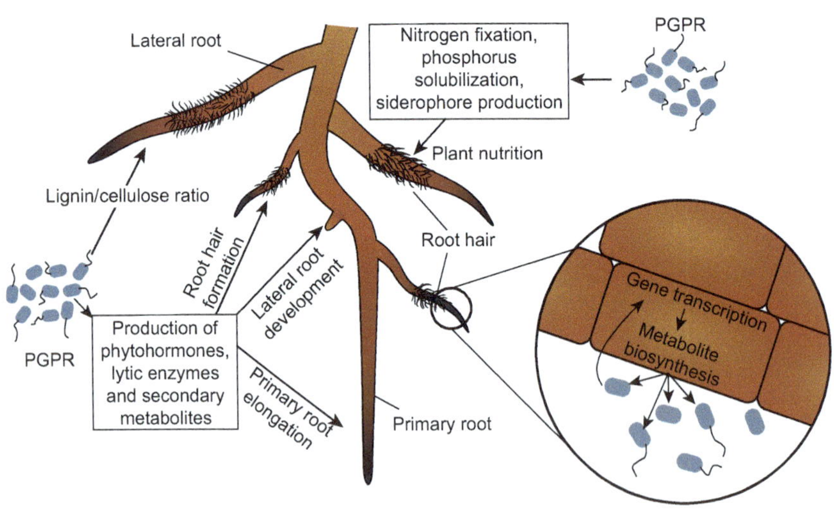

Usha Bishnoi, Figure 1 The possible mode of actions used by plant growth promoting rhizobacteria (PGPR) towards growth promotion in plants. *Adapted from Vacheron et al. (2013).*

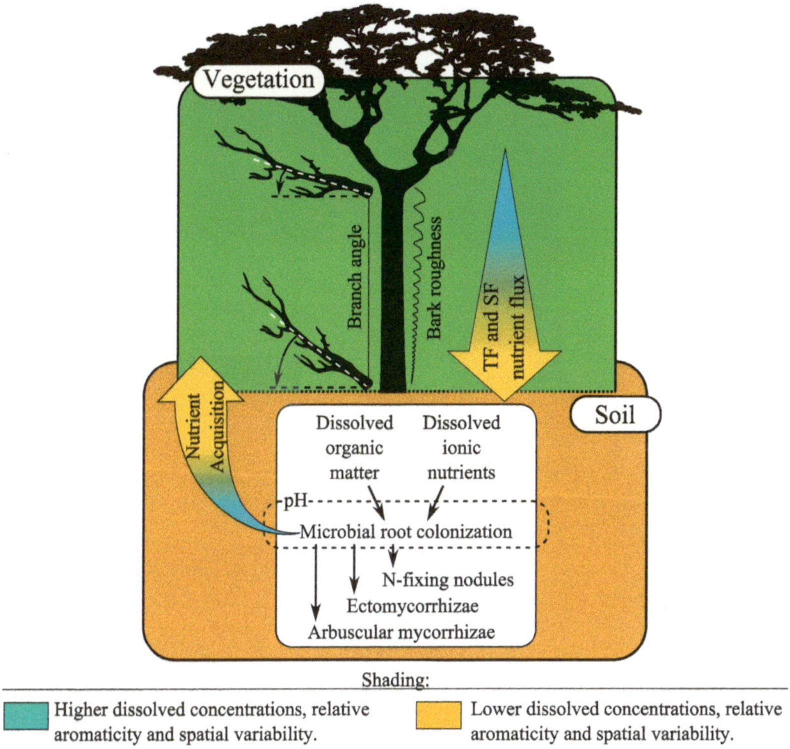

Carl L. Rosier *et al.*, Figure 1 The potential changes in edaphic soil conditions produced by contrasting net precipitation (i.e. throughfall and stemflow) effects on symbiotic microbial community.

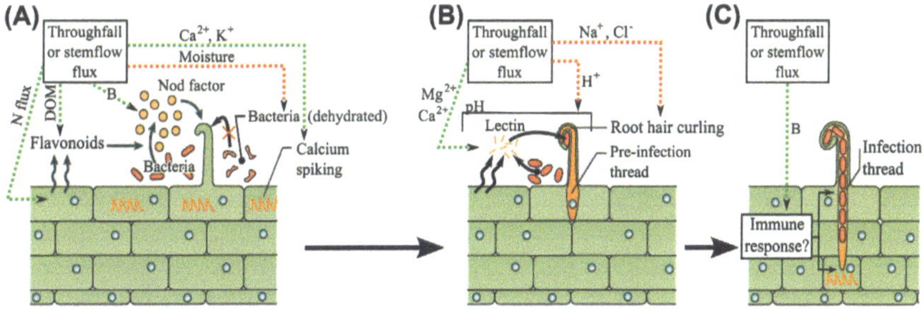

Carl L. Rosier *et al.*, Figure 2 The potential mechanisms for throughfall and stemflow to influence the signalling pathways of bacteria and host root during (A) initiation of symbiosis, (B) attachment of nodule-forming bacteria to root hairs, (C) termination of nodule development. *Figure adapted from Oldroyd (2013).*

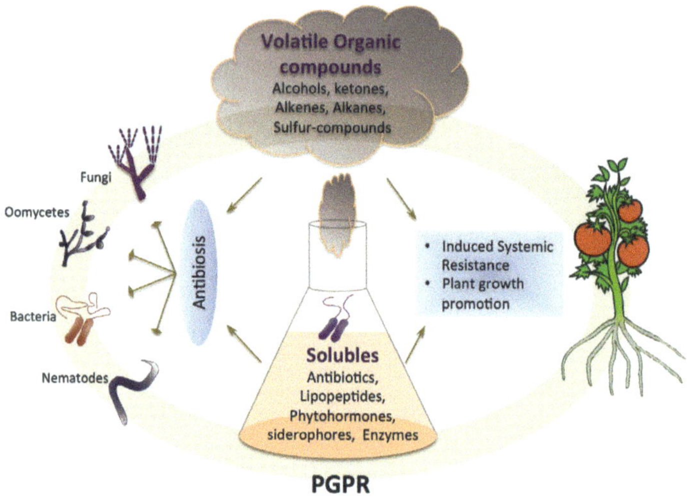

Gul Shad Ali *et al.*, Figure 1 Overview of soluble and volatile organic compounds produced by plant growth-promoting rhizobacteria (PGPRs).

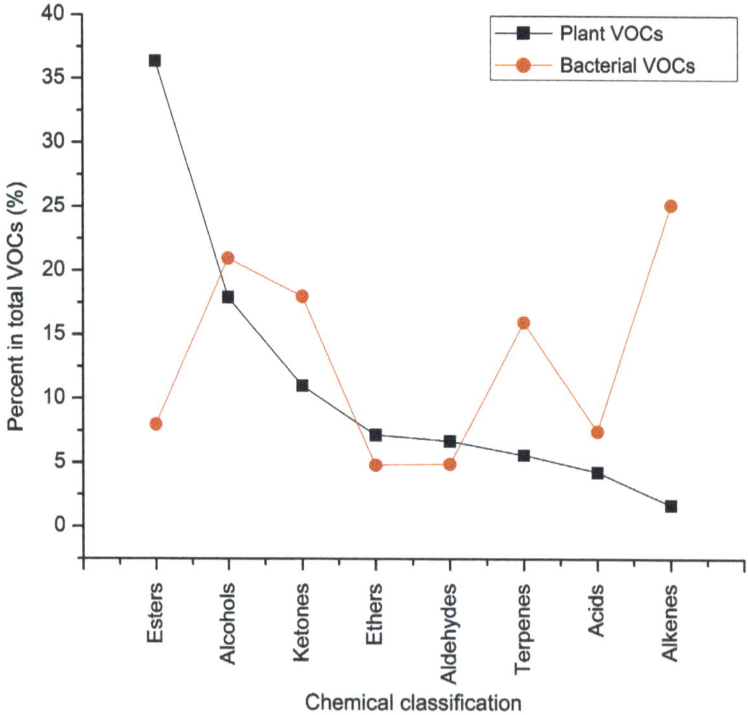

Gopinath Selvaraj, Figure 1 Difference between plant and bacterial VOCs in terms of chemical classification *(data obtained from Superscent and mVOC)*. Note: Only major chemical classes were considered for calculation, having at least ≥5% in either plant or bacterial volatiles.

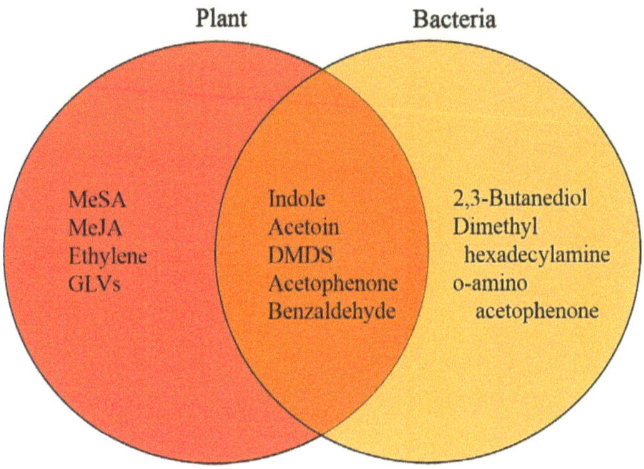

Gopinath Selvaraj, Figure 2 Venn diagram representing volatiles released during plant–bacteria interaction by both partners. These VOCs represent exclusive plant origin (plant volatilome) inside the pink circle, exclusive bacterial origin (bacterial volatilome) inside the sandal circle and VOCs common to both plant and bacteria in the overlapping region.

FSC
www.fsc.org
MIX
Papier | Fördert gute Waldnutzung
FSC® C083411

Kolibri 360 GmbH
Ferdinand-Jühlke-Straße 7
99095 Erfurt, Deutschland
produktsicherheit@kolibri360.de